Study Guide

to accompany

EIGHTH EDITION **LIFE** The Science of Biology

Sadava • Heller • Orians • Purves • Hillis

EDWARD M. DZIALOWSKI
University of North Texas

BETTY MCGUIRE
Cornell University

LINDSAY GOODLOE
Cornell University

NANCY GUILD
University of Colorado at Boulder

JON C. GLASE
Cornell University

Sinauer Associates, Inc.

W. H. Freeman and Company

Cover photograph © Max Billder.

Study Guide to accompany *Life: The Science of Biology,* **Eighth Edition**

Address editorial correspondence to:
Sinauer Associates, Inc.
23 Plumtree Road
Sunderland, MA 01375 U.S.A.
Fax: 413-549-1118
Internet: www.sinauer.com; publish@sinauer.com

Address orders to:
W.H. Freeman and Company
VHPS/W.H. Freeman & Co. Order Department
16365 James Madison Highway, U.S. Route 15
Gordonsville, VA 22942 U.S.A.
Internet: www.whfreeman.com

ISBN 978-0-7167-7893-6
Printed in U.S.A.

4 3 2 1

To the Student

Biology is an incredibly exciting field of study, but in order to appreciate new discoveries and discussions, it is necessary to have a firm grasp of the underlying concepts and ideas. Your textbook is designed to give you a comprehensive overview of important biological phenomena. It will also serve as a resource to you in future studies. Together with your instructor, your textbook will provide you with important information for beginning your study of biology.

This Study Guide is designed to supplement, not to replace, your textbook and your instructor. It was written for you, the student, in language that you can understand, but it does emphasize proper usage of biological terminology. Important concepts and ideas have been synthesized into short, easy-to-read summaries that provide an overview of the biological phenomena discussed in your textbook. Each Study Guide chapter includes four review elements: Important Concepts, The Big Picture, Common Problem Areas, and Study Strategies—these can help you preview a chapter before reading it, check your understanding, and review the chapter later.

Each Study Guide chapter also includes a series of questions, which have been grouped into two categories. Knowledge and Synthesis Questions are designed to determine if you have retained information from a chapter, and if you can put together various concepts in order to answer questions. Application Questions ask you to apply the knowledge you have gleaned from a chapter to answer questions that are more open-ended. The latter type of questions require you to have assimilated several concepts and to think beyond what you have just read. Your instructor may ask questions similar to these on exams, or may use an entirely different approach to assess your knowledge, but these questions will be a good check of how well you understand the material.

Brief answers to the questions are provided at the end of each chapter. These answers are not exhaustive, but are instead designed to point to the correct concepts in the textbook. Because of the nature of many of the Application Questions, your answers should be more expansive than the short explanations given in this Study Guide.

Strategies for Studying Biology

Each individual has his or her own unique study pattern. However, there are some successful study strategies that are universal. We recommend that you first preview a chapter in your textbook. In this initial preview, it is important to note the organization of the chapter and the main points, and to go over the chapter summary at the end. By referring to the Study Guide at this point you will further understand the organization of the material. If you follow these steps, you might have some specific questions in your mind that you will expect your reading to answer.

Reading a textbook is an active process. We recommend that you always have a pencil and paper available. Jotting notes in margins also serves as an excellent mental trigger when it comes time to review. As you read each section of your textbook, see if you can summarize it in your own words. Compare your summaries to those provided at the end of each section and at the end of the chapter. You want to assure yourself that the main points you are noting match those that the author has selected. Refer back to those questions that came up as you previewed the chapter. You should be able to answer your own questions by the time you have finished your reading.

As you are reading, take time to review all the figures and tables. Your textbook makes use of figures to illustrate points, describe pathways, and give visual representation to complex topics. Often these figures are more helpful than the paragraphs of written explanation. In places where figures might be beneficial to you but are not provided in the textbook, try to draw them yourself. This is especially important when attempting to understand structures and pathways. You will find that the Study Guide also refers you to specific tutorials and activities on the textbook's Companion Website (www.thelifewire.com). Each of these visual tools will aid in your understanding of the material.

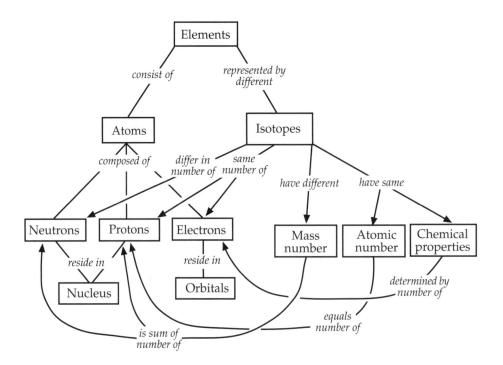

Once you have read the material, summarized it in your own words, and reviewed all of the figures and tables, it is time to do another quick review. Scan the summary in your textbook again and read over the important concepts in this Study Guide. We recommend further that you construct a "concept map" of the main points of a chapter. This will assure you that you see how concepts are interrelated and that you understand the organization of the material. An example of a concept map is shown above. Main concepts are in boxes, and arrows are used to connect concepts and show relationships.

Finally, when you are comfortable with chapter content, move to the questions section of this Study Guide. We recommend working through all of the questions before checking your answers. Any questions that you do not understand or that you answer incorrectly should be flagged as material you need to review. Remember that your goal is to understand the material completely, not just to answer these particular questions.

Experiments have suggested that the average attention span of a reader is approximately 12 minutes. You cannot expect to sit down and master an entire chapter in one sitting. Break up your study into short segments of approximately 30 minutes each. This will give you time to get down to business, maximize your attention, and learn without becoming frustrated or drained. At the end of this time, move to some other activity or area of study. When you return to biology, you will find your mind is ready to absorb more.

This entire process should take place well in advance of your exams. You cannot learn even one chapter adequately the day before an exam. Mastering biology requires daily study and review. If you keep up with learning the concepts as your course proceeds, you will find that reviewing the textbook summaries and the Study Guide is adequate preparation. The key to learning biology well is slow and diligent daily work.

Doing Well on Biology Exams

Your instructors are your best resource for doing well on exams. They are there to assist you in learning the material that is presented in your textbook. They are experts in their fields and understand how the concepts presented are vital to your biology education. Follow their lead in preparing for exams. Exam performance is directly linked to classroom attendance and daily study. Sit front and center in your lecture hall and pay close attention to what your instructor writes and provides in presentations or overhead displays. Your instructor will give you guidance about the most important points to study. You should take careful notes and ask for clarification when necessary.

You should always compare your classroom and lecture notes to your textbook and the Study Guide. Mark in the books those points your instructor emphasizes. Were there specific questions that the instructor brought up in class? Do these correspond to the textbook end-of-chapter questions or to those in the Study Guide? If so, chances are you will see them again on the exam. This practice should be part of your daily study regimen.

Approach the exam itself with confidence. Read the directions carefully and read the questions completely. One of the biggest mistakes students make on exams has to do with not following instructions. We recommend that you read carefully through the entire exam even before you begin to answer the first question. You may find that early questions are answered partially in later ones, and that an overview of the whole exam helps to trigger your memory of the material.

If your exam contains multiple choice questions, treat each of the answer options as if it were a separate true/false question. In other words, mentally fill in the question with each answer and ask yourself if the resulting statement is true or false. Be sure to read each of the options, even if the first one strikes you as the correct one. You may be dealing with a question that has multiple correct answers or asks you to select "all of the above" or "none of the above." If you are unsure about how to answer a question, begin by ruling out answers that you know are incorrect. By eliminating some possibilities, your chance of selecting the right one is greatly improved! If you narrow the choices down to two possible answers and just can't decide between them, go with your gut response. You just may be right. And when you go over your answers, change your response only if you know you answered a question incorrectly. If you aren't sure, it may be best to leave the answer as is. Your first instincts are often correct.

Essay questions require you not only to think through the concepts, but also to decide how you will present them. Organized, concise essay answers are always preferred to rambling answers with little direction. Take a look at the point value of each question. This is often an indication of how many "points" you need to make in your discussion. A 10-point question rarely can be answered with a single sentence. Look carefully at what your instructor is asking. If you are required to "discuss" a concept or a problem, do not present a list or some scattered phrases. Write in complete sentences and paragraphs unless you are asked specifically to list or itemize the material. Be sure that you address all points in the question. Essay questions frequently have multiple parts to test your understanding of the links between various concepts.

Always be aware of the time you have to complete your exam. Answer the questions you know swiftly, and allow yourself time to go back and really think about those you are struggling with. Be sure to concentrate on the questions with the highest point value. Those are the questions testing the most critical concepts and they are the ones that will have the most influence on your grade.

Once you think you have completed the exam, take a few minutes to go back over it. If there are any questions about which you are still unclear, be sure to ask your instructor for clarification.

Learn from Your Mistakes

After your instructor returns your exam or posts the answers, review the mistakes you have made. Take this opportunity to clarify misunderstandings and re-learn material when necessary. You will find this helpful not only for the final exam, but also as you approach upper level courses. Material reviewed, corrected, and re-learned is more likely to be retained in the long run.

Get Help before It Is Too Late

Many students come to college without adequate study habits and/or are ill prepared for the rigor of biology at the college level. But this is a problem that can be solved fairly easily. Many colleges and universities have academic centers that assist with locating tutors, teaching study habits, and acquainting you with a variety of other resources. Make good use of these facilities. They are there for you.

Good luck in your study of introductory biology!

Contents

Part Six ■ The Evolution of Diversity

Part Seven ■ Flowering Plants: Form and Function

Part Eight ■ Animals: Form and Function

Part Nine ■ Ecology

CHAPTER 1 Studying Life

Important Concepts

Biology is the study of living things.

- Living things are made up of cells. They can be composed of one cell; unicellular, or multiple cells; multicellular. Cell theory states that cells are the basic structural and physiological units of living things and can work alone or together to make up more complex organisms. This theory also states that all cells are made of the same things, they come from other cells, they are the site of the chemical reactions of life, and they contain the genetic material that is passed on from cell to cell.

- Charles Darwin proposed that all living things are related to one another and that species have evolved through the process of natural selection. A species can be defined as a group of organisms that are similar and can successfully breed and produce offspring. Natural selection works because individuals vary in their traits, and the characteristics of some traits possessed by some of the individuals may provide them with a better chance of survival and reproduction. Natural selection can produce adaptations, which are structural, physiological, or behavioral traits that enhance the chance of survival of an organism in a given environment.

- The genome of a cell contains all the genetic information in the form of DNA. Segments of DNA make up genes, which are used to produce the proteins necessary for life.

- All cells require nutrients for survival. These nutrients are used to carry out cellular work, such as building the cell or mechanical work.

- Cells continuously carry out multiple chemical reactions and thus must carefully regulate their internal environment. Multicellular organisms must also regulate the extracellular fluid that bathes each cell. This requires the specialization of the cells making up multicellular organisms. A group of similar cells that work together are known as a tissue. A number of different tissues can be organized into an organ, and multiple organs work together to form an organ system.

- Ecology is the study of how different species interact with each other. Organisms from one species living together make up a population. The many populations of different species in an area make up a community.

- Biologists like to use model species to study biological processes.

Certain evolutionary results had to occur before others.

- All species on Earth share a common ancestor. The study of the evolution of life on Earth involves both the fossil record and modern molecular methods.

- Life arose approximately 4 billion years ago. The random aggregation of complex chemicals led to the existence of the first biological molecules. These initial simple molecules led to molecules that could reproduce themselves and act as templates for larger, more complex molecules.

- The first known cells are thought to be approximately 3.8 billion years old. At this time, the cells became membrane-enclosed, allowing all the necessary complex biological molecules to interact. Early cells were simple prokaryotes with no internal membrane-enclosed compartments that lived in the oceans.

- Metabolism is the sum of all chemical reactions that occur within a cell and requires a source of energy to occur. Approximately 2.5 billion years ago, the ability to carry out photosynthesis arose. Photosynthesis involves taking light energy from the sun and converting it to chemical energy with oxygen as a by-product. Large numbers of photosynthetic organisms increased atmospheric oxygen levels, allowing for the evolution of aerobic metabolism. Photosynthetic organisms also contributed to the insulating ozone layer that shields Earth from radiation and modifies temperature. This shield eventually made life on land possible.

- Eukaryotic cells arose when larger prokaryotic cells engulfed and assimilated smaller cells, resulting in membrane-enclosed compartments known as organelles that carry out specific cellular functions.

- Multicellularity evolved about 1 billion years ago. Multicellularity makes it possible for cells to specialize and organism size to increase.

All organisms are descendants of a single unicellular organism.

- Different species come about when two groups of organisms in one species are isolated from each other so that they can no longer mate and produce viable offspring.

- There are as many as 30 million different species on Earth today. This is due to the splitting of populations during speciation events. The evolutionary relationship between all organisms can be described using an evolutionary tree.

- Organisms are referred to by their genus and species names (e.g., humans are *Homo sapiens*).

- Scientists are determining the branching pattern of the Tree of Life using the molecular evidence found in organisms. All life is placed into one of three major domains: Archaea, Bacteria, and Eukarya.

- Plantae, Fungi, and Animalia are three major groups of the Eukarya.

- Organisms that carry out photosynthesis and produce their own energy from the sun are known as autotrophs, and those that rely on other sources for energy are heterotrophs.

Biologists use observation and experimentation to investigate life.

- Biologists have always used observation as a means to study life. As we move forward, many new tools are being developed to help in our ability to observe life.

- The scientific method or hypothesis–prediction approach is used in most scientific investigations. This method involves making observations and asking questions. These questions lead to the formation of hypotheses, which answer the questions. Predictions are made in regard to the hypotheses, and then they are tested.

- Hypotheses are tested by either comparative experiments or by conducting controlled experiments. Comparative experiments involve gathering data to examine the patterns found in nature. Controlled experiments involve isolating the variables of interest, while keeping the other variables that may influence the outcome as steady as possible. Observations are then made to test the hypothesis.

- Statistical methods are used to determine if the results of an experiment are significant. Typically, these statistical tests start with a null hypothesis stating that there are no differences.

- Science depends on a hypothesis that is testable and can be rejected by direct observation and experiments.

Biological science has an impact on every human being and is the basis for many public policy decisions.

- The environmental stresses of population growth, the emergence of agricultural and medical technologies, and other modern challenges have increased the need for everyone to understand biological information.

The Big Picture

- Biology is the study of living things composed of cells. Cells contain genetic material, require nutrients for energy, maintain a constant internal environment, and interact with each other. All available evidence points to an origin of life some 4 billion years ago. Natural selection has led to the current host of organisms inhabiting the planet. There are specific characteristics of life, and the evolutionary roots of these characteristics can be traced through time.

- All life on Earth is organized in the Tree of Life based on molecular evidence. When studying organisms from an evolutionary perspective, scientists are concerned with evolutionary relatedness and common ancestors.

- Science follows a specific hypothesis–prediction approach. The necessity of a testable and falsifiable hypothesis sets science apart from other methods of inquiry. Control of variables and repeated testing lends credibility to the method.

Common Problem Areas

- Many students have trouble remaining open-minded enough to begin to understand evolutionary biology. Neither your instructors nor this book are asking you to set aside your beliefs. We are asking you to learn science. We encourage you to examine the evidence supporting evolution without prejudice, as you would any other scientific subject.

- Refrain from learning the hypothesis–prediction approach merely as a series of steps. Think about how each step in the process follows from the preceding step. Two characteristics of science set it apart from other modes of inquiry. One is the setting of hypotheses. The other is the continual process by which it is carried out. With each conclusion leading to a new set of hypotheses, science has the capacity to alter our overall base of knowledge.

Study Strategies

- Many students find that understanding the hypothesis–prediction approach makes understanding some of the evidence behind evolution easier. If you find you are having trouble, work on that section first.

- Drawing a time line or examining the calendar model of Figure 1.9 will help you understand the chronology of the major evolutionary events discussed in your textbook.

- Review the following activities on the Companion Website/CD:
  Activity 1.1 The Hierarchy of Life
  Activity 1.2 The Major Groups of Organisms

Test Yourself

Knowledge and Synthesis Questions

1. Life arose on Earth approximately how many years ago?
   a. 3.8 billion
   b. 3.8 million
   c. 4,000
   d. 1.5 billion
   Textbook Reference: 1.2 How Is All Life on Earth Related? p. 10

2. _____ is/are the differences among organisms that allow for life in such a wide variety of environments on Earth.
   a. Prokaryotic cells
   b. Eukaryotic cells
   c. Homeostasis
   d. Adaptations
   Textbook Reference: 1.1 What Is Biology? p. 6

3. Which of the following is *not* a characteristic of most living organisms?
   a. Regulation of internal environment
   b. Composed of one or more cells
   c. Ability to produce biological molecules
   d. None of the above
   Textbook Reference: 1.1 What Is Biology? p. 2

4. Photosynthesis was a major evolutionary milestone because
   a. photosynthetic organisms contributed oxygen to the environment, which led to the evolution of aerobic organisms.
   b. photosynthesis led to conditions that allowed life to arise on land.
   c. photosynthesis is the only metabolic process that can convert light energy to chemical energy.
   d. All of the above
   Textbook Reference: 1.2 How Is All Life on Earth Related? p. 11

5. The fact that all cells come from preexisting cells is part of _____ theory.
   a. animal
   b. genetic
   c. cell
   d. plant
   Textbook Reference: 1.1 What Is Biology? p. 5

6. A group of cells that work together to carry out a similar function is known as
   a. a tissue.
   b. an organ system.
   c. a unicellular organism.
   d. protein.
   Textbook Reference: 1.1 What Is Biology? p. 8

7. Which of the following are necessary for speciation to occur?
   a. Reproductive isolation of two groups of organisms
   b. A large amount of mutation within a population
   c. A reduction in the number of individuals in a population
   d. None of the above
   Textbook Reference: 1.1 What Is Biology? p. 6

8. Which of the following is *not* a domain in the Tree of Life?
   a. Archaea
   b. Plantae
   c. Eukarya
   d. Bacteria
   Textbook Reference: 1.2 How Is All Life on Earth Related? p. 12

9. The information needed to produce proteins is contained in
   a. nutrients.
   b. genes.
   c. evolution.
   d. organs.
   Textbook Reference: 1.2 How Is All Life on Earth Related? p. 7

10. Ecology is the study of how
    a. genes code for proteins.
    b. species interact with each other and the environment.
    c. organ systems function.
    d. None of the above
    Textbook Reference: 1.1 What Is Biology? p. 9

11. In an example experiment, researchers subjected frogs to various levels of UV-B radiation while keeping all other variables constant. This is an example of a _____ experiment.
    a. controlled
    b. repeated
    c. laboratory
    d. None of the above
    Textbook Reference: 1.3 How Do Biologists Investigate Life? p. 15

12. For a hypothesis to be scientifically valid, it must be _____ and it must be possible to _____ it.
    a. testable; prove
    b. testable; reject
    c. controlled; prove
    d. controlled; reject
    Textbook Reference: 1.3 How Do Biologists Investigate Life? p. 16

13. Eukaryotic cells differ from prokaryotic cells in the fact that they have
    a. genes.
    b. proteins.
    c. organelles.
    d. All of the above
    Textbook Reference: 1.2 How Is All Life on Earth Related? p. 11

14. Organisms are named using first their _____ and then their _____.
    a. species; genus
    b. genus; domain
    c. domain; genus
    d. genus; species
    Textbook Reference: 1.2 How Is All Life on Earth Related? p. 12

Application Questions

1. Discuss how the process of scientific inquiry is different from other forms of inquiry. Include in your discussion a description of the hypothesis–prediction approach.
   Textbook Reference: 1.3 How Do Biologists Investigate Life? p. 16

2. Look over a recent newspaper to see how many articles are directly related to biology. Select one article and discuss how the researchers followed or did not follow the hypothesis–prediction approach.
   Textbook Reference: 1.3 How Do Biologists Investigate Life? p. 13

3. What is one example of how biology influences public policy in your life?
   Textbook Reference: 1.4 How Does Biology Influence Public Policy? p. 16

Answers

Knowledge and Synthesis Answers

1. **a.** Available evidence puts the beginning of life at about 3.8 billion years ago.

2. **d.** Adaptations are the differences found in organisms that allow them to live in an environment.

3. **d.** Most living organisms are composed of cells, have the ability to make biological molecules, and can regulate their internal environment.

4. **d.** Photosynthesis caused the accumulation of oxygen in the atmosphere and contributed to the formation of an ozone layer. The presence of oxygen made the evolution of aerobic organisms possible, and the ozone layer shielded Earth from harmful radiation. These two phenomena contributed to the evolution of terrestrial life.

5. **c.** The cell theory states that all life is based on the cell.

6. **a.** Multicellular organisms have tissues that are formed from many similar cells.

7. **a.** For speciation to occur, reproductive isolation, such that two groups can no longer reproduce and produce viable offspring, is necessary.

8. **b.** Plantae is a kingdom found in the domain Eukarya.

9. **b.** Genes are specific sequences of DNA that contain the information used to make proteins.

10. **b.** Ecology is the study of how species interact with other species and the environment.

11. **a.** For experiments to be scientifically valid, they must be controlled.

12. **b.** Scientific hypotheses are set apart from mere conjecture by being testable and falsifiable.

13. **c.** Both eukaryotic and prokaryotic cells contain genes and proteins. Only the eukaryotes have organelles.

14. **d.** An example of this is *Homo sapiens.*

Application Answers

1. The process of scientific inquiry is different because a hypothesis must be testable, and it must be possible to reject it. The hypothesis–prediction approach begins with observations that lead to questions. From the questions, hypotheses are formed that are probable explanations for the phenomena being observed. Predictions are formed from the hypotheses and tested. Conclusions are drawn from the tests. These conclusions may, in turn, lead to additional hypotheses.

2. Often an article reporting on science news does not provide enough information for the reader to know whether a particular study followed the hypothesis–prediction approach. It pays to be a smart reader and consumer in this respect.

3. Many public policies that influence your life rely on sound biological findings. Examples include medical issues, environmental quality issues, and ecology of local populations of wild animals.

CHAPTER 2 The Chemistry of Life

Important Concepts

All matter is composed of the same chemical elements.

- Atoms combine to form all matter. The nucleus of atoms contains a defined number of positively charged protons and neutral neutrons. Negatively charged electrons move in their outer shells. An element is comprised of only one type of atom. There are more than 100 elements, and they are grouped according to certain characteristics in the periodic table.

- Six elements compose the majority of every living organism: carbon, hydrogen, oxygen, nitrogen, phosphorus, and sulfur.

- The number of protons in an element determines its type and is known as its atomic number. Neutrons are found in every element except hydrogen. The total number of protons and neutrons equals the mass number of the atom.

- Isotopes of an element have the same number of protons but differ in the number of neutrons present. Radioisotopes are unstable isotopes that give off energy. This decay eventually causes the atom to change to a new element. Radioisotopes are important in the medical field.

- Protons and neutrons have mass and contribute to atomic weight. Atomic weights on a periodic chart are averages of all isotopes of an element.

Biologists are interested in chemical reactions and the interactions of electrons.

- Interactions between atoms involve the associations of their electrons. Electrons continuously orbit the nucleus of an atom in a defined space.

- The orbital of an atom is the space in which an electron is found 90 percent of the time. These patterns of orbitals compose a series of electron shells or energy levels, each with a specific number of electrons. The first shell (s orbital) can contain up to two electrons. The second shell can have as many as four orbitals, each containing two electrons. The number of electrons in the outermost shell determines how the atom interacts with other atoms. Many biologically important atoms, such as carbon and nitrogen, are stable when they have eight electrons in the outermost shell. This phenomenon is referred to as the octet rule.

- Two or more linked atoms compose a molecule resulting in stabilization of the outer electron shell of the atoms.

Chemical bonds hold molecules together.

- Atoms share pairs of electrons to stabilize their outer shells in covalent bonding. This type of bond is very stable and strong and can be broken only with a great deal of energy. All molecules have a three-dimensional shape, and the interactions between a given pair of atoms always have the same length, angle, and direction. This shape of a molecule affects how it behaves.

- Multiple covalent bonds may exist. Although a single covalent bond involves one pair of shared electrons, double bonds involve two pairs of shared electrons. Triple bonds share three pairs, but are rare.

- Covalent bonding between atoms of the same element results in equal sharing of electrons. The attractive force of an atom on electrons is known as electronegativity. Two atoms that have the same electronegativity form nonpolar covalent bonds.

- Bonds between different elements generally result in an unequal sharing of the electrons. Unequal sharing of electrons results in partial (δ) charges in molecules because one nucleus is more electronegative than the other, attracting the electrons more strongly. One atom of a molecule may be partially negative, whereas the other is partially positive. This balance of partial charges results in a "polar" molecule with a δ^- pole and δ^+ pole.

- Partial charges allow hydrogen bonding of molecules. This occurs often between molecules of water. The δ^+ of H is attracted to the δ^- of another molecule. This is a weak bond that is easily broken because no electrons are shared, but this type of bond is important in stabilizing the three-dimensional shape of large molecules such as proteins and DNA.

- Ions are formed when atoms lose or gain electrons, resulting in a net positive or negative charge.
  - Cations are positively charged and have fewer electrons than protons.
  - Anions are negatively charged and have more electrons than protons.
- More than one electron can be gained or lost from an atom or molecule. A group of covalently bonded atoms may also gain or lose electrons to form complex ions.
- Ionic bonds form between ions of opposite charge. Ionic bonds are not as strong as covalent bonds. Ions can also attract the partial charges of polar molecules.
- Polar molecules are hydrophilic, or "water loving," because of the partial charges. Nonpolar molecules are hydrophobic, or "water hating." Attractions of nonpolar molecules are enhanced by van der Waals forces. Polar molecules tend to aggregate with other polar molecules, and nonpolar molecules aggregate with other nonpolar molecules.

Chemical reactions occur when atoms combine and form new bonding partners.

- Reactions involve reactants that are altered to produce products. Matter is neither created nor destroyed, but changed. Energy is the capacity to cause change. Some reactions produce energy, whereas others require the input of energy.
- Chemical bonds hold potential energy. When these bonds are broken, release of this energy is important to living systems.

Water is a biologically important molecule with many unique and crucial properties.

- Water is a polar molecule with the ability to form hydrogen bonds. Due to the four pairs of electrons in the outer shell of oxygen, water has a tetrahedral shape.
- Ice floats because of the less-dense pattern of hydrogen bonding that results when water molecules are frozen. Ice insulates ponds and lakes in the winter, allowing life to continue beneath the frozen layer.
- Much energy is required to break the hydrogen bonds when molecular ice melts. In contrast, much energy is lost during the freezing of water.
- Large bodies of water help moderate temperature on land and in the atmosphere. The amount of energy needed to raise the temperature of 1 gram of something by 1°C is known as the specific heat. Liquid water has a high specific heat, which provides it with a high heat capacity and the ability to moderate temperatures.
- Water has a high heat of vaporization and thus requires the input of energy to break the hydrogen bonds and go from a liquid to a gaseous state. This makes evaporating water an effective coolant.
- The polar nature of water and the formation of hydrogen bonds contribute to the cohesive strength of water and its surface tension, both of which are biologically important.
- Solutions are formed when a substance is dissolved in water or another liquid. An aqueous solution has water as the solvent.

Quantitative terms are important to understand for both biology and chemistry.

- Concentration is the amount of a substance in a given amount of solution.
- A mole is the amount of an ion or compound in grams whose weight is numerically equal to its molecular weight.
- Avogadro's number relates the number of molecules of any substance to its weight and is 6.02×10^{23} molecules per mole. A 1 molar solution is made by dissolving 1 mole of a substance into 1 liter of water.

Acids and bases are substances that dissolve in water and are biologically important.

- Acids dissolve in water and release H^+ ions.
- Bases dissolve in water and attach to H^+ ions.
- Acids and bases do not all ionize in the same way. Some ionize easily and completely and are called strong acids or bases. Others only partially or reversibly ionize and are called weak acids or bases.
- When strong acids and bases are ionized, the reaction is irreversible. Ionization of weak acids and bases can be reversible reactions.
- Water is a very weak acid because two molecules are required for ionization. Though ionization of water is uncommon, it is biologically important.
- Solutions are acidic or basic, whereas compounds and ions are bases or acids.
- The pH of a solution refers to the concentration of H^+ ions and is measured as the $\log_{10}$ of the H^+ concentration, $pH = -\log_{10}[H^+]$. Lower pH indicates a higher H^+ concentration or a more acidic solution. A higher pH indicates a lower H^+ concentration or a more basic solution.
- Homeostasis is the maintenance of a constant internal environment. A buffer is a mixture of a weak acid and its corresponding base. Buffers prevent fluctuations in the pH of a solution and play an important role in maintaining homeostasis.

The Big Picture

- Understanding the chemical building blocks of all matter is essential to understanding the biology of organisms. Atoms, containing protons, neutrons, and electrons, form elements or combine to form molecules. Molecules may be held together with covalent or ionic bonds and stabilized in three-dimensional conformations by hydrogen bonds. Chemical reactions change reactants to products without creating or destroying matter.

- Water has unique properties that make it biologically important. An essential property is its ability to act as a solvent. How a substance ionizes in water determines if it is acidic or basic. Acids have a high concentration of H^+, whereas bases have a low concentration of H^+. The pH scale allows us to compare the concentrations of H^+ in a variety of solutions. Weak acids acting as buffers are necessary for an organism to maintain homeostasis.

Common Problem Areas

- pH tends to confuse many people. Remember that it is a concentration of H^+ ions. Those that have a higher concentration are acidic and have a low pH, and those that have a lower concentration are basic with a high pH. Also remember that pH is a logarithmic scale. This means that if you decrease the pH from 8 to 7, you increase the concentration of H^+ ions by a factor of 10.

- Learning quantitative terminology and how solutions are made is often difficult. Completing practice problems is the best way to learn this material.

- People are sometimes overwhelmed by the periodic table. Consult Figure 2.2 in your book to make sure you understand the components and how they are arranged. Do not try to memorize the table, but keep it handy as a reference.

- Use Table 2.1 to understand the differences between the various types of bonds and interactions.

Study Strategies

- The best way to learn the quantitative terms and pH material is to work on many practice problems. Also take advantage of any laboratory experience making solutions. Every good biologist makes many solutions over time.

- Use the figures in your book to help you visualize what atoms and molecules look like and how they interact.

- Be careful not to get lost in terminology. Think about what each term means, and focus on understanding the concept rather than memorizing the term.

- Review the following animated tutorial and activity on the Companion Website/CD:
  Tutorial 2.1 Chemical Bond Formation
  Activity 2.1 Electron Orbitals

Test Yourself

Knowledge and Synthesis Questions

1. The atomic number of an element refers to the number of _____ in an atom.
   a. protons and neutrons
   b. protons
   c. electrons
   d. neutrons
   Textbook Reference: 2.1 What Are the Chemical Elements That Make Up Living Organisms? p. 23

2. Which of the following statements concerning electrons is *not* correct?
   a. Electrons orbit the nucleus of an atom in defined orbitals.
   b. The outer shell of all atoms must contain eight electrons.
   c. An atom may have more than one valence shell.
   d. Electrons are negatively charged particles.
   Textbook Reference: 2.1 What Are the Chemical Elements That Make Up Living Organisms? pp. 23–24

3. The element with which of the following atomic numbers would be most stable?
   a. 1
   b. 3
   c. 12
   d. 18
   Textbook Reference: 2.1 What Are the Chemical Elements That Make Up Living Organisms? p. 25

4. How can you differentiate between an element and a molecule?
   a. Molecules may be composed of different types of atoms, whereas elements are always composed of only one type of atom.
   b. Molecules are composed of only one type of atom, whereas elements are composed of different types of atoms.
   c. Molecules are elements.
   d. Molecules always have larger atomic weights than elements.
   Textbook Reference: 2.1 What Are the Chemical Elements That Make Up Living Organisms? p. 25

5. The strongest chemical bonds occur when
   a. two atoms share electrons in a covalent bond.
   b. two atoms share electrons in an ionic bond.
   c. hydrogen bonds are formed.
   d. van der Waals forces are in effect.
   Textbook Reference: 2.2 How Do Atoms Bond to Form Molecules? p. 27

6. You have discovered that a molecule is hydrophilic. What else do you know about this molecule?
   a. It cannot form hydrogen bonds.
   b. It is a polar molecule.
   c. It has a partial positive region and a partial negative region.
   d. Both b and c
   Textbook Reference: 2.2 How Do Atoms Bond to Form Molecules? p. 29

7. The stability of the three-dimensional shape of many large molecules is dependent on
   a. covalent bonds.
   b. ionic bonds.
   c. hydrogen bonds.
   d. van der Waals attractions.
   Textbook Reference: 2.2 How Do Atoms Bond to Form Molecules? p. 27

8. The molecular weight of glucose is 180. If you added 180 grams of glucose to a 0.5 liter of water, what would be the molarity of the resulting solution? (See Figure 2.2 for a periodic table.)
   a. 18
   b. 1
   c. 9
   d. 2
   Textbook Reference: *2.4 What Properties of Water Make It So Important in Biology? p. 33*

9. Why does ice float in water?
   a. Ice is less dense than water.
   b. There are no hydrogen bonds in ice.
   c. Ice is denser than water.
   d. Water has a higher heat capacity than ice.
   Textbook Reference: *2.4 What Properties of Water Make It So Important in Biology? p. 31*

10. Cola has a pH of 3; blood plasma has a pH of 7. The hydrogen ion concentration of cola is _____ than the hydrogen ion concentration of blood plasma.
    a. 4 times greater
    b. 4 times lesser
    c. 400 times greater
    d. 10,000 times greater
    Textbook Reference: *2.4 What Properties of Water Make It So Important in Biology? p. 34*

11. If solution A has a pH of 2 and solution B has a pH of 8, which of the following statements is true?
    a. A is basic and B is acidic.
    b. A is acidic and B is basic.
    c. A is a base and B is an acid.
    d. A has a greater $[OH^-]$ than B.
    Textbook Reference: *2.4 What Properties of Water Make It So Important in Biology? p. 34*

12. One mole of glucose ($C_6H_{12}O_6$) weighs
    a. 180 grams.
    b. 42 atomic mass units.
    c. 96 grams.
    d. 342 grams.
    Textbook Reference: *2.4 What Properties of Water Make It So Important in Biology? p. 33*

13. The role of a buffer is to
    a. allow the pH of a solution to vary widely.
    b. make a solution basic.
    c. maintain pH homeostasis.
    d. disrupt pH homeostasis.
    Textbook Reference: *2.4 What Properties of Water Make It So Important in Biology? p. 34*

14. Which of the following statements about water is correct?
    a. Water has a low heat of vaporization.
    b. Water has a high specific heat.
    c. When water freezes, it gains energy from the environment.

d. None of the above
Textbook Reference: *2.4 What Properties of Water Make It So Important in Biology? p. 32*

15. What occurs in a chemical reaction?
    a. The bonding partners of atoms remain constant during the reaction.
    b. All reactions release energy as they proceed.
    c. The bonding partners of atoms changes during the reaction.
    d. Matter is either created or destroyed.
    Textbook Reference: *2.3 How Do Atoms Change Partners in Chemical Reactions? p. 30*

Application Questions

1. In the following figures, place electrons in the appropriate shells based on the atomic numbers of the elements carbon ($_6$C), nitrogen ($_7$N), sodium ($_{11}$Na), chlorine ($_{17}$Cl), argon ($_{18}$Ar), and oxygen ($_8$O).

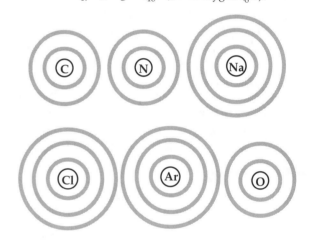

Textbook Reference: *2.1 What Are the Chemical Elements That Make Up Living Organisms? p. 24*

2. Calcium has an atomic number of 20. Draw structures for Ca and Ca^{2+}. What is different about these structures? Why is the most common ion of calcium Ca^{2+}? Why is the most common ion of lithium Li^+?
   Textbook Reference: *2.2 How Do Atoms Bond to Form Molecules? p. 28*

3. Nitrogen atoms can form triple bonds with each other. What is a triple bond? How many electrons are shared between two N atoms?
   Textbook Reference: *2.2 How Do Atoms Bond to Form Molecules? p. 27*

4. Water is a polar molecule. This property contributes to cohesion and surface tension. Draw six water molecules. In your drawing, indicate how hydrogen bonding between molecules contributes to cohesion and surface tension. (Be sure to include appropriate covalent bonds in each molecule.)
   Textbook Reference: *2.2 How Do Atoms Bond to Form Molecules? p. 29*

5. Rank the following solutions in order from the most acidic to the most basic.
   Lemon juice, pH = 2
   Mylanta, pH = 10
   Sprite, pH = 3
   Drain cleaner, pH = 15
   Seawater, pH = 8
   Of the preceding, which has the highest concentration of H^+ ions? Which has the lowest concentration of H^+ ions?
   Textbook Reference: 2.4 *What Properties of Water Make It So Important in Biology? p. 34*

6. If you have 12 moles of a substance, how many molecules do you have of that substance? Suppose the substance has a molecular weight of 342. How many grams of that substance would you have to dissolve in a liter of water to make a 12 M solution?
   Textbook Reference: 2.4 *What Properties of Water Make It So Important in Biology? p. 33*

Answers

Knowledge and Synthesis Answers

1. **b.** The atomic number refers to the number of protons in an atom. The atomic weight can vary with the isotope and is the total of the number of protons and neutrons in an atom.

2. **b.** The first energy shell requires only two electrons to be full; subsequent shells may hold eight or more electrons.

3. **d.** The element with atomic number 18 (Argon) is the most stable because its outermost electron shell is full.

4. **a.** An element is always composed of only one type of atom. Molecules contain many types of atoms that are bonded together.

5. **a.** Covalent bonds, resulting from the sharing of a pair (or more) of electrons, form the strongest types of bonds. Ionic bonds, where electrons are donated or received, are the second strongest. Hydrogen bonds are the weakest bonds and result from the attraction of partial charges to one another. Van der Waals forces are not bonds but attractions between nonpolar molecules.

6. **d.** Hydrophilic molecules are polar and have a partial positive and a partial negative pole. These molecules are said to be hydrophilic because they easily form hydrogen bonds with water.

7. **c.** Hydrogen bonds, though weak individually, are quite effective in large numbers and are responsible for maintaining the structural integrity of many large molecules such as proteins and DNA.

8. **d.** The molarity of the solution is 2. A 1 M solution is made by adding 180 grams to 1 liter of water. Adding the same amount to half a liter would result in a solution with twice the molarity.

9. **a.** When water forms ice, it becomes more structured. However, the resulting form is less dense than water in the liquid state allowing ice to float on water.

10. **d.** pH is a logarithmic scale. Each increase or decrease in pH value is by a factor of 10. Therefore, a difference between pH 3 and pH 7 would be a difference of $10 \times 10 \times 10 \times 10$, or 10,000.

11. **b.** A solution with a pH of 2 is acidic, and one with a pH of 8 is basic. The lower the pH value, the higher the concentration of H^+ ions.

12. **a.** Each carbon atom weighs 12 atomic mass units, each hydrogen atom is 1 atomic mass unit, and each oxygen is 16 atomic mass units. Adding these yields a molecular weight of 180 atomic mass units. Because one mole is equal to the atomic weight in grams, one mole of glucose equals 180 grams.

13. **c.** Buffers are solutions that contain a weak acid (carbonic acid) and the corresponding base (bicarbonate ions). Buffers help to maintain the pH of a solution when either an acid or base is added.

14. **b.** Water has both a high heat of vaporization and a high specific heat. Additionally, when water freezes it gives off a lot of energy to the environment.

15. **c.** Chemical reactions involve the combining or changing of bonding partners of the reactants to produce the products. During this process matter is neither created or destroyed. These reactions can either release energy or require energy.

Application Answers

1.

2.

Calcium (Ca) has two electrons in its outer shell.

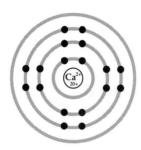

Calcium ion (Ca^{2+}) has lost its two outer electrons and therefore has a positive charge of 2 because it has two more protons than electrons. Lithium has only one electron in its outer shell; when that electron is lost, the ion gains a positive charge of only 1.

3. Three pairs of electrons are shared between two N atoms in a triple bond.

4. The partially positive hydrogens of one water molecule are attracted to the partially negative oxygens of another molecule of water. This attraction tends to lead to water molecules "sticking" together and creating surface tension.

5. Lemon juice, pH = 2 (highest concentration of H+ ions)
   Sprite, pH = 3
   Seawater, pH = 8
   Mylanta, pH = 10
   Drain cleaner, pH = 15 (lowest concentration of H+ ions)

6. You would multiply Avogadro's number (6.02×10^{23}) times 12 to determine the number of molecules. This gives you 7.224×10^{24} molecules of this substance. A 12 M solution would be made by multiplying the formula weight of the molecule times 12 and adding that many grams of the substance to one liter of water. The formula weight of this molecule is 342, so you would need 4104 g.

CHAPTER 3 Macromolecules and the Origin of Life

Important Concepts

Molecules have specific properties that enhance their biological activity.

- Functional groups are small groups of atoms that have specific properties and attach to larger molecules. The functional groups attach to a carbon skeleton or chemical groups that are noted with the letter R for "remainder" of the molecule. Each functional group confers different properties to the biological molecules.

- Isomers are molecules that have the same number and kind of atoms but are arranged differently. Structural isomers have a different order and binding of the atoms. Optical isomers occur when carbon has four different groups or atoms attached to it in differing orders.

Macromolecules are very large polymers with a wide range of biological functions.

- There is a biochemical unity among all organisms due to the fact that they are composed of four major biological macromolecules: proteins, nucleic acids, carbohydrates, and lipids. The specific types of macromolecules share similar roles throughout all of life that are dependent on the chemical properties of their component monomers. Macromolecules provide energy storage, structural support, protection, catalysis, transport, defense, regulation, movement, and information storage.

- Polymers are made by the covalent bonding of monomers in a process called condensation. See Figure 3.4A in your text.

- Polymers may be broken down into their constituent monomers through hydrolysis. See Figure 3.4B in your text.

- Both condensation and hydrolysis reactions require an input of energy.

- The size and structure of each polymer is related to its biological function.

Polymers of amino acids are called proteins and function in structural support, protection, transport, catalysis, defense, regulation, and movement.

- Proteins range in size from small proteins with 50 to 60 amino acids to large proteins with more than 4,600 amino acids. Proteins may be made of single chains of amino acids or multiple interacting chains.

- Proteins are composed of 20 amino acids. Each of the 20 amino acids has a central carbon atom that has a hydrogen atom, amino group, and carboxyl group attached to it. The specific amino acids differ in the unique side chain attached to the central carbon, providing them with specific properties.

- The shape and structure of proteins are influenced by the properties of the side chains of the constituent amino acids. Five amino acids have charged side chains, which attract water; five have polar, hydrophilic side chains that form hydrogen bonds; seven have hydrophobic, nonpolar side chains; and three have special hydrophobic functions.

- The side chain of cysteine is –SH and can form disulfide bridges with other cysteines and influences protein chain folding.

- Glycine is small and fits into folds of proteins.

- Proline has limited bonding ability and often contributes to looping in proteins.

- Amino acids are chained or polymerized together during condensation. The resulting bonds between the amino and carboxyl groups are called peptide bonds or linkages. See Figure 3.6 in your text. Peptide chains always begin with the N terminus and end with the C terminus.

- The C—N peptide linkages are relatively rigid, limiting the extent of protein folding. Further, hydrogen bonds are favored within the folded proteins, helping maintain the form and function of proteins.

Proteins have four levels of structural organization.

- Primary structure of a protein refers to the basic sequence of amino acids. All higher levels of structure are due to the specific amino acid sequence in the primary structure of the protein.

- Secondary structure refers to regular repeated spatial patterns of the amino acid chain that are stabilized by hydrogen bonds. The α helix consists of a right-handed coiling of a single polypeptide chain and is found in fibrous structural proteins. The β pleated sheet occurs when two or more chains form hydrogen bonds between themselves.

- Tertiary structure refers to bending and folding of an amino acid chain into a more complex pattern. This structure is stabilized by disulfide bridges, hydrophobic interactions, van der Waals forces, ionic interactions, and hydrogen bonds. The sequence of R groups in the primary structure is responsible for this level of folding.

- Quaternary structure refers to the three-dimensional interactions of multiple protein subunits.

Protein shape and surface chemistry influence function.

- The ability of proteins to bond noncovalently to other molecules makes them biologically important. This property allows them to function as cell junctions, carrier molecules, receptors, enzymes, and defensive proteins.

- Binding sites in proteins are of a specific shape, allowing for selective and specific interactions with other molecules. The particular shape and the ability of chemical groups to interact are a result of the protein's three-dimensional shape and the properties of its constituent amino acids.

- Because a protein's three-dimensional structure is stabilized by relatively "weak" bonds, environmental conditions such as temperature, pH changes, or altered salt concentrations can change the shape of the protein into an inactive form. This is called denaturation.

- Chaperonins are a class of proteins that prevent other proteins from denaturing by inhibiting inappropriate bonding.

Sugar polymers called carbohydrates serve the roles of primary energy source and carbon skeleton in biological systems.

- Monosaccharides are simple sugar monomers that are used in the synthesis of complex carbohydrates. More complex arrangements include disaccharides (2 monosaccharides), oligosaccharides (3–20 monosaccharides), and polysaccharides (large, complex carbohydrates).

- Simple sugars (monomers) are biologically important as an energy source (glucose) and as a structural backbone for RNA and DNA (ribose and deoxyribose).

- Simple sugars typically have three to six carbon atoms and may exist as different isomers.

- Dehydration synthesis between monomers results in glycosidic linkages.

- Polysaccharides are very large chains of monomers that provide energy storage or structural function. The specific structure of polysaccharides contributes to their function. For instance, starch and cellulose are both chains of glucose, but the glycosidic bonds are in different orientations and yield very different chemical properties.

- Added functional groups, such as a carboxyl group, give carbohydrates altered functions. Sugar phosphates and amino sugars, such as chitin, have significant biological activity.

Lipids are chemically diverse and provide many biological functions. They are all water insoluble.

- The insolubility of lipids in water is due to many nonpolar covalent bonds. Lipids aggregate due to their nonpolar nature and van der Waals forces. Lipids play a wide role in the biology of an animal, including energy storage, structural roles, capturing light, hormones, and insulators.

- Fats and oils, composed of fatty acids and glycerol, are also known as triglycerides or simple lipids and function primarily in the storage of energy.

- Each triglyceride contains a glycerol bound to three fatty acids with carbon atoms that are single-bonded (saturated fatty acids) or double-bonded (unsaturated fatty acids). The characteristics of the carbon bonds, saturated versus unsaturated, influence the shape of the molecule and determine its melting point.

- Phospholipids compose cellular membranes. The replacement of a hydrophobic fatty acid with a hydrophilic phosphate group allows phospholipids to have a hydrophilic "water-loving" head and hydrophobic "water-hating" tails. In an aqueous environment, the hydrophobic tails of the phospholipids tend to aggregate together with the phosphate heads facing out. This bilayer effect allows for the establishment of a hydrophobic inside surrounded by an aqueous environment.

- Carotenoids and steroids are lipids that do not have the glycerol-fatty acid structure. Carotenoids trap light energy during photosynthesis. Steroids serve as signaling molecules and are synthesized from cholesterol.

- Some vitamins are lipids and are not synthesized in the human body. Vitamins A, D, E, and K are all lipid derived.

- The nonpolar nature of lipids makes them an excellent water repellent. Wax coatings on hair, feathers, and leaves are all lipid derived.

The nucleic acids (DNA and RNA) store, transmit, and use genetic information.

- The primary role of nucleic acids is to store and transmit hereditary information.

- Nucleic acids are made up of nucleotide monomers. Each nucleotide monomer consists of a pentose (five-carbon) sugar, a phosphate group, and a nitrogenous base. DNA has one less oxygen atom in the ribose sugar than RNA and is thus called *deoxyribonucleic acid*. DNA and RNA also differ in one of their nitrogenous bases.

- The backbone of DNA consists of alternating sugars and phosphates with the bases projecting outward. In both RNA and DNA the nucleotides are joined by phosphodiester linkages that link the nucleotide with the phosphate group. RNA is a single strand of nucleotides, whereas DNA is a double strand. The two strands in DNA are stabilized by hydrogen bonds.

- Only four types of nucleotides are found in DNA: two purines (adenine and guanine) and two pyrimidines (thymine and cytosine).

- Nucleotide bases are capable of complementary base pairing. In DNA, adenine and thymine pair, and cytosine and guanine pair. In RNA, thymine is replaced by uracil. Complementary base pairing is facilitated by the size and shape of the molecules and by hydrogen bonding. DNA replication and the production of RNA from a DNA template are both made possible by complementary base pairing.

- The uniformity of the DNA molecules makes it suited for carrying information. The sequence of the nitrogenous bases provides the code. The regularity of the molecule allows it to be "read" in any cell.

- Similarities in base sequences of DNA can be used to help determine evolutionary relationships among organisms because it is a conserved and regular molecule.

- Nucleotides have additional biological functions. ATP and GTP serve as energy sources. cAMP serves as a messenger in many cellular processes.

Two major theories exist for the beginning of life on Earth.

- One theory of the origin of life on Earth is based on the idea that water and some molecules characteristic of life may have come to Earth from space. A meteorite from Mars, ALH 84001, may have had internal conditions that would allow for living organisms to survive a long space journey.

- Another theory of the origin of life on Earth states that the conditions on Earth billions of years ago led to the emergence of biological molecules. This theory is known as chemical evolution. Stanley Miller and Harold Urey produced a primitive atmosphere containing hydrogen gas, ammonia, methane gas, and water and using electrical charges were able to produce the initial building blocks of life.

- Specific RNA, ribozymes, can function as catalysts and may have been important in the origin of life. RNA can also pass along information due to the bases present.

- The conditions on Earth are no longer suitable for the generation of new life forms. All life comes from another living organism.

The Big Picture

- Macromolecules provide for specific biological activity and provide energy storage, information storage, structural components, catalytic activity, and other functions within the cell. Biologically important macromolecules fall within four basic classes: proteins, carbohydrates, nucleic acids, and lipids. All macromolecules are syn-thesized from smaller molecules through condensation and can be broken down through hydrolysis.

- Macromolecules all have regular primary structures held by covalent bonds, but many, especially proteins, can be folded into complex shapes held by ionic or hydrogen bonds or attractive forces.

- Chemically active side chains or functional groups contribute to the specific functions of macromolecules.

Common Problem Areas

- Many students are overwhelmed by the different types of macromolecules important to biological systems. Remember that each macromolecule is synthesized from smaller components that give each class of molecule its specific function.

Study Strategies

- Make a table of macromolecule characteristics. This will help you see the patterns of similarity and the differences.

- To avoid being overwhelmed, remember that each macromolecule is synthesized from smaller components that give the molecules its specific function.

- Avoid the temptation to memorize structures. If you understand condensation and hydrolysis and know the basic components of the macromolecules, memorizing the structures of the large macromolecules is unnecessary.

- Review the following animated tutorials and activities on the Companion Website/CD:
  Tutorial 3.1 Macromolecules
  Tutorial 3.2 Synthesis of Prebiotic Molecules in an Experimental Atmosphere
  Tutorial 3.3 Pasteur's Experiment
  Activity 3.1 Chemical Functional Groups
  Activity 3.2 Features of Amino Acids
  Activity 3.3 Forms of Glucose
  Activity 3.4 Nucleic Acid Building Blocks
  Activity 3.5 DNA Structure

Test Yourself

Knowledge and Synthesis Questions

1. Which of the following statements concerning polymers is *not* true?
   a. Polymers are synthesized from monomers during condensation.
   b. Polymers are synthesized from monomers during dehydration.
   c. Polymers consist of at least two types of monomers.
   d. Both b and c
   Textbook Reference: *3.1 What Kinds of Molecules Characterize Living Things? p. 39*

2. You are a biochemist and have recently discovered a new macromolecule. Studies of the bond types found in this macromolecule reveal many hydrogen bonds and peptide bonds. You most likely have found what type of macromolecule?

a. Carbohydrate
b. Lipid
c. Protein
d. Nucleic acid
Textbook Reference: *3.2 What Are the Chemical Structures and Functions of Proteins? p. 45*

3. An α helix is an example of which level of protein structure?
a. Primary
b. Secondary
c. Tertiary
d. Quaternary
Textbook Reference: *3.2 What Are the Chemical Structures and Functions of Proteins? p. 45*

4. You have isolated a monomer with the following components: a phosphate group, a sugar, and a nitrogen-containing base. Polymers synthesized from this monomer belong to what class of macromolecule?
a. Carbohydrate
b. Lipid
c. Protein
d. Nucleic acid
Textbook Reference: *3.5 What Are the Chemical Structures and Functions of Nucleic Acids? p. 57*

5. Cellulose and starch are composed of the same monomers. Which of the following results in their being structurally and functionally different?
a. They have different types of glycosidic linkages.
b. They have different numbers of glucose monomers.
c. They are held together by different bond types.
d. None of the above
Textbook Reference: *3.3 What are the Chemical Structures and Functions of Carbohydrates? p. 51*

6. DNA utilizes the bases guanine, cytosine, thymine, and adenine. In RNA, _____ is replaced by _____.
a. adenine; arginine
b. thymine; uracil
c. cytosine; uracil
d. cytosine; arginine
Textbook Reference: *3.5 What Are the Chemical Structures and Functions of Nucleic Acids? p. 59*

7. The pairing of purines with pyrimidines to create a double-stranded DNA molecule is called
a. complementary base pairing.
b. phosphodiester linkages.
c. antiparallel synthesis.
d. dehydration.
Textbook Reference: *3.5 What Are the Chemical Structures and Functions of Nucleic Acids? p. 57*

8. Triglycerides are synthesized from
a. glycerol and amino acids.
b. fatty acids and glycerol.
c. steroid precursors and starch.
d. cholesterol and glycerol.
Textbook Reference: *3.4 What Are the Chemical Structures and Functions of Lipids? p. 55*

9. Amino acids are linked together into proteins by which of the following bond types?
a. Noncovalent bonds
b. Peptide bonds
c. Phosphodiester bonds
d. Both a and b
Textbook Reference: *3.2 What Are the Chemical Structures and Functions of Proteins? p. 44*

10. Which of the following characteristics differentiate carbohydrates from other macromolecule types?
a. Carbohydrates are constructed of monomers that always have a ring structure.
b. Carbohydrates never contain nitrogen.
c. Carbohydrates consist of a carbon bonded to hydrogen and a hydroxyl group.
d. None of the above
Textbook Reference: *3.3 What Are the Chemical Structures and Functions of Carbohydrates? p. 49*

11. Which of the following statements about carbohydrates is *not* true?
a. Monomers of carbohydrates have six carbon atoms.
b. Monomers of carbohydrates are linked together during dehydration.
c. Carbohydrates are energy storage molecules.
d. None of the above
Textbook Reference: *3.3 What are the Chemical Structures and Functions of Carbohydrates? p. 50*

12. What would you expect to be true of the R groups of amino acids located on the surface of protein molecules found within the interior of biological membranes?
a. The R groups would be hydrophobic.
b. The R groups would be hydrophilic.
c. The R groups would be polar.
d. The R groups would be able to form disulfide.
Textbook Reference: *3.2 What Are the Chemical Structures and Functions of Proteins? p. 48; 3.4 What Are the Chemical Structures and Functions of Lipids? p. 56*

13. What characteristic of phospholipids allows them to form a bilayer?
a. They have a hydrophilic fatty acid tail.
b. They have a hydrophobic head.
c. They have a hydrophobic fatty acid tail.
d. All of the above
Textbook Reference: *3.4 What Are the Chemical Structures and Functions of Lipids? p. 55*

14. The first biological catalyst may have been _____.
a. glucose
b. DNA
c. a protein
d. a ribozyme
Textbook Reference: *3.6 How Did Life on Earth Begin? p. 63*

15. The double helix formation of DNA is due to _____.
a. ionic bonds
b. covalent bonds
c. hydrogen bonds

d. hydrophobic side chains
Textbook Reference: 3.5 What Are the Chemical Structures and Functions of Nucleic Acids? p. 59

Application Questions

1. Differentiate between primary, secondary, tertiary, and quaternary structures as they relate to proteins. If a protein is immersed in a pH solution far from its optimum, which structures are most likely to disassociate first, and why?
Textbook Reference: 3.2 What Are the Chemical Structures and Functions of Proteins? pp. 45–46

2. Dietary guidelines encourage people to stay away from saturated fats. What is meant by the term "saturated fat"? Why is this fat type of more concern than unsaturated fats in the diet? How are the two fats structurally different?
Textbook Reference: 3.4 What Are the Chemical Structures and Functions of Lipids? p. 55

3. Draw a phospholipid. What characteristics of phospholipids make them perfectly suited for membranes? What do you think might happen if phospholipids did not form a bilayer? How might they arrange themselves in an aqueous environment?
Textbook Reference: 3.4 What Are the Chemical Structures and Functions of Lipids? pp. 55–56

4. Complex carbohydrates should be a mainstay of your diet. What properties make them excellent food sources?
Textbook Reference: 3.3 What are the Chemical Structures and Functions of Carbohydrates? p. 51

5. Discuss how a protein's three-dimensional structure makes it perfect for acting as a carrier and receptor molecule. Why are proteins uniquely suited for this function, whereas other macromolecules are not?
Textbook Reference: 3.2 What Are the Chemical Structures and Functions of Proteins? p. 47

6. What properties of a DNA structure make it well suited for its function as an informational molecule?
Textbook Reference: 3.5 What Are the Chemical Structures and Functions of Nucleic Acids? p. 59

7. Amino acid R groups have a variety of chemical properties. How do these different properties contribute to the final three-dimensional shape of the molecule?
Textbook Reference: 3.2 What Are the Chemical Structures and Functions of Proteins? p. 46

8. You have isolated a protein with the following amino acid sequence: RSCFLA. Using Table 3.2 in your book, draw this protein. In your drawing, label the N terminus and the C terminus and show all peptide linkages. How many water molecules were generated in the synthesis of this protein?
Textbook Reference: 3.2 What Are the Chemical Structures and Functions of Proteins? p. 43

9. Consider the following triglycerides (*A* and *B*) in answering a–d.

A   *B*

a. In *B*, circle the remnant of the glycerol portion of the triglyceride.
b. Which triglyceride (*A* or *B*) is probably a solid at room temperature? _____ Explain your answer.
c. Which triglyceride (*A* or *B*) is probably derived from a plant? _____ Explain your answer.
d. How many water molecules result from the formation of triglyceride *B* from glycerol and three fatty acids?
Textbook Reference: 3.4 What Are the Chemical Structures and Functions of Lipids? p. 55

10. Use the base-pairing rules for DNA and RNA to label the complementary strand of RNA (right) to the single strand of DNA (left) shown following, where C = cytosine, G = guanine, A = adenine, T = thymine, and U = uracil.

Circle and label an example of a nucleotide and a nucleoside in the figure showing the double-stranded DNA and RNA hybrid. Based on the orientation of the sugar molecules, label the four ends of the molecule as 3′ or 5′ in the figure showing the double-stranded DNA and RNA hybrid.

Textbook Reference: *3.5 What Are the Chemical Structures and Functions of Nucleic Acids? p. 58*

Answers

Knowledge and Synthesis Answers

1. **d.** Polymers are synthesized via condensation and broken down via hydrolysis. Polymers may be made of a single type of monomer or from multiple types of monomers.

2. **c.** Proteins consist of amino acids joined by peptide bonds. Higher levels of structure are stabilized by hydrogen bonds.

3. **b.** α helices are examples of secondary structure and are maintained by hydrogen bonds between the amino acid residues.

4. **d.** These are all constituents of a nucleotide, the monomer that makes up nucleic acids.

5. **a.** Starch has β-glycosidic linkages; whereas cellulose has α-glycosidic linkages. This leads to structural and functional differences between the two macro-molecules.

6. **b.** RNA and DNA differ by one oxygen molecule in their ribose sugar and in the substitution of uracil (RNA) for thymine (DNA).

7. **a.** Complementary base pairing results from the attraction of charges and the ability to pair purines with pyrimidines through hydrogen bond formation.

8. **b.** Triglycerides are formed from one glycerol and three fatty acid molecules.

9. **b.** The peptide bond that links amino acids to form proteins is a type of covalent bond.

10. **c.** Carbohydrates always have carbons bonded to hydrogens and hydroxyl groups. They may have a variety of other associate molecules in addition to these.

11. **a.** Carbohydrate monomers may have different numbers of carbon atoms. Five-carbon monomers (pentoses) and six-carbon monomers (hexoses) are both common.

12. **a.** The interior of the plasma membrane is hydrophobic; therefore, an imbedded protein would have hydrophobic residues.

13. **c.** Phospholipids are composed of a hydrophilic head and a hydrophobic tail. When placed in water the hydrophobic tails come together in the interior of the bilayer, surrounded by the hydrophilic heads facing outward.

14. **d.** The first catalysts may have been ribozymes. These molecules can catalyze reaction on many cellular substances.

15. **c.** The double-helix structure of DNA is due to the hydrogen bonding.

Application Answers

1. See Figure 3.7 in your text for diagrams of primary, secondary, tertiary, and quaternary structures. The quaternary structure is the least stable and breaks down first in nonoptimum conditions. Protein structures continue to denature by unfolding the tertiary structures, then secondary structures. Primary structure is maintained by covalent bonds and is the last to break down. Disruptions in tertiary and quaternary structures are often reversible. Disruptions in primary or secondary structures are generally irreversible.

2. Saturated fats contain only single bonds and are "saturated" with hydrogen. This allows saturated fat molecules to pack together densely. This characteristic is easily seen in that most saturated fats are solid at room temperature. Unsaturated fats contain double bonds that affect the shape of the molecule. They are not "saturated" with hydrogen. This characteristic keeps them from packing together tightly, and they tend to be liquid at room temperature. Saturated fats are dangerous because of this "packing" ability.

3. See Figures 3.20 in your book for drawings of phospholipids. The hydrophilic head and the hydrophobic tail of phospholipids allow them to have an "inside" that resists an aqueous environment and an "outside" that can reside in such an environment. When they exist as a bilayer, the hydrophobic tails aggregate together. If they did not exist in two layers, the tails would still try to aggregate. This would result in a spherical aggregation of phospholipids called a micelle in which the tails are arranged toward the center of the sphere, away from the aqueous environment, and the heads are immersed in the aqueous environment.

4. Complex carbohydrates are easily broken down into glucose monomers, which provide nearly all cellular energy. By storing glucose monomers in large carbohydrates, the osmotic strain on any given cell is reduced without sacrificing availability of energy.

5. The three-dimensional nature of proteins allows them to form binding sites. These binding sites are uniquely shaped to interact with other molecules.

6. DNA structure is highly regular and conserved, which makes it usable by all cells, but the different bases allow it to encode specific genetic information.

7. The size of the R group, the charge of the R group, and any special binding properties all contribute to the final orientation of a protein molecule.

8. Five water molecules are generated.

9. a.

b. Triglyceride **A** is probably solid at room temperature. Its fatty acid chains are saturated (no double bonds) and relatively long, both characteristics of solid, animal-derived triglycerides.

c. Triglyceride **B** is probably derived from a plant. Its fatty acid chains are unsaturated (double bonds) and relatively short, both characteristics of liquid, plant-derived triglycerides.

d. **Three** water molecules will result. A water molecule results for each of the three fatty acids added to glycerol by a condensation reaction.

10.

CHAPTER 4 Cells: The Working Units of Life

Important Concepts

The cell is the building block of life.

- The cell theory states that the cell is the basic unit of life, that all organisms are composed of cells, and that all cells come from preexisting cells.

- Cell size is limited by surface area-to-volume ratio. Very large cells are not feasible because as volume increases, surface area increases at a slower rate. The capacity of a cell for chemical activity is related to the cell volume. However, cells with large volumes are unable to take up enough material across their surface to support the metabolic activity of the volume.

- Because cells are small, either light microscopes or electron microscopes must be used to visualize them. Light microscopes, using lenses and light, resolve objects as small as 0.2 μm. Electron microscopes, using electron beams, can resolve objects as small as 0.2 nm. Staining techniques are used to assist in visualizing cellular components.

- All cells are surrounded by a plasma membrane composed of a phospholipid bilayer with many embedded and protruding proteins. The plasma membrane and the associated proteins are selectively permeable, permitting some small molecules to pass through but preventing others from doing so. The membrane allows for maintenance of the internal cellular environment and is responsible for communicating and interacting with other cells.

Cells are classified as either prokaryotic or eukaryotic.

- Prokaryotic cells do not have internal membrane compartments and are generally smaller than eukaryotic cells. Prokaryotic are one-celled organisms found only in domains Archaea and Bacteria.

- Prokaryotic cells have a plasma membrane and a nucleoid containing the genetic material that is not membrane-enclosed. The cytoplasm of the cell consists of the liquid cytosol and insoluble particles that act as subcellular machinery (i.e., ribosomes for protein synthesis).

- Some prokaryotic cells have special feature:
  - Cell walls: This rigid structure is outside the cell membrane and is composed of peptidoglycans. Cell wall type is frequently used as a characteristic to identify bacteria.
  - Outer membrane: This membrane is outside the cell wall. It is polysaccharide rich, and some of the polysaccharides are disease-causing toxins.
  - Capsule: This slime layer is outside the wall and outer membrane. Though it provides protection, it is not necessary for the cell's survival.
  - Internal membranes: Photosynthetic prokaryotes have stacks of folded membranes on which photosynthesis is carried out. Other prokaryotes have membrane folds, which function in energy reactions and cell division.
  - Flagella: These tiny protein machines containing the protein flagellin cause motion of the cell.
  - Pili: These are protein extensions that assist in adherence.
  - Cytoskeleton: Actin-like proteins may help the cell maintain cell shape.

- Eukaryotic cells are larger than prokaryotic cells and are distinguished by having membrane-enclosed internal components such as the nucleus, mitochondrion, endoplasmic reticulum, Golgi apparatus, lysosome, vacuole, and chloroplast.

- Each organelle (membrane-enclosed compartment) has its own internal environment uniquely suited to its function. The membrane assists in regulating that environment. The function and study of organelles can be carried out by microscopy or by cell fractionation and organelle isolation.

Some organelles function to store and process information.

- The nucleus stores DNA and is the site of DNA duplication and DNA regulation of cellular activity. The nucleolus region of the nucleus is involved in ribosome assembly and RNA synthesis.

- The nucleus is bounded by a double membrane called the nuclear envelope. Small openings in the nuclear envelope called nuclear pores allow passage of RNA and ribosomes to the cell cytoplasm. The outer membrane of the nuclear envelope is continuous with the endoplasmic reticulum.

- Inside the nucleus, DNA molecules and proteins make up chromatin. At cell division, chromatin condenses to form chromosomes. Chromatin is organized on the nuclear matrix and the nuclear lamina. The lamina helps maintain nuclear shape.

- Ribosomes found in the cytoplasm, on the surface of the endoplasmic reticulum, and in the mitochondria act as information transcription centers and guide the synthesis of proteins from nucleic acid blueprints.

The endomembrane system functions in manufacturing and packaging cellular products.

- The endomembrane system consists of endoplasmic reticulum (ER), Golgi apparatus, and vesicles that shuttle between the two.

- The endoplasmic reticulum is a complex of membrane sacs throughout the cell and is continuous with the nuclear membrane. The ER has a large surface area due to the large amount of folding of the membrane.

- The endoplasmic reticulum is classified into two types based on the presence or absence of attached ribosomes.

  - Rough endoplasmic reticulum (RER) is studded with active ribosomes. Because ribosomes actively aid in the synthesis of new proteins, the RER is important for storage, transport, and modification of the proteins. Carbohydrate groups are added to proteins to make glycoproteins in the RER.

  - Smooth endoplasmic reticulum also functions in protein modification and transport but, more importantly, functions in the modification of chemicals taken in by the cell and as the site of hydrolysis of glycogen and synthesis of steroids and lipids.

- The Golgi apparatus, composed of flattened membrane stacks called cisternae and vesicles, is important in protein modification, packaging, and polysaccharide synthesis in plants.

- Different regions of the Golgi have different enzymes and functions. The *cis* region is closest to the nucleus or RER. The *trans* region is closest to the cell surface. Proteins are released from the ER in a vesicle and are transported to the *cis* region, where the vesicle membrane fuses with Golgi membrane to release the contents into the Golgi. Vesicles containing proteins in turn pinch off of the *trans* Golgi for transport away from the Golgi.

- Lysosomes are "digestion centers" within a cell that break down proteins, polysaccharides, nucleic acids, and lipids into their monomer components. They contain a host of powerful enzymes in a slightly acidic environment that break down engulfed molecules and cellular waste. In the process of phagocytosis, the cell takes up material in a phagosome, which fuses with a primary lysosome to make a secondary lysosome. Through the process of autophagy, lysosomes digest organelles such as mitochondria, breaking them down to monomers for reuse in new organelles.

Mitochondria and plastids function in energy transformation.

- During cellular respiration, mitochondria convert the potential chemical energy stored in glucose into adenosine triphosphate (ATP), a form usable by the cell.

- Mitochondria have two membranes—an outer smooth membrane and a highly folded internal membrane. The folds are called cristae, and the remaining internal space is called the matrix. The matrix holds ribosomes and DNA. Protein complexes used during cellular respiration are embedded in the cristae.

- Plastids are not found in animal cells, but in plants and protists. Plant cells have several different types of plastids, each with unique functions.

- Chloroplasts are the site of photosynthesis where light energy is converted to chemical energy. Chloroplasts are green because they contain the photosynthetic pigment chlorophyll.

- Chloroplasts have two membranes. The innermost membrane contains circular compartments, or thylakoids, that are folded into stacks called grana and hold chlorophyll and enzymes for photosynthesis. The liquid content of the chloroplast is called stroma and contains ribosomes and DNA.

- Other plastid types include chromoplasts, which contain pigments involved in flower color, and starch-storing leucoplasts.

Other specialized organelles are found in some but not all cells.

- Peroxisomes function to break down harmful peroxide by-products.

- Glyoxysomes are found in young plants, and they convert lipids to carbohydrates.

- Vacuoles, found in plants and protists, function in storage of waste products, structural support, reproduction, food store, or water regulation, depending on the organism.

The cytoskeleton provides support, shape, and movement for the cell.

- Microfilaments, composed of the protein actin, stabilize cell shape and assist with contraction of the cell. These filaments are involved in muscle cell contraction, cell division, cytoplasmic streaming, and cell shape. In muscle cells the protein myosin interacts with actin to produce muscle contraction.

- Intermediate filaments are found only in multicellular organisms and function in stabilizing structure and resisting tension. These make up the lamina, form desmosomes, and anchor the nucleus.

- Microtubules are long hollow tubes of the dimer protein tubulin that contribute to the rigidity of the cell and act as a framework of movement for motor proteins. They have a very specific structure that can be quickly added to or reduced. Motor proteins use microtubules as tracks from one area of the cell to another.

- Cilia and flagella of eukaryotes are powered by micro-tubules. Though cilia and flagella differ in size and function, they have the same basic structure: a "9 + 2" arrangement of microtubules (see Figure 4.22).

- Movement of cilia and flagella occurs when the micro-tubules slide past one another. The sliding is caused by an ATP-driven shape change in molecules of dynein while bound to the microtubules (see Figure 4.23). The motor proteins called kinesin are involved in the transfer of vesicles along microtubules within the cell.

- Centrioles are also made up of microtubules and are involved in cell division.

Extracellular structures are outside the plasma membrane and provide support, protection, and anchoring for the cell.

- Plants have semirigid cell walls composed of cellulose, which function to support and protect the cell. Plasmodesmata are small holes in the cell walls that allow connections between plant cells.

- The collagen- and glycoprotein-containing extra-cellular matrix of some animal cells functions to hold cells together, filter materials, orient cell movement, and assist with chemical signaling. Some cells, such as bone cells, secrete an elaborate and rigid matrix.

The endosymbiotic theory explains the origin of mitochondria and chloroplasts.

- Mitochondria and chloroplasts have their own DNA and are enclosed in a double membrane. They also reproduce within the cell. In other words, they act very much like cells within cells, but they are still under nuclear control.

- The endosymbiotic theory suggests that a small photo-synthetic prokaryote was engulfed and retained by a larger cell, with each getting mutual benefit from the other (endosymbiosis). Mitochondria may have origins as an engulfed respiring prokaryote (see Figure 4.26).

- There is evidence to support this theory, including DNA moving between organelles, similarities in DNA and biochemistry between modern-day bacteria and chloroplasts, and the potential for the initial formation of the mitochondria and chloroplasts' outer membrane from the engulfing cells' plasma membrane.

The Big Picture

- All living things are composed of cells, and all cells come from preexisting cells. There are two basic cell forms: the small prokaryotic cell with no organelles and the eukaryotic cell with many membrane-enclosed organelles.

- The organelles of eukaryotic cells have a variety of functions, including information storage and transmission, modification and packaging of cellular products, and energy transformation. See Figure 4.7 of the textbook to help you understand the organization, structure, and function of each component.

- The cytoskeleton of the cell provides support, structure, and protection. Components of the cytoskeleton also provide movement of the cell and within the cell.

Common Problem Areas

- People often are overwhelmed by the number of organelles and their diverse functions in a eukaryotic cell. Remember each organelle has a distinct location within the cell and structure, which are related to the function.

Study Strategies

- Terminology is the most difficult part of this material. It is easier to learn if you understand the concepts before you memorize the terms.

- Use Figure 4.7 while learning the organelles and their components. Visualizing the jobs each must do and how their functions relate to the structure makes learning easier.

- Review the following animated tutorials and activities on the Companion Website/CD:
Tutorial 4.1 Eukaryotic Cell Tour
Tutorial 4.2 The Golgi Apparatus
Activity 4.1 The Scale of Life
Activity 4.2 Know Your Techniques
Activity 4.3 Lysosomal Digestion

Test Yourself

Knowledge and Synthesis Questions

1. You have found a mass of cells in the sediment surrounding a thermal vent in the ocean floor. The salinity in the area is quite high. Upon microscopic examination of the cells, you find no evidence of membrane-enclosed organelles. How would you classify this cell?
   a. As a eukaryotic cell
   b. As a prokaryotic cell
   c. As a member of domain Archaea or Bacteria
   d. Both b and c
   Textbook Reference: *4.2 What Are the Characteristics of Prokaryotic Cells? pp. 72–73*

2. Centrifugation of a cell results in the rupture of the cell membrane and the contents compacting into a pellet in the bottom of the centrifuge tube. Bathing this pellet with a glucose solution yields metabolic activity, including the production of ATP. One of the contents of this pellet is most likely which of the following?
   a. Cytosol

b. Mitochondria

c. Lysosomes

d. Golgi bodies

Textbook Reference: *4.3 What Are the Characteristics of Eukaryotic Cells? p. 82*

3. Eukaryotic cells are thought to be derived from prokaryotic cells that underwent phagocytosis without digestion of the phagocytized cell. This mutualistic relationship is explained by the _____ theory.

a. endosymbiotic

b. cell

c. evolutionary

d. parasite

Textbook Reference: *4.5 How Did Eukaryotic Cells Originate? p. 92*

4. Though science fiction has produced stories like "The Blob," we don't see very many large single-celled organisms. Which of the following tends to limit cell size?

a. The ability to maintain a continuous large membrane

b. The ability to reproduce a large cell

c. Surface area-to-volume ratios

d. All of the above

Textbook Reference: *4.1 What Features of Cells Make Them the Fundamental Unit of Life? p. 70*

5. Microscopes are used to resolve images that cannot be seen with the unaided eye. Electron microscopes use _____ to resolve images, whereas light microscopes use _____ to resolve images.

a. light and lenses; diffraction of electron beams

b. diffraction of electron beams; light and lenses

c. lasers; light and lenses

d. None of the above

Textbook Reference: *4.1 What Features of Cells Make Them the Fundamental Unit of Life? pp. 70–72*

6. What is the correct cellular function of the RER?

a. DNA synthesis

b. Photosynthesis

c. Cellular respiration

d. Protein synthesis

Textbook Reference: *4.3 What Are the Characteristics of Eukaryotic Cells? p. 79*

7. Photosynthesis occurs in the _____.

a. chloroplast

b. mitochondria

c. Golgi apparatus

d. nucleus

Textbook Reference: *4.3 What Are the Characteristics of Eukaryotic Cells? p. 83*

8. Lysosomes are involved in _____.

a. DNA synthesis

b. breakdown of phagocytized material

c. protein folding

d. pigment production

Textbook Reference: *4.3 What Are the Characteristics of Eukaryotic Cells? p. 81*

9. The packaging of proteins to be used outside the cell occurs in the _____.

a. nucleus

b. SER

c. Golgi apparatus

d. chromoplast

Textbook Reference: *4.3 What Are the Characteristics of Eukaryotic Cells? pp. 80–81*

10. Which of the following organelles are enclosed in double membranes?

a. Nucleus

b. Chloroplast

c. Mitochondrion

d. All of the above

Textbook Reference: *4.3 What Are the Characteristics of Eukaryotic Cells? pp. 75, 82–83*

11. Movement of cells is accomplished in both prokaryotes and eukaryotes with which of the following structures?

a. Cilia

b. Pili

c. Dynein

d. Flagella

Textbook Reference: *4.3 What Are the Characteristics of Eukaryotic Cells? pp. 88–89*

12. Which of the following statements is true regarding mitochondria and chloroplasts?

a. Animal cells produce chloroplasts.

b. Mitochondria and chloroplasts may be found in the same cell.

c. Mitochondria and chloroplasts are not found in the same cell.

d. Chloroplasts can revert to mitochondria in certain conditions.

Textbook Reference: *4.3 What Are the Characteristics of Eukaryotic Cells? pp. 82–83*

13. Which of the following best describes ribosomes?

a. Ribosomes guide protein synthesis.

b. Ribosomes are found only in the nucleus or on the RER.

c. There are no ribosomes in the mitochondria.

d. All of the above

Textbook Reference: *4.3 What Are the Characteristics of Eukaryotic Cells? p. 79*

14. Nuclear DNA exists as a complex of proteins called _____ that condenses into _____ during cellular division.

a. chromosomes; chromatin

b. chromatids; chromosomes

c. chromophors; chromatin

d. chromatin; chromosomes

Textbook Reference: *4.3 What Are the Characteristics of Eukaryotic Cells? p. 79*

15. Rough endoplasmic reticulum and smooth endoplasmic reticulum differ

a. only by the presence or absence of ribosomes.

b. both in the presence or absence of ribosomes and in their function.

c. only in microscopic appearance.

d. None of the above

Textbook Reference: *4.3 What Are the Characteristics of Eukaryotic Cells? pp. 79–81*

Application Questions

1. The role of a certain cell in an organism is to secrete a protein. Trace the production of that protein from the nucleus through all necessary organelles to the point of release from the cell.

   Textbook Reference: *4.3 What Are the Characteristics of Eukaryotic Cells? pp. 79–81*

2. Explain how microtubules and dynein function to make cilia and flagella move.

   Textbook Reference: *4.3 What Are the Characteristics of Eukaryotic Cells? pp. 88–90*

3. Compare and contrast prokaryotic and eukaryotic cells.

   Textbook Reference: *4.2 What Are the Characteristics of Prokaryotic Cells? p. 72*

4. Explain the significance of organelles. What are the costs and benefits of having large compartmentalized cells?

   Textbook Reference: *4.3 What Are the Characteristics of Eukaryotic Cells? p. 75*

5. What is the primary function of a cell membrane? What characteristics of membranes allow them to contribute to metabolic activity?

   Textbook Reference: *4.1 What Features of Cells Make Them the Fundamental Unit of Life? p. 72*

6. The organelles that contain their own DNA are all enclosed in double membranes. Relate this observation to the endosymbiotic theory.

   Textbook Reference: *4.5 How Did Eukaryotic Cells Originate? p. 92*

7. Vacuoles in plants hold waste material. Why do plants tend to retain cellular wastes whereas animal cells do not?

   Textbook Reference: *4.3 What Are the Characteristics of Eukaryotic Cells? pp. 84–86*

8. There are structural similarities between mitochondria and chloroplasts. Using the idea that form follows function, why are there similarities in these two organelles?

   Textbook Reference: *4.3 What Are the Characteristics of Eukaryotic Cells? pp. 82–83*

Answers

Knowledge and Synthesis Answers

1. **b.** Several characteristics suggest this is a prokaryote. It survives in high salinity and high heat, but the sure indication is that it contains no membrane-enclosed organelles.

2. **b.** The pellet is undergoing cellular respiration, a function that occurs in the mitochondria. You can also assume that if the single membrane of the cell itself is ruptured, other organelles enclosed in single membranes would be ruptured as well.

3. **a.** The endosymbiotic theory is used to explain the presence of DNA in mitochondria and chloroplasts as well as the presence of two membranes around these organelles.

4. **c.** As volume increases, the surface area available for exchange does not increase proportionally. There comes a point at which the surface is not large enough for maintenance of the metabolic activity of the cell.

5. **b.** In electron microscopy, a concentrated beam of electrons is focused on an object, allowing resolution of structures as small as 2 nm. Light microscopy, using lights and lenses, can only resolve objects down to approximately 0.2 µm.

6. **d.** The RER is the site of protein synthesis.

7. **a.** The chloroplasts are the organelles involved in photosynthesis.

8. **b.** Lysosomes are organelles that contain digestive enzymes that are used to breakdown macromolecules taken in by phagocytosis.

9. **c.** The Golgi apparatus packages proteins for both internal use and external use.

10. **d.** The nucleus, the mitochondria, and the chloroplasts are the only organelles enclosed in double membranes.

11. **d.** Though the flagella are of different structures, they serve the same role in prokaryotes and eukaryotes.

12. **b.** Mitochondria and chloroplasts may be found in the same cell. Almost all eukaryotic cells contain mitochondria.

13. **a.** Ribosomes, found in the nucleus, the cytosol, and in organelles such as the mitochondria, RER, and chloroplasts, complex with RNA to guide protein production.

14. **d.** The complex of proteins and DNA is called chromatin. Chromatin only takes the form of chromosomes during cell reproduction.

15. **b.** Both the structure and the function of RER and SER differ.

Application Answers

1. The protein could be synthesized on ribosomes in the nucleus, or synthesis could take place in the cytosol or on the RER. From there it most likely would be transported along the RER for further modification before being packaged in a bit of ER membrane and transported to the *cis* region of the Golgi. Further modification might take place in the cisternae of the Golgi before the protein is encased in a final vesicle. When the vesicle fuses with the plasma membrane of the cell, its contents are released to the outside.

2. Dynein molecules bind to pairs of microtubules in the flagella or cilia. With the addition of cellular energy, the dynein molecules undergo a conformational change that results in microtubules sliding past one another, resulting in a whiplike motion of the flagella.

3. Prokaryotic Cells
    Small in size
    No membrane-enclosed organelles
    Found only in domains Archaea and Bacteria
    DNA is in nucleoid

    Eukaryotic Cells
    10 or more times greater in size
    Membrane-enclosed organelles
    Found in all domains other than Archaea and Bacteria
    DNA is in nucleus

4. Organelles allow different metabolic environments to exist in the same cell. This partitioning of jobs allows for greater specialization but comes at an energy cost. Eukaryotic cells are more energy expensive.

5. A cell membrane exists to form an inside and an outside of a cell. The presence of an inside and an outside allows the establishment of different environments. In addition, membranes hold integral proteins with a variety of chemical properties and activities. This allows for the enzymatic activity associated with membranes. Stacks of membranes, such as those in mitochondria and chloroplasts, increase the amount of chemical activity in an area.

6. See Figure 4.26 for a description of the origin of double membranes from endosymbiosis.

7. The presence of cell walls in plants limits what can be expelled outside the plant cell. Holding wastes in an internal compartment circumvents this problem.

8. Both mitochondria and chloroplasts are involved in energy-transformation activities that require many enzymes. The stacking or folding of membranes provides enzymatic activity centers for these reactions.

CHAPTER 5 The Dynamic Cell Membrane

Important Concepts

Biological membranes have a specific content and structure.

- Biological membranes are composed of lipids, proteins, and carbohydrates. Lipids provide the structure and barrier functions for the membrane. Proteins are involved in creating channels and transporting materials across the lipid barrier. Carbohydrates are involved in signaling and adhesion, and they are found on the outside of the plasma membrane. The fluid mosaic model describes how lipids interact to produce a fluid membrane on which the proteins and carbohydrates float.

- The majority of a biological membrane is composed of lipids, with phospholipids being the most abundant component. Phospholipids have both hydrophilic ("water-loving") heads and hydrophobic ("water-hating") tails. They arrange themselves into a bilayer with the hydrophobic tails touching and the heads extending into the aqueous environment inside and outside the cell. This arrangement allows for the fluid movement of the two layers on top of one another and sealing of any disruptions in the membrane.

- Though the basic structure of a bilayer is always the same, the inner and outer halves differ in lipid composition and thus have slightly different properties. Phospholipids differ in their length, degree of unsaturation, and degree of polarity. Additionally, cholesterol can make up 25 percent of the lipid content in the membrane. The amount of cholesterol present and the degree of fatty acid saturation influences the fluidity of the membrane.

- Proteins may be embedded in or extend across membranes. Regions or domains of a membrane protein with hydrophobic amino acid side chains tend to be found in the hydrophobic environment of the membrane. The regions of a membrane protein containing hydrophilic amino acid side chains extend out from the membrane. The phospholipids and proteins are independent of one another, allowing for movement throughout the membrane. This happens for the most part because the constituents interact only noncovalently.

- Proteins that penetrate the phospholipid bilayer are called integral proteins. Special types of integral proteins, transmembrane proteins, span the entire bilayer. Those not embedded are referred to as peripheral proteins. Proteins are distributed in membranes asymmetrically "as needed," and the numbers of proteins and their placement vary greatly based on cell type. The "inside" and "outside" of a membrane often have different properties due to the different characteristics of the transmembrane proteins on the two sides.

- Some membrane proteins may be anchored to cytoskeletal components or bound to lipid rafts. Such proteins have very specific functions.

- Membranes are in a constant state of flux. Phospholipids are produced on the surface of the smooth endoplasmic reticulum and distributed to membranes throughout the cell. Vesicles fuse with the plasma membrane in the process of exocytosis. Endocytosis balances this by extracting membrane from the plasma membrane.

- Membrane carbohydrates serve as recognition sites for other cells and molecules. Carbohydrates are frequently bound to lipids or proteins, forming glycolipids or glycoproteins, respectively. Because of the nature of carbohydrates, they contribute to cell adhesion.

Cells must be able to recognize and adhere to other cells to form tissues, organs, and organisms.

- Cells have the ability to recognize and adhere to other cells allowing for the grouping of cells in an organism. Cell adhesions are specific to the organism and to the cell type. A sponge is an animal composed of many cells that exhibits species-specific cell adhesion.

- A cell's ability to recognize and adhere to other cells is due to the presence of recognition proteins on the cell surface. Often the proteins on two cells that adhere to one another are the same; these molecules and the bonds they form are referred to as homotypic. Occasionally the two proteins are different, but have complementary binding sites. In this case the molecules and the bonding are called heterotypic.

Cells must make connections to other cells to communicate and stabilize tissues.

- Cell junctions are sites where cell recognition proteins from two cells form a connection. The three main types of cell junctions are tight junctions, desmosomes, and gap junctions.

- Tight junctions are found specifically in epithelial cells and function to prevent substances from moving through the spaces between cells. They also restrict migration of membrane proteins and phospholipids that would otherwise be free to float around in the "fluid" plasma membrane. Tight junctions help tissues move materials in specific directions.

- Desmosomes are structural connections between cells that hold them together. Desmosomes involve plaques on the plasma membrane that attach to special cell adhesion molecules. Because a desmosome involves the keratin fibers of the cytoskeleton and crosses the cell membrane, the connections have integrity and are not easily broken.

- Gap junctions facilitate communication between cells. They are made of protein connexons that span adjacent plasma membranes and act as channels.

Transport of substances across membranes can be a passive process.

- Plasma membranes are not permeable to all substances, but are selectively permeable. Passive transport is the movement of a substance across a lipid bilayer through channel proteins or by carrier molecules that does not require an input of energy. Passive transport can either occur by simple diffusion or facilitated diffusion.

- Diffusion is a process of random movement toward equilibrium and is the net movement of a substance from areas of greater concentration to lesser concentration. The rate of diffusion depends on the size of diffusing substance, electrical charge of diffusing substance, temperature of solution, and concentration gradient.

- Diffusion within small areas such as single cells may occur rapidly, but it occurs more slowly with increasing distance.

- Membranes alter the rate of diffusion. A membrane is said to be permeable to those substances that can pass through it and impermeable to those that cannot. If a substance can pass through a membrane, it will diffuse until concentrations on either side of the membrane are equal. At equilibrium, the molecules of the substance continue to move across the membrane, but the net movement of molecules in both directions is equal. If a substance cannot permeate the membrane, it cannot diffuse across it.

- In simple diffusion, small molecules pass through a membrane. The more lipid soluble the molecule is, the faster it diffuses. Lipid-soluble molecules readily pass through biological membranes. In contrast, charged and polar molecules do not readily pass through a membrane due to the formation of many hydrogen bonds with water and the hydrophobic nature of the internal layer of the membrane. The exception to this rule is water.

- Osmosis is the diffusion of water across a membrane. Water will move across a membrane to equalize solute concentrations on either side of the membrane if the solute cannot move. Water moves from areas of low solute concentration to high solute concentration in an attempt to equalize solute concentrations on both sides of the membrane.

- Solute concentrations separated by a membrane are classified as isotonic, hypertonic, or hypotonic. Isotonic solutions have the same solute concentrations on both sides of a membrane. A hypertonic solution has a higher solute concentration than the solution on the other side of the membrane. A hypotonic solution has a lower solute concentration than the solution with which it is being compared. Environmental and cellular solute concentrations dictate the direction of osmosis in living cells.

- Turgor pressure is the pressure within cells and is due to the uptake of water by osmosis. Turgor pressure can build up only if there is a cell wall to limit cell expansion.

Facilitated diffusion uses proteins that span the membrane.

- Facilitated diffusion is the passive movement of a substance across a membrane with the help of membrane-bound proteins that act as channels or carriers.

- A substance may cross the membrane by facilitated diffusion through protein channels running through the plasma membrane. The pores of these proteins have polar amino acids that allow polar molecules and ions to cross the membrane.

- The best studied channels are ion channels. Most ion channels are either ligand-gated or voltage-gated, allowing the passage of ions to be controlled depending on the cellular environment. Movement of ions through a voltage-gated channel depends on both the concentration gradient for the ion and the electrochemical gradient. Cells have an unequal distribution of ions between their inside and outside. There is typically a K^+ imbalance, with a larger amount of K^+ inside than outside. This imbalance produces a membrane potential that can be calculated with the Nernst equation. The membrane potential of a cell is typically around −70 mV.

- Ion channels are very specific for the ions that they allow to pass through their pore. Specific channels called aquaporins allow water to cross the membrane. Facilitated diffusion through protein channels is still governed by factors affecting rates of diffusion.

- Facilitated diffusion may be aided by carrier proteins that bind a substance and transport it across the membrane. Diffusion in this case is not only limited by the concentration gradient, but also by the number of available membrane carrier proteins. When all of the carrier proteins are bound to the substance, they are saturated, limiting the rate of diffusion.

Active transport is directional, works against a concentration gradient, and requires an input of energy.

- Three types of protein systems are involved in active transport. Uniport transporters move a single solute across a membrane in one direction. Symport transporters move two solutes in the same direction across a membrane. Antiport transporters move two solutes in opposite directions.

- Primary active transport requires the direct participation of ATP to move ions against their concentration gradient. The sodium–potassium pump is an important antiporter that uses ATP to transport two K^+ ions and three Na^+ ions.

- Secondary active transport uses ATP indirectly by coupling solute transport with the ion concentration gradient established by primary transport. It typically moves amino acids and sugars across the membrane with the help of an ion like Na^+.

Endocytosis and exocytosis move large molecules into and out of cells.

- Endocytosis is the process by which a cell brings large substances into the cell. This is accomplished by the cell membrane, which folds around the substance to form a vesicle. See Figure 5.16.

- Large substances and even entire cells are engulfed in the process of phagocytosis. Once inside the cell, the vesicle fuses with a lysosome for digestion. The cell takes up liquids in small vesicles from the outside in the process of pinocytosis.

- Animal cells use receptor-mediated endocytosis to capture specific macromolecules such as cholesterol from the environment. Receptors for specific macromolecules cluster together on the cell surface in coated pits containing the protein clathrin. Upon binding of the specific molecule to the receptors, the coated pit invaginates to form a vesicle. The resulting vesicle becomes clathrin coated until it is well inside the cell, where it loses its coat and fuses with a lysosome. Animal cells take up cholesterol by the process of receptor-mediated endocytosis.

- Exocytosis moves materials out of the cell. Binding proteins found on the surface of the cell-produced vesicles bind with receptor proteins of the cytoplasmic side of the cell membrane. The vesicle membrane fuses with the cell membrane, and the contents of the vesicle are released outside the cell membrane.

- Membranes have additional multiple biological functions. Though the primary membrane functions are to act as barriers and binding sites, membranes also have roles in information processing, energy transformation, and the organization of chemical reactions.

- In order to carry out these functions, membranes are necessarily dynamic. They are continually being formed, modified, and degraded.

The Big Picture

- Cellular membranes are a dynamic composition of a phospholipid bilayer, integral and peripheral proteins, and carbohydrates. The nature of the constituent phospholipids allows the formation of a barrier that is semipermeable. Small hydrophilic molecules can traverse the membrane via simple diffusion, water may cross the membrane by osmosis, small charged ions may pass through via protein channels, and some other molecules may be shepherded through via carrier molecules. Transport of other molecules must be active and requires energy input, either directly from ATP or as coupled to ATP-driven transport. Larger substances depend on endocytosis and exocytosis for transport into and out of the cell.

- In multicellular organisms, adhesion and cell-to-cell connections are important for structure and communication. Tight junctions allow multiple cells to form a sheet that directs movement of substances in a single direction. Desmosomes enhance structural support of sheets of cells that are prone to abrasion. Gap junctions allow communication by connecting adjacent cells.

Common Problem Areas

- The biggest challenge when studying membranes is in understanding diffusion and osmosis. It is very easy to get the terminology confused, especially the terms "hypertonic" and "hypotonic," when discussing osmosis. Looking at the Latin roots of the words is helpful. "Hyper-" generally means excess, and "hypo-" generally means " less than." "Tonic" refers to solute concentration. Therefore, "hypertonic" means excess solutes, and "hypotonic" means fewer solutes.

- Secondary active transport also tends to be confusing. Remember that secondary active transport does not use ATP directly, but is tightly coupled to ion transport that does require ATP.

Study Strategies

- When studying the fluid mosaic model, think about the properties of the constituent molecules. This will make understanding the membrane's structure much easier. Draw a cartoon of the plasma membrane and all the potential components.

- Draw diagrams of the movement of water and solutes when working with diffusion and osmosis. Diagrams help you to visualize what is happening across a membrane.

- Review the following animated tutorials and activities on the Companion Website/CD:
  Tutorial 5.1 Passive Transport
  Tutorial 5.2 Active Transport
  Tutorial 5.3 Endocytosis and Exocytosis
  Activity 5.1 The Fluid-Mosaic Model
  Activity 5.2 Animal Cell Junctions

Test Yourself

Knowledge and Synthesis Questions

1. Which of the following statements regarding cellular membranes is *not* true?
   a. The hydrophobic nature of the phospholipid tails limits the migration of polar molecules across the membrane.
   b. Integral proteins and phospholipids move fluidly throughout the membrane.
   c. Membrane phospholipids flip back and forth from one side of the bilayer to the other.
   d. Glycolipids and glycoproteins serve as recognition sites on the cell membrane.
   Textbook Reference: 5.1 What Is the Structure of a Biological Membrane? pp. 98–101

2. Which of the following contributes to differences in the two sides of the cell membrane?
   a. Differences in peripheral proteins
   b. Different domains expressed on the ends of integral proteins
   c. Differences in phospholipid types
   d. All of the above
   Textbook Reference: 5.1 What Is the Structure of a Biological Membrane? p. 99

3. Which of the following cell membrane components serve as recognition signals for interactions between cells?
   a. Cholesterol
   b. Glycolipids or glycoproteins
   c. Phospholipids
   d. All of the above
   Textbook Reference: 5.1 What Is the Structure of a Biological Membrane? p. 101

4. Which of the following types of junctions are responsible for communication between cells?
   a. Tight junctions
   b. Desmosomes
   c. Gap junctions
   d. None of the above
   Textbook Reference: 5.2 How Is the Plasma Membrane Involved in Cell Adhesion and Recognition? p. 104

5. You are monitoring the diffusion of a molecule across a membrane. Which of the following will result in the fastest rate of diffusion?
   a. An internal concentration of 5 percent and an external concentration of 60 percent
   b. An internal concentration of 60 percent and an external concentration of 5 percent

   c. An internal concentration of 35 percent and an external concentration of 40 percent
   d. Both a and b
   Textbook Reference: 5.3 What Are the Passive Processes of Membrane Transport? p. 105

6. If a red blood cell with an internal salt concentration of about 0.85 percent is placed in a saline solution (salt solution) that is 4 percent, which of the following will most likely happen?
   a. The red blood cell will loose water and shrivel.
   b. The red blood cell will gain water and burst.
   c. The turgor pressure in the cell will greatly increase.
   d. The cell will remain the same.
   Textbook Reference: 5.3 What Are the Passive Processes of Membrane Transport? p. 107

7. In which of the following is solution X hypotonic relative to solution Y?
   a. Solution X has a greater solute concentration than solution Y.
   b. Solution X has a lower solute concentration than solution Y.
   c. Solution X and solution Y have the same solute concentration.
   d. None of the above
   Textbook Reference: 5.3 What Are the Passive Processes of Membrane Transport? p. 106

8. Which of the following statements regarding osmosis is *not* true?
   a. Osmosis refers to the movement of water along a concentration gradient.
   b. In osmosis, water moves to equalize solute concentrations on either side of the membrane.
   c. If osmosis occurs across a membrane, then diffusion is not occurring.
   d. The movement of water across a membrane can affect the turgor pressure of some cells.
   Textbook Reference: 5.3 What Are the Passive Processes of Membrane Transport? pp. 106–107

9. Channel proteins allow ions that would not normally pass through the cell membrane to pass through via the channel. What properties of the proteins are responsible for this?
   a. The channels are often composed of polar amino acid groups.
   b. The channels are often composed of hydrophobic amino acid groups.
   c. Both a and b
   d. None of the above
   Textbook Reference: 5.3 What Are the Passive Processes of Membrane Transport? p. 109

10. Which of the following limits the movement of molecules when carrier-mediated facilitated diffusion is involved?
    a. Concentration gradient
    b. Availability of carrier molecules
    c. Temperature

d. All of the above

Textbook Reference: 5.3 What Are the Passive Processes of Membrane Transport? p. 110

11. Active transport differs from passive transport in that active transport
    a. requires energy.
    b. never requires direct input of ATP.
    c. moves molecules with a concentration gradient.
    d. Both a and c

    Textbook Reference: 5.4 How Do Substances Cross Membranes against a Concentration Gradient? p. 111

12. Single-celled animals like amoebas engulf entire cells for food. Which of the following represents the manner in which amoebas "eat"?
    a. Exocytosis
    b. Phagocytosis
    c. Facilitative transport
    d. Active transport

    Textbook Reference: 5.5 How Do Large Molecules Enter and Leave a Cell? p. 113

13. Sodium–potassium pumps are common in many cells. Which of the following are necessary for the pumps to work?
    a. ATP
    b. A channel protein
    c. No concentration gradient
    d. All of the above

    Textbook Reference: 5.4 How Do Substances Cross Membranes against a Concentration Gradient? p. 111

14. Bacterial cells are often found in very hypotonic environments. Which of the following characteristics keeps them from continuing to take on water from their environment?
    a. The presence of a cell wall allows a buildup of turgor pressure that prevents any more water from entering the cell.
    b. The presence of a cell wall allows a buildup of tonic pressure that prevents any more water from entering the cell.
    c. The cell expels water as fast as it takes it up.
    d. None of the above

    Textbook Reference: 5.3 What Are the Passive Processes of Membrane Transport? p. 107

15. Which of the following may affect the rate of diffusion?
    a. Temperature
    b. Molecule size
    c. Concentration gradient
    d. All of the above

    Textbook Reference: 5.3 What Are the Passive Processes of Membrane Transport? p. 105

Application Questions

1. Draw a diagram of a cell membrane and label the phospholipids, integral proteins, peripheral proteins, and carbohydrates. Using your diagram, describe the fluid mosaic model.

    Textbook Reference: 5.1 What Is the Structure of a Biological Membrane? p. 98

2. Proteins for export are frequently modified in the endoplasmic reticulum, passed to the Golgi apparatus, and then exported from the cell. Explain the role of membranes in a protein's journey from the ER to the extracellular environment.

    Textbook Reference: 5.1 What Is the Structure of a Biological Membrane? p. 101

3. A marathon runner has just come into your emergency room with severe dehydration. You must decide what type of solution to pump into his veins so that he survives. Your choices are: pure water, 0.9 percent saline, and 1.5 percent saline. You cannot remember which solution is correct, so you treat blood samples that you have placed on microscope slides with each solution and observe what happens. (Hint: Blood cells are approximately 0.9 percent saline.) Describe what you think will happen to the blood cells when exposed to each solution. Which solution you would choose to rehydrate the runner?

    Textbook Reference: 5.3 What Are the Passive Processes of Membrane Transport? p. 107

4. Compare and contrast active and passive transport.

    Textbook Reference: 5.4 How Do Substances Cross Membranes against a Concentration Gradient? p. 114

5. Barrier formation is only one function of the cell membrane. Describe some other functions of the membrane and discuss how the membrane is suited for those functions.

    Textbook Reference: 5.6 What Are Some Other Functions of Membranes? p. 114

Answers
Knowledge and Synthesis Answers

1. **c.** Because of the hydrophobic tails and hydrophilic head of the phospholipids, it is impossible for them to flip back and forth from one side of the membrane bilayer to the other.

2. **d.** The cell membrane is asymmetric and has different properties and functions on the cytoplasmic side versus the extracellular side. These properties arise from differences in the constituents of the membrane.

3. **b.** Both glycolipids and glycoproteins serve as recognition signals.

4. **c.** Gap junctions are involved in chemical and electrical signaling between cells.

5. **d.** Diffusion may take place in either direction across a membrane and always follows a concentration gradient. The larger the gradient, the faster diffusion will occur.

6. **a.** The cell will lose water as solute concentrations on both sides of the membrane equalize.

7. **b.** Hypotonic solutions have lower solute concentrations than the solution they are being compared to.

8. **c.** Diffusion and osmosis are not mutually exclusive and may take place at the same time.

9. **a.** The charged or polar lining of the channel proteins allows passage of polar and charged molecules.

10 **d.** Anything that affects the rate of diffusion will affect carrier-mediated facilitated diffusion. In addition to those effects, carrier-mediated facilitated diffusion also relies on the availability of carrier molecules.

11. **a.** Active transport works against a concentration gradient and requires energy to do so. That energy does not always have to be directly supplied in the form of ATP.

12. **b.** Cells carry out cellular eating by phagocytosis.

13. **a.** Sodium–potassium pumps are forms of primary active transport and require energy in the form of ATP.

14. **a.** Turgor pressure limits osmosis, and once a cell is turgid, no more water may be taken on.

15. **d.** Temperature, molecule size, molecule charge, and concentration gradients all affect the rate at which diffusion takes place.

Application Answers

1. Refer to Figure 5.1 in your text.

2. The protein leaves the ER encased in a membrane vesicle. Signal molecules on the surface of the membrane bind to receptor molecules on the membrane of the Golgi. The membranes fuse, and the protein contents of the vesicle enter the Golgi for modification. Upon leaving the Golgi, the protein is encased in membrane, which again, with the appropriate signals and receptors, binds to the cytoplasmic surface of the cell membrane. The vesicle membrane and the cell membrane fuse, and the contents of the vesicle are released to the extracellular side of the membrane.

3. In pure water, the blood cells will take on water through osmosis, swell, and eventually rupture. In 0.9 percent saline, the cells should neither gain nor lose a significant amount of water and should remain the same. In a 1.5 percent saline solution, the cells should lose water and shrivel. You might jump to the conclusion that for quick rehydration the patient needs a hypotonic solution. But remember that you will be adding this substance directly to the bloodstream—the patient would end up with ruptured cells. You want to use a solution isotonic to the patient's blood cells, 0.9 percent, to avoid rupture of cells.

4. The main difference between active and passive transport is that active transport goes against a concentration gradient and requires energy, whereas passive transport diffuses passively and does not require energy.

5. Membranes function in processing energy transformation and in the organization of chemical reactions. Integral and peripheral proteins contribute to these functions. The membrane serves as a holding site for the catalytic enzymes associated with these processes.

CHAPTER 6 Energy, Enzymes, and Metabolism

Important Concepts

Energy is the capacity to do work.

- No living cell manufactures energy. The energy needed for life must be obtained from an environmental energy source and transformed to a usable form.

- Energy transformations are linked to chemical transformations in cells.

- Kinetic energy is energy in motion and does work. Potential energy is stored energy. Energy can be stored biologically in chemical bonds of fatty acids and other molecules. Breaking the bonds converts the energy to kinetic energy.

- Metabolism is all of the chemical reactions occurring in a living organism. Anabolic reactions or anabolism link simple molecules to create complex molecules and store energy in the resulting bonds. Catabolic reactions or catabolism break down complex molecules and release stored energy.

- First law of thermodynamics: Energy is neither created nor destroyed. Energy can be converted from one form to another.

- Second law of thermodynamics: Not all energy can be used. When energy is converted from one form to another, some becomes unusable.

  - Total energy (enthalpy, H) = usable energy (free energy, G) + unusable energy (entropy, S) × absolute temperature (T).

- We cannot measure absolute energy; we can only measure the change (Δ) in energy. We determine the change in usable energy as follows: $\Delta G = \Delta H - T\Delta S$

  - If ΔG is negative, free energy is released, and the reaction is exergonic.

  - If ΔG is positive, free energy is required, and the reaction is endergonic.

- With each conversion, entropy (S) tends to increase; therefore, processes tend toward disorder and randomness.

- All biochemical reactions must either release ($-\Delta G$) or take up ($+\Delta G$) usable energy.

- Reactions that are exergonic in one direction are endergonic in the other. The direction of the reaction depends on concentrations of reactants versus products. The point at which the forward reaction occurs at the same rate as the reverse reaction is termed chemical equilibrium.

- The further toward completion a reaction's equilibrium lies, the more free energy is given off (larger $-\Delta G$). A larger $+\Delta G$ means equilibrium favors the reverse reaction (equilibrium falls toward the reactant). A ΔG near zero indicates that both products and reactants have free energy.

ATP is the primary energy currency of living cells.

- ATP (adenosine triphosphate) is needed to capture, transfer, and store free energy (usable energy) in cells to do work. Additionally, ATP can donate a phosphate group to different molecules by phosphorylation.

- Hydrolysis of ATP to ADP and P_i releases a large amount of free energy ($\Delta G = -7.3$ kcal/mol). The equilibrium falls toward ADP production.

- The formation of ATP is endergonic, requiring as much energy as is released by ATP hydrolysis. This results in an "energy-coupling cycle" of the endergonic and exergonic reactions. ATP is formed when coupled with the exergonic reactions of cellular respiration. ATP is hydrolyzed to fuel endergonic reactions like protein synthesis (see Figure 6.6).

Enzymes catalyze biochemical reactions by lowering activation energy.

- Catalysts speed up reactions without being consumed in the process. Most biological catalysts are protein enzymes. The protein enzyme acts as scaffolding upon which the reaction takes place.

- Energy barriers exist between reactants and products that slow down reactions. Energy must be added to get past this barrier. You may think of it as a little shove to get the reaction going. This is called activation energy (E_a) (see Figure 6.8).

- For reactions to take place, the reactants must be slightly destabilized and turned into "transition-state species" with higher free energy than that of either the reactants or the products. This is accomplished with the help of a catalyst.

- Enzymes are highly specific biological catalysts that act on specific reactants or substrates. They have specific substrate-binding areas on their surfaces called active sites. The enzyme's specificity comes from the shape of its active site. The names for enzymes typically end in the suffix –ase. Only specific substrates can fit in and bind with an enzyme's active site. Once bound, the site is referred to as an enzyme–substrate complex (ES).

- The enzyme alters the conformation of the substrate, thus lowering the activation energy for the reaction (see Figure 6.11). From this, the product is formed, and the enzyme remains unchanged. Such reactions can be represented as follows:

$$E + S \rightarrow ES \rightarrow E + P$$

- Enzymes do not alter the equilibrium of a reaction. Both the forward and reverse reactions are sped up.

The structure of an enzyme determines its function.

- Enzymes speed up reactions by orienting substrates for maximum chemical interactions, by inducing strain on the substrate, or by adding charges to substrates.

- Enzymes are very large macromolecules with small active sites for binding small substrates.

- Binding of small molecules to the active sites depends on H-bonds, charge interactions, and hydrophobic interactions.

- Binding of substrate by an enzyme can change the enzyme's shape. The enzyme's shape can be altered by the substrate to produce an "induced fit." The large size of an enzyme helps it position the correct amino acids at the active site and helps regulate shape and allow for the induced fit.

- Cofactors, coenzymes, and prosthetic groups all contribute to an enzyme's activity. Cofactors are inorganic ions such as zinc, copper, or iron. Coenzymes are organic molecules that bind in the active site and are considered a substrate. Coenzymes do not change form during the chemical reaction. Prosthetic groups are bound to the enzyme.

- The rate of a reaction increases as the substrate concentration increases, until all enzyme active sites are occupied. At that point, no amount of additional substrate will increase the reaction rate because no more enzyme is available for catalysis. Enzyme efficiency is measured in turnover number, or how fast an enzyme can convert substrate to product and free up its active site.

Homeostasis requires that enzyme activity be regulated.

- All life must maintain stable internal conditions or homeostasis, and this is accomplished by the regulation of enzymes. The chemical reactions occurring in an organism are organized into specific metabolic pathways that are catalyzed by specific enzymes at each step.

- Inhibitors are substances that bind with enzymes to inhibit their function. Irreversible inhibition occurs when an inhibitor forms a covalent bond to an enzyme and permanently destroys the active site.

- Reversible inhibition occurs when an inhibitor binds to and alters the enzyme active site but the inhibition is reversible, depending on the concentration of the inhibitor and substrate. Some reversible inhibitors are called competitive inhibitors because they compete with the substrate for the active site. Others are called noncompetitive inhibitors because they bind elsewhere but alter the active site so that the substrate cannot bind, or the rate of binding is reduced.

Allosteric enzymes are regulated by other molecules.

- Allostery is the regulation of an enzyme by a molecule that binds to the enzyme at a site other than the active site, resulting in a change in the enzyme shape. Noncompetitive inhibitors are an example of inhibitors that work allosterically.

- Most allosteric enzymes exist naturally in an active form and an inactive form, and they can switch back and forth between the two forms. When an allosteric regulator binds to the enzyme, the enzyme becomes locked in either its active form or inactive form.

- Most allosteric enzymes are proteins with quaternary structures, where each subunit may have a different regulatory job. Catalytic subunits contain the active site for a specific substrate, whereas the regulatory subunit contains the site for the activator or inhibitor.

- Allosteric interactions help control metabolism. The first step in an enzyme-mediated pathway is referred to as the commitment step. Once initiated, the pathway is followed to completion. Frequently, the final product allosterically inhibits the enzyme of the commitment step. This prevents overproduction of the final product. This process is called end-product or feedback inhibition.

Enzyme function is influenced by the cellular environment.

- Changes in pH influence charges of amino and other groups on the enzyme or substrate. This may change the folding of the enzyme or its ability to interact with the substrate and drastically alter its catalytic ability. Enzymes typically have an ideal pH at which they function best.

- Increases in temperature may aid in reduction of activation energy by adding kinetic energy; however, this may also result in denaturing of enzymes. Each enzyme tends to have an optimal temperature. Organisms may produce different forms of the same enzyme, called isozymes, which have different optimal temperatures.

The Big Picture

- Cells need energy to carry out their functions. Metabolism is a series of energy-transferring reactions that fuel the processes of the cell. All reactions either give off energy (exergonic) or require energy to proceed (endergonic). Energy may be stored in chemical bonds (potential energy) and released when the bonds are broken to do work (kinetic energy). ATP serves as the energy shuttle for many metabolic processes.

- Enzymes aid biological reactions by lowering the activation energy required to start the reaction. Each enzyme has a specific three-dimensional conformation that interacts specifically with the substrate. Interaction between the enzyme and substrate results in optimum orientation for a reaction to take place. Enzymes are proteins and are affected by temperature, pH, concentrations, reactants, and inhibitors.

- Metabolism is regulated by enzymes. Most metabolic pathways are under allosteric control. Enzyme complexes allow for interactions and regulation of adjacent active sites. Often the final product of a specific pathway regulates the commitment step of the pathway itself.

Common Problem Areas

- The chemistry and physics involved in this material are often overwhelming, and you may question why you are learning this in biology. Remember that biological processes are based on physical and chemical properties. To understand the function of organisms, it is necessary to understand the basics of chemistry and physics as they pertain to energy transfer.

- Allosteric regulation can be confusing because the terms "allosteric regulation" and "allosteric enzyme" are related but slightly different. Remember that an allosteric enzyme is an enzyme with more than one binding site: an active site (for binding and acting on substrates) and one or more allosteric sites. When allosteric regulators are bound to the allosteric site, they control whether or not the enzyme can bind substrate.

Study Strategies

- With this chapter, the best strategy is to take "small bites." You may find the concepts to be very unfamiliar; therefore, make sure you understand each section before proceeding. This chapter contains a large amount of terminology. Making a terminology/vocabulary list may be helpful.

- When studying the energy equations, be sure that you understand how any change in either free energy, unusable energy, or absolute temperature will affect the total energy of the system. Also focus on what a larger or smaller ΔG value means in terms of how reactions proceed and whether they require or release energy.

- When thinking of enzyme-mediated reactions, be sure to consider that altering anything on the left side of the equation is going to lead to changes in the amount of product formed. Always remember that enzymes are conserved across the equation.

- Review the following animated tutorials and activities on the Companion Website/CD:
  Tutorial 6.1 Enzyme Catalysis
  Tutorial 6.2 Allosteric Regulation of Enzymes
  Activity 6.1 ATP and Coupled Reactions
  Activity 6.2 Free Energy Changes

Test Yourself

Knowledge and Synthesis Questions

1. ATP is necessary for the conversion of glucose to glucose 6-phosphate. Splitting ATP into ADP and P_i releases energy into what form?
   a. Potential
   b. Kinetic
   c. Entropic
   d. Enthalpic
   Textbook Reference: 6.2 What Is the Role of ATP in Biochemical Energetics? p. 123

2. Before ATP is split into ADP and P_i, it holds what type of energy?
   a. Potential
   b. Kinetic
   c. Entropic
   d. Enthalpic
   Textbook Reference: 6.1 What Physical Principles Underlie Biological Energy Transformations? p. 120

3. Which of the following statements concerning energy transformations is true?
   a. Increases in entropy reduce usable energy.
   b. Energy may be created during transformation.
   c. Potential energy increases with each transformation.
   d. Increases in temperature decreases total amount of energy available.
   Textbook Reference: 6.1 What Physical Principles Underlie Biological Energy Transformations? p. 121

4. A reaction has a ΔG of –20 kcal/mol. This reaction is
   a. endergonic, and equilibrium is far toward completion.
   b. exergonic, and equilibrium is far toward completion.
   c. endergonic, and the forward reaction occurs at the same rate as the reverse reaction.
   d. exergonic, and the forward reaction occurs at the same rate as the reverse reaction.
   Textbook Reference: 6.1 What Physical Principles Underlie Biological Energy Transformations? p. 122

5. ATP hydrolysis releases energy to fuel cellular functions. ATP hydrolysis is
   a. endergonic.
   b. exergonic.
   c. chemoautotrophic.
   d. None of the above
   Textbook Reference: 6.2 What Is the Role of ATP in Biochemical Energetics? p. 124

6. Enzymes are biological catalysts and function by
   a. increasing free energy in a system.
   b. lowering activation energy of a reaction.
   c. lowering entropy in a system.
   d. increasing temperature near a reaction.
   Textbook Reference: *6.3 What Are Enzymes? p. 125*

7. Which of the following contribute to the specificity of enzymes?
   a. Each enzyme has a narrow range of temperature and pH optima.
   b. Each enzyme has a specific active site that interacts with a particular substrate.
   c. Substrates themselves may alter the active site slightly for optimum catalysis.
   d. All of the above
   Textbook Reference: *6.4 How Do Enzymes Work? p. 128*

8. Coenzymes and cofactors, as well as prosthetic groups, assist enzyme function by
   a. stabilizing three-dimensional shape and maintaining active sites.
   b. assisting with the binding of enzyme and substrate.
   c. Both a and b
   d. None of the above
   Textbook Reference: *6.4 How Do Enzymes Work? p. 129*

9. Which of the following are characteristics of enzymes?
   a. They are consumed by the enzyme-mediated reaction.
   b. They are not altered by the enzyme-mediated reaction.
   c. They raise activation energy.
   d. All of the above
   Textbook Reference: *6.3 What Are Enzymes? pp. 126–127*

10. Ascorbic acid, found in citrus fruits, acts as an inhibitor to catecholase, the enzyme responsible for the browning reaction in fruits such as apples, peaches, and pears. One possibility for its function could be that ascorbic acid is very similar in size and shape to catechol, the substrate of the browning reaction. If this is true, then this inhibition is most likely an example of _____ inhibition.
    a. competitive
    b. indirect
    c. noncompetitive
    d. None of the above
    Textbook Reference: *6.5 How Are Enzyme Activities Regulated? pp. 131–132*

11. Refer to question 10. Suppose further studies indicate that ascorbic acid is not similar to catechol in size and shape but that the pH of the ascorbic acid solution is altering the protein folding of catecholase. If this is true, then this inhibition is most likely an example of _____ inhibition.
    a. competitive
    b. irreversible
    c. noncompetitive

    d. None of the above
    Textbook Reference: *6.5 How Are Enzyme Activities Regulated? pp. 131–132*

12. Metabolism is organized into pathways. The pathway is linked in which of the following manners?
    a. All cellular functions feed into a central pathway.
    b. All steps in the pathway are catalyzed by the same enzyme.
    c. The product of one step in the pathway functions as the substrate in the next step.
    d. Products of the pathway accumulate and are secreted from the cell.
    Textbook Reference: *6.5 How Are Enzyme Activities Regulated? p. 133*

13. Which of the following represents an enzyme-catalyzed reaction?
    a. $E + P \rightarrow E + S$
    b. $E + S \rightarrow E + P$
    c. $E + S \rightarrow P$
    d. $E + S \rightarrow E$
    Textbook Reference: *6.3 What Are Enzymes? p. 127*

14. Which of the following graphs of enzyme-mediated reactions represents an allosteric enzyme?

    a.

    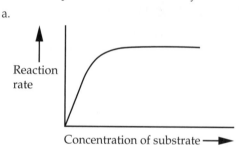

    b.

    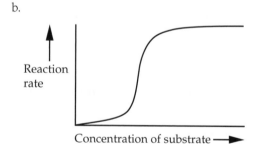

    c.

    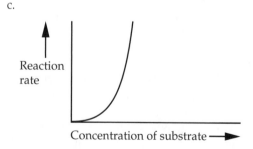

    Textbook Reference: *6.5 How Are Enzyme Activities Regulated? p. 133, Figure 6.19*

15. You are studying a new species never before studied. It lives in acidic pools in volcanic craters where temperatures reach 100°C. You determine that it has a surface enzyme that catalyzes a reaction leading to its protective coating. You decide to study this enzyme in the laboratory. Under what conditions would you most likely find optimum activity of this enzyme?
    a. 0 °C
    b. 37 °C
    c. 55 °C
    d. 95 °C
    Textbook Reference: 6.5 How Are Enzyme Activities Regulated? p. 131–133

Application Questions

1. It is estimated that approximately 90 percent of energy that passes between levels in a food web is "lost" at each level. Explain the first law of thermodynamics and discuss why this apparent loss of energy does not contradict the law.
   Textbook Reference: 6.1 What Physical Principles Underlie Biological Energy Transformations? p. 119

2. Amylase is a digestive enzyme that breaks down starch and is secreted in the mouth of humans. Amylase functions well in the mouth but ceases to function once it hits the acidic stomach environment. Explain why amylase does not function in the stomach.
   Textbook Reference: 6.5 How Are Enzyme Activities Regulated? p. 131

3. The ultimate goal of metabolism is to drive ATP synthesis. ATP is considered the energy currency of the cell. Discuss how ATP couples endergonic and exergonic reactions and why it is so important in cellular functions.
   Textbook Reference: 6.2 What Is the Role of ATP in Biochemical Energetics? p. 123

4. Figure 6.18 shows the behavior of an allosteric enzyme that has binding sites for a negative regulator. Describe the behavior of an enzyme with binding sites for a positive regulator instead of a negative regulator.
   Textbook Reference: 6.5 How Are Enzyme Activities Regulated? p. 131

5. Explain how substrate concentration affects the rate of an enzyme-mediated reaction.
   Textbook Reference: 6.4 How Do Enzymes Work? p. 128

6. Explain how free energy, total energy, temperature, and entropy are related.
   Textbook Reference: 6.1 What Physical Principles Underlie Biological Energy Transformations? p. 119

7. Use a graph to explain how temperature affects enzyme activity.
   Textbook Reference: 6.5 How Are Enzyme Activities Regulated? p. 131

Answers
Knowledge and Synthesis Answers

1. **b.** The released energy is available to do work; therefore, it is kinetic energy.

2. **a.** Potential energy is energy held within chemical bonds that may be converted to working kinetic energy.

3. **a.** Total energy = Free energy + entropy * temperature. Any increase in entropy is necessarily going to reduce free energy.

4. **b.** A negative ΔG indicates an exergonic reaction with energy being liberated. A large ΔG indicates that equilibrium lies toward completion.

5. **b.** ATP hydrolysis is exergonic, resulting in a ΔG of -7.3 kcal/mol.

6. **b.** Enzymes reduce activation energy and speed up reactions.

7. **d.** Enzymes are specific to particular substrates that may actually "adjust" the fit of the active site. They also function in specific, narrow optimum ranges of pH and temperature.

8. **c.** Cofactors, coenzymes, and prosthetic groups assist with the maintenance of an enzyme's three-dimensional shape and the conformation of the active site.

9. **b.** Enzymes are not consumed or altered in any way during an enzyme-mediated reaction, and they function to lower the activation energy of a reaction.

10. **a.** Competitive inhibitors compete for the active site with the substrate.

11. **b.** Denaturing an enzyme alters its three-dimensional structure, often irreversibly.

12. **c.** Within a given pathway, the products of the preceding step act as substrates for subsequent steps.

13. **b.** Substrate is converted to product and the enzyme is unchanged.

14. **b.** Allosteric enzymes produce sigmoid curves on reaction rate graphs. See Figure 6.19B.

15. **d.** Enzymes typically have a temperature range over which they work at a maximal rate. This optimal temperature tends to be correlated with the body temperature of the organism.

Application Answers

1. The first law of thermodynamics states that energy cannot be created or destroyed, but that it may be converted from one form to another. In the transfer between levels of a food web, approximately 90 percent of the energy is converted to unusable heat energy. There is a net loss of usable energy during each conversion, but the total amount of energy (usable and unusable) remains the same.

2. The pH optimum of amylase is approximately 7. At that pH, the protein has the three-dimensional shape to

allow starch to bind to its active site and catalyze its hydrolysis. When it is at the stomach pH (approximately 2), the protein is denatured, and its three-dimensional shape and active site are lost; therefore, it can no longer catalyze the reaction.

3. The conversion of ATP to ADP and P_i releases approximately 7.3 kcal/mol of energy. This energy release fuels (endergonic) reactions in the cell. Equilibrium of the reaction is far to the right and favors the formation of ADP. In the converse, the formation of ATP from ADP and P_i is energy intensive and can be coupled to highly exergonic reactions within the cell. Thus ATP functions as an energy shuttle between endergonic and exergonic reactions. The small size of the molecule and its ubiquitousness allow it to be available and move freely within the cell.

4. A positive regulator would stabilize the enzyme in its active form. In the absence of the regulator, the enzyme would alternate between its inactive and active forms. When it encountered substrate, it would bind it only if it happened to be in its active form. When bound to the regulator, the enzyme would be fixed in its active form, and it would bind substrate at a greater rate.

5. Increasing substrate concentration will result in an increased rate of reaction until all available active sites are occupied. At that point, no amount of substrate increase will increase the rate of reaction (see Figure 6.14).

6. Total energy = free energy + unusable energy × absolute temperature.

7. See Figure 6.22.

CHAPTER 7 Pathways That Harvest Chemical Energy

Important Concepts

Glucose is the energy source most often used by living organisms.

- All living organisms require a source of energy to survive. Most organisms use glucose ($C_6H_{12}O_6$) as a metabolic fuel source. Cells oxidize glucose for energy: $C_6H_{12}O_6 + 6 O_2 \rightarrow 6 CO_2 + 6 H_2O$ + energy. The energy is typically captured in ATP and is used to carry out cellular work.

- The metabolism of glucose involves up to three metabolic processes. In the first pathway; glycolysis, glucose is converted to pyruvate with a small net energy release. In the second pathway; cellular respiration, the pyruvate from glycolysis is converted to CO_2 and energy in the form of ATP. Cellular respiration occurs only in the presence of O_2 and is thus an aerobic process. In the absence of O_2, pyruvate is converted to lactic acid or ethanol with a small net energy release in the process of fermentation. Both fermentation and glycolysis occur in the absence of O_2, making them anaerobic processes.

- Complete conversion of glucose to CO_2 in the presence of O_2 releases 686 kcal/mol of energy. Incomplete breakdown through fermentation produces substantially less energy.

Redox reactions are the electron transferring mechanism of metabolism.

- Energy is transferred from compound to compound through transfer of electrons in redox reactions.

- The gain of electrons or hydrogen atoms is called reduction, and the loss of electrons or hydrogen atoms is called oxidation. Oxidation and reduction are always coupled. In order for a material to lose an electron or a hydrogen atom, another material must accept it.

- Oxidizing agents accept electrons and become reduced. In metabolism, O_2 is the oxidizing agent.

- Reducing agents donate electrons and become oxidized. In metabolism, glucose is the reducing agent.

- The net ΔG of a redox reaction is negative; therefore, energy is liberated as heat.

Just as ATP is an energy shuttle in metabolism, NAD is an electron shuttle in the redox reactions of metabolism.

- NAD (nicotinamide adenine dinucleotide) may exist as either NAD^+ or $NADH + H^+$ and thus acts as an electron carrier.

- Two electrons are transferred in the reduction of NAD^+ to $NADH + H^+$. The oxidation of NADH by O_2 is exergonic and liberates 52.4 kcal/mol. FAD (flavin adenine dinucleotide) is another electron carrier used in glucose metabolism.

Glycolysis is the process of converting glucose to pyruvate and usable energy.

- Glycolysis is a fundamental metabolic pathway occurring in the cytosol of the cell. It occurs during both aerobic and anaerobic cell respiration.

- During glycolysis, glucose is converted to two molecules of pyruvate with a net production of two ATP and two NADH.

- Though glycolysis results in a net gain of two ATP, the initial steps are endergonic and require energy input.

- Refer to Figure 7.5 in your text to follow each of the ten enzyme-catalyzed reactions involved in the conversion of glucose to two molecules of pyruvate. The first five steps are endergonic and require energy input. Subsequent steps are exergonic.

- The following are key points to remember about glycolysis:

  - Glycolysis begins with a 6-C sugar that is ultimately converted to two 3-C pyruvate molecules.

  - During the initial three energy-intensive steps of glycolysis, two phosphate groups from two ATP molecules are added to the 6-C sugar to produce the 5-C fructose-1,6-biphosphate (FBP). During steps 4 and 5 the carbon ring is opened and split into two 3-C phosphorylated molecules: glyceraldehyde 3-phosphate (G3P).

  - The subsequent energy-producing steps occur when the phosphate groups are transferred to ADP to make ATP, and NAD^+ is reduced.

- The conversion of glyceraldehyde 3-phosphate (G3P) to 1,3-bisphosphoglycerate (BPG) during step 6 is significant because it is an oxidation process, and energy is stored in two NADH + H$^+$ after this conversion. The cell has only small amounts of NAD, so it must be recycled for glycolysis to continue.

- During steps 7 and 10, the phosphate groups on BPG (step 7) and Phosphoenolpyruvate (PEP, step 10) are transferred to ADP to produce ATP in the process of substrate-level phosphorylation. Phosphorylation is the term to describe the addition of a phosphate group to a molecule.

- During glycolysis, two ATPs are used and four ATPs are produced, resulting in a net gain of two ATPs.

Aerobic conditions allow the oxidation of pyruvate to acetate, which is combined with Coenzyme A to form acetyl CoA.

- Pyruvate oxidation to acetyl CoA occurs on the inner mitochondrial membrane with the help of the pyruvate dehydrogenase complex.

- CO_2 is liberated, and NADH + H$^+$ and acetyl CoA are produced.

The citric acid cycle oxidizes acetate to CO_2 and forms FADH$_2$, NADH + H$^+$, and ATP.

- The citric acid cycle consists of eight reactions. Figure 7.8 details the reactions of the citric acid cycle. In the initial step of the citric acid cycle, citrate is formed by the combination of acetyl CoA and oxaloacetate. During this step CoA is liberated and recycled for combination with another pyruvate.

- The concentrations of the intermediate molecules involved in the citric acid cycle are maintained at a constant or steady state.

- In the third and fourth reactions of the cycle, one CO_2 molecule and one NADH + H$^+$ are formed at each step. Reaction five involves the conversion of GDP + P$_i$ into GTP, which is then used to produce ATP. Reaction six yields one FADH$_2$. The last reaction in the cycle results in the formation of one NADH + H$^+$ and oxaloacetate, which reenters the cycle at the beginning. So one turn of the cycle produces two CO_2 molecules, three NADH + H$^+$, one FADH$_2$, and one ATP.

- Remember that each glucose molecule yields two acetyl CoA that enter the cycle, so the yield-per-glucose is doubled.

- The majority of the citric acid cycle takes place in the mitochondrial matrix.

- The NADH produced by glycolysis and the citric acid cycle must be reoxidized during fermentation in the absence of oxygen or oxidative phosphorylation in the presence of oxygen.

In the absence of O_2, ATP is generated via fermentation.

- Fermentation occurs in the cell cytosol. The purpose of fermentation is to reduce pyruvate (or a metabolite)

and return NAD for use in glycolysis in the absence of O_2. This allows the cell to continue to produce small amounts of ATP by glycolysis.

- There are two types of fermentation based on the final end products produced. In lactic acid fermentation, pyruvate serves as the electron acceptor and lactic acid is produced. In yeast and some plant cells, alcoholic fermentation occurs with ethyl alcohol as the end product.

Energy is liberated from reduced NAD$^+$ and FAD through electron transport in the respiratory chain.

- The process of oxidative phosphorylation involves the passing of electrons through a series of membrane-associated electron carriers in the mitochondria.

- The respiratory chain is composed of four enzyme complexes (I, II, III, IV) plus cytochrome c and ubiquinone (Q). The first two complexes shuttle the electrons of NADH + H$^+$ and FADH$_2$ to Q. The third complex moves electrons from Q to cytochrome c. The final complex passes the electrons on to O_2, the ultimate electron acceptor, resulting in H_2O as a by-product. Figures 7.11 and 7.12 detail the respiratory chain.

- The purpose of the electron shuttling is to pump protons (H$^+$) across the mitochondrial membrane against a concentration gradient. This movement of electrons by these proton pumps results in the establishment of a proton gradient across the inner mitochondria membrane. This also polarizes the membrane by creating a charge differential across the membrane. These coupled effects are called the proton-motive force.

ATP is produced as protons diffuse across the mitochondrial membrane.

- An ATP synthase in the inner mitochondrial membrane couples the movement of protons with the synthesis of ATP in the chemiosmotic mechanism. ATP synthase is a protein made of two parts, the F$_0$ unit spans the membrane and acts as the channel for H$^+$, and the F$_1$ unit is the site of ATP synthesis.

- Because the ATP synthase reaction may go in either direction, ATP is removed from the mitochondrial matrix immediately to favor ATP formation. Also, the proton gradient is maintained to favor ATP synthesis.

- If the proton flow is uncoupled from ATP synthase, the resulting energy is lost as heat. This phenomenon is used in thermoregulation.

Fermentation and cellular respiration have different energy yields.

- Fermentation yields a net total of two ATP for every one glucose molecule, regardless of whether it occurs through lactic acid fermentation or alcoholic fermentation.

- Aerobic respiration yields a total of 32 ATP for every one glucose molecule. Two ATP are produced in glycolysis, two from the citric acid cycle, and 28 from the respiratory chain.

- The general formula for aerobic cellular respiration is:

$$C_6H_{12}O_6 + 6\,O_2 \rightarrow 6\,O_2 + 6\,H_2O + 32\,ATP$$

Glycolysis and cellular respiration occur while other metabolic processes are occurring.

- Components of the metabolic pathways are also components of other metabolic processes including catabolism and anabolism.

- Catabolism is the breaking down of molecules to release energy. Polysaccharides, lipids, and proteins all feed into the metabolic pathways at different points. Polysaccharides are broken down into glucose and enter at glycolysis. Lipids are broken down and can enter at glycolysis or the citric acid cycles as acetyl CoA. Proteins enter glycolysis and the citric acid cycle as amino acids.

- Just as intermediates can be broken down (catabolized) to release energy, they can also be synthesized in anabolism for storage or use by the cell. Gluconeogenesis is the anabolic process by which glucose is produced from the intermediates of glycolysis.

- Metabolism, both anabolism and catabolism, is regulated by enzymes to maintain stable concentrations of intermediates for metabolic homeostasis.

Metabolic enzymes are allosterically controlled to regulate the production of ATP.

- Metabolism responds to both positive and negative feedback provided by intermediates and end products of the process.

- Each process has a specific control point for regulation. The control point for glycolysis is phosphofructokinase, which is inhibited by ATP. This allows glycolysis to speed up during fermentation and slow down during cellular respiration.

- The main control point for the citric acid cycle is isocitrate dehydrogenase. $NADH + H^+$ and ATP inhibit the enzyme to slow down the process, and NAD^+ and ADP act as activators.

- If the citric acid cycle slows due to abundant ATP, glycolysis slows as well.

- If excess acetyl CoA is produced and ATP is abundant, acetyl CoA can be shuttled to fatty acid synthesis for storage.

The Big Picture

- Metabolic processes occur in pathways and are regulated by enzymes. These pathways may be further controlled by compartmentalization into organelles in eukaryotic cells. Because metabolic processes are energy-transferring reactions, ATP and NAD are necessary for shuttling energy and electrons between steps of the pathways. Redox reactions are the basis of

metabolism. As one compound is oxidized, another is reduced.

- Glucose provides cellular energy. It may be metabolized in the absence of oxygen through glycolysis and fermentation. In the presence of oxygen, it is metabolized through glycolysis and cellular respiration in the form of the citric acid cycle, electron transport, and oxidative phosphorylation. Anaerobic respiration produces significantly less cellular energy (in the form of ATP) than does aerobic respiration.

- Each step of the catabolic pathway is regulated by the products of subsequent pathways and depends on oxygen being present as the final electron acceptor. This level of control is achieved through allosteric regulation.

Common Problem Areas

- The biggest mistake you can make in learning glucose metabolism is to focus on memorizing the pathways without understanding what is happening. When you begin your study, do not focus on the pathways themselves. Focus on understanding the beginning and end products of each pathway and why the pathway does what it does. Once you conceptually understand why the pathway is present, move on to learning the pathway itself.

- Students often fail to see how regulation of the pathways occurs. By understanding how the presence or absence of intermediates affects the entire pathway, you can better understand the pathways themselves.

Study Strategies

- Remember that ATP and NAD are energy currencies. They move energy from pathway to pathway. Think about what roles ATP and NAD have in redox reactions to completely understand the pathways.

- Do not memorize pathways. Make sure you *understand* the pathways and how they are connected. Also understand how affecting one pathway can alter another pathway.

- This chapter is very visual in nature. You should spend a significant amount of time with the diagrams.

- Review the following animated tutorials and activities on the Companion Website/CD:
  Tutorial 7.1 Electron Transport and ATP Synthesis
  Tutorial 7.2 Two Experiments Demonstrate the Chemiosmotic Mechanism
  Activity 7.1 Glycolysis and Fermentation
  Activity 7.2 Energy Pathways in Cells
  Activity 7.3 The Citric Acid Cycle
  Activity 7.4 Electron Transport Chain
  Activity 7.5 Energy Levels
  Activity 7.6 Regulation of Energy Pathways

Test Yourself

Knowledge and Synthesis Questions

1. Which of the following cellular metabolic processes can occur in the presence *or* the absence of oxygen?
   a. The citric acid cycle
   b. Electron transport
   c. Glycolysis
   d. Fermentation
   Textbook Reference: *7.3 How Is Energy Harvested from Glucose in the Absence of Oxygen? p. 147*

2. Which of the following statements regarding glycolysis is true?
   a. A 6-C sugar is broken down to two 3-C molecules.
   b. Two ATP molecules are consumed.
   c. A net sum of two ATP molecules is generated.
   d. All of the above
   Textbook Reference: *7.2 What Are the Aerobic Pathways of Glucose Metabolism? p. 143, Figure 7.5*

3. During which process is most ATP generated in the cell?
   a. Glycolysis
   b. The citric acid cycle
   c. Electron transport coupled with chemiosmosis
   d. Fermentation
   Textbook Reference: *7.5 Why Does Cellular Respiration Yield So Much More Energy Than Fermentation? p. 153*

4. One purpose of the electron transport chain is to
   a. cycle $NADH + H^+$ back to NAD^+.
   b. use the intermediates from the citric acid cycle.
   c. break down pyruvate.
   d. All of the above
   Textbook Reference: *7.2 What Are the Aerobic Pathways of Glucose Metabolism? p. 147*

5. Cellular respiration is allosterically controlled. Which of the following act as inhibitors at the various control points?
   a. ATP
   b. $NADH + H^+$
   c. Both a and b
   d. None of the above
   Textbook Reference: *7.6 How Are Metabolic Pathways Interrelated and Controlled? p. 156*

6. Which of the following describes the role of the mitochondrial membrane?
   a. The membrane acts as an anchor for the membrane-associated enzymes of cellular respiration.
   b. The membrane allows for the establishment of a proton-motive force.
   c. Both a and b
   d. None of the above
   Textbook Reference: *7.4 How Does the Oxidation of Glucose Form ATP? p. 151, Figure 7.13*

7. In a redox reaction between G3P and NAD^+ yielding BPG and $NADH + H^+$, _____ is oxidized and _____ is reduced.
   a. G3P; NAD^+
   b. BPG; $NADH + H^+$
   c. G3P; $NADH + H^+$
   d. NAD^+; $NADH + H^+$
   Textbook Reference: *7.2 What Are the Aerobic Pathways of Glucose Metabolism? p. 143, Figure 7.5*

8. Which of the following is true regarding redox reactions?
   a. Oxidizing agents accept electrons.
   b. A molecule that accepts electrons is said to be reduced.
   c. Redox reactions involve electron transfers.
   d. All of the above
   Textbook Reference: *7.1 How Does Glucose Oxidation Release Chemical Energy? p. 140*

9. Cyanide poisoning inhibits aerobic respiration at cytochrome *c* oxidase. Which of the following is not a result of cyanide poisoning at the cellular level?
   a. Oxygen is reduced to water.
   b. ATP cannot be synthesized in the mitochondria because electron transport is never completed.
   c. Cells (with the exception of brain cells) must switch to anaerobic respiration.
   d. All of the above
   Textbook Reference: *7.4 How Does the Oxidation of Glucose Form ATP? p. 151, Figure 7.13*

10. Which of the following is correctly matched with its catabolic product?
    a. polysaccharides → amino acids
    b. lipids → glycerol and fatty acids
    c. proteins → glucose
    d. polysaccharides → glycerol and fatty acids
    Textbook Reference: *7.6 How Are Metabolic Pathways Interrelated and Controlled? p. 154*

11. The main purpose of cellular respiration is to
    a. convert energy stored in the chemical bonds of glucose to an energy form that the cell can use.
    b. destroy energy in the cell.
    c. convert kinetic to potential energy.
    d. create energy in the cell.
    Textbook Reference: *7.1 How Does Glucose Oxidation Release Chemical Energy? p. 139*

12. Which of the following statements concerning the synthesis of ATP in the mitochondria is true?
    a. ATP synthesis cannot occur without the presence of ATP synthase.
    b. The proton-motive force is the establishment of a charge and concentration gradient across the mitochondrial membrane.
    c. The proton-motive force is not necessary to drive protons back across the membrane through channels established by the ATP synthase channel protein.
    d. The ATP synthase protein is composed of two units.
    Textbook Reference: *7.4 How Does the Oxidation of Glucose Form ATP? p. 150*

13. Which of the following does not occur in the mitochondria of eukaryotic cells?
    a. Fermentation
    b. Oxidative phosphorylation
    c. Citric acid cycle
    d. Electron transport chain
    Textbook Reference: 7.1 How Does Glucose Oxidation Release Chemical Energy? p. 141, Table 7.1

14. The largest change in free energy during glycolysis occurs at what reaction?
    a. Reaction 2: G6P → F6P
    b. Reaction 5: DAP → G3P
    c. Reaction 6: G3P → BPG + NADH
    d. Reaction 7: BPG → 3PG + ATP
    Textbook Reference: 7.2 What Are the Aerobic Pathways of Glucose Metabolism? p. 144

15. Which of the following is recycled and reused in cellular metabolism?
    a. ADP
    b. NAD
    c. FAD
    d. All of the above
    Textbook Reference: 7.1 How Does Glucose Oxidation Release Chemical Energy? pp. 141–142

Application Questions

1. Why is oxygen necessary for aerobic respiration?
   Textbook Reference: 7.4 How Does the Oxidation of Glucose Form ATP? p. 148

2. Glycolysis yields two molecules of pyruvate, two ATP, and two NADH + H[+] regardless of whether oxygen is present or not. What are the fates of these molecules in the absence of oxygen? What would happen if NADH + H[+] was not recycled?
   Textbook Reference: 7.3 How Is Energy Harvested from Glucose in the Absence of Oxygen? pp. 147–148

3. Explain how the proton-motive force drives chemiosmosis.
   Textbook Reference: 7.4 How Does the Oxidation of Glucose Form ATP? p. 150

4. The fate of acetyl CoA differs according to how much ATP is present in the cell. Explain what happens to acetyl CoA when ATP is limited, and compare that to what happens when acetyl CoA is abundant. How do these processes help regulate metabolism?
   Textbook Reference: 7.6 How Are Metabolic Pathways Interrelated and Controlled? p. 157

5. Cellular respiration occurs simultaneously along with many other cellular processes. Describe generally how cellular respiration interacts with other cellular metabolic events.
   Textbook Reference: 7.6 How Are Metabolic Pathways Interrelated and Controlled? pp. 154–155

6. Compare and contrast energy yields from aerobic respiration and fermentation.
   Textbook Reference: 7.5 Why Does Cellular Respiration Yield So Much More Energy than Fermentation? p. 153

7. One of the by-products of aerobic cellular respiration is carbon dioxide. Assuming you begin with labeled glucose, trace the fate of that molecule until carbon dioxide is released.
   Textbook Reference: 7.2 What Are the Aerobic Pathways of Glucose Metabolism? p. 143, Figure 7.5; p. 146, Figure 7.8

8. Identify the controlling steps of glycolysis, the citric acid cycle, and electron transport. What regulators affect each of these steps?
   Textbook Reference: 7.6 How Are Metabolic Pathways Interrelated and Controlled? pp. 156–157

Answers

Knowledge and Synthesis Answers

1. **c.** Glycolysis proceeds during both fermentation and cellular respiration. Only in cellular respiration is oxygen needed as the terminal electron acceptor of the pathway.

2. **d.** During glycolysis, 6-C glucose is broken down into two 3-C pyruvate molecules. In the process, four total ATP are produced, but two are consumed, leaving a net production of two ATP molecules.

3. **c.** Most of the ATP produced during cellular respiration is produced as electron transport and chemiosmosis are coupled in oxidative phosphorylation.

4. **a.** The electron transport chain is responsible for oxidizing NADH + H[+] back to NAD[+].

5. **c.** Both ATP and NADH + H[+] allosterically control metabolism. ATP controls both phosphofructokinase and isocitrate dehydrogenase, which are commitment steps for glycolysis and the citric acid cycle respectively. NADH + H[+] controls isocitrate dehydrogenase.

6. **c.** The mitochondrial membrane is necessary for the anchoring of proteins as well as the establishment of a barrier across which a gradient can be established.

7. **a.** A molecule is oxidized when it loses electrons or protons and is reduced when it gains electrons or protons.

8. **d.** Oxidizing agents accept electrons and cause oxidation of another molecule. Reducing agents donate electrons and cause the reduction of another molecule.

9. **a.** Cyanide stops aerobic cellular respiration because cytochrome *c* oxidase loses the ability to reduce oxygen to water. Those cells that can switch to anaerobic respiration (fermentation) do so. Brain cells cannot make that switch and so they are killed.

10. **b.** Lipids are broken down into glycerol and fatty acids; polysaccharides are broken down into glucose; proteins are broken down into amino acids.

11. **a.** Cellular respiration is the cell's way of converting potential energy in the chemical bonds of glucose to potential energy that the cell ultimately can use.

12. **c.** The proton-motive force results in a concentration and charge gradient across the mitochondrial membrane. In order for that gradient to equalize, the protons must flow through a channel protein. If this channel protein has an associated ATP synthase, ATP is generated as protons flow through.

13. **a.** Fermentation occurs in the cytosol, whereas all the other processes occur in the mitochondria of eukaryotic cells.

14. **c.** The largest change in free energy occurs in reaction 6, with more than 100 kcal/mol released.

15. **d.** ADP, NAD, and FAD are all recycled and reused in the process of cellular respiration.

Application Answers

1. Oxygen acts as the terminal electron acceptor in the electron transport pathway. Without it, NADH + H$^+$ cannot be cycled back to NAD$^+$. The accumulated NADH + H$^+$ acts as an inhibitor to the citric acid cycle and effectively shuts it down. Therefore, in the absence of oxygen, a cell can only undergo glycolysis.

2. In the absence of oxygen, pyruvate is either reduced to lactate (in lactic acid fermentation) or it is metabolized and its metabolites are reduced to ethyl alcohol (in alcoholic fermentation). In either case, NADH + H$^+$ is the reducing agent, and it is oxidized back to NAD$^+$ in the process. The two ATP would be used as cellular energy. If NADH + H$^+$ was not oxidized to NAD$^+$, there would eventually be no NAD$^+$ available for glycolysis.

3. The proton-motive force results in a concentration and charge gradient across the mitochondrial membrane. For that gradient to equalize, the protons must flow through a channel protein. If this channel protein has an associated ATP synthase, ATP is generated as protons flow through.

4. If ATP is limited, acetyl CoA enters the citric acid cycle, and cellular respiration utilizes it to produce ATP. If ATP is abundant, acetyl CoA is shuttled to fatty acid synthesis, thus storing the energy in chemical bonds.

5. Consult Figure 7.17 to see where different metabolic pathways in the cell interact.

6. Fermentation yields only two ATP. Cellular respiration yields 32 ATP. Though this difference is great, organisms can survive quite well relying on fermentation because the rate at which glycolysis occurs is increased nearly tenfold.

7. Follow Figures 7.6 and 7.8. Carbon dioxide is liberated when α-ketoglutarate is oxidized to succinyl CoA.

8. This control point for glycolysis is phosphofructo-kinase, which is inhibited by ATP. This allows glycolysis to speed up during fermentation and slow down during cellular respiration. The control point for the citric acid cycle is isocitrate dehydrogenase. NADH + H$^+$ and ATP inhibit the enzyme, and NAD$^+$ and ADP are activators. Electron transport is controlled by the amount of NADH + H$^+$ fed in and by NADH-Q reductase.

CHAPTER 8 Photosynthesis: Energy from Sunlight

Important Concepts

The photosynthetic production of O₂ by green plants is necessary for most organisms to obtain the energy for life.

- Photosynthesis uses CO_2, water, and light to produce carbohydrate and O_2 in the reaction:

$$6\ CO_2 + 12\ H_2O \rightarrow C_6H_{12}O_6 + 6\ O_2 + 6\ H_2O$$

- In plants, water for photosynthesis must be acquired by the roots and transported to the leaves, where photosynthesis takes place. CO_2, O_2, and water vapor are exchanged through openings or stomata in the leaf's surface. Light from the sun is required for photosynthesis to occur. Photosynthesis in eukaryotes occurs in chloroplasts.

- Photosynthesis occurs in a two-pathway process. The light reactions produce ATP from light energy. The light-independent reactions use ATP and NADPH + H⁺ produced by the light reactions to trap CO_2 and produces sugars that act as an energy store. The light-independent pathways are the Calvin cycle, C_4 photosynthesis, and crassulacean acid metabolism. *Neither* pathway operates in the absence of light.

The interaction of light and pigment is fundamental to photosynthesis.

- Light is a form of electromagnetic radiation that comes in discrete packets called photons and has wavelike properties. Photons can be scattered, transmitted, or absorbed. The properties of light depend on its wavelength. For light to be biologically available, it must be absorbed by the receptive molecule and contain enough energy to carry out chemical work. When a molecule absorbs a proton, it goes from a grounded state to an excited state.

- Pigments are molecules that absorb wavelengths in the visible spectrum. The pigment chlorophyll appears green because blue and red light are absorbed and green light is reflected.

- Compounds have unique absorption spectra, depending on which wavelengths of light are absorbed (Figure 8.6). An action spectrum can be measured that relates biological activity to particular wavelengths.

- In plants, photosynthesis relies on chlorophyll *a*, chlorophyll *b*, and accessory pigments. Chlorophylls are more abundant, but they can only absorb photons with blue or orange-red wavelengths. Accessory pigments, such as carotenoids and phycobilins, absorb photons that chlorophylls cannot, and transfer energy to the chlorophylls.

In photosynthesis, a pigment molecule in the excited state passes the absorbed energy along.

- Once a pigment absorbs light, two things can happen; the energy can be released as fluorescence, or it can be passed to another pigment.

- Photosynthetic antennae systems function to pass the energy from light from one pigment to another until the reaction center is reached, where light energy is converted to chemical energy. The excitation energy is passed from one pigment that absorbs at a shorter wavelength to another pigment that absorbs at a longer wavelength. In plants, the reaction center is chlorophyll *a*. Excited chlorophyll acts as a reducing agent resulting in electron transport.

- Noncyclic electron flow results in the production of ATP and NADPH + H⁺. Water is oxidized to form O_2, H⁺, and electrons. Electrons move from water to chlorophyll, through electron carriers, and ultimately to NADP⁺. Release of free energy drives ATP synthesis through chemiosmosis.

- Noncyclic electron flow requires two photosystems in the thylakoid membrane to continuously absorb light. Photosystem II takes slightly higher energy (680 nm) than photosystem I (700 nm). These two photosystems interact together to pass electrons in a model called the Z scheme.

- Figure 8.9 shows the flow of electrons in the Z scheme from the splitting of water to the reduction of NADP⁺. The electron transport pumps protons actively into the thylakoid compartment. This proton gradient powers the formation of ATP. Thus, noncyclic flow produces equal amounts of ATP and NADPH + H⁺.

- The light-independent reactions require more ATP than NADPH + H⁺. Cyclic electron flow supplies the additional ATP by producing ATP without producing NADPH + H⁺.

- Cyclic electron transport involves only photosystem I and cycles back to the same chlorophyll molecule (see Figure 8.10). The amount of NADPH + H$^+$ present regulates whether cyclic or noncyclic electron flow occurs. If NADPH + H$^+$ is abundant, electrons are accepted by plastoquinone, which pumps two protons back across the thylakoid membrane.

- ATP synthesis in either cyclic or noncyclic electron flow is produced via chemiosmosis. Just as in the mitochondria, a proton-motive force coupled to ATP synthase is established in the chloroplast as electrons are passed along the transport chain. This results in the photophosphorylation of ADP to ATP.

The Calvin cycle incorporates CO$_2$ into sugars.

- In the Calvin cycle, ATP and NADPH from the light reactions are used to convert CO$_2$ to sugars for storage. Because ATP and NADPH are energy-rich coenzymes that cannot be stockpiled, the Calvin cycle reactions occur only in the light when ATP and NADPH can be made.

- The Calvin cycle occurs in the stroma of the chloroplasts.

- The cycle was revealed using radiolabeled CO$_2$ and observing where it was incorporated. Following this process, researchers revealed the pathway by which CO$_2$ is converted to sugar. See Figure 8.12 to understand how this was done.

- CO$_2$ is combined with ribulose 1,5-bisphosphate (RuBP), which is then split into two molecules of 3-phosphoglycerate (3PG) by an enzyme called rubisco. Rubisco is the world's most abundant protein, and it regulates the Calvin cycle.

- ATP and NADPH + H$^+$ from the light reactions are then used to convert the fixed CO$_2$ in 3PG into carbohydrate (glyceraldehyde 3-phosphate; G3P).

- Finally, ATP is used to regenerate RuBP so that additional CO$_2$ may be fixed.

- The resulting carbohydrate (G3P) is converted to starch or sucrose, either for storage or for conversion to glucose and fructose for cellular fuel.

- Light stimulates the Calvin cycle by inducing pH changes that result in the activation of Calvin cycle enzymes and by reducing disulfide bonds on four of the Calvin cycle enzymes.

Rubisco also mediates the process of photorespiration.

- Rubisco can act as an oxygenase and fix O$_2$ at the expense of CO$_2$ to produce phosphoglycolate. Phosphoglycolate enters membrane-enclosed peroxisomes and is converted to glycine. The glycine is then converted into glycerate and CO$_2$ in the mitochondrion.

- This process is known as photorespiration. ATP and NADPH from the light reaction are used in photorespiration, but CO$_2$ is released instead of used to make carbohydrate.

- Whether CO$_2$ or O$_2$ is fixed depends on relative concentrations of CO$_2$ and O$_2$. Excess O$_2$ (abundant, for example, on hot, dry days) forces photorespiration.

Photorespiration is metabolically expensive, so plants have evolved mechanisms to avoid it.

- To avoid photorespiration, the level of CO$_2$ around rubisco must be high. This is difficult to achieve because O$_2$ levels in the air are much higher than CO$_2$ levels. If hot conditions force the closing of stomata, CO$_2$ levels are quickly depleted.

- Plants such as roses, wheat, and rice that fix CO$_2$ to 3PG are called C$_3$ plants and are very sensitive to CO$_2$/O$_2$ levels. Photorespiration occurs frequently and limits the range of conditions under which C$_3$ plants can grow.

- Many tropical plants such as corn, sugarcane, and other tropical grasses are C$_4$ plants. They fix CO$_2$ to the acceptor phosphoenolpyruvate (PEP) using the enzyme PEP carboxylase to form oxaloacetate. PEP carboxylase has no oxygenase activity and fixes CO$_2$ at very low levels. The resulting four-carbon compound diffuses to the bundle sheath cells in the interior of the leaf, releases CO$_2$, and is recycled. The released CO$_2$ goes through the Calvin cycle including rubisco as normal. The early fixation process is a CO$_2$-concentrating mechanism around rubisco.

- Cacti, pineapples, and other succulents undergo crassulacean acid metabolism (CAM) that separates CO$_2$ fixation and the Calvin cycle. Plants that carry out CAM open their stomata only at night and store CO$_2$ for use during the day when light is present. CO$_2$ is fixed as oxaloacetate and converted to malic acid. CO$_2$ is stored in malic acid until it is transferred to the chloroplasts or photosynthetic cells when light is available.

Metabolic pathways in plants involve photosynthesis and respiration.

- Because plants are autotrophic, they can synthesize all necessary molecules to survive from CO$_2$, H$_2$O, phosphate, sulfate, and NH$_4$. All cells, whether photosynthetic or not, go through respiration to produce the ATP necessary for cellular processes, and they do so in the light and the dark.

- The Calvin cycle and the respiratory pathways are closely linked. G3P can be converted to pyruvate and enter respiration, or it can enter the gluconeogenic pathway to form sucrose. In order for plants to grow, the energy stored must exceed the energy used in respiration.

- Figure 8.19 illustrates how plant metabolism is interconnected.

The Big Picture

- Photosynthesis allows organisms with the appropriate pigments and metabolic processes to convert light energy from the sun to chemical energy that can be used in the cell or stored. This conversion process forms the basis of all food webs and is vital to life on Earth. Photosynthetic organisms fix CO_2 into sugars. This process requires water and light energy and produces the by-products O_2 and water.

- The process of photosynthesis occurs in two steps. The first step utilizes light energy to produce ATP and NADPH with the help of the electron transport system. The movement of electrons through the electron transport system generates cellular energy in the form of ATP that can be used in the fixing of CO_2. The fixing of CO_2 occurs in the Calvin cycle and is mediated by an enzyme called rubisco, and the reactions are CO_2 concentration dependent.

- Rubisco can also act as an oxygenase and allow a plant to go through a metabolically expensive process called photorespiration. Plants that grow in conditions favoring photorespiration have evolved a wide variety of systems to avoid photorespiration.

Common Problem Areas

- There is the temptation to learn pathways without understanding why they occur. Be sure to focus on why the plant has a particular pathway before attempting to learn the pathway itself.

- It is very common to think that the light reactions happen in the light and the Calvin cycle occurs in the dark. This is not the case! The light reactions must occur simultaneously with the Calvin cycle to supply the needed energy in the form of ATP and NADPH + H$^+$.

- Remember that plants respire as well as photosynthesize and that photorespiration and respiration are different processes. All plant cells go through cellular respiration. Cellular respiration takes the energy stored in photosynthesis and makes it available to drive other cellular processes. Photorespiration is energy expensive and is of little benefit to the plant. Photosynthesis occurs in specialized photosynthetic cells and happens at the same time as respiration.

Study Strategies

- The process of photosynthesis is easier to understand when you visualize what is happening. Use the figures in your book to understand where energy and carbon are flowing. Remember that what the plant is doing is taking light energy and converting it to chemical energy that can be stored. In the process it uses CO_2 and gives off O_2. Watch for where energy is flowing and how.

- The properties of chlorophyll and light itself greatly affect how well a plant photosynthesizes. Take some time to understand the properties of light and how pigments interact with light.

- Rubisco is the most abundant protein on the planet. Focus on how the relative concentrations of O_2 and CO_2 dictate how the enzyme functions. Also understand that C_4 and CAM plants have modifications to avoid photorespiration.

- Review the following animated tutorials and activities on the Companion Website/CD:
Tutorial 8.1 The Source of the Oxygen Produced by Photosynthesis
Tutorial 8.2 Photophosphorylation
Tutorial 8.3 Tracing the Pathway of CO_2
Activity 8.1 Calvin Cycle
Activity 8.2 C_3 and C_4 Leaf Anatomy

Test Yourself
Knowledge and Synthesis Questions

1. The main purpose of photosynthesis is to
   a. consume CO_2.
   b. produce ATP.
   c. convert light energy to chemical energy.
   d. produce starch.
   Textbook Reference: 8.1 What Is Photosynthesis? p. 161

2. Which of the following best represent the components that are necessary for photosynthesis to take place?
   a. Mitochondria, accessory pigments, visible light, water, and CO_2
   b. Chloroplasts, accessory pigments, visible light, water, and CO_2
   c. Mitochondria, chlorophyll, visible light, water, and O_2
   d. Chloroplasts, chlorophyll, visible light, water, and CO_2
   Textbook Reference: 8.1 What Is Photosynthesis? p. 163, Figure 8.3

3. Chlorophyll is suited for the capture of light energy because
   a. certain wavelengths of light raise it to an excited state.
   b. in its excited state chlorophyll gives off electrons.
   c. chlorophyll's structure allows it to attach to thylakoid membranes.
   d. All of the above
   Textbook Reference: 8.2 How Does Photosynthesis Convert Light Energy into Chemical Energy? p. 165

4. Plants give off O_2 because
   a. O_2 results from the incorporation of CO_2 into sugars.
   b. They do not respire; they photosynthesize.
   c. water is the initial proton donor, leaving O_2 as a photosynthetic by-product.
   d. All of the above
   Textbook Reference: 8.2 How Does Photosynthesis Convert Light Energy into Chemical Energy? pp. 166–167

5. Cyclic and noncyclic electron flow are used in plants to
   a. meet the ATP demands of the Calvin cycle.
   b. produce excess NADPH + H$^+$.
   c. unbalance ATP and NADPH + H$^+$ ratios in the chloroplast.

d. All of the above

Textbook Reference: 8.3 *How Is Chemical Energy Used to Synthesize Carbohydrates? p. 169*

6. Which of the following statements concerning the light reactions of photosynthesis is true?
   a. Photosystem I cannot operate independently of photosystem II.
   b. Photosystems I and II are activated by different wavelengths of light.
   c. Photosystems I and II transfer electrons and create proton equilibrium across the thylakoid membrane.
   d. All of the above

Textbook Reference: 8.2 *How Does Photosynthesis Convert Light Energy into Chemical Energy? p. 167*

7. ATP is produced during the light reactions via
   a. CO_2 fixation.
   b. chemiosmosis.
   c. reduction of water.
   d. All of the above

Textbook Reference: 8.2 *How Does Photosynthesis Convert Light Energy into Chemical Energy? p. 168*

8. Because of the properties of chlorophyll, plants need adequate _____ light to grow properly.
   a. green
   b. blue and red
   c. infrared
   d. ultraviolet

Textbook Reference: 8.2 *How Does Photosynthesis Convert Light Energy into Chemical Energy? p. 165, Figure 8.6*

9. Which of the following statements concerning the Calvin cycle is *not* true?
   a. Light energy is not required for the cycle to proceed.
   b. CO_2 is assimilated into sugars.
   c. RuBP is regenerated.
   d. It uses energy stored in ATP and NADPH + H$^+$.

Textbook Reference: 8.3 *How Is Chemical Energy Used to Synthesize Carbohydrates? p. 170*

10. Which of the following statements concerning rubisco is true?
    a. Rubisco is an enzyme.
    b. Rubisco catalyzes both the beginning steps of photorespiration and the Calvin cycle.
    c. Rubisco is the most abundant protein on Earth.
    d. All of the above

Textbook Reference: 8.3 *How Is Chemical Energy Used to Synthesize Carbohydrates? p. 169; 8.4 How Do Plants Adapt to the Inefficiencies of Photosynthesis? p. 172*

11. Which of the following begins the Calvin cycle that results in the entire pathway being carried out?
    a. 3PG is reduced to G3P using ATP and NADPH + H$^+$.
    b. RuBP is regenerated.
    c. CO_2 and RuBP join forming 3PG.
    d. As a cycle, it can start at any point.

Textbook Reference: 8.3 *How Is Chemical Energy Used to Synthesize Carbohydrates? p. 170*

12. The Calvin cycle results in the production of _____.
    a. glucose
    b. starch
    c. rubisco
    d. G3P

Textbook Reference: 8.3 *How Is Chemical Energy Used to Synthesize Carbohydrates? pp. 170–171*

13. Which of the following statements regarding photorespiration is true?
    a. Photorespiration is a metabolically expensive pathway.
    b. Photorespiration is avoided when CO_2 levels are low.
    c. Photorespiration increases the overall CO_2 that is converted to carbohydrates.
    d. All of the above

Textbook Reference: 8.4 *How Do Plants Adapt to the Inefficiencies of Photosynthesis? pp. 172–173*

14. The fixation of CO_2 by PEP carboxylase functions to
    a. concentrate O_2 for use in photosynthetic cells.
    b. allow plants to close stomata without having photorespiration occur.
    c. allow plants to photosynthesize in the dark.
    d. Both a and b

Textbook Reference: 8.4 *How Do Plants Adapt to the Inefficiencies of Photosynthesis? p. 174*

15. CAM plants differ from C$_4$ plants in that
    a. CO_2 is stored as malic acid.
    b. photosynthesis can occur at night in these plants.
    c. their stomata close during periods that favor photorespiration.
    d. they use PEP carboxylase to fix CO_2.

Textbook Reference: 8.4 *How Do Plants Adapt to the Inefficiencies of Photosynthesis? p. 175*

16. Which of the following statements is true regarding the relationship between photosynthesis and cellular respiration in plants?
    a. Photosynthesis occurs in specialized photosynthetic cells.
    b. Cellular respiration occurs in specialized respiratory cells.
    c. Cellular respiration and photosynthesis can occur in the same cell.
    d. Both a and c

Textbook Reference: 8.5 *How Is Photosynthesis Connected to Other Metabolic Pathways in Plants? pp. 175–176*

Application Questions

1. Plants consume CO_2 and give off O_2. How is this possible if plants must also undergo cellular respiration?

Textbook Reference: 8.5 *How Is Photosynthesis Connected to Other Metabolic Pathways in Plants? pp. 175–176*

2. Why do plants undergo both the light reactions of photosynthesis and the Calvin cycle? Why don't they

simply use the ATP produced in the light reactions of photosynthesis to drive cellular processes?
Textbook Reference: 8.3 How Is Chemical Energy Used to Synthesize Carbohydrates? p. 169

3. Why do plants undergo both photosynthesis and cellular respiration? Why don't they simply use the ATP produced in the light reactions of photosynthesis to drive cellular processes?
Textbook Reference: 8.5 How Is Photosynthesis Connected to Other Metabolic Pathways in Plants? pp. 175–176

4. Rubisco has both carboxylase and oxygenase activities. These processes compete with one another. What determines which function the enzyme has? What conditions favor photorespiration? What conditions favor photosynthesis?
Textbook Reference: 8.4 How Do Plants Adapt to the Inefficiencies of Photosynthesis? pp. 172–173

5. Compare and contrast C_3 and C_4 plants.
Textbook Reference: 8.4 How Do Plants Adapt to the Inefficiencies of Photosynthesis? pp. 173–174

6. The Calvin cycle was once referred to as the "dark" reactions of photosynthesis. Why is this a misnomer?
Textbook Reference: 8.1 What Is Photosynthesis? p. 162

7. Explain the differences between cyclic and noncyclic electron flow. Why are both processes necessary?
Textbook Reference: 8.2 How Does Photosynthesis Convert Light Energy into Chemical Energy? pp. 166–168

8. How do accessory pigments enhance photosynthetic activity in plants?
Textbook Reference: 8.2 How Does Photosynthesis Convert Light Energy into Chemical Energy? p. 165

9. Why are plants green?
Textbook Reference: 8.2 How Does Photosynthesis Convert Light Energy into Chemical Energy? pp. 163–164

Answers

Knowledge and Synthesis Answers

1. **c.** Photosynthetic organisms are the only life forms capable of trapping light energy and converting it to chemical energy. Because of this they form the basis of food chains.

2. **d.** Chloroplasts are the site of the photosynthetic reactions; chlorophyll is excited by photons of light and serve as reaction centers for the photosystems; visible light is necessary to excite chlorophyll and accessory pigments; water is the initial electron donor for the pathway; and CO_2 is necessary to make precursor molecules for energy storage.

3. **d.** The "tails" of chlorophyll molecules are associated with the thylakoid membranes of the chloroplasts. This close membrane association assists with establishing the proton-motive force that will drive ATP synthesis. When excited by light, the chlorophyll moves into an excited state and passes electrons to acceptor molecules. This begins to set up the proton gradient across the membrane that will drive ATP synthesis.

4. **c.** Water is split at photosystem II to donate electrons to the reaction center. The resulting protons are moved across the membrane to establish the proton-motive force, and O_2 is given off as a by-product.

5. **a.** ATP is required at higher levels in the Calvin cycle than NADPH + H$^+$ is; therefore, there must be a mechanism for producing additional ATP. Cyclic electron flow provides that mechanism. If noncyclic electron flow were to be sped up to meet ATP needs, an excess of NADPH + H$^+$ would result. Shifting between cyclic and noncyclic flow balances ATP/NADPH + H$^+$ ratios.

6. **b.** Photosystems I and II operate depending on whether electron flow is cyclic or noncyclic. Activity is controlled by the ATP levels in the chloroplast. Photosystem II is activated by light of a higher energy level than photosystem I. Both photosystems transfer electrons and create proton gradients across the thylakoid membranes.

7. **b.** In the light reactions, ATP synthesis occurs when protons flow through an ATP synthase channel protein in the thylakoid membrane. This is a chemiosmotically driven process.

8. **b.** Chlorophyll and accessory pigments absorb light in the blue and orange-red wavelengths of visible light. Green light is reflected; therefore, plants appear green.

9. **a.** Light energy is required for the Calvin cycle to proceed. ATP synthesis is dependent on light energy, and the Calvin cycle is dependent on ATP.

10. **d.** Rubisco, the most abundant enzyme on Earth, has both oxygenase and carboxylase activities.

11. **c.** The first step of the Calvin cycle is the fixation of CO_2 into 3PG. This is the regulatory step, and it requires ATP and NADPH + H$^+$.

12. **d.** The Calvin cycle produces only G3P. G3P can then be metabolized into storage products like sugars and starch.

13. **a.** Photorespiration uses as much ATP as photosynthesis, but results in no energy gains for the plant and reduces net carbon fixation by 25 percent compared with the Calvin cycle. If CO_2 is abundantly available, rubisco acts as a carboxylase rather than an oxygenase.

14. **b.** Plants do not photosynthesize in the dark. PEP carboxylase allows the fixation of CO_2 and the concentration of it at rubisco.

15. **a.** CAM plants functionally store CO_2 as malic acid.

16. **d.** Photosynthesis occurs only in cells that have the necessary structures, but cellular respiration occurs in *every* living cell.

Application Answers

1. Plant cells undergo cellular respiration in all cells. Therefore, all cells consume O_2. Photosynthesis occurs in specialized cells that consume both CO_2 and O_2. Because atmospheric O_2 levels are high, there is excess O_2 available for the plant to utilize; therefore, O_2 continues to be emitted from the plant.

2. The light reactions of photosynthesis produce ATP. ATP cannot be stored for use later (such as when light is not available); therefore, there has to be a mechanism for that energy to be stored. The Calvin cycle stores the energy in the chemical bonds of G3P, which can be incorporated into carbohydrates for longer-term storage.

3. Though photosynthesis produces all the necessary energy for a plant, a plant cannot be continuously photosynthetically active. Therefore, a plant stores energy in carbohydrates. Cellular respiration is necessary to break down stored carbohydrates.

4. Whether rubisco acts as a carboxylase or an oxygenase depends on the relative ratio of O_2 to CO_2. At higher CO_2 levels it acts as a carboxylase. At low CO_2 levels it acts as an oxygenase. Photorespiration is favored during hot, dry weather, which forces the closing of stomata and leads to increases in O_2 levels within the leaf. Photosynthesis is favored when stomata can remain open and light intensity is optimal.

5. Refer to Table 8.1 in your book.

6. Light is required for both the light reactions of photosynthesis and the Calvin cycle. The Calvin cycle depends on the ATP generated during the light-dependent reactions.

7. See Figures 8.9 and 8.10. See also the answer to Question 5 under "Knowledge and Synthesis Questions."

8. Accessory pigments allow utilization of light in all wavelengths of the visible spectrum. The energy absorbed is channeled to the reaction centers of the photosystems.

9. The primary pigments in plants are chlorophylls. Chlorophylls absorb blue and orange-red wavelengths of light and reflect green light, thus making plants appear green. See the absorption spectra and action spectra of chlorophyll in Figure 8.6.

9 Chromosomes, the Cell Cycle, and Cell Division

Important Concepts

Cell division

- Cell division requires a reproductive signal either from inside or outside the cell, subsequent DNA replication, segregation of the newly replicated chromosomes to opposite poles of the cell, and the division of the cytoplasm (cytokinesis) to form two daughter cells.

Prokaryotes divide by fission.

- Prokaryotic cells divide by increasing in size, replicating their DNA (a single circular chromosome), and dividing into two new cells through the process of fission.

- DNA replication in prokaryotic cells requires a number of different proteins that form a replication complex and special sites on the chromosome where replication begins (the *ori* site) and terminates (the *ter* site).

Eukaryotic cells divide by mitosis or meiosis.

- Cell division in multicellular eukaryotes occurs in response to the needs of the entire organism.

- Eukaryotic cells have more than one chromosome, and these chromosomes are linear.

- Eukaryotic chromosomes are located in the nucleus, which must also be divided into two new nuclei during cell reproduction.

- Eukaryotic cells replicate their chromosomes in response to a reproductive signal, segregate those chromosomes during mitosis (which are closely associated with each other as sister chromatids), and divide their cytoplasm during cytokinesis.

- During the cell cycle, DNA replication, mitosis, and cell division occur sequentially.

- There are two major phases of the cell cycle: interphase and mitosis. Interphase is divided into three sub phases, G1 (or Gap 1), S (or DNA synthesis), and G2 (or Gap 2). Mitosis (the M phase) includes cytokinesis.

  - G1 is the most variable phase of the cell cycle in which the decision and subsequent preparation for DNA synthesis occurs.

- The G1/S transition is where the commitment to cell division occurs.

- S is the phase of the cell cycle in which all the chromosomes are replicated; each replicated chromosome consists of two identical sister chromatids.

- During G2, the cell prepares for mitosis. During M (mitosis), the cell segregates the newly replicated chromosomes to opposite poles of the cell, and nuclear division occurs.

- In cells that produce gametes (eggs and sperm) nuclear division occurs via meiosis, which generates diversity.

Both kinases and cyclins play a key role in cell cycle transitions.

- Cdk, a cyclin-dependent kinase that phosphorylates other proteins in the cell, is catalytically active when it is bound to another protein, cyclin.

- Different cyclin–Cdk protein complexes act at various stages in the cell cycle: in the middle of G1 to help cells get past the restriction point (R), in G1 to stimulate DNA replication, and to initiate the transition from G2 to M.

- Other proteins, such as retinoblastoma protein (RB), regulate cyclin–Cdk complexes.

- Growth factors are external chemical signals that stimulate cell division.

Eukaryotic chromosome structure

- Eukaryotic chromosomes are composed of protein and DNA, collectively called chromatin. Cohesin is a protein that holds the sister chromatids together along their length.

- When eukaryotic chromosomes are replicated, the daughter chromosomes, termed sister chromatids, are still joined at the centromere, and condensin proteins make the chromosome more compact.

- In a eukaryotic nucleus, DNA molecules are wrapped around beadlike particles of protein called nucleosomes. Nucleosomes are composed of eight histone molecules, two from each class. Histones are very basic proteins.

- The nucleosomes are further condensed with the assistance of another histone, histone 1. During mitosis more compaction occurs, and the sister chromatids become visible (see Figures 9.7 and 9.8).
- Separation of the sister chromatids (which are held together by cohesin) into the daughter cells is facilitated by microtubules.

Mitosis

- During mitosis a single nucleus gives rise to two nuclei that are genetically identical to the parent nucleus.
- Centrosomes determine the plane of cell division. Centrosome duplication occurs during S phase. Centrosomes may contain centrioles, which consist of two hollow tubes of microtubules positioned at right angles to one another.
- Duplicated centrosomes migrate to opposite poles of the cell during the G2-to-M transition.
- Microtubules grow from a microtubule-organizing center, the centrosome. The microtubules form a spindle along which the chromosomes will move. Later in the cell cycle, some of these microtubules (the kinetochore microtubules) attach to the chromosomes and help separate them to opposite poles of the cell.
- Mitosis can be divided into five phases: prophase, prometaphase, metaphase, anaphase, and telophase.
- During prophase, the mitotic spindle forms and includes polar and kinetochore microtubules. The newly replicated chromosomes condense so that the sister chromatids are visible. Kinetochores assemble at the centromere of each newly replicated chromosome, and microtubules attach to the kinetochore.
- Prometaphase signals the disappearance of the nuclear envelope and the nucleoli, the attachment of kinetochore microtubules to the kinetochores of each newly replicated chromosome, and movement of the chromosomes to the equatorial plate.
- Metaphase signals the positioning of all of the centromeres at the equatorial (metaphase) plate.
- At the end of metaphase the cohesion proteins are hydrolyzed by the protease separase. Separase is normally bound to an inhibitory subunit called securin; when all the chromatids are connected to the spindle, securin is hydrolyzed, and separase becomes active.
- During anaphase, the chromosomes move to opposite poles of the spindle assisted by dynein motors acting at the kinetochores, by the shortening of the microtubules at the poles, and by the sliding of polar microtubules, which pushes the poles further apart.
- At telophase, the poles break down, the chromosomes begin to uncoil, and the nuclear envelope and nuclei coalesce, resulting in two identical nuclei.

Cytokinesis

- During cytokinesis in animal cells, the cell membrane contracts due to the interaction of actin and myosin microfilaments at the cell membrane, forming a contractile ring (see Figure 9.12A).
- In plant cell cytokinesis, a cell plate forms between the newly segregated chromosomes. This cell plate is derived from Golgi vesicles located at the equatorial region, which fuse to form a plasma membrane. The plasma membrane secretes plant cell wall materials into the cell plate to complete cell division.
- On completion of cytokinesis, there are two distinct cells, each with a full complement of chromosomes.

Cell reproduction

- There are two kinds of reproduction. The first is asexual (or vegetative reproduction) in which the replicated chromosomes are separated by mitosis. After cytokinesis the offspring are identical to the parent. The second kind of reproduction is sexual, in which gametes from two different parents fuse to form a zygote. This fusion is called fertilization and produces offspring that are different from both parents. Gametes each contain a single set of chromosomes (n) resulting from meiosis, and the zygote contains two sets of chromosomes ($2n$), one set from each parent.
- Somatic cells from diploid organisms contain pairs of homologous chromosomes, one set of chromosomes coming from each parent. Homologous chromosomes are similar in size and appearance and bear corresponding genetic information.
- Haploid cells are gametes and contain only one set of chromosomes.
- Mature organisms can exist primarily in the diploid or haploid state or can alternate between haploid and diploid states during their life cycles.
- Haplontic organisms are haploid as mature organisms. They produce spores that fuse to form a zygote, which is diploid. The zygote then undergoes meiosis to become haploid again. Those cells divide mitotically to become the mature organism.
- Most plants and some protists have mature forms that alternate between diploid and haploid stages (see Figure 9.14).
- Diplontic organisms are diploid in their mature state; only their gametes are haploid.
- Fusion of gametes to form a zygote increases diversity because the zygote's genetic makeup is derived from two different parents, each contributing one haploid set of randomly selected chromosomes.
- Mitotic chromosomes can be characterized by their shape, number, and size. Each organism has a distinctive set of chromosomes, called a karyotype.

Meiosis

- The purpose of meiosis is to reduce the chromosome number from diploid to haploid, to ensure haploid products have a complete set of chromosomes, and to promote genetic diversity in gametes and in the species.

- Meiosis is distinguished from mitosis by *two* nuclear divisions (meiosis I and II), but only *one* round of DNA replication, resulting in a reduction in the number of chromosomes from diploid to haploid.

- Each haploid cell has one complete set of chromosomes, and the combinations of alleles on the chromosomes of each gamete are different. The advantage of creating gametes with different combinations of alleles is genetic diversity in the future zygote, potentially increasing the resultant organism's ability to survive.

- There are five phases in each meiotic division: prophase, prometaphase, metaphase, anaphase, and telophase.

- In prophase of meiosis I, the newly replicated chromosomes pair with their homologs (synapsis). Because there are four chromatids (two from each homolog), these paired chromosomes are called tetrads or bivalents.

- The synaptonemal complex forms a scaffold of proteins that promotes the pairing of the chromatids during prophase. Crossing over between nonsister chromatids on homologous chromosomes occurs during prophase at special sites called chiasmata (see Figure 9.18).

- Prometaphase of meiosis I marks the disappearance of the nuclear envelope, the disaggregation of the nucleoli, the formation of the spindle, and the attachment of the microtubules to the kinetochore. Kinetochores of both chromatids in each chromosome become attached to the same half spindle. As a result, during anaphase, when the chromosomes are separated to opposite poles by microtubules attached to the kinetochore, the entire homolog is pulled to the pole.

- In some organisms the telophase completes meiosis I and is followed by an interphase termed interkinesis, whereas other organisms enter meiosis II immediately after meiosis I.

- Unlike mitosis, no DNA replication occurs before meiosis II begins, and the number of chromosomes on the equatorial plate is half the number in the mitotic nucleus (even though each chromosome still consists of two chromatids). Unlike the chromatids in mitosis, each chromatid is different from its sister chromatid, due to crossing over in meiosis I.

- Segregation of each chromatid occurs in anaphase II as in mitosis, with microtubules forming a spindle and the chromatids separating due to the breakdown of cohesion proteins.

- Each of the four haploid nuclei resulting from meiosis is genetically different due to crossing over in prophase I and due to the independent assortment of each chromatid to the daughter cells. (In humans there are 23 homologs and thus 2^{23} ways these chromosomes can segregate into the daughter nuclei.)

Abnormal numbers and kinds of chromosomes

- Aneuploidy, an abnormal number of chromosomes, occurs in cells whose chromosomes fail to segregate in meiosis I (both homologs fail to separate and go to the same pole) or meiosis II (both chromatids fail to separate and go to one pole; see Figure 9.20).

- One of the causes of nondisjunction in meiosis I may be a lack of cohesins, which orient chromosomes at the equatorial plate during metaphase I. In the absence of cohesins, homologs may line up at random and wind up in the wrong daughter cell.

- Aneuploid gametes can produce trisomies such as Down's individuals who have an extra chromosome 21. Monosomic zygotes have one copy of a homologous chromosome.

- Translocations, another chromosome abnormality, occur when part of one chromosome breaks away and becomes attached to another chromosome.

- Polyploids are cells that have increased numbers of complete sets of chromosomes. Polyploids with an even number of sets of chromosomes ($4n$, $6n$, $8n$) can undergo meiosis successfully, whereas odd-number polyploids ($3n$, $5n$, etc.) cannot, because each homolog needs to pair with is partner.

Cell death

- Cell death can occur in cells in two ways: by necrosis (cells poisoned or starved of oxygen or essential nutrients) or by apoptosis (programmed cell death). Apoptosis is a normal process during development.

- Signals for cell death include a lack of mitotic signals and recognition of DNA damage (see Table 9.2).

The Big Picture

- In the presence of an internal or external environmental signal, cells undergo a specific set of steps to replicate their DNA, segregate those newly replicated chromosomes, and divide their cytoplasm into two daughter cells. These steps ensure that the daughter cells receive a complete set of genetic instructions.

- Diploid organisms produce haploid gametes, which fuse during fertilization to form new diploid organisms.

- During the generation of these gametes, crossing over between sister chromatids and random segregation of those chromatids create genetically diverse gametes, leading to increased species diversity and survival.

Common Problem Areas

- Be sure you can distinguish the differences between mitosis and meiosis. In mitosis, the replicated chromosomes are segregated to the daughter cells, resulting in each daughter cell's having two sets of chromosomes (diploid), like its parent cell. After every DNA replication, there is one cell division. In meiosis, the replicated chromosomes undergo two cell divisions (for every one round of DNA replication), resulting in cells that have only one set of chromosomes (haploid). Furthermore, the homologous chromosomes pair in prophase of meiosis I (which is not the case in mitosis), and crossing over occurs.

Study Strategies

- Make several sequential drawings that show a cell with four chromosomes and what those chromosomes look like after DNA replication and during each phase of mitosis. Repeat these sequential drawings for meiosis I and II, using different colors for each homologous chromosome. In your drawings be sure to include crossing over in prophase I of meiosis.

- Diagram meiosis showing the effect of a nondisjunction in meiosis I and include the gametes that would result from this nondisjunction. Repeat this process for a nondisjunction in meiosis II. What kind of zygotes would be produced from these gametes?

- Draw a picture of the cell cycle and indicate the points at which the cyclin–Cdk complexes act during that cell cycle. Refer to Figure 9.6 as well as the information in your text.

- List the five phases of mitosis and describe what happens in each phase. Repeat this process for meiosis.

- Review the following animated tutorials and activities on the Companion Website/CD:
  Tutorial 9.1 Mitosis
  Tutorial 9.2 Meiosis
  Activity 9.1 The Mitotic Spindle
  Activity 9.2 Images of Mitosis
  Activity 9.3 Sexual Life Cycle
  Activity 9.4 Images of Meiosis

Test Yourself

Knowledge and Synthesis Questions

1. Which of the following is true of mitosis?
   a. The chromosome number in the resulting cells is halved.
   b. DNA replication is completed prior to the beginning of this phase.
   c. The chromosome number of the resulting cells is the same as that of the parent cell.
   d. Both b and c
   Textbook Reference: 9.3 What Happens during Mitosis? pp. 190–191

2. Which of the following is true of meiosis?
   a. The chromosome number in the resulting cells is halved.
   b. DNA replication occurs before meiosis I and meiosis II.
   c. The homologs do not pair during prophase I.
   d. The chromosome number of the resulting cells is the same as that of the parent cell.
   Textbook Reference: 9.5 What Happens When a Cell Undergoes Meiosis? p. 195

3. Which of the following is true of kinetochores on mitotic chromosomes?
   a. They are located at the centromere of each chromosome.
   b. They are the sites where microtubules attach to separate the chromosomes.
   c. They are organized so that there is one per sister chromatid.
   d. All of the above
   Textbook Reference: 9.3 What Happens during Mitosis? p. 189

4. Which of the following is true of the mitotic spindle?
   a. It is composed of actin and myosin microfilaments.
   b. It is composed of kinetochores at the metaphase plate.
   c. It is composed of microtubules, which help separate the chromosomes to opposite poles of the cell.
   d. It originates only at the centrioles in the centrosomes.
   Textbook Reference: 9.3 What Happens during Mitosis? p.189

5. Imagine that there is a mutation in the Cdk2 gene such that its gene product is nonfunctional. What kind of effect would this mutation have on a mature red blood cell?
   a. The cell would be unable to replicate its DNA.
   b. The cell would not be able to enter G1.
   c. The cell would be unable to reproduce itself.
   d. There would be no effect, because mature red blood cells do not enter the cell cycle.
   Textbook Reference: 9.2 How is Eukaryotic Cell Division Controlled? p. 186

6. Imagine that there is a mutation in the Cdk2 gene such that its gene product is nonfunctional. What kind of effect would this mutation have on a mammalian white blood cell?
   a. The cell would be unable to replicate its DNA.
   b. The cell would be unable to enter G1.
   c. The cell would be unable to reproduce itself.
   d. Both a and c
   Textbook Reference: 9.2 How is Eukaryotic Cell Division Controlled? p. 186

7. Which is true of DNA replication and cytokinesis in *Escherichia coli*?
   a. DNA replication occurs in the nucleus.
   b. Cytokinesis is facilitated by microfilaments of actin and myosin.
   c. Cell reproduction is initiated by reproductive signals, which result in DNA replication, DNA segregation, and cytokinesis.
   d. The *E. coli* chromosome is linear.
   Textbook Reference: 9.1 How do Prokaryotic and Eukaryotic Cells Divide? p. 182

8. Which of the following is true of chromatids?
   a. They are replicated chromosomes still joined together at the centromere.
   b. They are identical in mitotic chromosomes.

c. They are identical in meiotic chromosomes.

d. Both a and b

Textbook Reference: *9.3 What Happens during Mitosis? pp. 187–189*

9. Histones are positively charged because

a. the majority of the ions in the nucleus of the cell are negatively charged.

b. histones interact with acidic residues of proteins found in the nucleus.

c. the basic side chains of histone proteins interact with the negatively charged DNA.

d. histones have a majority of acidic residues in their protein sequence.

Textbook Reference: *9.3 What Happens during Mitosis? p. 187*

10. Chromosome movement during anaphase is the result of

a. the molecular motors at the kinetochores that move the chromosomes toward the poles.

b. molecular motors at the centrosome that pull the microtubules toward the poles.

c. shortening of the microtubules at the centrosome that pull the chromosomes toward the poles.

d. Both a and c

Textbook Reference: *9.3 What Happens during Mitosis? p.190*

11. Programmed cell death (apoptosis)

a. occurs in cells that have been deprived of essential nutrients.

b. occurs only in cells that have damaged DNA.

c. is a natural process during development.

d. is signaled by the initiation of mitosis.

Textbook Reference: *9.6 How Do Cells Die? p. 202*

12. What would happen to an *E. coli* cell if the *ori* site on its chromosome was deleted?

a. Nothing.

b. Replication would start but could not continue.

c. Replication could not start.

d. The chromosome would be replicated but the cell could not divide.

Textbook Reference: *9.1 How do Prokaryotic and Eukaryotic Cells Divide? p. 182*

13. Chiasmata

a. are sites where nonsister chromatids can exchange genetic material during meiosis.

b. are sites where sister chromatids can exchange genetic material during meiosis.

c. increase genetic variation among the products of meiosis.

d. increase genetic variation among the products of mitosis.

e. Both a and c

Textbook Reference: *9.5 What Happens When a Cell Undergoes Meiosis? p. 198*

14. The difference between asexual and sexual reproduction is

a. asexual reproduction only occurs in bacteria, and sexual reproduction occurs in plants and animals.

b. asexual reproduction results in an organism that is identical to the parent, whereas sexual reproduction results in an organism that is not identical to either parent.

c. asexual reproduction results from the fusion of two gametes; sexual reproduction produces clones of the parent organism.

d. asexual reproduction only occurs in haplontic organisms, and sexual reproduction occurs only in diplontic organisms.

Textbook Reference: *9.4 What Is the Role of Cell Division in Sexual Life Cycles? p. 192*

15. A chromatid is

a. one of the pairs of homologous chromosomes.

b. a homologous chromosome.

c. a newly replicated bacterial chromosome.

d. one half of a newly replicated eukaryotic chromosome.

Textbook Reference: *9.3 What Happens during Mitosis? pp. 188–189*

Application Questions

1. How is cell division different in prokaryotic cells and eukaryotic cells?

Textbook Reference: *9.1 How Do Prokaryotic and Eukaryotic Cells Divide? pp 181–183*

2. By administering a drug that inhibits cytokinesis, you have created peaches that are tetraploid. How many sets of chromosomes do these peaches have? (What is the ploidy of these chromosomes?) Will these peaches produce gametes that are fertile? What if the peaches were triploid?

Textbook Reference: *9.5 What Happens When a Cell Undergoes Meiosis? p.199*

3. How does cytokinesis differ in animal and plant cells?

Textbook Reference: *9.3 What Happens during Mitosis? p. 192*

4. Describe how two meters of DNA in a typical human cell can fit into the nucleus, which is 5 μm in diameter.

Textbook Reference: *9.3 What Happens during Mitosis? pp. 187–188, Figure 9.8*

5. Describe two ways that the genetic diversity of organisms is increased during meiosis.

Textbook Reference: *9.5 What Happens When a Cell Undergoes Meiosis? p. 199*

Answers

Knowledge and Synthesis Answers

1. **d.** Mitosis occurs after DNA replication and results in cells having the same number of chromosomes as the parent cell.

2. **a.** Meiosis occurs after one round of DNA replication. Homologous chromosomes pair during prophase I of

meiosis, and after meiosis II the resulting cells have half the number of chromosomes as the parent cell.

3. **d.** Kinetochores, one per sister chromatid, are assembled at the centromere of each chromosome and are the sites in which microtubules attach to segregate the chromosomes.

4. **c.** The mitotic spindle is composed of microtubules, not actin and myosin filaments. The spindle originates from the centrosome, which may or may not have centrioles.

5. **d.** Many cells, such as red blood cells, muscle cells, and nerve cells, lose their ability to divide as they mature.

6. **d.** Dividing cells do enter the cell cycle, and cyclin–Cdk complexes signal transitions in the cell cycle. The cyclin E–Cdk2 complex acts in the middle of G1, and the cyclin A–Cdk2 complex acts in S1 and stimulates DNA replication. Without functional Cdk2, no catalytically active cyclin–Cdk complexes can form. The cell will be unable to replicate its DNA and will not progress through the cell cycle to reproduce itself.

7. **c.** *Escherichia coli* is a prokaryote and lacks a nucleus, has a circular chromosome, and does not synthesize actin or myosin proteins. Cytokinesis in *E. coli* is a result of a reproductive signal that causes the DNA to be replicated and segregated and, finally, causes the cell to divide.

8. **d.** Chromatids are highly condensed, newly replicated chromosomes, which will be segregated to the daughter cells. After DNA replication, chromatids are still attached to one another at the centromere. Meiotic sister chromatids are different from one another due to crossing over in prophase of meiosis I. Mitotic sister chromatids are identical.

9. **c.** The positive charges on histone proteins are due to the large number of basic amino acid residues found in these proteins. These positive charges interact with the negatively charged phosphate sugar backbone of DNA during assembly of the DNA on the nucleosome.

10. **d.** Chromosomes are attached to the microtubules at their kinetochores. There are molecular motors at the kinetochores, which help move the chromosomes to opposite poles. Chromosomes are also pulled toward the poles by the shortening of the kinetochore microtubules.

11. **c.** Programmed cell death occurs during the development of many organisms (for instance, tadpoles lose their tails to become adult frogs). One of the stimuli for programmed cell death is DNA damage, but it is not the only cause of death. Necrosis (cell death that is not programmed) occurs when cells have been deprived of essential nutrients. The initiation of mitosis is part of the cell cycle, in which cells reproduce, and is not a step in programmed cell death.

12. **c.** Without the origin of replication, there would be no site for the replication proteins to bind to initiate DNA replication, so DNA synthesis would not start.

13. **e.** Chiasmata are sites where nonsister chromatids can exchange genetic material during meiosis, which increases genetic variation in the gametes (the products of meiosis).

14. **b.** Asexual reproduction produces cells that are identical to the parent and can occur in plants. Sexual reproduction can occur in haplontic organisms (such as fungi).

15. **d.** A chromatid is one half of a newly replicated eukaryotic chromosome, and is connected to the other (sister) chromatid at the centromere.

Application Answers

1. In most prokaryotic cells there is only one circular chromosome. As the cell enlarges to prepare for division, the newly replicated daughter chromosomes are separated at opposite sides of the cell. During fission, the cell membrane pinches in, and cell wall components are synthesized between the daughter cells. In eukaryotic cells, there are more chromosomes, and they are linear. The cell undergoes a sequential set of steps called the cell cycle, in which the chromosomes are replicated and then separated to opposite poles of the cell. Microtubules are used to segregate the chromosomes equally into the daughter cells, and actin filaments and myosin cause the cell membrane to form a contractile ring and separate to form two daughter cells.

2. Peaches that are tetraploid have four sets of chromosomes. Because there are an even number of chromosomes ($4n$), each replicated homologous chromosome will be able to find a replicated homolog to pair with at meiosis and will produce fertile gametes. These gametes will be diploid. Triploid cells will not be fertile because one of the three homologs will not find its pair during prophase of meiosis I, and the single homologs will be segregated randomly into the daughter cells.

3. In animal cells, cytokinesis results from the interaction of actin filaments and myosin, which causes the cell membrane to pinch in and divide the cytoplasm into two cells. In plant cells, a cell plate forms between the newly segregated chromosomes, and Golgi vesicles fuse at that site to form the new cell membranes. Cell wall components are then secreted between the plasma membranes to complete cytokinesis.

4. See Figure 9.8.

5. Genetic diversity is increased during crossing over of prophase I of meiosis so that each gamete has chromosomes with different combinations of alleles. During meiosis, each homologous chromosome is randomly segregated to one of the two poles, resulting 2^{23} different possible combinations of homologous chromosomes per gamete.

CHAPTER 10 Genetics: Mendel and Beyond

Important Concepts

Genes are units of inheritance.

- Mendel studied the inheritance of characters (such as flower color) and traits (a particular feature, such as a red, purple, or white color) as they were passed from one generation of pea plants to another.

- He used plants that were true-breeding (which meant that the observed trait must be the only form for many generations). To test if plants were true-breeding, Mendel allowed these plants to self-fertilize for many successive generations to determine if the trait was consistently expressed in every generation.

- In genetic crosses, the two individuals in the initial cross are the parents (the P generation) and their offspring are the first filial generation (the F_1 generation). The F_2 (second filial generation) are the offspring from a cross of the F_1 generation.

- Monohybrid crosses are matings between two true-breeding parents, each expressing one different trait (i.e., a cross between one true-breeding parent with a spherical seed and one true-breeding parent with a wrinkled seed). Mendel found that in these crosses, the F_1 generation expressed only one trait (spherical), and in the F_2 generation both the spherical and the wrinkled traits were expressed in ratios of 3:1 (spherical to wrinkled). He termed the F_1 trait (the spherical seed) dominant and the F_2 trait (the wrinkled seed) recessive.

- Mendel's theory stated that discrete particles of inherited traits existed in pea plants in pairs (one from each parent) and that these particles separated from each other and were assorted independently during gamete formation. Each of the gametes received one of these particles, whereas the zygote received two particles for a trait, one from each parent. This unit of inheritance is a gene.

- Alleles are different forms of the same gene. The gene for seed shape in peas has a spherical allele and a wrinkled allele.

- True-breeding parents are homozygous for that trait and have two copies of the same allele, one on each homologous chromosome.

- Heterozygous individuals have two different alleles for a gene, one on each homologous chromosome. For example, the F_1 generation of a monohybrid cross between pea plants with spherical seeds and pea plants with wrinkled seeds is heterozygous for seed shape, having one spherical allele and one wrinkled allele.

- The phenotype of an organism is the physical appearance of that organism (i.e., having seeds that are spherical or wrinkled).

Predicting the outcome of genetic crosses

- Mendel's theories have been restated as laws. Mendel's first law, the law of segregation, states that when any individual produces gametes, alleles separate so that each gamete receives only one allele for each gene.

- Punnett squares can be utilized to analyze the potential outcome of a genetic cross. All of the possible gametes from one parent are lined up on one side of the Punnett square, and all of the gametes of the other parent are lined up on the other side. Genotypes of the progeny are predicted by placing both alleles of the gametes in the square (see Figures 10.4, 10.6, and 10.7).

- Genes are located at particular sites on chromosomes called loci (locus, singular).

- Test crosses are used to determine if a given individual is heterozygous or homozygous for a dominant allele. That individual is crossed with an individual homozygous for the recessive trait (the test cross). If the individual from this test cross is homozygous for that trait, then all of the offspring will express the dominant trait. If the test individual is heterozygous, then approximately one-half of the offspring from this cross will express the dominant trait and one-half will express the recessive trait (see Figure 10.6).

- Mendel's second law, the law of independent assortment, states that all of the alleles of different genes assort independently during gamete formation. When parents that are heterozygous for two traits—for example, seed shape and seed color—are crossed (a dihybrid cross), all the alleles are expressed in the F_2 generation. In the F_2 progeny from this dihybrid cross, both yellow and green seed color are observed as well as wrinkled and spherical seeds. Review the Punnett square in Figure 10.7.

Probability calculations can be used to predict the outcome of genetic crosses.

- The product rule states that when predicting the outcome of two independent events happening together (a joint probability), that outcome is equal to the product of the probability of each of those individual events (see Figure 10.9).

- The sum rule is used to predict the probability of an event that can occur in two or more different ways. The probability that the event will occur in one of these ways is equal to the sum of the individual probabilities (see Figure 10.9).

Pedigrees can be used to predict how traits are inherited from generation to generation.

- If a trait is due to a rare dominant allele (see Figure 10.10A):
  - The affected person has an affected parent.
  - On average, one-half of the offspring of an affected parent will be affected.
  - The phenotype is seen equally in both sexes.
- If a trait is due to a rare recessive allele (see Figure 10.10B):
  - The affected individual has unaffected parents.
  - On average, one-fourth of the children from unaffected parents express the trait.
  - The phenotype occurs in both sexes.

Mutations are rare changes in the DNA sequence that are stable and passed on to the progeny (heritable).

- "Wild type" refers to traits that occur in most individuals in nature, whereas "mutant" refers to traits that are different from wild type.

- A polymorphic trait is a trait that has many different forms, with each individual phenotype present in less than 99 percent of the population.

Alleles interact in different ways.

- Multiple alleles: There can be more than two alleles for a particular trait (for example, rabbit fur color; see Figure 10.11).

- Incomplete dominance: Individuals that are heterozygotic for a particular trait appear as intermediate phenotypes of the parents. For example, a cross between a red-flowered plant and a white-flowered plant yields F_1 plants with pink flowers (see Figure 10.12).

- Codominance: Both alleles are expressed in the individual (for example, blood types in humans) (see Figure 10.13).

- Pleiotropic: A single allele can have more than one effect in an individual. For example, in the Siamese cat, a single allele can affect both pigmentation and crossed eyes.

Genes interact in different ways.

- Epistasis: The phenotypic expression of one gene is affected by another gene. For example, an allele for yellow coat color in Labrador retrievers can mask another allele for brown or black color (see Figure 10.14). Complementary gene action is a form of epistasis in which the expression of one gene depends on another.

- Hybrid vigor or heterosis: Individuals heterozygous for a particular trait are superior to the homozygote. For example, hybrid varieties of corn produce much better yields than their homozygous parents (see Figure 10.15).

- Environmental effects: Environmental conditions such as light, temperature, or nutrition can affect the expression of a particular trait. Both penetrance and expressivity are related to environmental effects (see Figure 10.16).
  - Penetrance is the proportion of individuals within a group with a particular genotype that actually show the expected phenotype.
  - Expressivity is the degree a particular genotype is expressed in an individual.

- Quantitative or continuous variation: More than one gene (quantitative trait loci) can affect a particular phenotype, with each allele intensifying or diminishing the phenotype (for example, human height). These traits can be affected by the environment.

- Linkage: Different alleles on the same chromosome do not assort independently, as first noted by Morgan in his studies on *Drosophila* (see Figure 10.18).

Genes on the same chromosome are linked.

- Linkage groups are the full set of loci on a given chromosome. Genes on the same chromosome are linked.

- Recombination frequencies between alleles that are linked (due to crossing over during prophase of meiosis I, see Figure 10.19) can be used to map positions of genes on a chromosome. Recombination frequencies are expressed in map units and are equal to the percent of recombinant progeny divided by the total progeny. A map unit (also referred to as a centimorgan) corresponds to a recombination frequency of 0.01 (see Figures 10.20, 10.21, 10.22, and 10.23).

Sex is determined in different ways in different species.

- Monoecious organisms (earthworms, pea plants, corn, etc.) produce both female and male gametes.

- Dioecious organisms (some plants and most animals) produce only male or only female gametes, resulting in two separate sexes.

- In many animals including humans, sex is determined by one or two sex chromosomes. The other chromosomes are called autosomes.

- In mammals, there are two X chromosomes in females and one X chromosome and one Y chromosome in males.

- Nondisjunction of the sex chromosomes produces gametes that can result in a Turner's female, XO, or a Klinefelter's male, XXY.

- The *SRY* (sex determining *r*egion on the arm of the *Y* chromosome) determines maleness in males. *SRY* is responsible for the primary sex determination in males and causes the embryo to develop sperm-producing tissue, the testes. Males without the *SRY* gene develop as females. The *DAX1* gene, which encodes the antitestis factor, is found on the X chromosome and inhibits the development of testis tissue. In XY individuals (males), the *SRY* gene product inhibits the effect of *DAX1*, and testes develop.

- Secondary sex determination, the outward manifestation of male and female characteristics, is determined by autosomal and X-linked genes that control the actions of hormones such as testosterone or estrogen.

- In *Drosophila*, sex is determined by the ratio of X chromosomes to autosomes: XX is female (as is XXY), and XY is male (as is XO). The Y chromosome is not necessary for sex determination but is required for male fertility.

- In birds, ZZ (XX) are male and ZW (XY) are female. In these species the egg (rather than the sperm) determines the sex of the offspring.

Sex-linked inheritance is governed by loci on the sex chromosome.

- In mammals and other organisms, including some insects, females have two copies of each X-linked gene, whereas males only have one copy of the X-linked genes.

- Because males are hemizygous for X-linked traits, all their X-linked traits will be expressed.

- If an allele is located on the X chromosome, reciprocal crosses do not have the same result. For instance, a cross between a homozygous red-eyed female *Drosophila* and a white-eyed male *Drosophila* will generate all red-eyed flies. The reciprocal cross between a white-eyed female (who must be homozygous, because the white-eyed allele is recessive) and a red-eyed male will produce an F_1 generation in which all female flies have red eyes and all male flies have white eyes (see Figure 10.23).

- X-linked recessive traits show the following characteristics.

  - The phenotype is much more common in males than in females.

  - Males with X-linked traits can only pass those traits to their daughters, but not their sons.

  - Females who are carriers of an X-linked trait pass that trait on average to one-half of their sons and one-half of their daughters. Because sons always receive the X chromosome from their mother, one-half of those sons, on average, will express the X-linked trait. Daughters who receive the X-linked trait from their mothers will be carriers (assuming they received the X chromosome from their father that carried the dominant allele).

- X-linked traits can skip a generation; a trait that is expressed in the grandfather will be carried by his daughter and may be expressed in his grandson.

- There are several dozen genes in humans that are Y-linked. These traits are always passed from father to son.

Cytoplasmic inheritance differs from the Mendelian pattern.

- Mitochondria and plastids contain genomes that are passed to the zygote from the gamete that makes the largest cytoplasmic contribution (in humans, the egg).

- These organelles are generally so numerous that genes for cytoplasmic traits are polyploid.

- Genes in organelles mutate faster than those in the nucleus, leading to multiple alleles for many nonnuclear genes.

- In plants, mutations in plastid genes affect proteins that assemble chlorophyll molecules into photosystems, resulting in white rather than green leaves.

- Mutations in mitochondrial genes affect the electron transport chain and the synthesis of ATP. These mutations have a strong effect on tissues that require high ATP concentrations, particularly the nervous and muscular systems and kidneys.

The Big Picture

- Genes are units of inheritance that are passed down to the progeny by the fusion of gametes to form a new organism.

- Meiosis generates gametes that contain only one allele for each gene, and those alleles are segregated independently (unless they are linked).

- By studying the expression of particular phenotypes in successive generations, one can begin to understand the alleles that gave rise to those phenotypes (i.e., rare recessive, rare dominant, sex-linked).

- Exceptions to Mendel's rules also help explain the interactions of gene products and the interaction of alleles that lead to the observed phenotype.

Common Problem Areas

- It is easy to be overwhelmed by all the exceptions to Mendel's rules. Initially, make sure you understand Mendel's laws of segregation and independent assortment. Examine the exceptions to these rules by thinking about the molecular role of these alleles. For example, what is the explanation for a codominant phenotype at the molecular level, such as the AB blood type in humans? Both gene products are synthesized by the cell and modify the cell surface glycoproteins, producing both A and B antigens. Or what is the molecular explanation for epistasis? The enzyme required to produce an initial substrate necessary for pigmentation is missing, even though the other gene products (further down the biochemical pathway) for producing a particular color are present in the cell.

Study Strategies

- Review the figures of meiosis in Chapter 9 to see how Mendel's laws of independent assortment and segregation apply. Draw the chromosomes with different alleles (*S* on one chromosome, *s* on the other) and follow how those alleles are segregated to the gametes during meiosis. Repeat this process, but draw two separate pairs of chromosome with each bearing a different allele (for instance *Ss* and *Yy*). Show how these alleles sort independently in the gametes (review Figure 10.8).

- Use a Punnett square to predict the outcome of different monohybrid and dihybrid crosses. Use the probability rules to predict the genotypes and phenotypes of these same crosses and compare the two methods.

- After reviewing Mendel's laws of independent assortment and allele segregation, make a list of the exceptions to those laws with specific examples for each exception.

- Review the linkage problem in Figure 10.22 to familiarize yourself with allele loci on a chromosome and how recombination frequencies affect the genotypes of the gametes.

- Review the following animated tutorials and activities on the Companion Website/CD:
  Tutorial 10.1 Independent Assortment of Alleles
  Tutorial 10.2 Alleles That Do Not Sort Independently
  Activity 10.1 Homozygous or Heterozygous?
  Activity 10.2 Concept Matching I
  Activity 10.3 Concept Matching II

Test Yourself

Knowledge and Synthesis Questions

1. In the beginning of Chapter 10, hemophilia is mentioned as a trait carried by the mother and passed to her sons. What is the pattern of inheritance for this trait?
   a. Hemophilia is an allele carried on one of the mother's autosomal chromosomes.
   b. Hemophilia is an allele carried on the Y chromosome because more males have this genetic disorder than females.
   c. Hemophilia is an allele carried on the X chromosome and can be directly inherited by the son from the father or the mother.
   d. Hemophilia is carried on the X chromosome and can only be inherited by the son if the mother is a carrier.

Textbook Reference: 10.4 What Is the Relationship between Genes and Chromosomes? p. 227

2. Originally, genetic inheritance was thought to be a function of the blending of traits from the two parents. Which exception to Mendel's rules is an example of blending?
   a. Polygenic inheritance
   b. Incomplete dominance
   c. Codominance
   d. Pleiotropism
   Textbook Reference: 10.2 How Do Alleles Interact? pp. 218–219

3. True-breeding plants
   a. produce the same offspring when crossed for many generations.
   b. result from a monohybrid cross.
   c. result from a dihybrid cross.
   d. result from crossing over during prophase I of meiosis.
   Textbook Reference: 10.1 What Are the Mendelian Laws of Inheritance? p. 209

4. What is the probability that a cross between a true-breeding pea plant with spherical seeds and a true-breeding pea plant with wrinkled seeds will produce F_1 progeny with spherical seeds?
   a. 1/2
   b. 1/4
   c. 0
   d. 1
   Textbook Reference: 10.1 What Are the Mendelian Laws of Inheritance? p. 211, Figure 10.3

5. What is the pattern of inheritance for a rare recessive allele?
   a. Every affected person has an affected parent.
   b. Unaffected parents can produce children who are affected.
   c. Unaffected mothers have affected sons and daughters who are carriers.
   d. None of the above
   Textbook Reference: 10.1 What Are the Mendelian Laws of Inheritance? p. 217

6. What is the pattern of inheritance for a rare dominant allele?
   a. Every affected person has an affected parent.
   b. Unaffected parents can produce children who are affected.
   c. Unaffected mothers have affected sons and daughters who are carriers.
   d. None of the above
   Textbook Reference: 10.1 What Are the Mendelian Laws of Inheritance? p. 217

7. What is the pattern of inheritance for a sex-linked allele?
   a. Every affected person has an affected parent.
   b. Unaffected parents can produce children who are affected.

c. Unaffected mothers have affected sons and daughters who are carriers.

d. None of the above

Textbook Reference: 10.4 *What Is the Relationship between Genes and Chromosomes? p. 227*

8. Penetrance and expressivity are related to
   a. the increased expression of a particular trait when a hybrid species is formed.
   b. quantitative traits that diminish or intensify a particular phenotype.
   c. the influence of environment on the expression of a particular genotype.
   d. the expression of one gene masking the effects of another gene.

   Textbook Reference: 10.3 *How Do Genes Interact? p. 221*

9. Sex determination in humans and *Drosophila* is similar because
   a. females are hemizygous.
   b. males have one X chromosome and females have two X chromosomes.
   c. all males from both species always have one Y chromosome
   d. the ratio of X chromosomes to sets of autosomes determines maleness or femaleness.

   Textbook Reference: 10.4 *What Is the Relationship between Genes and Chromosomes? pp. 225–226*

10. Linked genes are genes that
    a. assort independently.
    b. segregate equally in the gametes during meiosis.
    c. always contribute the same trait to the zygote.
    d. are found on the same chromosome.
    e. recombine during mitosis.

    Textbook Reference: 10.4 *What Is the Relationship between Genes and Chromosomes? p. 218*

11. Cytoplasmic inheritance
    a. results from polygenic nuclear traits.
    b. is the result of gametes contributing equal amounts of cytoplasm to the zygote.
    c. is determined by genes on DNA molecules in mitochondria and chloroplasts.
    d. follows Mendel's law of segregation.

    Textbook Reference: 10.5 *What Are the Effects of Genes Outside the Nucleus? pp. 228–229*

12. Epistasis is
    a. the degree a particular genotype is expressed in an individual.
    b. the proportion of individuals within a group with a particular genotype that show the expected phenotype.
    c. when a heterozygotic individual expresses an intermediate phenotype of the parents.
    d. when one gene masks the expression of another gene.

    Textbook Reference: 10.3 *How Do Genes Interact? pp. 219–220*

13. Quantitative traits are
    a. affected by the environment.
    b. traits that affect the same phenotype.
    c. traits where each allele intensifies or diminishes the phenotype.
    d. All of the above

    Textbook Reference: 10.3 *How Do Genes Interact? pp. 221–222*

14. A test cross
    a. is used to determine if an organism that is displaying a dominant trait is heterozygous or homozygous for that trait.
    b. is used to determine if an organism that is displaying a recessive trait is heterozygous or homozygous for that trait.
    c. results in the F_2 generation having a phenotypic ratio of 3/4 dominant to 1/4 recessive.
    d. results in the same alleles being transferred from generation to generation.

    Textbook Reference: 10.3 *What Are the Mendelian Laws of Inheritance? p. 213*

15. An individual has a karyotype that is XX but is phenotypically male. What could explain this result?
    a. The Y chromosome was not visible in the karyotype.
    b. Sex determination is determined by the autosomes and not the X chromosomes.
    c. A translocation has occurred placing the *SRY* gene on one of the X chromosomes.
    d. The DAX I protein is overproduced.

    Textbook Reference: 10.4 *What Is the Relationship between Genes and Chromosomes? p. 226*

Application Questions

1. Draw a pedigree for three generations in which the grandfather has red–green color blindness and his daughter is a carrier. This daughter has four sons. Predict how many of the sons will be color-blind.

   Textbook Reference: 10.4 *What Is the Relationship between Genes and Chromosomes? p. 228*

2. Draw a sample pedigree with three generations in which the paternal grandfather has a rare dominant autosomal trait. What is the probability that one of his children will have the disease? That one of his grandchildren will have the disease?

   Textbook Reference: 10.1 *What Are the Mendelian Laws of Inheritance? p. 216*

3. Draw a sample pedigree with three generations in which the maternal grandmother and paternal grandfather are carriers of a rare recessive autosomal trait. What is the probability that one of their children will be carriers of this trait? What is the probability that the grandchild with these grandparents would have the disease?

   Textbook Reference: 10.1 *What Are the Mendelian Laws of Inheritance? p. 216*

4. Cytoplasmic traits in certain species of trees are passed from the male plant to all of its progeny. Compare this observation to cytoplasmic inheritance in humans.
   Textbook Reference: 10.5 What Are the Effects of Genes Outside the Nucleus? p. 228

5. You are a genetics counselor who is working with a 21-year-old pregnant woman who has just discovered that her father has Huntington's chorea, a rare dominant autosomal trait. This disease usually develops in middle-aged individuals, so people carrying this trait do not find out they have this genetic disorder until midlife. What are the chances that the child she is carrying will develop Huntington's chorea? (Assume that her husband's family has no history of Huntington's chorea.) What is the chance that she has Huntington's chorea?
   Textbook Reference: 10.1 What Are the Mendelian Laws of Inheritance? pp. 216–217

Answers

Knowledge and Synthesis Answers

1. **d.** Hemophilia is an X-linked trait and can only be inherited by the son from his mother's X chromosome. The father contributes the Y chromosome to his son (not his X chromosome) and thus cannot pass any of his X-linked alleles to his son.

2. **b.** Incomplete dominance results in the progeny's expressing an intermediate form of the two parental alleles. (In a cross between red-flowered plants and white-flowered plants, the expression of pink-flowered plants would be a "blend" of the parental traits.) Codominance is not an example of blending because both alleles are fully expressed in the individual.

3. **a.** Monohybrid and dihybrid crosses produce heterozygous individuals; true-breeding individuals are always homozygous.

4. **d.** This is an example of a monohybrid cross. All of the F_1 progeny would have spherical seeds. (The F_1 generation would all have the genotype *Ss*, producing the phenotype of spherical seeds because the spherical allele, *S*, is dominant to the wrinkled allele, *s*.)

5. **b.** Rare recessive alleles can be carried by both parents but not expressed in those parents. The parents will be heterozygous for this allele (*Aa*). Their children will have a one-fourth probability of expressing that recessive allele (*aa*).

6. **a.** If an allele is dominant, every affected individual has at least one dominant allele. An affected individual must have received that allele from one of his or her parents. Because the allele is dominant, that parent must also be affected.

7. **c.** The most common sex-linked alleles are X-linked and are passed from a mother to her son (because the mother always donates one of her X chromosomes to her son, and the father always donates the Y chromosome to his son). Daughters can also receive the X-linked allele from their mothers, but the father donates the other X chromosome, so daughters can be carriers.

8. **c.** Penetrance and expressivity are related to the effects the environment has on a particular phenotype. Answer **a** refers to hybrid vigor, answer **b** refers to quantitative traits, and answer **d** refers to epistasis.

9. **b.** In these three species, females have two X chromosomes and males have one X chromosome. The ratio of X chromosomes to autosome sets is important for sex determination in *Drosophila*, but not in humans. In *Drosophila*, male flies can be XO (they are normally XY), and female flies can be XXY (they are normally XX).

10. **d.** Linked genes by definition are on the same chromosome and thus do not sort independently, do not contribute the same trait to the zygote, and do not recombine during mitosis or segregate equally to the gametes during meiosis.

11. **c.** The genes on the mitochondria and chloroplast chromosomes are passed on to all of the progeny from the gamete that contributes the majority of the cytoplasm.

12. **d.** When one gene masks the expression of another gene. Answer **a** refers to expressivity, answer **b** refers to penetrance, and answer **c** refers to incomplete dominance.

13. **d.** Quantitative traits are traits that are affected by the environment and can either diminish or intensify one phenotype.

14. **a.** A test cross is used to determine if an organism that is expressing a dominant trait is homozygous or heterozygous for that trait. A ratio of 3/4 dominant to 1/4 recessive in the F_2 generation results from a monohybrid cross. True-breeding individuals continue to express the same alleles generation after generation.

15. **c.** The *SRY* (sex-determining region) gene has been moved via a translocation to the X chromosome and in the presence of the SRY protein, the XX individual has developed sperm producing testes. SRY also inhibits the expression of the *DAX I* gene, which encodes a male inhibitor. If *DAX I* was overproduced during embryonic development, a female would be produced.

Application Answers

1. See Figure 10.24, generations II, III, and IV. One-half of her sons could be color-blind.

2. See Figure 10.10A. One-half of his children could get the disease; one-fourth of his grandchildren could get the disease.

3. See Figure 10.10B, generations III, and IV. One-half of the children of these grandparents could be carriers. One-sixteenth of the children could have the disease.

4. In humans, the gamete with the largest cytoplasmic contribution is the egg, so cytoplasmic inheritance is passed from the female parent to all her children. In certain tree species, the male gamete contributes the

majority of the cytoplasm to the zygote, so that all the mitochondria and chloroplasts in the zygote are inherited from the male parent.

5. She has a 50 percent chance that she will develop Huntington's chorea. Because the trait is an autosomal dominant allele, one-half of her father's gametes will contain the homologous chromosome carrying that allele and one-half of his gametes will contain the homologous chromosome that carries the wild-type allele. If she received the Huntington's allele, her child has a 50 percent chance of receiving this allele from her. The product rule is used to predict the probability that her child will inherit the Huntington's allele: 1/2 (the probability that she has the Huntington's allele) × 1/2 (the probability her child will inherit this allele from her) = 1/4 (the probability her child has the allele). Her child has a 25 percent chance of carrying the Huntington's chorea allele and thus of developing the disease. She has a 50 percent chance of carrying the Huntington's chorea allele.

CHAPTER 11 DNA and Its Role in Heredity

Important Concepts

Specific observations and discoveries revealed that DNA is the genetic material.

- Staining techniques showed that DNA is present in the nucleus.

- Different organisms showed varying amounts of DNA in their nuclei.

- Proportional amounts of DNA were found in somatic and germ cells.

- Griffith's transformation experiments indicated that some transforming principle (later identified as DNA) could cause a heritable change in living cells.

- Avery, MacLeod, and McCarty's work used purified DNA to demonstrate that the transforming factor that contained the genetic information was DNA.

- Hershey and Chase, using a virus that infected bacteria, demonstrated definitively that DNA carried the hereditary information.

- Transformation in eukaryotes (termed transfection) was the final proof that DNA was the genetic material.

A series of related yet independent experiments elucidated the structure of DNA.

- Franklin and Wilkins prepared X-ray crystallographic images, which were critical to understanding the structure of DNA.

- Chargaff provided rules for relative ratios of pyrimidines and purines, the nitrogenous bases of the nucleotides of DNA.

- Watson and Crick coupled results from previous experimentation and model building and revealed the three-dimensional, double-helical structure of DNA.

The form of DNA is tied to its function.

- DNA is a right-handed double helix formed by two antiparallel DNA strands. The DNA helix has a major and minor groove, and the bases in those grooves provide distinct surfaces for protein recognition. The two strands are aligned in the helix with the 5′ end of one strand pairing with the 3′ end of its complementary strand.

- The sugar–phosphate backbones of the polynucleotide chains are on the outside of the helix, and the nitrogenous bases point toward the center of the helix.

- Base pairs are complementary; they consist of a pyrimidine pairing with a purine via hydrogen bonding (A pairs with T, G pairs with C). Hydrophobic interactions between the base pairs help them to stack on top of one another in the center of the molecule, stabilizing the molecule and maintaining a constant diameter.

- The specific base sequences in DNA are used to store genetic information. Altering the base sequence in the DNA by mutation may cause a change in the genetic information. Faithful replication of the DNA molecule is precisely generated using complementary base pairing. The expression of genetic information in the DNA results in specific phenotypes.

DNA must be able to be replicated in order to be heritable.

- All of the following must be present for replication to take place: a template strand, all four deoxyribonucleoside triphosphates (dATP, dCTP, dGTP, and dTTP), a primer base-paired to the template DNA, and DNA polymerase.

- Meselson and Stahl's experiments demonstrated that DNA replication is semiconservative: each parental strand of the DNA duplex served as a template for the newly synthesized (daughter) DNA, and each new DNA molecule was composed of one parental (template) and one newly replicated (daughter) strand (see Figure 11.11).

- During replication, DNA is unwound, and the new strand is synthesized by complementary base pairing of each incoming nucleotide to the parental strand via hydrogen bonds. New nucleotides are always added at the 3′ (—OH) end of the elongating strand via covalent bonds.

- Replication complexes assemble at origins (*ori*) on the DNA to begin replication. Replication proceeds in both directions from the replication origin, forming replication forks. Another model suggests that DNA is threaded through the replication complex during replication. Both leading-strand and lagging-strand synthesis occur at replication forks (see Figure 11.18).

- Replication complexes are composed of many proteins: helicases unwind parental DNA, single-stranded binding proteins prevent parental strands from reassociating, primases generate RNA primers, DNA polymerases elongate daughter strands from the 3′ OH end of those primers, and DNA ligases seal the nicks in the sugar–phosphate backbone. Topoisomerases separate newly replicated chromosomes.

- Prokaryotes have one replication origin for their circular chromosome; most eukaryotes have many origins for their large linear chromosomes.

- New daughter strands must begin with a short RNA primer, formed using the parental strand as a template. Once a primer is synthesized, DNA polymerase then adds deoxyribonucleotides to the 3′ (—OH) end of the primer.

- There are several different DNA polymerases in cells, all replicate DNA, some remove primers, and some are involved in DNA repair.

Replication proceeds in a specified direction.

- Nucleotides can only be added to the 3′(—OH) end of a strand. Thus, the parental strand is read in the 3′-to-5′ direction, and the daughter strand elongates in the 5′-to-3′ direction.

- The leading strand elongates continuously in the "right" direction (5′ to 3′) as the replication fork opens up.

- The lagging strands (Okazaki fragments) are synthesized 5′ to 3′ in a direction opposite to opening of the replication fork. This process requires many primers, and the resulting Okazaki fragments that are generated are ligated (covalently joined by DNA ligase) to form a continuous strand. RNA primers are removed prior to this ligation by DNA polymerase I (see Figure 11.19).

- The sliding clamp increases the efficiency of DNA replication by keeping DNA polymerase tightly associated with the newly replicated strand (see Figure 11.20).

- Telomeres are repetitive sequences found at the ends of eukaryotic chromosomes, which, with the protein telomerase, help maintain the stability of those ends during each replicative cycle. These telomeres are gradually lost during each replicative cycle, and that loss can eventually lead to cell death.

Many mistakes are made during initial synthesis of DNA; therefore, DNA proofreading and repair mechanisms are essential.

- DNA polymerase proofreads and repairs base mismatches during replication.

- Mismatch repair systems scan for possible errors in newly synthesized DNA.

- Excision repair mechanisms replace abnormal bases with functional bases.

Advanced techniques in molecular biology have allowed us to make multiple copies of DNA from a single strand and to determine the sequence of bases in DNA.

- The polymerase chain reaction (PCR) allows the rapid amplification of short DNA sequences into many identical copies (see Figure 11.23).

- DNA sequencing is now a highly automated process in which modified bases are mixed with the normal substrates for DNA replication to generate a mixture of DNA fragments. These fragments are used to determine the base sequence of the DNA.

The Big Picture

- Knowing that DNA is the genetic material has allowed scientists to understand how hereditary information is passed on at the molecular level and how mutations can alter that hereditary information.

- Advances in our knowledge of how DNA is replicated provide new technologies for understanding genes, their function, their expression, and genetic relatedness in different organisms.

Common Problem Areas

- Students frequently get lost in the history of discovery. Do not focus on *who* and *when*, but instead on *how* the experiments provide evidence to prove that DNA is the genetic material.

- Understanding DNA replication is a highly visual process. Use the figures in your text to "see" what is happening. Make your own diagrams of these processes.

- 5′ and 3′ ends are often confused when comparing parental (template) strands and daughter (newly synthesized) strands. On both strands, the 3′ end corresponds to a site where a hydroxyl group (—OH) is (or was), and the 5′ end corresponds to a site where a phosphate tail is (or was).

- Take advantage of laboratory activities that simulate or actually allow you to experience sequencing or PCR. Nearly all molecular research labs utilize one or both of these techniques.

Study Strategies

- Focus on how scientific evidence proved that DNA was the genetic material and how Meselson and Stahl demonstrated that DNA replication was semiconservative.

- Draw a picture of the replication fork. Position the primers on that fork, and then draw in the leading and lagging strands. Label all of the 3′ and 5′ ends. Use an arrow to indicate where ligase will seal up the ends.

- Make a list of all the proteins involved in DNA replication. Describe the function of each protein.
- Review the following animated tutorials on the Companion Website/CD:
  Tutorial 11.1 DNA Replication, Part 1: Replication of a Chromosome and DNA Polymerization
  Tutorial 11.2 The Meselson-Stahl Experiment
  Tutorial 11.3 DNA Replication, Part 2: Coordination of Leading and Lagging Strand Synthesis

Test Yourself

Knowledge and Synthesis Questions

1. Griffith's experiments showing the transformation of R strain pneumococcus bacteria to S strain pneumococcus bacteria in the presence of heat-killed S strain bacteria gave evidence that
   a. an external factor was affecting the R strain bacteria.
   b. DNA was definitely the transforming factor.
   c. S strain bacteria could be reactivated after heat killing.
   d. All of the above
   Textbook Reference: 11.1 What Is the Evidence that the Gene Is DNA? p. 234

2. Experiments by Avery, MacLeod, and McCarty supported DNA as the genetic material by showing that
   a. both protein and DNA samples provided the transforming factor.
   b. DNA was not complex enough to be the genetic material.
   c. only samples with DNA provided transforming activity.
   d. even though DNA was molecularly simple, it provided adequate variation to act as the genetic material.
   Textbook Reference: 11.1 What Is the Evidence that the Gene Is DNA? pp. 234–235

3. Hershey and Chase used radioactive ^{35}S and ^{32}P in experiments to provide evidence that DNA was the genetic material. These experiments pointed to DNA because
   a. progeny viruses retained ^{32}P but not ^{35}S.
   b. presence of ^{32}P in progeny viruses indicated that DNA was passed on.
   c. absence of ^{35}S in progeny viruses indicated that proteins were not passed on.
   d. All of the above
   Textbook Reference: 11.1 What Is the Evidence that the Gene Is DNA? pp. 235–237

4. X-ray crystallography provides information about the _____ of DNA but is limited because of the _____ of DNA. The technique is based on the pattern of _____ off the atoms in the molecule.
   a. structure; difficulty of purification; light absorption
   b. dimensions; molecular weight; diffraction
   c. molecular weight; shape; diffraction
   d. dimensions; linearity; light absorption
   Textbook Reference: 11.2 What Is the Structure of DNA? p. 238

5. Chargaff observed that the amount of
   a. purines is roughly equal to the amount of pyrimidines in all tested organisms.
   b. A is roughly equal to the amount of T in all tested organisms.
   c. A + T is roughly equal to the amount of G + C in all tested organisms.
   d. Both a and b
   Textbook Reference: 11.2 What Is the Structure of DNA? p. 238

6. Watson and Crick's model allowed them to visualize
   a. the molecular bonds of DNA.
   b. how the purines and pyrimidines fit together in a double helix.
   c. that the two strands of the DNA double helix were antiparallel.
   d. All of the above
   Textbook Reference: 11.2 What Is the Structure of DNA? p. 238

7. A fundamental requirement for the function of genetic material is that it must be
   a. conserved among all organisms with very little variation.
   b. passed intact from one species to the next species.
   c. accurately replicated.
   d. found outside the nucleus.
   Textbook Reference: 11.2 What Is the Structure of DNA? p. 241

8. Evidence indicating that DNA replication was semiconservative came from
   a. DNA staining techniques.
   b. DNA sequencing.
   c. density gradient studies using "heavy" nucleotides.
   d. None of the above
   Textbook Reference: 11.3 How Is DNA Replicated? pp. 241–243

9. Current evidence indicates that replication complexes are attached to stationary nuclear components and that DNA is threaded through these complexes. Which of the following best describes the role of the replication complex?
   a. The complex acts as an enzymatic center for DNA replication.
   b. The complex binds specifically to replication origins, then controls the rate at which replication occurs.
   c. The complex is the initiating site of replication forks.
   d. All of the above
   Textbook Reference: 11.3 How Is DNA Replicated? pp. 243–244

10. The primary function of DNA polymerase is to
    a. add nucleotides to the growing daughter strand.
    b. seal nicks along the sugar–phosphate backbone of the daughter strand.

    c. unwind the parent DNA double helix.

    d. prevent reassociation of the denatured parent DNA strands.

Textbook Reference: *11.3 How Is DNA Replicated? p. 244, Figure 11.12; p. 245*

11. The lagging daughter strand of DNA is synthesized in what appears to be the "wrong" direction. This synthesis is accomplished by

    a. synthesizing short Okazaki fragments in a 5′-to-3′ direction.

    b. synthesizing multiple short RNA primers to initiate DNA replication.

    c. using DNA polymerase I to remove RNA primers from Okazaki fragments.

    d. All of the above

Textbook Reference: *11.3 How Is DNA Replicated? pp. 246–247*

12. RNA primers are necessary in DNA synthesis because

    a. DNA polymerase is unable to initiate replication without an origin.

    b. the DNA polymerase enzyme can only catalyze the addition of deoxyribonucleotides onto the 3′ (—OH) end of an existing strand.

    c. RNA primase is the first enzyme in the replication complex.

    d. All of the above

Textbook Reference: *11.3 How Is DNA Replicated? p. 245*

13. Proofreading and repair occur

    a. at any time during or after synthesis of DNA.

    b. only before DNA methylation occurs.

    c. only in the presence of DNA polymerase.

    d. only in the presence of an excision repair mechanism.

Textbook Reference: *11.4 How Are Errors in DNA Repaired? pp. 249–250*

14. DNA replication is an _____ process and _____ energy.

    a. exergonic; does not require

    b. endothermic; does require

    c. endergonic; does require

    d. endodontic; does not require

Textbook Reference: *11.3 How Is DNA Replicated? p. 243*

15. *T. aquaticus* DNA polymerase is not denatured during the heat cycling required to denature DNA. This property led to advances in what technique?

    a RFLP analysis

    b. PCR

    c. Sequencing

    d. EPA

Textbook Reference: *11.5 What Are Some Applications of Our Knowledge of DNA Structure and Replication? p. 251*

16. Thirty percent of the bases in a sample of DNA extracted from eukaryotic cells is adenine. What percentage of cytosine is present in this DNA?

    a. 10 percent

    b. 20 percent

    c. 30 percent

    d. 40 percent

Textbook Reference: *11.2 What Is the Structure of DNA? p. 238*

17. Which of the following represents a bond between a purine and a pyrimidine (in that order)?

    a. C–T

    b. G–A

    c. G–C

    d. T–A

Textbook Reference: *11.2 What Is the Structure of DNA? pp. 239–240*

18. Which of the following statements about DNA replication is *false*?

    a. Okazaki fragments are synthesized as part of the leading strand.

    b. Replication forks represent areas of active DNA synthesis on the chromosomes.

    c. Error rates for DNA replication are often less than one in every billion base pairings.

    d. Ligases and polymerases function in the vicinity of replication forks.

Textbook Reference: *11.3 How Is DNA Replicated? pp. 243, 246–247; 11.4 How Are Errors in DNA Repaired? p. 249*

19. Which of the following would *not* be found in a DNA molecule?

    a. Purines

    b. Ribose sugars

    c. Phosphates

    d. Sulfur

Textbook Reference: *11.2 What Is the Structure is DNA? p. 240, Figure 11.9*

20. If a nucleotide lacking a hydroxyl group at the 3′ end were added to a PCR reaction, what would be the outcome?

    a. No additional nucleotides would be added to a growing strand containing that nucleotide.

    b. Strand elongation would proceed as normal.

    c. Nucleotides would only be added at the 5′ end.

    d. *T. aquaticus* DNA polymerase would be denatured.

Textbook Reference: *11.5 What Are some Applications of Our Knowledge of DNA Structure and Replication? p. 251*

Application Questions

1. Given the following parent strand sequence, what would the daughter strand sequence look like?

    5′ – G C T A A C T G T G A T C G T A T A A G C T G A – 3′

Textbook Reference: *11.2 What Is the Structure of DNA? p. 239*

2. Diagram the double helix. Be sure to label those properties that make it most suited as the genetic material.

Textbook Reference: *11.2 What Is the Structure of DNA? p. 240*

3. Diagram a replication fork as it would be seen in a replicating segment of DNA. In your diagram label the 5′ and 3′ ends of each parent strand and daughter strand. Indicate which new strand is the leading strand and which is the lagging strand of the daughter DNA.
   Textbook Reference: 11.3 How Is DNA Replicated? p. 247

4. Based on your diagram in Question 3, construct a flowchart that represents what occurs during DNA replication. Divide your chart into three parts: initiation, elongation, and termination of replication. In your chart, indicate the roles of helicase, DNA polymerase, single-stranded DNA binding proteins, nucleotides, parental (template) DNA strands, DNA ligase, RNA primase, and RNA primers.
   Textbook Reference: 11.3 How Is DNA Replicated? pp. 242–248

5. Explain the difference between conservative and semiconservative models of DNA replication. What results supported the semiconservative model? What would the results have looked like had the conservative model of DNA replication been accurate? Are there any other potential hypotheses?
   Textbook Reference: 11.3 How Is DNA Replicated? pp. 241–242

6. Explain the role of Okazaki fragments in the synthesis of the lagging strand.
   Textbook Reference: 11.3 How Is DNA Replicated? p. 246

7. Differentiate between proofreading, mismatch repair, and excision repair. Which of these repair mechanisms is responsible for repairing a mutation that occurs in an adult cell from overexposure to the sun? Explain your answer.
   Textbook Reference: 11.4 How Are Errors in DNA Repaired? pp. 249–250

8. Explain how modified nucleotides are used in DNA sequencing.
   Textbook Reference: 11.5 What Are Some Applications of Our Knowledge of DNA Structure and Replication? pp. 251–253

Answers to Questions

Knowledge and Synthesis Answers

1. **a.** The experiments showed only that something was causing the R strain pneumococcus to develop the S strain virulence. Additional experiments were necessary to provide evidence that the transforming factor was indeed DNA.

2. **c.** These researchers were able to isolate nearly pure DNA samples. It was only these samples that provided transformation activity.

3. **d.** Only DNA incorporates radioactive P. Radioactive S is incorporated into proteins. The fact that progeny viruses retained radioactive P and not radioactive S indicated that DNA is heritable and protein is not.

4. **b.** X-ray crystallography allows researchers to measure the distances between atoms and determine the physical dimension of a molecule. The technique becomes quite difficult with molecules that have high molecular weights. Wilkins's contributions were significant because he was able to isolate nearly perfect strands of the molecule. The resulting images are patterns of how the X-rays are diffracted off the constituent atoms of the molecule.

5. **d.** Chargaff found that in most DNA sampled, the amount of A equaled the amount of T and the amount of G equaled the amount of C. It follows that the amount of purines (A + G) equals the amount of pyrimidines (T + C). The same does not hold for the amount of A + T versus the amount of G + C.

6. **d.** Model building by Watson and Crick created a three-dimensional visualization of the size, bond angles, base pairings, and overall structure of the DNA molecule. The information contributing to their model came from a variety of experiments performed by other scientists.

7. **c.** Replication of DNA is a fundamental requirement for its function as the genetic material. DNA must be correctly replicated in each cell of an organism and must be passed from parent to offspring or from cell to daughter cell. DNA is not passed from one species to a different species. There is variation in DNA sequences from different organisms. DNA is found outside the nucleus in eukaryotic cells (specifically in the mitochondria and, in plants, chloroplasts), but this DNA is inherited in a non-Mendelian fashion (see Chapter 10).

8. **c.** Density gradient labeling allowed Meselson and Stahl to track parental versus daughter strands of DNA. Their research showed that DNA replication was a semiconservative process with each newly synthesized DNA molecule containing one parental (template) strand base-paired to one daughter (newly synthesized) strand.

9. **d.** Replication complexes bind to replication origins. The enzymes contained in the complex are responsible for unwinding the parental DNA, for initiating replication, and for the actual synthesis of the new strands of daughter DNA.

10. **a.** DNA polymerase adds nucleotides to an existing nucleotide strand.

11. **d.** Okazaki fragments are small segments of newly synthesized DNA that have been added to short RNA primers. The RNA is removed from these small fragments by DNA polymerase I and replaced with DNA. The remaining DNA fragments are ligated (covalently bound) together to form a continuous newly synthesized DNA strand.

12. **b.** DNA polymerase cannot initiate synthesis of a nucleotide strand, it can only add onto an existing strand of RNA primer or DNA. Primers provide DNA polymerase with the 3′ (—OH) end required for the

addition of deoxyribonucleotides. The origin is also required for DNA synthesis, but it is the site where the replication complex is initially assembled. Primase is not the first enzyme in the replication complex.

13. **a.** The mismatch repair mechanism operates before the newly synthesized DNA strand is methylated, but for the integrity of DNA to be maintained, repair mechanisms must be active during synthesis, modification, and utilization of DNA.

14. **c.** DNA replication is an energy-consuming process that must have an input of energy to proceed. Energy is provided by the breaking of the phosphodiester bonds of each incoming deoxyribonucleoside triphosphate (see Figure 11.10).

15. **b.** PCR depends on continuous cycles of heating, in which DNA is denatured, followed by cooling, which then allows primer annealing and DNA synthesis to occur. The original limitation of the technique was that DNA polymerases were also denatured at the high temperature along with the DNA itself. Because *T. aquaticus* polymerase is active at high temperatures, it is not denatured during the heating cycle of PCR. This allows the heat cycles of the PCR reaction to be repeated continuously without any addition of extra polymerase enzyme.

16. **b.** If 30 percent of DNA is adenine, then by Chargaff's rule, 30 percent will be thymine. The remaining 40 percent of the DNA is cytosine and guanine. Because the ratio of cytosine to guanine must be equal, the percentage of cytosine in this DNA must be 20 percent.

17. **c.** Guanine is a purine, and its paired pyrimidine is cytosine.

18. **a.** Okazaki fragments are involved in synthesis of the lagging strand.

19. **d.** Sulfur is a constituent of many protein molecules, but is not found in DNA.

20. **a.** A hydroxyl group at the 3′ position of a nucleotide is necessary for the binding of any additional nucleotides. If this hydroxyl group were absent, no other nucleotides could be added to a growing strand.

Application Answers

1. 3′ – C G A T T G A C A C T A G C A T A T T C G A C T – 5′

2. See Figure 11.6 in the text.

3–4. See Figure 11.15 and 11.16 in the text.

5. In the conservative model of DNA replication, the parent DNA remains intact, and a newly synthesized molecule consists of two newly replicated daughter strands. Had this been DNA's method of replication, Meselson and Stahl would not have seen the intermediate density band in the first generation of replication. They would have seen a heavy band corresponding to the parental DNA and a light band corresponding to the daughter DNA. Because they saw an intermediate band, they knew one strand was heavy and one was light; therefore, replication was semiconservative, with each new molecule consisting of one parental strand and one daughter strand. A third hypothesis was the dispersive model, with each new molecule containing bits and pieces of both old and new strands.

6. See Figure 11.17 in the text.

7. Proofreading occurs during synthesis of the DNA molecule, mismatch repair occurs after synthesis, and excision repair removes any abnormalities in "mature" DNA. DNA damage due to UV exposure is generally repaired via excision repair mechanisms.

8. Refer to Figure 11.21 in your textbook for an explanation of DNA sequencing.

CHAPTER 12 From DNA to Protein: Genotype to Phenotype

Important Concepts

One gene encodes one enzyme.

- Beadle and Tatum studied special mutant (auxotroph) and wild-type (prototroph) strains of *Neurospora*, a bread mold. Auxotrophs are mutant strains that cannot grow on minimal media without the addition of an external mineral, vitamin, or nutrient, because auxotrophs are missing a particular enzyme needed to synthesize these nutrients. Prototrophs, also known as "wild type" strains, can grow on minimal media because in these cells all the enzymes needed to synthesize these nutrients are functional.

- Auxotrophs for a particular nutrient are missing the enzymes needed to synthesize that nutrient from a biochemical intermediate. For the amino acid arginine, for example, each of these enzymes defines a step in the biochemical pathway of arginine synthesis. These mutant strains can be characterized by supplying them with chemical intermediates in the arginine pathway to determine which enzymes are nonfunctional (see Figure 12.1).

- Beadle and Tatum concluded from their experiments that one gene encoded one enzyme. Because enzymes can consist of several different polypeptides, this conclusion has been modified to the theory that one gene encodes one polypeptide.

- Genes also code for some RNAs (such as ribosomal RNA, transfer RNA, and regulatory RNA) that do not become translated into polypeptides.

RNA differs from DNA in three ways.

- Unlike DNA, which consists of two polynucleotide strands, RNA is generally a single polynucleotide strand. A strand of RNA can base-pair with a single strand of DNA. A strand of RNA may also fold over to base-pair with itself.

- In RNA, the sugar molecule is ribose rather than deoxyribose, as seen in DNA.

- The bases A, G, and C in DNA also occur in RNA, but T is replaced by U in RNA. In RNA, G still base-pairs with C, but A base-pairs with U.

The central dogma of molecular biology says that information flows in only one direction.

- The central dogma of molecular biology states that information flows from DNA to RNA to protein.

- In a eukaryotic cell, a message (the messenger RNA or mRNA) is needed to carry the genetic information (the DNA) from the nucleus to the cytoplasm. The process of synthesizing mRNA from DNA is called transcription.

- An adapter (the transfer RNA or tRNA) is needed to translate the nucleic acid (in the messenger RNA) into proteins. Translation is the process of synthesizing protein from mRNA.

- Some viruses use RNA as their genome and during infection make more RNA (using viral RNA as a template). These viruses include tobacco mosaic virus, poliovirus, and influenza virus. Other viruses, such as HIV, make a DNA copy of their RNA and use the DNA to synthesize more RNA. These viruses are exceptions to the central dogma.

Transcription is the process of synthesizing RNA from a DNA template.

- RNA polymerase is the enzyme that synthesizes RNA using one strand of the DNA as a template. The resulting RNA molecules can be messenger RNAs (which are used to synthesize proteins) or they can become transfer RNA, ribosomal RNA, or regulatory RNA.

- RNA polymerase initiates transcription at special sequences on the DNA called promoters. When RNA polymerase binds the promoter, it unwinds the DNA. The promoter tells the RNA polymerase where to start transcription, which strand of DNA to read, and which direction to move on the DNA to begin synthesizing RNA.

- New RNA is synthesized in a 5'-to-3' direction, antiparallel to the template DNA strand. New nucleotides are added to the 3' end of the growing strand. No primer is required for initiation of RNA synthesis.

- RNA polymerase unwinds about 10 base pairs in the DNA as it reads and synthesizes the complementary RNA strand.
- RNA polymerase terminates transcription at special sequences in the DNA.
- In prokaryotic cells, translation of the mRNA can begin before transcription terminates.
- In eukaryotic cells, the mRNA is processed in the nucleus and must be transported from the nucleus to the cytoplasm before translation can begin.

Translation is the process of synthesizing proteins from an mRNA template.

- The genetic code provides the key to "reading" mRNA sequences.
  - Genetic information in the messenger RNA is read three bases at a time; each three-base sequence is called a codon.
  - There are four possible bases that can be used in a codon, so there are 4^3, or 64, different codons.
  - Sixty-one codons specify specific amino acids, so the genetic code is redundant, meaning that there is more than one codon for many of the 20 amino acids.
  - The codon used to start translation is AUG, which also specifies the amino acid methionine.
  - Three codons, the stop codons UAG, UAA, and UGA, are used to terminate translation.
  - The genetic code for nuclear genes is almost universally used in all living cells. There are slight differences in the genetic code of one group of protists and in mitochondria and chloroplasts.
  - The genetic code was first determined by experiments in which artificial RNAs (poly U, for example) were added to reaction mixtures containing all of the necessary components needed for protein translation (see Figure 12.7). The resulting polypeptides were analyzed to determine what the protein-synthesizing instructions were for these RNA sequences. When the synthetic RNA poly U was used, polyphenylalanine was made. Thus, one of the codons for phenylalanine was UUU.
- Transfer RNA (tRNA) is the link between a codon and its amino acid.
  - tRNA is a small RNA molecule (75–80 nucleotides) that base-pairs with itself to form a characteristic shape, the "cloverleaf," found in all tRNAs (see Figure 12.8).
  - The anticodon of the tRNA (found about at the midpoint in the molecule) base-pairs with the codon of the mRNA in an antiparallel fashion. This base pairing occurs on the ribosome during translation.
  - Base pairing between the anticodon and the codon is always complementary (A pairs with U, C pairs with G) in the first two positions. At the third position of the codon (and the first position of the anticodon) base pairing does not have to be exact. This position is called the wobble position and allows the cell to use one tRNA to base-pair with up to four different codons.
  - An amino acid is covalently attached to the 3' end of the tRNA; thus the tRNA (with an anticodon of AAA) has phenylalanine covalently attached to its 3' end. The enzymes that catalyze the covalent attachment of the amino acids to the correct tRNAs are tRNA synthetases (see Figure 12.9).
- The ribosome is the site in the cell where protein translation occurs.
  - Ribosomes are composed of a large subunit and a small subunit; both subunits consist of different ribosomal proteins and RNA molecules (ribosomal RNA).
  - The important sites on a ribosome are the A site (where the anticodon of the charged tRNA binds the codon of the mRNA), the P site (where the tRNA with the growing polypeptide chain is located), and the E site (where the tRNA exits from the ribosome) (see Figure 12.10).
- Initiation of translation requires the small subunit of the ribosome, the initiator tRNA, the mRNA with the start codon AUG positioned correctly on the ribosome, and the initiation factors (see Figure 12.11).
- When all these components are in place, the large subunit of the ribosome joins the small subunit of the ribosome, and the incoming charged tRNA is positioned at the A site. The anticodon of this tRNA base-pairs with the codon on the mRNA that is at the A site of the ribosome.
- The large subunit of the ribosome catalyzes the formation of a peptide bond between the amino acid on the tRNA in the P site and the amino acid on the tRNA in the A site. This peptidyl transferase activity can be attributed to the RNA (not to any of the proteins) in the large ribosomal subunit.
- Once the peptide bond is formed, the ribosome shifts down the messenger RNA so that the next codon is placed at the A site. The tRNA that was in the P site is moved to the E site and will exit the ribosome. The tRNA that was in the A site is shifted to the P site.
- During elongation, the messenger RNA is read 5' to 3', with each new codon being moved into the A site during elongation so that the next incoming charged tRNA can try to base-pair with it (see Figure 12.12).
- Elongation factors are proteins that assist in this phase of translation.
- Termination occurs when a stop codon is moved into the A site of the ribosome. Release factor binds the stop codon, and peptidyl transferase hydrolyzes the bond between the amino acid (which will be the C terminus) that is covalently attached to the tRNA in the P site.

The polypeptide is released from the ribosome and the small and large subunits of the ribosome dissociate (see Figure 12.13).

- In prokaryotes, one messenger RNA is usually bound by many translating ribosomes, forming a polysome (see Figure 12.14).
- Within the amino acid sequence of some proteins in eukaryotic cells are signal sequences that serve as addresses for the protein's final cellular destination. These signal sequences are recognized by receptor proteins at their destination (in the case of nuclear, mitochondrial, plastid, or peroxisomal proteins (see Figure 12.15), or by signal recognition particles in the cytoplasm that direct the partially translated protein to the endoplasmic reticulum for further addressing and completion of translation (see Figure 12.16). The signal recognition particle will bind the docking protein in the endoplasmic reticulum and protein translation continues. Proteins synthesized in the endoplasmic reticulum may be secreted or modified further.
- Proteins that lack signal sequences remain in the cytoplasm.
- Protein modification after translation includes proteolysis, glycosylation, and phosphorylation.
  - Proteolysis is the process of cutting proteins into smaller pieces. The removal of the signal sequence from a protein in the ER is an example of proteolysis.
  - Glycosylation is the addition of sugars to proteins. Glycosylation in the Golgi apparatus marks proteins as destined for lysosomes, the vacuole (in plants), or the plasma membrane.
  - Phosphorylation is the addition of phosphate groups to proteins. Phosphorylation contributes to the tertiary structure of a protein.

Mutations are heritable changes in genetic information.

- Mutations can occur in germ-line cells and thus be passed on to progeny, or they can occur in somatic cells.
- Conditional mutants are a class of mutations that are only observable under altered conditions, such as high temperature. The inability of certain mutants to grow at high temperature can be related to the unstable tertiary structure of the protein whose gene sequence has been altered.
- Point mutations are changes of one or two base pairs and include silent mutations, missense mutations, nonsense mutations, and frame-shift mutations.
- Silent mutations are mutations that do not change the meaning of the codon. For example, the change of an A to a C in the third position of a codon would change a CCA to a CCC, but both codons specify proline.
- Missense mutations are single-base changes that change one codon to another (GAU, which codes for asp, to GUU, which codes for val).

- Nonsense mutations are single-base changes that change a codon to a stop codon (UGG, which codes for trp, to UAG, stop). Nonsense mutations result in a shortened polypeptide.
- Deletions or insertions of one or two bases are frame-shift mutations, which alter the reading frame of the genetic message.
- Chromosome mutations are larger deletions, duplications, inversions, or translocations that occur in the chromosome.
- Spontaneous mutations can be caused by tautomerization of the bases, deamination of the bases, errors during DNA replication, and errors during meiosis (leading to unequal crossovers or unequal separation of the chromosomes).
- Induced mutations are caused by chemicals and radiation.
- Mutations for the most part are deleterious, but some mutations may actually benefit the organism by creating an altered gene product that may improve the survival of the organism.

The Big Picture

- Every protein and RNA in the cell has a DNA blueprint (the gene sequence) that specifies the amino acid or nucleotide sequence of that gene product. Those sequences determine how that gene product will fold up three-dimensionally and ultimately function in the cell.
- Alterations in those gene products, caused by mutations in the genes, help us to understand how those proteins and RNAs function in the cell.

Common Problem Areas

- Students often try to understand every detail of gene expression without comprehending the larger picture: accessing genetic information in the DNA and expressing it in the cell so that the cell can function. Familiarize yourself with the details of the central dogma, but then take a step back and understand that all of these details describe the steps required to synthesize gene products.
- Students often confuse transcription and translation. Be able to distinguish between the two processes, and for each one carefully review the template, the product, and the sites of initiation and termination.

Study Strategies

- Make diagrams of the processes of transcription and translation. Be sure to orient the 5′ and the 3′ ends of the nucleic acid and the N and the C terminus of the protein.
- Pick a hypothetical gene that encodes a peptide of four amino acids. Draw a flowchart describing how that gene in the chromosome is made into protein. Now follow the details of gene expression: the synthesis of the RNA

from the DNA, and the synthesis of the protein from the mRNA. Be sure to include the start and stop signals in each of these processes. Alter one of those nucleotides in the DNA sequence. Transcribe and translate the gene. Is the protein sequence changed? Repeat this process using a different mutation in the DNA.

- Review the following animated tutorials and activity on the Companion Website/CD:
  Tutorial 12.1 Transcription
  Tutorial 12.2 Deciphering The Genetic Code
  Tutorial 12.3 Protein Synthesis
  Activity 12.1 Genetic Code

Test Yourself

Knowledge and Synthesis Questions

1. Transcription in prokaryotic cells
   a. occurs in the nucleus, whereas translation occurs in the cytoplasm.
   b. is initiated at a start codon with the help of initiation factors and the small subunit of the ribosome.
   c. is initiated at a promoter and uses only one strand of DNA, the template strand, to synthesize a complementary RNA strand.
   d. is terminated at stop codons.
   Textbook Reference: 12.3 How Is the Information Content in DNA Transcribed to Produce RNA? p. 261

2. Which of the following about RNA polymerase is *not* true?
   a. It synthesizes mRNA in a 5'-to-3' direction reading the DNA strand 3'-to-5'.
   b. It synthesizes mRNA in a 3'-to-5' direction reading the DNA strand 5'-to-3'.
   c. It binds at the promoter and unwinds the DNA.
   d. It does not require a primer to initiate transcription.
   Textbook Reference: 12.3 How Is the Information Content in DNA Transcribed to Produce RNA? p. 262

3. Translation of messenger RNA into protein occurs
   a. in a 3'-to-5' direction and from N terminus to C terminus.
   b. in a 5'-to-3' direction and from N terminus to C terminus.
   c. in a 3'-to-5' direction and from C terminus to N terminus.
   d. in a 5'-to-3' direction and from C terminus to N terminus.
   Textbook Reference: 12.4 How Is RNA Translated into Proteins? pp. 269–270, Figure 12.12

4. If a codon were read two bases at a time instead of three bases at a time, how many different possible amino acids could be specified?
   a. 16
   b. 64
   c. 8
   d. 32
   Textbook Reference: 12.3 How Is the Information Content in DNA Transcribed to Produce RNA? p. 264

5. Translate the following mRNA:
   3'-G A U G G U U U U A A A G U A- 5'
   a. NH$_2$ met—lys—phe—leu—stop COOH
   b. NH$_2$ met—lys—phe—trp—stop COOH
   c. NH$_2$ asp—gly—phe—lys—val COOH
   d. NH$_2$ asp—gly—phe—lys—stop COOH
   Textbook Reference: 12.3 How Is the Information Content in DNA Transcribed to Produce RNA? p. 264

6. What would happen if a mutation occurred in DNA such that the second codon of the resulting mRNA was changed from UGG to UAG?
   a. Nothing. The ribosome would skip that codon and translation would continue.
   b. Translation would continue, but the reading frame of the ribosome would be shifted.
   c. Translation would stop at the second codon, and no functional protein would be made.
   d. Translation would continue, but the second amino acid in the protein would be different.
   Textbook Reference: 12.6 What Are Mutations? pp. 275–276

7. If the following synthetic RNA were added to a test tube containing all the components necessary for protein translation to occur, what would the amino acid sequence be?
   5'-A U A U A U A U A U A U - 3'
   a. Polyphenylalanine
   b. Isoleucine-tyrosine-isoleucine-tyrosine
   c. Isoleucine-isoleucine-isoleucine-isoleucine
   d. Tyrosine-tyrosine-tyrosine-tyrosine
   Textbook Reference: 12.3 How Is the Information Content in DNA Transcribed to Produce RNA? pp. 264–265

8. What part of the tRNA base-pairs with the codon in the mRNA?
   a. The 3' end, where the amino acid is covalently attached
   b. The 5' end
   c. The anticodon
   d. The promoter
   Textbook Reference: 12.4 How Is RNA Translated into Proteins? p. 266

9. Peptidyl transferase is an
   a. enzyme found in the nucleus of the cell that assists in the transfer of mRNA to the cytoplasm.
   b. enzyme found in the large subunit of the ribosome that catalyzes the formation of the peptide bond in the growing polypeptide.
   c. RNA molecule that is catalytic.
   d. Both b and c
   Textbook Reference: 12.4 How Is RNA Translated into Proteins? p. 269

10. Termination of translation requires
    a. release factor, initiator tRNA, and ribosomes.
    b. initiation factors, the small subunit of the ribosome, and mRNA.
    c. elongation factors and charged tRNAs.

d. a stop codon positioned at the A site of the ribosome, peptidyl transferase, and release factor.
Textbook Reference: 12.4 How Is RNA Translated into Proteins? p. 270

11. Which of the following mutations would probably be the most deleterious?
a. A missense mutation in the second codon
b. A frame-shift mutation in the second codon
c. A nonsense mutation in the last codon
d. A silent mutation in the second codon
Textbook Reference: 12.6 What Are Mutations? pp. 275–276

12. If the DNA encoding a nuclear signal sequence were placed in the gene for a cytoplasmic protein, what would happen?
a. The protein would be directed to the lysosomes.
b. The protein would be directed to the nucleus.
c. The protein would be directed to the cytoplasm.
d. The protein would stay in the endoplasmic reticulum.
Textbook Reference: 12.5 What Happens to Polypeptides after Translation? p. 273

13. Auxotrophs are mutant strains that
a. can grow on a minimal medium.
b. require the addition of an essential nutrient to grow on a minimal medium.
c. behave like wild-type strains.
d. can only grow if arginine is added to the growth medium.
Textbook Reference: 12.1 What Is the Evidence that Genes Code for Proteins? p. 258

14. The central dogma of molecular biology states that _____ is transcribed into _____, which is translated into _____.
a. genes; polypeptides; gene product
b. protein; DNA; RNA
c. DNA; mRNA; tRNA
d. DNA; RNA; protein
Textbook Reference: 12.2 How Does the Information Flow from Genes to Proteins? p. 260

15. A gene product can be a(n)
a. enzyme.
b. polypeptide.
c. RNA.
d. All of the above
Textbook Reference: 12.1 What Is the Evidence that Genes Code for Proteins? p. 260

16. The enzyme that catalyzes the synthesis of RNA is
a. DNA polymerase.
b. tRNA synthetase.
c. ribosomal RNA.
d. RNA polymerase.
Textbook Reference: 12.3 How Is the Information Content in DNA Transcribed to Produce RNA? p. 262

Application Questions

1. What would be the effect of a deletion of the DNA encoding the targeting sequence for that gene product? (Imagine that this protein was targeted to go to the mitochondria and the signal sequence was removed as a result of this deletion.)
Textbook Reference: 12.5 What Happens to Polypeptides after Translation? pp. 272–274

2. What would happen if the tRNA synthetase for tryptophan actually added a phenylalanine to the trp tRNAs instead of tryptophan?
Textbook Reference: 12.4 How Is RNA Translated into Proteins? pp. 267–268

3. Mutations can be very harmful to an organism, yet without them life as we know it today would not exist. Explain.
Textbook Reference: 12.6 What Are Mutations? pp. 277–278

4. Suppose that two different mutant strains of a bacterium are unable to grow on a minimal medium without the addition of the amino acid lysine. Explain how this phenotype might be caused by different mutations in each strain, perhaps on the same gene and perhaps in two different genes.
Textbook Reference: 12.1 What Is the Evidence that Genes Code for Proteins? p. 258

5. When fed the same biochemical intermediate, two auxotrophic mutants that had been isolated were able to grow. According to the experiments of Beadle and Tatum, the mutations in each of these auxotrophs should be in the same gene, because they were blocked at the same step in a biochemical pathway. Yet, these two auxotrophs had mutations that mapped in different genes. How do you explain this?
Textbook Reference: 12.1 What Is the Evidence that Genes Code for Proteins? p. 258

Answers
Knowledge and Synthesis Answers

1. **c.** Answer **a** describes transcription and translation in eukaryotic cells; answers **b** and **d** describe translation.

2. **b.** RNA polymerase binds at a promoter, unwinds the DNA, synthesizes mRNA in a 5′-to-3′ (not 3′-to-5′) direction, and does not require a primer to synthesize the RNA.

3. **b.** Translation of messenger RNA occurs 5′ to 3′, and the polypeptide is synthesized from N terminus to C terminus.

4. **a.** Four possible bases read two at a time gives 4^2, or 16, different codons.

5. **b.** See the codon table, Figure 12.6. Recall that translation occurs in the 5′-to-3′ direction.

6. **c.** UAG is a stop codon, so translation would terminate at that site.

7. **b.** See the codon table, Figure 12.6.

8. **c.** Neither the 3′ end nor the 5′ end of the tRNA is part of the anticodon. The promoter is a DNA sequence, to which RNA polymerase binds to initiate transcription.

9. **d.** Peptidyl transferase is the enzyme that catalyzes the formation of the peptide bond, and it is located in the large subunit of the ribosome. Its catalytic activity is due to ribosomal RNA found in the large subunit of the ribosome.

10. **d.** Termination of translation requires a stop codon to be positioned in the A site of the ribosome, and peptidyl transferase and release factor. Peptidyl transferase adds water across the last amino acid attached to the tRNA in the P site, creating the C terminus.

11. **b.** The mutation that is potentially the most dangerous is the frame-shift in the second codon. All of the other codons would be out of register and no functional protein could be made. A missense mutation in the second codon could be disastrous or minimal, depending on how crucial that second amino acid was to the folding and functioning of the protein. A nonsense mutation at the end of the coding sequence would have less effect on protein function because almost every amino acid in the protein would have been correctly synthesized. Again, this would depend on how important that last codon was to the proper folding and functioning of the protein. A silent mutation would have no effect on protein function because the amino acid inserted at that codon would be the same.

12. **b.** The nuclear sequence would direct this protein to be delivered to the nucleus.

13. **b.** Answer **d** is true only for arginine auxotrophs.

14. **d.** Genes are not transcribed into polypeptides, protein is not transcribed into DNA, and messenger RNAs are not translated into tRNAs.

15. **d.** Gene products can be RNAs (such as rRNA and tRNA) as well as enzymes and other polypeptides. Messenger RNA is translated into a gene product, protein.

16. **d.** DNA polymerase catalyzes the synthesis of DNA, tRNA synthetase covalently attaches amino acids to tRNAs, and ribosomal RNA (in the large subunit) catalyzes the formation of the peptide bond during translation.

Application Answers

1. Because the protein lacked its targeting sequence, it would no longer be moved to the mitochondria and would remain in the cytoplasm after it had been translated.

2. If the tRNA synthetase for tryptophan added phenylalanine to the trp tRNAs, everytime a tryptophan codon was read by these trp tRNAs, phenylalanine would be added to the polypeptide. This would create proteins that were nonfunctional, and the cell would die.

3. Mutations very often have damaging effects on a gene, rendering that gene product nonfunctional. Some mutations actually create gene products that are better for the cell or organism and may increase its ability to survive. Mutations over the course of evolutionary time have created new organisms that could survive in different environments or compete more effectively for limited resources. Without mutations evolution does not occur.

4. The mutations in these two strains of bacteria apparently interfere with lysine synthesis. The mutations might both be in the same gene coding for an enzyme necessary for lysine synthesis, but one could be a nonsense mutation in the fifth codon and the other a frame-shift mutation in the twenty-third, for example (the number of mutations that can disable a gene is enormous). If lysine synthesis in this bacterium requires more than one enzyme (as is likely), the two mutations could be in different genes coding for different enzymes. In this case, the phenotypes would not be strictly identical; it should be possible to distinguish the two by trying to grow them on minimal media to which different intermediates in the synthesis of lysine have been added (see Figure 12.1).

5. These two genes must encode different polypeptides that are both subunits for the same enzyme in this biochemical pathway.

CHAPTER 13 The Genetics of Viruses and Prokaryotes

Important Concepts

Viruses and prokaryotes are useful model organisms for genetic studies.

- The genome size of bacterial viruses and bacteria are small (about 1/1000th the size of human DNA), making identification and characterization of genes and gene function in these organisms relatively simple.

- Prokaryotic cells and the viruses that infect them (also known as bacteriophage) are easy to grow, have a short generation time, and can produce large numbers of individuals (10^9 E. coli per milliliter of growth medium).

Viruses are intracellular parasites that can reproduce only within living cells.

- Viruses are acellular, do not regulate the transport in and out of cell membranes and perform no metabolic function. They are obligate intracellular parasites; they develop and reproduce only within the cells of specific hosts.

- Outside the host cell, viruses exist as individual particles called virions. These virions have a central core of DNA or RNA enclosed in a protein capsid.

- Viruses are classified by several characteristics: whether their genome is RNA or DNA, whether their nucleic acid is single-stranded or double-stranded, whether their shape is simple or complex, and whether their virion is surrounded by membrane or not.

- Viruses can be categorized by the type of organism they infect and their infection life cycle.

- In prokaryotic cells, viral infection is initiated when proteins on the viral capsid bind receptor proteins or carbohydrates on the host bacterium's cell wall. Viral nucleic acid is injected into the cytoplasm through the viral tail assembly.

- Bacterial viruses (bacteriophage) can have two different kinds of life cycles: the lytic cycle, during which the virus multiplies and lyses the cell, and the lysogenic cycle, during which the viral genetic material is harbored in the host chromosome for many generations. When environmental conditions are altered, the viral genome can excise itself from the host chromosome, multiply, and lyse the cell (see Figure 13.3).

- Some bacteriophage (the virulent viruses) are only able to undergo a lytic cycle. Other bacteriophage (the temperate viruses) can alternate between lytic and lysogenic life cycles.

- During the early phase of the lytic cycle, the phage injects its nucleic acid into the host cytoplasm, and viral genes are transcribed and translated. These early gene products shut down host transcription, degrade host DNA, and stimulate viral genome replication and viral gene transcription. Late gene products include viral capsid proteins and proteins that lyse the cell at the end of the lytic cycle (see Figure 13.3). The lytic cycle takes about 30 minutes and produces hundreds of bacteriophage per cell.

- Lysogeny occurs when the viral genome inserts itself into the host genome. The virus is called a prophage at this stage. The prophage prevents the host from being infected by other bacteriophage, providing the host immunity to infections by the same virus. Lysogenesis usually occurs when environmental conditions are not limited and the host cell is growing rapidly. When the cell encounters stressful environmental conditions, lysogenic viruses respond with a change in gene expression, resulting in excision from the host chromosome and completion of the lytic cycle.

- The use of bacteriophage to treat bacterial infections has become an active area of research as antibiotic-resistant bacteria are becoming more common.

Animal viruses are diverse.

- Animal viruses infect a wide variety of vertebrates, but among invertebrates only arthropods (crustaceans and insects) are commonly infected. Arboviruses infect insects but do not adversely affect them. Infected insects are vectors for the viruses because they transmit the virus to vertebrates via an insect bite.

- Like bacteriophage, animal viruses consist of a nucleic acid core (either DNA or RNA) surrounded by protein. Some animal viruses (enveloped viruses) have a membrane surrounding the viral protein core, which is

- derived from the host plasma membrane. The genome of an animal virus usually encodes only a few proteins.

- The lytic cycle of animal viruses, like bacterial viruses, also has early and late stages. Transcription and translation of early genes includes the synthesis of viral proteins that replicate the viral genome and turn off host transcription. Late viral gene products include proteins that package the virus and lyse the cell (see Figure 13.4).

- Animal viruses lacking a membrane are taken into the host cell via the endocytic pathway (refer to the discussion on endocytosis, Chapter 5). Inside the cell, the membrane of the endocytic vesicle is broken down, releasing the virus to the cytoplasm. Enzymes in the cytoplasm degrade the protein capsid of the virus and liberate the viral nucleic acid.

- Enveloped viruses bind host cells using viral glycoproteins (embedded in the viral envelope) that bind specific protein receptors in the host cell membrane. These viruses can be taken into the host via the endocytic pathway and released from the endocytic vesicle into the cytoplasm (see Figure 13.5). Alternatively, some enveloped viruses release the virion into the cytoplasm of the host cell by fusion of their viral envelope with the cell membrane of the host cell. Enveloped viruses leave the host cell by budding out of the plasma membrane, carrying part of that membrane with them (see Figure 13.6).

- The influenza virus, an RNA virus, is taken into the host cell by endocytosis. Inside the cell, the viral envelope (the provirus) fuses with the endocytic vesicle, releasing the RNA into the host cell's cytoplasm. The viral RNA is used to make mRNA for viral proteins and to make more copies of the viral genome (see Figure 13.5).

- HIV is an RNA virus that enters the host cell by fusion of its viral membrane with the host plasma membrane. HIV I is a retrovirus, copying its RNA genome into DNA using reverse transcriptase. The DNA copy of the viral genome is then integrated into the host cell's chromosome forming a provirus (see Figure 13.6).

- The provirus is transcribed as mRNA and translated into protein by the host cell's translation machinery. Viruses are assembled at the host cell membrane, and mature virions bud from the cell membrane.

Regulation of gene expression in viruses is controlled by viral proteins and operator sites.

- Viral proteins activate a switch in response to environmental factors, which can determine if temperate phage (such as lambda, λ) will make a decision to lyse or lysogenize.

Viruses that infect plants need to penetrate the cell wall.

- Plant viruses can infect horizontally (plant to plant) or vertically (plant to offspring).

- In a horizontal infection, the virus is most commonly transmitted from one plant to another by a vector, usually an insect. Virus may also be transmitted by physical contact between a bruised leaf of an infected plant and an uninfected plant. Once inside the cell, the virus reproduces and spreads to other cells in the plant.

- Viral infection of other plant cells is facilitated by the plasmodesmata, which are special plant structures that form cytoplasmic bridges between cells.

Genetic recombination in prokaryotes is not tied to reproduction.

- Unlike viruses, prokaryotes are living cells. They reproduce asexually, generating genetically identical cells (clones).

- New genetic material can be introduced into bacterial cells in several ways, including conjugation, transformation, and transduction.

- Bacteria can pass some of their genetic material to other bacterial cells through the process of conjugation (first characterized by Lederberg and Tatum using two different bacterial auxotrophs, see Figure 13.10). Donor cells synthesize a thin extension called a sex pilus, which eventually contacts a recipient cell, forming a conjugation tube (see Figure 13.11). DNA is transferred through the conjugation tube to the recipient cell, where it can recombine with the chromosome of the recipient cell (see Figure 13.12).

- During transformation, external DNA can be introduced to cells, as demonstrated by experiments in which rough nonvirulent pneumococci bacteria were transformed to smooth virulent bacteria (see Chapter 11). Once the DNA enters the living cell it can recombine with the bacterial chromosome (see Figure 13.13A).

- Some viruses can pick up bacterial genes from one cell, pack them in their viral capsid, and transfer them to the next bacterial cell they infect. This process of gene transfer using a viral vector is called transduction (see Figure 13.13B).

Plasmids are small circular DNA molecules.

- Plasmids are found in the cytoplasm of some bacterial cells. These DNA molecules contain a few dozen genes plus an origin for DNA replication (the *ori* site).

- Plasmids can move from one bacterial cell to another during conjugation.

  - Some plasmids contain genes that encode enzymes with unusual metabolic functions, such as enzymes that digest oil.

  - Some plasmids contain genes for mating factors, such as the proteins that form the pilus and the conjugation tube. These plasmids are called fertility factors or F factors.

  - Some plasmids (R plasmids) contain genes whose products destroy or inactivate antibiotics and heavy metals. These factors are called R factors because they impart resistance to drugs or heavy metals. Some bacteria are resistant to more than one antibiotic due to the presence of multiple R factors.

Transposable elements and transposons carry genes that help them insert into other chromosomes.

- Transposable elements are special DNA sequences that can readily move to other sites on the same chromosome or to another chromosome.

- Transposable elements carry genes whose products can copy or excise the transposable elements and insert them elsewhere in the chromosome. Some transposable elements are copied and inserted at another position, whereas others are simply excised and placed at a different chromosomal site without DNA replication (see Figure 13.15).

- Longer transposable elements that have additional genes are called transposons. Each resistance gene in an R plasmid contains parts of transposons, suggesting that transposition is the mechanism by which plasmids originally gained antibiotic resistance.

Prokaryotes regulate gene expression in response to changing environments.

- Gene regulation in prokaryotes is primarily transcriptional—gene activity is controlled by allowing or blocking the transcription of different genes.

- Some regulation occurs at the posttranscriptional level, including hydrolysis of messenger RNA, blocking of translation, protein hydrolysis, or protein inhibition.

- Genes can be expressed in response to specific environmental conditions (inducible) or constitutively (all the time).

- Enzyme activity can be repressed by end-product feedback (also known as feedback inhibition, see Figure 13.17) or by regulating gene expression.

Repressors block transcription when bound to operators.

- In prokaryotic cells, one promoter can control more than one gene.

- Transcription in prokaryotes can be regulated by proteins that bind operator sites on the DNA.

- Operons are units of transcription in prokaryotes that consist of structural genes and the DNA sequences that control their transcription. The control sequences include the promoter and the operator.

- RNA polymerase binds to the promoter of the operon to initiate transcription. If it cannot bind to the promoter, it cannot transcribe the structural genes.

- The operator is a DNA sequence immediately adjacent to the promoter in the operon. If a repressor molecule is bound to the operator, it prevents RNA polymerase from binding to the promoter.

- The *lac* operon is an inducible operon. This means that the repressor is bound to the operator, blocking transcription of the structural genes, unless another molecule (the inducer) binds to the repressor. When the inducer (in this case, lactose) binds the repressor, the repressor becomes inactive and cannot bind the operator, allowing transcription to proceed.

  - The structural genes for the metabolism of lactose are adjacent to one another in the *lac* operon and are all transcribed from one promoter (review transcription, Chapter 12). There are three structural genes, which code for three enzymes necessary for the cell to use lactose (see Figure 13.18).

  - As long as lactose is present in sufficient quantities, it will bind the repressor and allow production of the proteins that metabolize lactose. As lactose levels drop, lactose molecules separate from the repressor, allowing it to bind the operator again and prevent unnecessary production of proteins for lactose metabolism (see Figure 13.19).

  - The repressor for the *lac* operon is encoded by the *i* gene. The *i* gene is constitutive (it does not have an associated repressor, so it is made at a constant rate). The promoter for the *i* gene does not bind RNA polymerase very well, so very few repressor molecules are synthesized. There are generally only about 10 molecules of *lac* repressor per cell, but this is sufficient for proper regulation of the *lac* operon.

  - To summarize, inducible operons have the following features: operators and promoter are bound by regulatory protein to control transcription; in the absence of the inducer, the operon is off; the repressor prevents transcription of the operon; and adding the inducer changes the repressor so transcription of the operon occurs.

- The *trp* operon is a repressible operon. This means that its repressor is *not* bound to the operator unless another molecule, the corepressor (tryptophan in this case), first binds to the repressor. When the corepressor binds to the repressor, the repressor becomes active and binds to the operator, preventing transcription of the structural genes.

  - The *trp* operon has five structural genes (*e, d, c, b,* and *a*) that encode five enzymes necessary for synthesis of the amino acid tryptophan (see Figure 13.20).

  - Recall that tryptophan is the corepressor of the *trp* repressor. As tryptophan levels rise in the cell, tryptophan binds to the *trp* repressor, making it active. The repressor then binds to the *trp* operator, preventing transcription of the *trp* operon's structural genes (and thereby preventing the synthesis of additional enzymes for making tryptophan). In this way, the cell avoids making more tryptophan than necessary.

- In inducible systems (like the *lac* operon) the substrates of these pathways bind the repressor, allowing transcription to occur. The presence of the substrate thus turns *on* production of the machinery necessary to catabolize the substrate.

- In repressible systems (like the *trp* operon) the product of the metabolic pathway (the corepressor) interacts with the regulatory protein enabling it to bind the promoter and block transcription. The presence of the product thus turns *off* the transcriptional machinery necessary to produce more of it.

Transcription can be positively regulated by increasing promoter efficiency.

- The cAMP receptor protein (CRP), when bound to cAMP, will cause the transcription of certain operons to increase by helping RNA polymerase bind the promoter more efficiently (see Table 13.2 for an outline of positive and negative regulation of transcription).

- The binding site for the CRP protein is just upstream 5′ of the promoter. Glucose reduces the levels of cellular cAMP. When glucose levels are high in the cell, cAMP levels are low and the CRP protein is not active. Under these conditions, transcriptional activation by the CRP protein does not occur. As glucose levels fall, cAMP levels increase, and the CRP protein binds the available cAMP molecules and activates transcription (see Figure 13.21).

Viruses also need to control transcription.

- In bacteriophage λ (lambda), a lysogenic phage, there are two λ proteins, cI and Cro, which determine whether the phage will complete a lytic or a lysogenic cycle. These two regulatory proteins control a genetic switch.

- Both cI and Cro compete for two operator promoter sites in the λ DNA that control genes involved in lytic and lysogenic functions. If the bacteriophage has infected a host cell that is experiencing rich environmental resources, then lysogeny is favored, and the cI-controlled switch is active. If the infected cells are damaged by mutagens or other environmental stress, Cro synthesis is high, the Cro-controlled switch is favored, and the bacteriophage undergoes a lytic cycle (see Figure 13.8).

Many viral and bacterial genomes have been completely sequenced.

- Genomic sequencing of viruses and bacteria have revealed open reading frames, amino acid sequences of known gene products, and gene regulatory sequences.

- Functional genomic analysis has been used to assign function to gene products (annotation).

- Comparative genome analysis has been used to compare genomes from different organisms to understand the physiology of different organisms.

- Genomic analysis for prokaryotes has revealed useful information, including new genes for ATP synthesis in *Chlamydia*, virulence genes in *Rickettsia*, new species of prokaryotes in the Sargasso Sea, lipid metabolism genes in *Mycobacteria*, new antibiotic resistance genes in *Streptomyces*, unique genes in pathogenic strains of *E. coli*, *Salmonella*, and *Shigella*, and methane-producing

and -oxidizing genes in methanococcus and methlyococcus. Some of the genome information can be used to develop new vaccines against pathogens. Genomic analysis in simple organisms has also been used to determine the minimal number of genes required for cellular life and to begin to create artificial life in a test tube.

The Big Picture

- Prokaryotic cells and viruses have small genomes. These have been probed by molecular biologists to understand gene expression, function, regulation, and viral life cycles. During viral infection, sequential gene expression occurs early and late to facilitate successful viral production.

- Prokaryotic cells use operons to coordinately regulate the expression of several genes. In response to environmental signals (the presence of lactose or the absence of tryptophan, for example) the repressor protein releases the operator site, leaving the promoter accessible to RNA polymerase to initiate transcription. In bacteriophage λ, other proteins bind promoters and control whether lysis or lysogeny will occur.

- New genes can be introduced into prokaryotic cells during transformation, conjugation, and transduction. Transposable elements and transposons are mobile genetic elements that move from one part of the chromosome to another. Plasmids are small circular DNA molecules that can contain genes for metabolic factors, drug resistance, heavy metal resistance, and conjugation proteins.

- Genomics is used to study and compare genomes from different organisms and to identify genes, potential genes, and their regulatory sequences. Information from these studies has been used to design new medicines for pathogens and to understand the minimal requirements for a cell to sustain life.

Common Problem Areas

- It is easy to confuse inducible operons (the *lac* operon) with repressible operons (the *trp* operon) because they both use repressors to regulate gene expression. Review the environmental conditions that must exist for each repressor to bind its operator site and what causes each repressor to release the operator site. (In the presence of lactose, the *lac* repressor cannot bind the *lac* operator site; in the presence of tryptophan, the *trp* repressor binds the *trp* operator site.) Review gene expression (Chapter 12) to see how, once transcription is initiated, proteins are produced that will act on those environmental signals.

- Although bacteria reproduce asexually, there are several ways that new genetic material is introduced into bacterial cells, including conjugation, transformation, and transduction. Review transposable genetic elements and plasmids, keeping in mind the advantages that mobile DNA elements provide to their host cells.

- It may seem puzzling at first that some viruses insert their chromosome into the host chromosome, because the goal of viral infection seems to be rapid multiplication and infection of adjacent cells. However, lysogeny allows the viral genome to persist while environmental resources are abundant. When resources are depleted or cell damage occurs, the virus can enter the lytic cycle to infect other cells.

Study Strategies

- Gene expression often occurs in response to environmental signals. Make a list of the environmental signals that a prokaryote receives and then outline the steps the cell initiates to respond to those signals. Include the following signals: lactose in the cell, glucose and lactose in the cell, high levels of glucose in the cell, low levels of tryptophan in the cell, high levels of tryptophan in the cell, bacteriophage λ infection under poor environmental conditions, and bacteriophage λ infection under rich environmental conditions.

- Outline the progress of different animal virus life cycles (review Figures 13.5 and 13.6) from the start of infection through multiplication of virus to viral release. Compare these life cycles to the lytic and lysogenic life cycles of bacteriophage (see Figure 13.3).

- Review the following animated tutorials and activity on the Companion Website/CD:
  Tutorial 13.1 The *lac* Operon
  Tutorial 13.2 The *trp* Operon
  Activity 13.1 Concept Matching

Test Yourself

Knowledge and Synthesis Questions

1. Viruses consist of
   a. a protein core and a nucleic acid capsid.
   b. a cell wall surrounding nucleic acid.
   c. RNA and DNA enclosed in a membrane.
   d. a nucleic acid core surrounded by a protein capsid and in some cases a membrane.
   Textbook Reference: 13.1 How Do Viruses Reproduce and Transmit Genes? p. 284

2. Bacterial cells that are resistant to viruses
   a. lack a cell surface receptor that the virus must bind to infect the cell.
   b. harbor a prophage in their chromosome, making the bacterial cell immune to further viral infection.
   c. cannot be lysed by the bacteriophage.
   d. All of the above
   Textbook Reference: 13.1 How Do Viruses Reproduce and Transmit Genes? p. 286

3. Lytic bacterial viruses
   a. infect the cell, replicate their genomes, and lyse the cell.
   b. infect the cell, replicate their genomes, transcribe and translate their genes, and lyse the cell.

c. infect the cell, replicate their genomes, transcribe and translate their genes, package those genomes into viral capsids, and lyse the cell.
   d. infect the cell, translate their RNA, and lyse the cells.
   Textbook Reference: 13.1 How Do Viruses Reproduce and Transmit Genes? p. 285, Figure 13.3

4. Animal viruses that integrate their DNA into the host chromosome
   a. are RNA viruses.
   b. are prophages.
   c. copy their RNA genome into DNA using reverse transcriptase.
   d. Both a and c
   Textbook Reference: 13.1 How Do Viruses Reproduce and Transmit Genes? p. 288

5. During conjugation,
   a. DNA from one bacterial cell is transferred to another bacterial cell using a bacteriophage.
   b. mutants that are auxotrophic for one nutrient can be converted to prototrophs when mixed with mutants that are auxotrophic for another nutrient.
   c. a pilus is synthesized, and DNA is transferred from one bacterium across the conjugation tube to the recipient bacterium.
   d. Both b and c
   Textbook Reference: 13.3 How Do Prokaryotes Exchange Genes? pp. 291–292

6. Plasmid DNA may contain genes that can
   a. confer drug resistance to the host cell.
   b. regulate conjugation.
   c. confer resistance to heavy metals.
   d. All of the above
   Textbook Reference: 13.3 How Do Prokaryotes Exchange Genes? pp. 294–295

7. An operon
   a. is regulated by a repressor binding at the promoter.
   b. has structural genes that are all transcribed from same promoter.
   c. has several promoters, but all of the structural genes are related biochemically.
   d. is a set of structural genes all under the same translational regulation.
   Textbook Reference: 13.4 How Is Gene Expression Regulated in Prokaryotes? p. 297

8. If the gene encoding the *lac* repressor is mutated so that the repressor can no longer bind the operator, will transcription of that operon occur?
   a. Yes, but only when lactose is present.
   b. No, because RNA polymerase is needed to transcribe the genes.
   c. Yes, because RNA polymerase will be able to bind the promoter and transcribe the operon.
   d. No, because cAMP levels are low when the repressor is nonfunctional.
   Textbook Reference: 13.4 How Is Gene Expression Regulated in Prokaryotes? pp. 297–298

9. If the gene encoding the *trp* repressor is mutated such that it can no longer bind tryptophan, will transcription of the *trp* operon occur?
   a. Yes, because the *trp* repressor can only bind the *trp* operon and block transcription when it is bound to tryptophan.
   b. No, because this mutation does not affect the part of the repressor that can bind the operator.
   c. No, because the *trp* operon is repressed only when tryptophan levels are high.
   d. Yes, because the *trp* operon can allosterically regulate the enzymes needed to synthesize the amino acid tryptophan.

   Textbook Reference: *13.4 How Is Gene Expression Regulated in Prokaryotes? p. 298*

10. Transcriptional regulation in prokaryotes can occur by
    a. a repressor binding an operator and preventing transcription.
    b. an activator binding upstream from a promoter and positively affecting transcription.
    c. different promoter sequences binding RNA polymerase more tightly, resulting in more effective transcriptional initiation.
    d. All of the above

    Textbook Reference: *13.4 How Is Gene Expression Regulated in Prokaryotes? pp. 297–299*

11. Functional genomics
    a. assigns function to the products of genes.
    b. assigns functions to regulatory sequences.
    c. compares genes in different organisms to see how the how those organisms are related physiologically.
    d. All of the above

    Textbook Reference: *13.5 What Have We Learned from the Sequencing of Prokaryotic Genomes? pp. 301–303*

12. Comparative genomics
    a. assigns function to the products of genes.
    b. assigns functions to regulatory sequences.
    c. compares genes in different organisms to see how the how those organisms are related physiologically.
    d. All of the above

    Textbook Reference: *13.5 What Have We Learned from the Sequencing of Prokaryotic Genomes? pp. 301–303*

13. A transposon is used to inactivate genes in a bacterium. If the inactivated gene is essential the bacterium will
    a. live.
    b. die.
    c. be a prototroph.
    d. be resistant to viral infection.

    Textbook Reference: *13.5 What Have We Learned from the Sequencing of Prokaryotic Genomes? pp. 301–303*

14. In order to infect a plant, plant viruses must
    a. pass through the cell wall as well as the plasma membrane.
    b. utilize an insect vector to travel from plant to plant.
    c. spread through the plasmodesmata between cells.
    d. All of the above

Textbook Reference: *13.1 How Do Viruses Reproduce and Transmit Genes? p. 289*

15. For the bacteriophage λ, the decision to become a prophage is made
    a. if environmental resources (i.e. food) for the host are limiting.
    b. when the Cro protein binds the promoter.
    c. when the CI protein binds the promoter.
    d. when bacterial lysis occurs.

    Textbook Reference: *13.2 How Is Gene Expression Regulated in Viruses? p. 290*

Application Questions

1. Animal viruses are obligate parasites. What does this mean?

   Textbook Reference: *13.1 How Do Viruses Reproduce and Transmit Genes? pp. 284–285*

2. Why are antibiotics useless in combating animal viral infections?

   Textbook Reference: *13.1 How Do Viruses Reproduce and Transmit Genes? p. 284*

3. Genomics has revealed that the bacterium that causes tuberculosis has over 250 genes that metabolize lipids. What does this finding suggest about the bacterium and about medicinal approaches to this disease?

   Textbook Reference: *13.5 What Have We Learned from the Sequencing of Prokaryotic Genomes? p. 302*

4. A cell has a mutation that deletes the gene encoding the repressor for a certain operon. A plasmid is introduced into the host cell that carries a wild-type copy of the gene for the repressor. Is normal regulation of this operon restored in the presence of this plasmid?

   Textbook Reference: *13.4 How Is Gene Expression Regulated in Prokaryotes? p. 297*

5. A cell has a mutation that deletes the gene encoding the operator for a certain operon. A plasmid is introduced into the host cell that carries a wild-type copy of the operator. Is normal regulation of this operon restored in the presence of this plasmid?

   Textbook Reference: *13.4 How Is Gene Expression Regulated in Prokaryotes? p. 297*

6. Review some of the auxotrophic bacterial strains that Lederberg and Tatum were studying. Describe three different ways that an auxotrophic bacterium can be converted to a prototrophic bacterium.

   Textbook Reference: *13.3 How Do Prokaryotes Exchange Genes? pp. 291–293*

Answers

Knowledge and Synthesis Answers

1. **d.** Nucleic acids do not form capsids; cell walls are found in bacterial and plant cells, not viruses. Viruses are organized so that the nucleic acid is surrounded by protein (not membranes), and the protein capsid may be surrounded by a membrane.

2. **d.** Bacteriophage must bind to cell surface receptors to initiate their infective cycle. In bacteriophage λ, a prophage (a viral chromosome inserted into the host chromosome) prevents other λ phage from infecting that bacterium. Bacteria that are resistant to viruses cannot be infected by them and thus cannot be lysed.

3. **c.** This sequence includes the most complete details of the viral life cycle.

4. **d.** Answer **b** describes a provirus, which is a bacterial virus that has inserted its genome into a host chromosome.

5. **d.** Answer **a** describes transduction, **b** describes one possible result of conjugation, and **c** describes what happens during conjugation.

6. **d.** All of these genes can be found on plasmid molecules.

7. **b.** An operon is a set of genes that are all transcribed from the same promoter. Answer **a** is not correct. The repressor binds at the *operator* site, which overlaps the promoter. Answer **d** is not correct because the operon is regulated transcriptionally, not translationally.

8. **c.** If the *lac* repressor is nonfunctional, it cannot bind the operator site, and transcription of the *lac* operon will occur at all times, whether or not lactose is present.

9. **a.** If the repressor can no longer bind tryptophan, then it cannot bind the operator, and transcription of the *trp* operon will always be *on*, whether tryptophan levels in the cell are high or low.

10. **d.** Answer **a** refers to the *lac* and *trp* repressors, answer **b** to the CRP protein, and answer **c** to promoters that have different transcriptional efficiencies.

11. **a.** Functional genomics assigns functions to genes, not the regulatory sequences. Comparative genomics compares genes between different organisms to see what genes one organism has that another is missing. These comparisons can be related to the physiology of the organisms that are being compared.

12. **c.** Comparative genomics compares genes between different organisms to see what genes one organism has that another is missing. These comparisons can be related to the physiology of the organisms that are being compared. Functional genomics assigns functions to genes, not the regulatory sequences.

13. **b.** If an essential gene in a bacterial cell is inactivated by the insertion of a transposon, that bacterium will die.

14. **d.** In order for viruses to infect plant cells, they must be able to get through the cell walls. They are transmitted most often by insects and can move from cell to cell through cytoplasmic bridges called plasmodesmata.

15. **c.** The decision to become a prophage by bacteriophage λ is made when cI binds the promoter. This decision is made if the host is experiencing rich nutrient conditions, and results in lysogeny, not cell lysis.

Application Answers

1. Animal viruses cannot express their genes, replicate their genomes, or multiply unless they are in the cytoplasm of the host cell. They use the host cell's components (ribosomes, ATP) to grow and reproduce.

2. Most antibiotics attack bacterial cells by inhibiting prokaryotic translation. Animal viruses use the host translation machinery, which is different enough from the bacterial translation machinery that it is unaffected by antibiotics. Drugs that would inhibit host translation would kill the host cell (by preventing protein synthesis) as well as the animal virus.

3. The tuberculosis bacterium must use lipids as a source of energy-rich compounds; inhibiting lipid synthesis in this bacterium may inhibit the growth of this bacterium.

4. Yes. The repressor gene can be transcribed and translated from the plasmid DNA, and normal regulation will be restored.

5. No. The operator site on the plasmid cannot restore regulation unless it recombines with the host operator site in such a way that it replaces the mutant operator on the host chromosome. The DNA site on the plasmid would bind repressor, but because that site is not adjacent to the promoter or the structural genes on the chromosome, normal regulation of those genes cannot occur.

6. Converting an auxotroph to a prototroph would require the introduction of a wild-type copy of the gene that was nonfunctional in the auxotroph (for example, introducing a gene that encoded an enzyme necessary for leucine biosynthesis to a leucine auxotroph lacking that enzyme). The wild-type copy of this gene could be introduced to the recipient cell by conjugation, transformation, or transduction.

CHAPTER 14 The Eukaryotic Genome and Its Expression

Important Concepts

Genes in eukaryotic cells differ from those of prokaryotic cells in terms of content and organization.

- The eukaryotic genome is larger (more DNA) and more complex (more genes) than the prokaryotic genome. For example, the size of the human genome is 6 billion base pairs (bp) per cell, whereas bacteria such as *E. coli* have 4.5 million bp per cell, and viral genomes can have as few as 10 thousand bp.

- Among eukaryotes, genome size is not related to relative complexity; cells of the lily plant have 18 times more DNA than human cells.

- Eukaryotic genomes have more regulatory sequences and more regulatory proteins than prokaryotes.

- Much of the eukaryotic genome is noncoding and includes noncoding regions within genes.

- The eukaryotic genome is located on several linear chromosomes.

- Each chromosome has three essential parts: an origin (the recognition sequence for DNA polymerase to initiate replication), a centromere (which holds the replicated chromosomes together before mitosis and meiosis is completed), and telomeres (the ends of the chromosome).

- In eukaryotes, transcription occurs in the nucleus and translation occurs in the cytoplasm. Pre-mRNAs have to be processed before they are transported to the cytoplasm for translation.

Model eukaryotes are used to study eukaryotic gene expression and development.

- *Saccharomyces cerevisiae*
  - *Saccharomyces cerevisiae*, budding yeast, is a simple eukaryote that has 12 million bp on 16 chromosomes (per haploid genome).
  - *S. cerevisiae* has 5,800 genes including many genes whose products are involved in protein targeting and organelle function.
  - The genome of *E. coli* and the yeast genome have the same number of genes for basic functions of cell

survival. Other genes that are present in yeast and other eukaryotes but absent from prokaryotes include histone genes, cyclin-dependent kinases, and RNA processing genes.

- *Caenorhabditis elegans*
  - The nematode *Caenorhabditis elegans* is a multicellular organism used to study development.
  - The worm's body is transparent, and its growth from a fertilized egg to a thousand-celled differentiated adult organism takes just 3 days.
  - The genome of *C. elegans* is eight times larger than that of yeast (97 million bp) and has four times the number of protein-coding genes (19,099 proteins versus 6,000 proteins). Genome comparisons have revealed that 3,000 of these genes in *C. elegans* have direct homologs in yeast.
  - Other gene products in *C. elegans* include proteins used for holding cells together to form tissues, for cellular differentiation, and for intracellular communication.

- *Drosophila melanogaster*
  - The fruit fly *Drosophila melanogaster* has ten times more cells than *C. elegans*. Its genome has 180 million more bp.
  - Genome analysis has revealed that the fly has 13,449 genes (fewer than in *C. elegans*).
  - In *Drosophila* there are more mRNAs (18,941) than there are genes (13,449), which means that the fly's genome codes for more proteins than it has genes.
  - There are 514 genes that do not code for proteins. These gene products include tRNAs, rRNAs, and small nuclear RNAs.
  - The fly genome has 177 genes with sequences known to be similar to human disease genes, including some genes involved in cancer and neurological disease.

- *Arabidopsis thaliana* and *Oryza sativa*
  - *Arabidopsis thaliana* is a plant that has a small genome (119 million bp) with 26,000 protein-coding genes. Many of these genes are duplicates of each

other. When the duplicates are removed, only 15,000 genes remain.

- *Arabidopsis* contains genes unique to plants, including genes involved in photosynthesis, in water transport into the root, in cell wall synthesis, in uptake and metabolism of inorganic substances, and in defense against herbivores.

- Most of the genes in *Arabidopsis* can be found in various species of rice, *Oryza sativa*.

- Rice has many more genes than *Arabidopsis*. These additional genes include genes that make it possible for the plant to grow submerged in water and genes that provide resistance to certain diseases.

Eukaryotic genomes contain many repetitive sequences.

- There are two types of highly repetitive sequences: minisatelllites and microsatellites.

- Minisatellites are 10–40 bp long and are repeated several thousand times. The number of repeats of minisatellites varies in individual organisms, and this variation provides unique molecular markers that can be used to identify individuals.

- Microsatellites are short 1–3 bp sequences that occur in small clusters of 15–100 copies scattered all over the genome.

- Moderately repetitive DNA includes telomeres, tRNA and rRNA genes, *Alu* elements, and transposable elements.

- Multiple copies of the tRNA and rRNA genes are needed to provide the cell with high concentrations of components needed for protein translation.

- In mammals there are four different rRNA molecules: 18S, 5.8S, 28S, and 5S. The 18S, 5.8S, and 28S rRNA genes are all transcribed from one promoter, producing a large precursor RNA, which is enzymatically trimmed to yield the mature rRNA molecules (see Figure14.3).

- In humans there are 280 copies of these rRNA gene clusters, located on five different chromosomes.

- Transposons, or transposable elements, are moderately repetitive sequences that are not stably integrated into the DNA but can move from place to place in the genome.

- There are four kinds of transposable elements: SINEs, LINEs, retrotransposons, and DNA transposons.

- SINEs are short *int*erspersed *e*lements of up to 500 bp; they are transcribed but not translated.

- LINEs are *l*ong *int*erspersed *e*lements of up to 7000 bp; some are transcribed and translated. LINES make up about 17 percent of the human genome and include the 300 bp *Alu I* element.

- SINEs and LINEs (greater than 100,000 copies/genome) make an RNA copy of DNA; this RNA is then used as a template for DNA synthesis, which is inserted at a new site in the genome.

- Retrotransposons are transposons that make an RNA copy when they move to a different site in the genome. These transposable elements make up about 8 percent of the human genome.

- DNA transposons do not replicate when they move to a new site on the chromosome nor do they use RNA as an intermediate (see Figure 14.4).

- Transposons adversely affect the cell by inserting their DNA into different genes and inactivating them.

- Transposons can replicate themselves and adjacent chromosomal genes, resulting in gene duplication. Transposons can also pick up adjacent genes during transposition and move them to a new site on the chromosome, potentially creating a new gene that will be advantageous to the organism's survival.

- Transposons help explain why some mitochondrial and chloroplast genes are located in the nucleus, whereas other organelle genes remain inside the organelle.

Many protein-coding genes are single-copy DNA sequences that include structural genes.

- Single-copy genes contain noncoding regions, including the promoter, the transcriptional terminator, and introns. The promoter is the site where RNA polymerase binds to initiate transcription. The terminator sequence signals the end of transcription. The stop codon on the mRNA signals the ribosome to terminate translation.

- Some eukaryotic genes are part of gene families, groups of structurally and functionally related genes within the genome.

- Introns are noncoding internal sequences that are transcribed in the pre-mRNA and removed in the nucleus. During intron removal, the exons, which contain the coding sequence for the gene, are spliced together.

- Nucleic acid hybridization experiments clearly demonstrated the presence of introns in pre-mRNA molecules (see Figure 14.6).
  - Denatured β-globin DNA was hybridized to mature β-globin mRNA, and loops of single-stranded DNA representing noncoding intron sequences were observed.
  - Hybridizations with β-globin DNA and pre-mRNA for β-globin revealed no single-stranded DNA loops, indicating that introns were still present in the pre-mRNA.

- In DNA, introns separate gene-coding sequences into distinct parts called exons. All of the exons are in order on the gene. In some cases, different exons code for different functional domains in the final protein.

- Gene families are sets of duplicate or related genes (such as the β-globin genes).

- Gene families provide the organism with a functional gene while allowing mutations in the other members of the gene family. Some of these mutations may create new genes that are advantageous to the organism.

- In humans, the globin gene family consists of three α-globin genes and five β-globin genes (see Figure 14.8).

- Different globins are expressed at different times in development. γ-globin, which is expressed in the fetus, binds oxygen more tightly to ensure oxygen transfer from the mother across the placenta to the fetus.

- Many gene families include pseudogenes, which are inexact copies of genes. Pseudogenes are nonfunctional because they lack promoters and/or recognition sites for intron removal.

Modifications of pre-mRNA (RNA processing) occur in the nucleus of eukaryotic cells.

- A cap of modified GTP (a G cap) is added to the 5′ end of eukaryotic mRNA to facilitate binding of mRNA to the ribosome in the cytoplasm and to protect the mRNA from degradation in the cytoplasm.

- The pre-mRNA is first cut by an enzyme that recognizes the sequence A A U A A A at the 3′ end of the mRNA, and the poly A tail (from 100–300 nucleotides) is added to the 3′ end.

- The poly A tail assists in the transport of the mRNA from the nucleus to the cytoplasm and increases the mRNA stability in the cytoplasm.

- Introns are "spliced" or removed from pre-mRNAs by small nuclear ribonucleoprotein particles (snRNPs).

- snRNPs bind consensus sequences at the 3′ and 5′ boundaries of the introns using complementary base pairing.

- Other proteins bind the snRNPs to form the spliceosome, which uses the energy of ATP hydrolysis to cut the RNA, release the intron, and rejoin the RNA to form the mature mRNA (see Figure 14.11).

- Some forms of β-thalassemia (in which there is an inadequate amount of β-globin, resulting in severe anemia) are the result of a mutation in the consensus sequence causing inappropriate splicing of β-globin RNA and a nonfunctional gene product.

- Mature RNA is transported from the nucleus to the cytoplasm through the nuclear pores.

The expression of eukaryotic genes must be precisely regulated.

- Multicellular organisms must regulate their genes precisely so that the correct genes are expressed at the appropriate time and development can proceed normally.

- Differentiated cells have different functions and must express genes specific for those functions.

- Gene regulation in eukaryotes occurs both at the transcriptional and the translational level (see Figure 14.12).

- Some regulatory mechanisms alter chromosome function and structure, making the DNA more accessible for the transcription complex. Genes may be selectively replicated to provide extra copies for transcription or genes may be rearranged on the chromosome.

- Genes in specialized cells that are specific to cell type can be selectively transcribed.

- All cells must express housekeeping genes, such as proteins involved in glycolysis and protein translation.

- In prokaryotes, operons often include genes that code for several related gene products, all of which are transcribed as a unit. In eukaryotic cells, each gene is usually associated with its own promoter. Eukaryotic genes that code for related gene products must have similar control sequences near their promoters to allow those genes to respond to the same signal.

- There are three RNA polymerases in eukaryotic cells:
  - RNA polymerase I transcribes rRNAs.
  - RNA polymerase II transcribes mRNAs.
  - RNA polymerase III transcribes tRNAs and small nuclear RNAs.

- Eukaryotic promoters are diverse and have several kinds of regulatory sequences in front of those genes, including promoters, enhancers, silencers, and regulators. These sequences are bound by proteins that affect transcriptional efficiency.

- Initiation of transcription in eukaryotic cells requires regulatory proteins called transcription factors. These are proteins that bind to the promoter and form a transcription complex that RNA polymerase can bind in order to initiate transcription. This is in contrast to prokaryotic promoters, which have conserved sequences that RNA polymerase binds to initiate transcription.

- Most eukaryotic genes have a TATA sequence in their promoters as well as other recognition sequences that can be bound by other transcription factors.

- TFIID, a transcription factor, binds the TATA box, changing the shape of the DNA and the transcription factor. Other transcription factors then bind the promoter, and finally, RNA polymerase binds to initiate transcription.

- Regulator sequences, which bind proteins that activate transcription (see Figure 14.14), occur just upstream of the promoter.

- Enhancer sequences are located farther away from the gene they affect (some as far as 20,000 bp); activator proteins bind enhancer sequences and stimulate the transcription complex.

- Silencer sequences are bound by repressor proteins and negatively regulate transcription.

- The rate of transcription of any eukaryotic gene is determined by the combination of the following regulatory factors: transcription factors, repressors, and activators.

Proteins that bind DNA (activators, transcription factors, regulators, and repressors) have characteristic protein motifs.

- DNA protein-binding motifs include helix-turn-helix, zinc finger, leucine zipper, and helix-loop-helix (see Figure 14.15).

- These proteins can interact with the DNA by fitting into the major or minor groove, using amino acid side chains that can project into the interior of the helix and hydrogen bond with the bases.

- Coordinated gene regulation is achieved by positioning similar regulator sequences near the promoters of each of those genes. Proteins bind to these regulatory sequences and stimulate RNA synthesis. In plants, regulatory elements known as stress response elements (SREs; see Figure 14.16) are found near the promoters of genes that respond to drought.

For gene expression to occur, the DNA must be made accessible to the transcription complex.

- The DNA in eukaryotic chromosomes is associated with nucleosomes and other chromosomal proteins, which can inhibit the initiation and elongation steps of transcription. Chromatin remodeling must occur to make the DNA accessible to transcription complexes.

- The first step of chromatin remodeling is the binding of protein complexes upstream of the initiation site. These protein complexes help the nucleosomes become disaggregated so that the transcription complex can bind the DNA and transcription can be initiated.

- The second step in chromatin remodeling occurs when other protein complexes bind the DNA to help the transcription complex move through the nucleosomes.

- Nucleosomes become disassembled and reassemble with the help of acetylation, methylation, and phosphorylation.

- Transcriptionally active chromatin (euchromatin) is less condensed and takes up less nuclear stain than the transcriptionally inactive, highly condensed chromatin (heterochromatin).

- The genotype for the sex chromosomes in mammals is XY in males and XX in females. However the expression of X-linked genes is the same in both sexes, due to X-inactivation in females.

  - Early in embryonic development, one of the two X chromosomes is randomly inactivated, producing a highly condensed heterochromatic chromosome, called a Barr body.

  - The number of Barr bodies is equal to the number of X chromosomes minus one (XXX has two Barr bodies and normal females (XX) have one Barr body).

  - The DNA of the inactive X is hypermethylated (methyl groups are added to the 5′ carbon of cytosine).

  - Transcription of the *XIST* gene occurs on the inactive X chromosome.

  - The *XIST* RNA (also known as interference RNA) binds the X chromosome from which it was transcribed and inactivates it (see Figure 14.19).

Specific genes can be amplified to meet the needs of the cell, as seen in the amplification of the ribosomal RNA genes.

- Egg cells in frogs and fishes selectively amplify the ribosomal RNA gene clusters so that there are greater than a million copies of ribosomal DNA per egg cell (see Figure 14.20).

- Gene amplification is also seen in cancer cells, specifically in oncogenes (genes that cause cancer) and in genes encoding protein receptors for cancer drugs.

Posttranscriptional modifications to RNA include alternate splicing, modulating mRNA stability, and RNA editing.

- Different RNAs can be made from the same gene by alternate splicing.

- In mammals, tropomyosin pre-mRNA is spliced five different ways to generate five different forms of this gene product, each for a different tissue: skeletal muscle, smooth muscle, fibroblast cells, liver, and brain (see Figure 14.2).

- When an mRNA arrives in the cytoplasm, ribonucleases degrade it. This degradation can be affected by other cytoplasmic factors. The exosome is a ribonuclease complex that breaks down AU sequences in mRNAs.

- MicroRNAs are small RNAs (about 20 bp) that bind specific mRNAs and block their translation.

- The sequence of mRNAs can be altered by RNA editing (adding additional nucleotides for example, U's) or altering individual nucleotides (changing C's to U's).

Translational and posttranslational regulation occurs in many eukaryotic genes.

- Messenger RNAs in the cytoplasm are not always translated, because other factors regulate the accessibility of these mRNAs to the translational machinery. For example, if the G of the 5′ cap is not modified, as seen in the embryo of the hornworm moth, the mRNA is not translated. After fertilization, this G cap is modified, and the mRNA is translated.

- Ferritin is a storage protein for iron. The mRNA necessary for making ferritin is present in cells at steady levels, regardless of the concentration of iron ions in the cell. When the iron level in the cell is low, a repressor protein binds to ferritin mRNA, preventing its translation into ferritin. As the iron level rises in the cell, some iron ions bind to the repressor protein, altering its shape so that it can no longer bind the ferritin mRNA. This permits translation of ferritin from the ferritin mRNA; the new ferritin is then available for storage of iron ions.

- The translation of the globin mRNA increases in the presence of excess heme in the cell. Heme removes a block to translational initiation of globin mRNA.

Protein modification includes glycosylation, phosphorylation, and the removal of signal sequences. Protein degradation regulates the longevity of a protein in the cell.

- Proteins that are targeted for degradation are first linked to ubiquitin. Subsequently, more ubiquitin chains attach, forming a ubiquitin complex. This ubiquitin complex binds to a proteasome, which is a hollow cylinder that has ATPase activity.

- When a protein-ubiquitin complex enters the proteasome, ubiquitin is removed, the protein is unfolded, and three proteases digest the protein (see Figure 14.24).

- Human papilloma virus targets p53 (a cellular protein that inhibits cell division) for proteasome degradation, leading to unregulated cell division and cancer.

The Big Picture

- Eukaryotic genomes are larger and more complex than prokaryotic genomes. The linear chromosomes in eukaryotic cells are multiple and located in the nucleus.

- Eukaryotic DNA contains many genes and many noncoding regions, including nongene sequences, moderately repetitive sequences, highly repetitive sequences, nontranscribed regions, and introns. Some of the repetitive DNA sequences, such as centromeres and telomeres, have a structural function for the chromosome.

- DNA is highly condensed in the nucleus due to its association with nucleosomes and other chromosomal proteins. Gene expression is a two-step process: The chromatin must be remodeled to expose the DNA containing the genes so that the transcriptional complexes (which consist of transcription factors, regulatory proteins, activators, and RNA polymerases) can access those DNA sites to regulate transcription of that gene. Transcription factors and other regulatory proteins have particular motifs that facilitate their binding to DNA.

- In eukaryotic cells, transcription is separated from translation by the nuclear membrane. RNAs are processed in the nucleus before being transported to the cytoplasm. Posttranscriptional regulatory elements affect gene expression, including different mRNA stabilities, alternate splicing, RNA editing, translation regulation, protein modification, and protein degradation.

Common Problem Areas

- Sometimes it is difficult to understand why there are so many different ways gene expression is regulated in eukaryotic cells. Examine a picture of a eukaryotic cell to get a feeling for all of the specialized parts of the cell and how gene expression must be tailored for each of these parts. Now think about a multicellular organism and how different genes must be expressed in all of the differentiated tissues.

- Grasping the different events that are involved in the regulation of gene expression in eukaryotic cells may be overwhelming. Begin by examining how transcription of a eukaryotic gene occurs by reviewing chromatin structure, the promoter, RNA polymerase, and transcription factors. Next examine how regulators, enhancers, and silencers, when bound by proteins, can modulate the levels of transcription of a eukaryotic gene. To complete your analysis, review the details of posttranscriptional regulation.

Study Strategies

- Make a list of the sites on the DNA that affect transcriptional initiation. Next, list the names of the proteins that bind those DNA sites and describe how they affect transcription.

- Draw a simple diagram of a eukaryotic gene. Include the flanking 5′ and 3′ sequences of the gene that are important for regulation and gene expression.

- Draw a diagram of the pre-mRNA and indicate sites on the RNA where processing occurs.

- Make another diagram showing the mature mRNA as it appears in the cytoplasm. Indicate sites on the mRNA where posttranscriptional regulation might occur.

- Finally, describe ways that proteins can be modified to regulate gene expression.

- Give examples of moderately repetitive and highly repetitive sequences in the DNA and describe their function.

- Review model organisms described in this chapter and describe two advantages for studying gene regulation in each of these organisms.

- Review the following animated tutorials and activities on the Companion Website/CD:
Tutorial 14.1 RNA Splicing
Tutorial 14.2 Initiation of Transcription
Activity 14.1 Eukaryotic Gene Expression: Structural Features
Activity 14.2 Eukaryotic Gene Expression: Control Points

Test Yourself
Knowledge and Synthesis Questions

1. Eukaryotic chromosomes
   a. are circular and contain origins and terminator sequences.
   b. are linear and have origins and telomeres.
   c. contain coding and noncoding sequences.
   d. Both b and c
   Textbook Reference: *14.1 What Are the Characteristics of the Eukaryotic Genome? pp. 307–308*

2. Model eukaryotic organisms have helped biologists understand
   a. genes involved in development.
   b. gene families.
   c. genes encoding proteins that are essential for all cells.
   d. All of the above

   Textbook Reference: *14.1 What Are the Characteristics of the Eukaryotic Genome? pp. 308–311*

3. Moderately repetitive DNA includes
   a. only coding sequences.
   b. only noncoding sequences.
   c. coding and noncoding sequences.
   d. satellites, minisatellites, and microsatellites.

   Textbook Reference: *14.1 What Are the Characteristics of the Eukaryotic Genome? p. 311*

4. Transposable genetic elements
   a. always affect the cell adversely, because when they move, they inactivate genes.
   b. are retroviruses.
   c. provide a mechanism for moving genetic material from organelle genomes to the nuclear genome.
   d. always replicate their DNA when they move.

   Textbook Reference: *14.1 What Are the Characteristics of the Eukaryotic Genome? pp. 311–312*

5. Introns are DNA sequences that
   a. code for functional domains in proteins.
   b. are removed from pre-mRNA by spliceosomes.
   c. allow one gene to make different gene products, depending on which introns are removed during splicing.
   d. Both b and c

   Textbook Reference: *14.2 What Are the Characteristics of Eukaryotic Genes? pp. 313–315*

6. Globin genes are
   a. introns that are visible using nucleic acid hybridization with DNA and globin mRNA.
   b. translationally regulated by excess heme.
   c. part of a family of globin genes, with different genes expressed at different times in development.
   d. All of the above

   Textbook Reference: *14.2 What are the Characteristics of Eukaryotic Genes? p. 315; 14.5 How Is Eukaryotic Gene Expression Regulated After Translation? p. 326*

7. Pre-mRNAs must be processed in the nucleus in order to
   a. increase their stability in the cytoplasm.
   b. allow RNA polymerase to initiate transcription.
   c. permit coding sequences to be joined to adjacent noncoding sequences.
   d. facilitate ribosome recognition in preparation for DNA synthesis.

   Textbook Reference: *14.3 How Are Eukaryotic Gene Transcripts Processed? p. 316*

8. Coordinated regulation of genes in eukaryotic cells
   a. is the result of positioning the same regulatory sequence in front of each gene.
   b. results from all of those genes being under the control of one promoter.
   c. occurs because related genes all have the same operons.
   d. occurs because enhancers cause DNA to bend.

   Textbook Reference: *14.4 How Is Eukaryotic Gene Transcription Regulated? pp. 321–322*

9. The transcription complex includes _____ and _____.
   a. transcription factors; promoters
   b. regulator proteins; regulators
   c. repressor proteins; silencers
   d. Both a and b

   Textbook Reference: *14.4 How Is Eukaryotic Gene Transcription Regulated? p. 320*

10. DNA binding proteins
    a. have distinct three-dimensional structures that allow them to bind to the DNA.
    b. can be transcription factors.
    c. can help condense the DNA in the nucleus.
    d. All of the above

    Textbook Reference: *14.4 How Is Eukaryotic Gene Transcription Regulated? pp. 320–321*

11. Chromatin structure must be altered for gene expression to occur because
    a. condensed chromatin is replicated but not transcribed.
    b. condensed chromatin makes most DNA sequences inaccessible to the transcription complex.
    c. decondensed chromatin has more nucleosomes per DNA molecule.
    d. heterochromatin is actively transcribed and euchromatin is not transcribed.

    Textbook Reference: *14.4 How Is Eukaryotic Gene Transcription Regulated? p. 322*

12. When DNA sequences are moved to new sites on a chromosome,
    a. new genes can be transcribed.
    b. genes can be inactivated.
    c. new genes can be created.
    d. All of the above

    Textbook Reference: *14.1 What Are the Characteristics of the Eukaryotic Genome? p. 312*

13. rRNA gene copies are amplified to
    a. ensure that gene expression in somatic cells will continue.
    b. make more copies of DNA origins for replication.
    c. ensure that there is enough translational machinery for the rapid development of the embryo.
    d. help heterochromatin become more accessible to the transcription complex.

    Textbook Reference: *14.4 How Is Eukaryotic Gene Transcription Regulated? p. 324*

14. Posttranscriptional regulation can include
    a. binding of repressor on silencer regions.
    b. insertion and alteration of nucleotides.
    c. decreasing mRNA stability in the cytoplasm.
    d. Both b and c
    Textbook Reference: 14.5 How Is Eukaryotic Gene Expression Regulated After Transcription? p. 325

15. Genes can be inactivated by
    a. inaccurate removal of introns.
    b. transposable genetic elements.
    c. movement of genes to heterochromatic regions of the chromosome.
    d. All of the above
    Textbook Reference: 14.1 What Are the Characteristics of the Eukaryotic Genome? p. 312; 14.4 How Is Eukaryotic Gene Transcription Regulated? p. 323; 14.5 How Are Eukaryotic Genes Processed? p. 317

Application Questions

1. The cells of the lily plant have 18 times more DNA than human cells, yet human DNA is more complex. Explain this phenomenon.
   Textbook Reference: 14.1 What Are the Characteristics of the Eukaryotic Genome? p. 311

2. When normal cells become transformed into cancer cells, gene regulation and chromosome stability in those cells is altered. Describe three different ways that cancer cells are different from normal cells.
   Textbook Reference: 14.1 What Are the Characteristics of the Eukaryotic Genome? p. 312; 14.4 How Is Eukaryotic Gene Transcription Regulated? p. 324; 14.6 How Is Gene Expression Controlled During and After Translation? p. 327

3. A mutation occurs in the promoter of a eukaryotic gene, eliminating the TATA box. How will this mutation affect the transcription of this gene?
   Textbook Reference: 14.4 How Is Eukaryotic Gene Transcription Regulated? p. 319

4. A different mutation eliminates the enhancer sequence on the DNA for this gene. How will this mutation affect the transcription of this gene?
   Textbook Reference: 14.4 How Is Eukaryotic Gene Transcription Regulated? p. 320

5. Describe one posttranscriptional regulatory event and one posttranslational regulatory event that controls the level of gene expression in eukaryotic cells.
   Textbook Reference: 14.5 How Is Eukaryotic Gene Expression Regulated After Transcription? pp. 324–326; 14.6: How Is Eukaryotic Gene Expression Controlled During and After Translation? pp. 326–327

Answers

Knowledge and Synthesis Answers

1. **d.** Eukaryotic chromosomes are linear, contain origins, telomeres and coding and non coding sequences. (Prokaryotic chromosomes are circular and contain termination sequences.)

2. **d.** Important developmental genes and gene families have been found in several different eukaryotic organisms. Comparisons of genes in eukaryotes and prokaryotes have revealed many similar genes that provide the same essential functions to those cells.

3. **c.** Moderately repetitive sequences include the ribosomal gene clusters as well as noncoding sequences such as telomeres and *Alu* sequences. Satellites, minisatellites, and microsatellites do not contain genes.

4. **c.** Transposable elements do not always have an adverse effect on genes; sometimes their insertion can create new genes. Retrotransposons are like retroviruses, but lack the genes for the viral capsid proteins. DNA transposons do not replicate their DNA before they move to a different site on the chromosome.

5. **d.** Introns are removed by spliceosomes, and differential splicing can produce different gene products as seen in the tropomyosin example (review Figure 14.20).

6. **d.** The introns in the β-globin gene are visible as loops in hybridizations between β-globin DNA and the mRNA for β-globin (see Figure 14.6). Globin genes are translationally regulated by heme, and they are part of the family of globin genes.

7. **a.** Pre-mRNAs are processed in the nucleus for many reasons (see Figures 14.10 and 14.11), including to increase their stability in the cytoplasm. Transcription by RNA polymerase *precedes* pre-mRNA processing. Noncoding sequences (introns) are removed from the pre-mRNA, not joined to coding sequences (exons). Ribosome recognition does not lead to DNA synthesis.

8. **a.** Each eukaryotic gene is transcribed from a separate promoter, and eukaryotic genes do not have operons to regulate them.

9. **d.** Transcription complexes include transcription factors bound to the promoter and regulator proteins bound to regulators. (When repressor proteins are bound to silencers, transcription is inhibited.)

10. **d.** DNA binding proteins must have three-dimensional structures appropriate for binding DNA. Some DNA-binding proteins are transcription factors, some are telomerases, which bind the ends of chromosomes, and some are nucleosomes, which bind the DNA and help pack it into the nucleus.

11. **b.** Nucleosomes condense the DNA. Heterochromatin is transcriptionally inactive, whereas euchromatin is transcriptionally active.

12. **d.** Transposable elements can be transcribed if they move adjacent to a promoter. Transposable elements can inactivate a gene if they insert into the middle of a coding sequence. Transposable elements can move parts of genes with them to new chromosomal sites, potentially creating new genes.

13. **c.** Ribosomal RNA amplification occurs in the eggs of fishes and frogs to prepare for the rapid development and protein synthesis that immediately follows fertilization. Amplified rRNA is not needed in somatic cells, DNA origins or for accessibility to the chromosome.

14. **d.** Repressors that bind at silencer sequences regulate transcription and thus are not posttranscriptional mechanisms of regulation.

15. **d.** Inaccurate removal of introns can create mRNAs that are missing coding sequence or that have extra noncoding sequences. Transposable genetic elements can move into the coding regions of genes, inactivating those gene products. Moving a gene to a heterochromatic (transcriptionally inactive) region of a chromosome results in that DNA being inaccessible to the transcription complex.

Application Answers

1. The DNA of the lily plant must contain more noncoding and repetitive sequences and less coding sequence than human DNA.

2. Genetic changes can make a normal cell into a cancerous cell in the following ways: a transposable element can move into the coding sequence of a gene that regulates cell division, inactivating that gene, amplification of oncogenes can transform normal cells into cancer cells, targeting proteins that control cell division for proteasome degradation (as seen in human papilloma virus and p53) will cause cells to become cancerous.

3. If the TATA box of the promoter for a eukaryotic gene is deleted, transcription factor TFIID will not bind the promoter. The transcription complex will not assemble at the promoter and there will be no initiation of transcription.

4. If the enhancer for a gene is deleted, transcription for that gene will still occur, but at a reduced rate.

5. Posttranscriptional regulatory events can include events that affect splicing, mRNA stability and inhibition of translation by microRNAs. Posttranslational events can include modifying the G cap so that the mRNA will be translated and protein degraded.

15 Cell Signaling and Communication

Important Concepts

Cells respond to specific environmental signals by changing their cellular function. This process of receiving a signal and communicating it to the cell is called signal transduction.

- Cells receive signals from their environment (chemicals, light, temperature, touch, or sound) and from other cells (usually chemicals).

- Multicellular organisms receive signals from their environment, from other cells, or extracellular fluid.

- Autocrine signals are local signals that affect the cells that make them (see Figure 15.1A).

- Paracrine signals are local signals that affect nearby cells (see Figure 15.1A).

- Hormones are circulatory signals that travel through the circulatory system and affect distant cells (see Figure 15.1B).

- To transmit a signal a cell must receive it, must respond to it, and that response must have some effect on the function of the cell.

- Signal transduction includes the following elements:

  - A receptor that binds the signal, altering its conformation.

  - A responder that is activated in response to the conformational change of the receptor (such as a kinase enzyme that adds a phosphate to another protein).

  - An amplification of the signal (such as a single enzyme repeatedly catalyzing the addition of a phosphate to its target protein as long as the signal is bound to the receptor).

  - Further activation of the signal transduction pathway such as transcription factor activation, which causes gene products to be expressed and alter the cell's activity.

- Response to osmotic change in E. coli is an example of signal transduction (see Figure 15.2). An environmental signal (a change in solute concentration in the inner membrane space) causes a conformational change in a receptor protein (EnvZ), which is a transmembrane protein in the bacterium's plasma membrane. Signal transduction then proceeds along the following pathway: EnvZ picks up a phosphate from ATP and transfers it to OmpR (a protein in the cytoplasm). The OmpR protein changes shape so that it can bind to the promoter for the gene that encodes the protein OmpC, thereby increasing production of OmpC protein. The OmpC protein blocks pores in the outer membrane, preventing solutes from entering the intermembrane space.

Receptors are specific. Not all cells have the same receptors, so a given signal can have different effects (or no effect) on different cell types.

- There are two classes of receptors: plasma membrane receptors, which bind large and/or polar ligands that cannot cross the plasma membrane, and cytoplasmic receptors, which bind small nonpolar ligands that diffuse across the plasma membrane.

- Binding to the receptor requires a fit of the ligand to the receptor binding site.

- Binding must cause a conformational change on the receptor to have an effect.

- The ligand is not changed in the binding process, and binding of the signal to the receptor is reversible.

- Binding sites may be inhibited by competing chemicals.

- There are three main types of plasma membrane receptors: ion channels, protein kinases, and G protein-linked receptors.

  - Ion channel receptors: Ligand binding causes a conformational change to open "gates" that allow ions (Na^+, K^+, Ca^+, or Cl^-) to pass through (see Figure 15.5). These receptors respond to sensory stimuli and chemical ligands.

  - Protein kinases: Ligand binding stimulates the transfer of a phosphate group from ATP to a target protein (see Figure 15.6).

  - G protein-linked receptors: These receptors have seven-transmembrane-spanning G protein-linked receptors. Ligand binding to this transmembrane receptor protein changes its shape so that a G protein on the cytoplasmic side can bind. When the

G protein is activated (by binding to the cytoplasmic side of the G linked receptor), it binds GTP and activates an effector protein, changing cellular function. After binding the effector protein, GTP is hydrolyzed to GDP causing the G protein to be inactivated until the receptor binds its ligand again (see Figure 15.7).

- Cytoplasmic receptors are in the cell cytoplasm.
- Ligand binding causes a conformational change in the receptor that allows passage of the receptor-signal complex into the cell nucleus where it can function as a transcription factor (see Figure 15.8).

Signal transduction may be direct (occurring at ligand binding) or indirect (requiring second messenger molecules).

- Direct signal transduction is a function of the receptor itself (see Figure 15.9A).
  - Protein kinase receptors are examples of direct signal transduction (see Figure 15.10). They can initiate protein kinase cascades where several different kinases are activated after the initial signal is bound. Protein kinase cascades can result in the transcriptional activation of genes. They are useful transducers because at each step, the signal is amplified, the signal is communicated to the nucleus, and multiple steps in the pathway provide specificity.
- Indirect signal transduction utilizes a second messenger to mediate the response between receptor binding and cellular response.
- Secondary messengers are released to the cytoplasm after ligand binding and function to amplify the signal. They act as cofactors or allosteric regulators of target enzymes and allow the cell to have multiple responses to a single signal. Common secondary messengers include the following:
  - cAMP: This secondary messenger can affect ion channels in the cell or can activate protein kinases.
  - IP_3 and DAG: These end products of the hydrolysis by phospholipase C of PIP2 can interact with protein kinase C or Ca^+ channels in the endoplasmic reticulum.
  - Free ions: Calcium ions frequently act as secondary messengers either independently or through calmodulin.
  - Nitric oxide: This gas stimulates the formation of cGMP, which can initiate a protein kinase cascade.

Signal transduction is highly regulated.

- Nitric oxide breaks down rapidly.
- Ca^+ concentration is controlled by membrane pumps and ion channels.
- Protein phosphatases remove phosphate groups from phosphorylated proteins.
- GTPases convert GTP on G proteins to GDP, thereby inactivating G proteins.

- cAMP levels are affected by phosphodiesterase, which converts cAMP to AMP, which has no secondary messenger activity.
- To regulate secondary messengers, cells can synthesize, break down, activate, or inhibit enzymes that generate those messengers.

Signal effects are changes in cell function.

- Ion channels open (as seen in sensing a particular odor).
- Enzymes are inhibited or activated (as seen with epinephrine stimulation).
- Transcription is stimulated (as seen with lipid soluble hormones).

Signals are often transduced between cells.

- Cell-to-cell signaling in animal cells can be facilitated by gap junctions. Gap junctions are protein-lined (connexon protein) channels that allow the passage of small molecules, including signal molecules and ions between adjacent cells (see Figure 15.19).
- Plasmodesmata are membrane-lined channels connecting adjacent plant cells. They are filled with endoplasmic reticulum-derived tubules called desmotubules. Desmotubules allow small metabolites and ions to move between plant cells. The plant and plant viruses can stimulate the synthesis of "movement proteins" to increase pore size and permit proteins, mRNAs and viruses to move through plasmodesmata (see Figure 15.20).

The Big Picture

- Signal transduction is the means by which cells receive information from the environment or other cells and react to those signals. Transduction is a highly regulated series of events that depends on the binding of a signal ligand to a receptor protein, which causes a responder protein to initiate events in the cell that change its function. The signal binding must cause a change in the shape of the receptor protein, which results in a cascade of intracellular events that leads to a change in cell function. The effects of signals are often mediated by secondary messengers.

Common Problem Areas

- It's easy to get overwhelmed by the different examples of signal transduction. Try to simplify your studying approach to signal transduction by looking at particular details of the systems. For instance, distinguish between direct and indirect signal transduction. Compare plasma membrane receptors with cytoplasmic receptors. List the three kinds of secondary messengers for signal transduction. Then, pick one of the signal transduction examples and make a table that includes the signal, the receptor, the transduction (responders and amplification), and the effect in the cell. Expand your table to include other examples of signal transduction.
- G protein action frequently gives students trouble. Remember that the G protein interacts with the

receptor, it binds GTP, and then it interacts with the effector protein.

- Gap junctions and plasmodesmata are physical connections between adjacent cells, and although they are similar in function, they are fundamentally different in structure.

Study Strategies

- Make a flow chart or a diagram of the signal transduction pathways.
- There are several different secondary signals. List each one and give an example of how that signal was activated and what it affects.
- Review the figures in your textbook to clarify the different examples of signal transduction.
- Review the following animated tutorial and activities on the Companion Website/CD:
  Tutorial 15.1 Signal Transduction Pathway
  Activity 15.1 Signal Transduction
  Activity 15.2 Concept Matching

Test Yourself

Knowledge and Synthesis Questions

1. Which of the following does *not* occur during signal transduction?
   a. Ligand binding to receptor
   b. Conformational change to the receptor protein
   c. Conformational change of the signal
   d. Alteration of cellular activity
   Textbook Reference: *15.1 What Are Signals, and How Do Cells Respond to Them? pp. 334–336*

2. When looking at how *E. coli* cells cope with osmotic changes, the ultimate goal of the *E. coli* cell is
   a. to change the permeability of the membrane.
   b. to change what DNA is transcribed.
   c. phosphorylation of OmpR.
   d. binding of the ligand to EnvZ.
   Textbook Reference: *15.1 What Are Signals, and How Do Cells Respond to Them? pp. 334–335*

3. What do the following receptors all have in common: G protein-linked receptors, protein kinases, and ion channels?
   a. Ligand binding
   b. Conformational change once the ligand is bound
   c. Amplification of the signal
   d. All of the above
   Textbook Reference: *15.2 How Do Signal Receptors Initiate a Cellular Response? pp. 337–338*

4. How does signal binding to a receptor differ from enzyme–substrate binding?
   a. The signal is not altered in any way during the process.
   b. The process is reversible.
   c. Cell receptors can become saturated by signal molecules.

   d. None of the above
   Textbook Reference: *15.2 How Do Signal Receptors Initiate a Cellular Response? p. 336*

5. Secondary messengers function to
   a. amplify the signal.
   b. bind to the active site of the receptor.
   c. result in multiple effects from a single signal.
   d. Both a and c
   Textbook Reference: *15.3 How Is a Response to a Signal Transduced through the Cell? pp. 340–344*

6. Caffeine is a stimulant that works because it acts as _____ to the adenosine receptors in a person's brain, and stimulates _____ in that person's heart and liver that increases blood flow and blood glucose.
   a. an allosteric effector; a pathway
   b. an inhibitor; a cascade pathway
   c. an inhibitor; ligands
   d. a signal; inhibitors
   Textbook Reference: *What Are Signals, and How Do Cells Respond to Them? pp. 334–333*

7. Cytoplasmic receptors only bind
   a. small signals that can diffuse through the plasma membrane.
   b. secondary messengers like cAMP.
   c. hydrophilic molecules.
   d. All of the above
   Textbook Reference: *15.2 How Do Signal Receptors Initiate a Cellular Response? pp. 336, 339*

8. Protein kinase cascades are significant because
   a. amplification can occur at each step in the path.
   b. information at the plasma membrane is communicated to the nucleus.
   c. the multiple steps allow for specificity of the process.
   d. All of the above
   Textbook Reference: *15.3 How Is a Response to a Signal Transduced through the Cell? p. 340*

9. Signal transduction is regulated in which of the following ways?
   a. The amount of signal present can be regulated.
   b. Enzymes in the pathway convert active forms of proteins to inactive forms.
   c. Signals are denatured.
   d. None of the above
   Textbook Reference: *15.4 How Do Cells Change in Response to Signals? p. 346*

10. Which of the following statements regarding receptors is true?
    a. Only one type of receptor is utilized throughout a cascade.
    b. Cascades may involve many types of receptors including ion channel receptors, protein kinase receptors, and G protein-linked receptors.
    c. Receptors amplify signals in a one-to-one ratio.
    d. All of the above
    Textbook Reference: *15.3 How Is a Response to a Signal Transduced through the Cell? pp. 340–341*

11. Gap junctions and plasmodesmata differ in that
    a. gap junctions are connected by protein tubules called connexons, and plasmodesmata are connected by extensions of the plant's plasma membrane.
    b. gap junctions allow much larger molecules to pass through them.
    c. gap junctions have no real physical connection but are the space between adjacent cell membranes.
    d. one is of animal origin and the other is of plant origin, otherwise they are physically the same.
    Textbook Reference: 15.5 How Do Cells Communicate Directly? pp. 348–349

12. Estrogen is an example of which kind of signal?
    a. Autocrine
    b. Paracrine
    c. Hormone
    d. Plasma membrane receptor
    Textbook Reference: 15.1 What Are Signals, and How Do Cells Respond to Them? p. 334; How Is a Response to a Signal Transduced through the Cell? p. 339

13. Which of the following signals directly results in an enzyme being activated?
    a. Acetylcholine binding to its receptor
    b. Insulin binding its receptor
    c. Cortisol binding its receptor
    d. Fertilization of an egg by a sperm cell
    Textbook Reference: 15.3 How Is a Response to a Signal Transduced through the Cell? p. 340–342

14. Select the order that is correct for signal transduction:
    a. Binding of a signal, secondary messenger released, receptor conformation altered, cellular function altered
    b. Binding of a signal, secondary messenger released, receptor conformation altered, transcription of gene
    c. Binding of a signal, responder activates target protein, receptor conformation altered, secondary messenger released transcription of gene
    d. Binding of a signal, receptor conformation altered, responder activates target protein, cellular function altered
    Textbook Reference: 15.2 How Do Signal Receptors Initiate a Cellular Response? pp. 337–338

15. Which of the following is not a response to a signal binding its receptor?
    a. Channel opening
    b. G protein exchanging GDP for GTP
    c. Intracellular nitrous oxide concentration increases
    d. The diffusion of solutes through the porous outer membrane of *E. coli*
    Textbook Reference: 15.1 What are Signals and How do Cells Respond to Them? pp. 334–335; 15.2 How Do Signal Receptors Initiate a Cellular Response? pp. 337–338; 15.3 How Is a Response to a Signal Transduced through the Cell? p. 344

Application Questions

1. Based on your knowledge of prokaryotic and eukaryotic cell structure and function, how might signal transduction differ in prokaryotes and eukaryotes?
   Textbook Reference: 15.1 What Are Signals, and How Do Cells Respond to Them? pp. 334–335

2. Describe how protein kinases and G protein-linked receptors may interact in a signal transduction cascade.
   Textbook Reference: 15.2 How Do Signal Receptors Initiate a Cellular Response? pp. 337–338; 15.3 How Is a Response to a Signal Transduced through the Cell? pp. 340–343

3. Diagram the Ras signaling pathway, and identify the components by receptor type. Be sure to identify the signal, the receptors, and the resulting effect on the cell.
   Textbook Reference: 15.3 How Is a Response to a Signal Transduced through the Cell? pp. 340–341

4. Discuss the role of secondary messengers in a signaling pathway. How are they different from signals and receptors? What roles do they have in common with signals?
   Textbook Reference: 15.2 How Do Signal Receptors Initiate a Cellular Response? pp. 337–338; 15.3 How Is a Response to a Signal Transduced through the Cell? pp. 340–344

5. True or false: A receptor may act as a signal. Explain your answer.
   Textbook Reference: 15.3 How Is a Response to a Signal Transduced through the Cell? pp. 340–341

6. Trace how changes in osmotic pressure in the plasma membrane of *E. coli* result in changes in gene transcription. What would be the outcome if OmpR could not bind to the *ompC* promoter?
   Textbook Reference: 15.1 What Are Signals, and How Do Cells Respond to Them? pp. 334–335

Answers

Knowledge and Synthesis Answers

1. **c.** For signal transduction to take place, the ligand must bind to the receptor, the receptor must undergo a conformational change, and the activity of the cell must be altered.

2. **a.** The ultimate goal of any signaling pathway is to alter the function of the cell in response to the signal.

3. **d.** All types of receptors bind ligands, undergo conformational change, and amplify the signal.

4. **a.** Substrates are altered in enzyme–substrate complexes. Signals are not altered when they bind to receptors.

5. **d.** Secondary messengers both amplify the signal and have multiple effects within a cell.

6. **b.** Caffeine and adenosine bind to the same receptor protein. The binding of caffeine prevents adenosine from binding in the brain. In other organs, a series of cascades begin in response to caffeine binding.

7. **a.** To bind to a cytoplasmic receptor, the ligand must be able to pass through the plasma membrane.

8. **d.** Amplification, regulation, and specificity are all roles of protein kinase cascades.

9. **b.** Active forms of proteins may be converted to inactive forms via enzyme action.

10. **b.** A given cascade may utilize ion channel receptors, protein kinase receptors, and G protein-linked receptors.

11. **a.** Both plasmodesmata and gap junctions provide physical connections between the cytoplasms of adjacent cells.

12. **c.** The hormone estrogen is made in the ovaries of female mammals and travels through the circulatory system to its target cells. Autocrine signals affect the cells that make them and paracrine cells affect nearby cells. A receptor in the plasma membrane binds hydrophilic signals.

13. **b.** Insulin binds and activates a protein kinase receptor. Acetylcholine leads to a channel opening, cortisol binds to a cytoplasmic receptor and activates transcription, and fertilization leads to an increase in intracellular calcium concentrations.

14. **d.** Signals must bind their receptors, the conformation of the receptor is altered, a responder activates target protein, and cell function changes.

15. **d.** The diffusion of solutes through the porous outer membrane of *E. coli* is a signal; the other answers are responses to signals.

Application Answers

1. Signal transduction in prokaryotes is limited to only one cell; there is no nuclear membrane to cross, and cascades tend to be simpler with fewer intermediate steps.

2. G protein-linked receptors frequently expose the protein kinase activities of effector molecules.

3. Refer to Figure 15.10 in your text.

4. Secondary messengers function to amplify and spread signals. They do not bind to receptors and do not act like signals in the cascade. Signals result in a change in cell function; secondary messengers are part of the pathway that results in the change in cell function.

5. Receptors may act as signals for the next step in a given cascade. See Figure 15.10 in your text.

6. Refer to Figure 15.2 in your text. If OmpR could not bind to the *ompC* promoter, then the *ompC* gene would not be transcribed and translated, and the *E. coli* cell would be unable to prevent further solute entry into the inner membrane space.

CHAPTER 16 Recombinant DNA and Biotechnology

Important Concepts

Restriction endonucleases are bacterial enzymes that cut double-stranded DNA at specific sites.

- Bacteria defend themselves against viral infection by synthesizing enzymes known as restriction endonucleases. Restriction endonucleases cut viral DNA but leave their own DNA unharmed. Bacteria use their own enzymes (methylases) to add a methyl group to the bases of the DNA, making their DNA unable to be cut by their own restriction enzymes.

- Restriction endonucleases cut the backbone of the DNA at specific sites (restriction sites) between the 3′ OH of one nucleotide and the 5′ PO$_4$ of the next nucleotide. For example, EcoRI is a restriction endonuclease that cuts the sequence GAATTC and its complementary strand CTTAAG between the G and the A on each strand. This sequence is a palindrome; it reads the same in both directions.

- EcoRI cuts on average about 1 in every 4,000 base pairs in prokaryotic genomes, but sizes of the DNA fragments vary, depending on the actual location of these EcoRI sites in the chromosome.

DNA fragments can be separated using gel electrophoresis.

- A gel is a porous molecular sieve that allows smaller DNA molecules to move through it faster than larger DNA molecules. An electric field is applied to the gel, causing the negatively charged phosphates of the DNA to be pulled toward the positively charged electrode.

- DNA fragments of known sizes are also separated on the gel as reference.

- DNA fragments containing specific sequences can be identified on a gel using a blot procedure and subsequent hybridization with a radioactively labeled probe. Denatured DNA fragments are transferred to a nylon filter (the blotting part of the procedure, also known as a Southern blot) and then exposed to a single-stranded DNA probe that has been radioactively labeled. The radioactive DNA probe will hybridize to complementary base sequences on the DNA fragments on the filter (see Figure 16.2).

- The identified DNA fragments can then be removed from the gel and purified.

Purified DNA fragments can be mixed with other DNA molecules that have been cut with the same restriction endonuclease or that have complementary sticky ends to form recombinant DNA.

- When restriction endonucleases cut DNA, the often leave ends that have 5′ or 3′ overhangs of single-stranded DNA. These ends are called "sticky ends" and can form complementary base pairs with other DNA molecules that have the same sticky ends (see Figure 16.8).

- Some restriction endonucleases cut between the same bases on both DNA strands, producing blunt ends.

- DNA ligase (the enzyme that covalently joins the Okazaki fragments during DNA replication and mends broken DNA; see Chapter 11) is used to form a covalent bond on each DNA strand of the recombinant molecule (see Figure 16.4).

- DNA fragments cut with the same restriction enzyme can be joined together by ligase, even if they are from different species.

How are genes inserted into cells?

- Recombinant DNA is inserted (transfected) into a host cell creating a transgenic cell or organism.

- Reporter genes are often added to recombinant DNA to identify host cells that have the clone of interest. These reporter genes serve as genetic markers for the sequence of interest.

DNA fingerprinting uses electrophoresis and restriction digests.

- Genes that are used for DNA fingerprinting are highly polymorphic and include single nucleotide polymorphisms (SNPs) or short tandem repeats (STRs) (see Figure 16.4).

- Sometimes DNA samples need to be amplified by PCR before they are used in fingerprint analysis.

- DNA fingerprinting is used in forensic cases to identify suspects and to determine relatedness between skeletal remains and living relatives (i.e., the Romanoffs, see Figure 16.5)

- Scientists have proposed a DNA barcode using a short sequence from a highly conserved gene to classify all organisms on Earth. This project can help us understand evolution and species diversity and detect undesirable microbes in food.

Genes can be cloned into prokaryotes or eukaryotes.

- Prokaryotes have been used to clone many genes. However, they lack splicing machinery and protein modification enzymes that may be required for eukaryotic gene expression.

- Yeast, plants, mice, and even humans have been used to clone eukaryotic genes.

- Yeast cells have a rapid cell division cycle (2–8 hours), are easy to grow, have a small genome size (12 million base pairs), and have been used to clone many eukaryotic genes.

- Many plant cells are totipotent. They can be grown in culture, coaxed to take up recombinant DNA, and manipulated by alterations in the growth medium to form an entire new transgenic plant containing the recombinant DNA molecule.

Once the recombinant DNA has been successfully introduced into the cell, it must be replicated to be maintained.

- Recombinant DNA can be inserted into the host chromosome, where it is replicated when the chromosome is replicated. Such insertion might happen randomly after the DNA is introduced into the cell. Alternatively, recombinant DNA molecules can enter the host cell by being part of a vector (plasmid, virus, or artificial chromosome).

- Vectors must be able to replicate independently and have restriction sites, reporter genes, and a small size.

- Plasmid vectors in *E. coli* are small (from 2,000 to 6,000 base pairs) circular DNA molecules. They have a single set of unique restriction sites, where the cut DNA is inserted, and a drug resistance marker allowing the investigator to confirm the presence of the plasmid in the *E. coli* cell. Plasmids synthesize their DNA independently from their own origins of replication, making many copies of plasmid DNA per cell.

- Both prokaryotic and eukaryotic virus vectors are used to clone larger DNA sequences, especially eukaryotic genes, which have introns, exons, and flanking sequences. Genetically engineered bacteriophage λ can accommodate up to 20,000 base pairs of inserted DNA. For example, viruses infect cells naturally allowing easy entry of cloned sequences into the cytoplasm of the cell.

- Yeast artificial chromosomes (YACs) can be used to clone genes in a eukaryote, the yeast cell (see Figure 16.9B), where normal eukaryotic DNA replication and gene expression occur.
  - YACs have a yeast origin of replication, telomeres, a centromere, marker reporter genes, and restriction endonuclease sites where foreign DNA can be inserted. They are only about 10,000 base pairs in size, but they can accommodate insertions of 50,000 to 1,500,000 base pairs of DNA.

- *Agrobacterium tumefaciens,* a bacterium that causes crown gall in plants, harbors a Ti plasmid. This plasmid contains a transposon, T DNA, which inserts copies of itself into the host cell's DNA (see Figure 16.9C).

Reporter genes and genetic markers (drug-resistance) can be used to determine if host cells contain recombinant DNA.

- If foreign DNA has been inserted into the *Tet* gene (which confers resistance to tetracycline) of a pBR322 plasmid, that gene will be inactivated, and the cell bearing this plasmid will be tetracycline sensitive. The plasmid also has an *Amp* gene, which confers ampicillin resistance to the host cell (see Figure 16.10). To determine if the plasmid is present in the host cell, the investigator will have to screen three kinds of cells: cells that did not receive the plasmid (amp^s, tet^s), cells that received the plasmid but without any inserted DNA (amp^r, tet^r), and cells that received the plasmid with the DNA insert (amp^r, tet^s).

- Other marker genes that are commonly used for detection of recombinant DNA molecules in host cells include the *lac* operon and the gene for green fluorescent protein (GFP).

DNA molecules can be isolated from genetic libraries, which include random pieces of DNA, from DNA copies of messenger RNA (cDNAs), or from artificially synthesized DNA.

- Genetic libraries of DNA are usually cloned into bacteriophage λ because each virus can accommodate 20,000 base pairs of DNA.

- To make cDNA (complementary DNA), the messenger RNA must first be isolated from the cell. Molecules of oligo dT are hybridized to the poly A tail of the messenger RNA and used as primers on the messenger RNA for reverse transcriptase. A complementary DNA molecule is synthesized by reverse transcriptase, and the new molecule can then be cloned. (see Figure 16.10).

- cDNA clones are used to compare gene expression in different tissues at different stages of development. One-third of all genes in an animal are expressed only during prenatal development.

- To make synthetic DNA, the investigator must know the amino acid sequence of the gene. Sequences for transcriptional and translational initiation and termination can be added to the gene sequence. Regulatory regions and preferred codons may also be added to the synthetic DNA.

Other techniques used to manipulate DNA include knockout experiments, gene silencing, and DNA chips.

- Mutant genes can be synthesized and compared to wild-type genes to analyze gene function.

- Gene function can be eliminated by creating knockout genes, which are the result of a homologous recombination event in the cell in which a normal gene is replaced by an inactivated form of the gene. The inserted DNA carries a reporter gene so its expression can be followed in the host's cells.

- In constructing transgenic mice, a plasmid containing the inactivated marker gene is transfected into a mouse stem cell (see Figure 16.13). If recombination occurs, the reporter gene will be expressed. The transfected stem cell is then transplanted into an early mouse embryo, and the resulting phenotype is analyzed.

DNA chip technology allows screening of thousands of DNA sequences at the same time.

- DNA chips have been used to study transcription patterns in cells in different physiological states, to study gene expression in different tissues at different times in an organism, and to detect all the possible variants of a particular gene.

- DNA chips are small glass chips containing thousands of copies of different DNA sequences per chip. These sequences (greater than 20 base pairs long) are attached to the chip in a precise order and can contain up to 60,000 different DNA sequences.

- For example, cDNA copies of cellular mRNA can be made (this technique is called real time PCR or RT-PCR) and amplified using PCR (see Chapter 11). Those cDNA molecules are coupled with fluorescent dyes and allowed to hybridize with the DNA chips. The chip is then exposed to fluorescent light to determine which sequences formed a hybrid with the cDNA (see Figure 16.15).

- Chip technology has been used to look at gene expression from different breast cancers to predict the prognosis of the patients.

Antisense messenger RNA and RNAi can prevent the translation of specific genes.

- Antisense RNA will form a double-stranded RNA molecule in the cell's cytoplasm, which cannot be translated and will be degraded by the cell.

- Micro RNAs, known as interference RNAs (RNAi), are short (about 20 nucleotides) double-stranded RNA molecules that bind specific mRNAs and target them for degradation.

- Small inhibitory RNAs are more effective at inhibiting translation than antisense RNAs and show promise as chemotherapeutic agents.

Biotechnology is the use of living cells to produce useful materials for people, including food, medicines, and chemicals.

- For cells to produce a cloned gene product, the vector (an expression vector) must have DNA sequences that allow the cloned gene to be expressed.

- Prokaryotic expression vectors require a promoter, a termination site for transcription, and a ribosome-binding site.

- Eukaryotic expression vectors require the same elements as well as a poly A addition site, transcription factor binding sites, and enhancers.

- Modifications of expression vectors include adding inducible promoters (which respond to a specific signal), tissue-specific promoters, and signal sequences.

- Medically useful products that have been cloned in expression vectors include tissue plasminogen activator, human insulin, and vaccine proteins (see Table 16.1).

- Recombinant DNA offers breeders the opportunity to choose specific genes that will be incorporated into an organism, to introduce any gene into a plant or animal species, and to generate new organisms quickly.

- Transgenic plants have been created that express toxins for insect larva, have resistance to herbicides, produce extra nutrients, or can grow in high-salt conditions.

- Human genes, such as blood-clotting factors, antibodies to colon cancer, and elastase inhibitors have been cloned in sheep and goats next to the lactoglobulin promoter, causing those animals to produce human proteins in their milk, a technique called "pharming."

- Creating transgenic plants has raised some concerns about these crops being unsafe for human consumption, that it is unnatural to interfere with nature, and that transgenes could escape into other noxious plants. Transgenic plants are extensively field tested, and scientists are examining these issues and proceeding cautiously.

The Big Picture

- The ability to isolate DNA from any organism, ligate it to vector DNA, introduce that DNA into host cells, and propagate those cells has had an enormous impact on our understanding of genetics, molecular biology, and cell function and development. These techniques have been used to elucidate evolutionary relationships between different organisms and to understand gene regulation and function in greater depth. Applications of these techniques have been used to develop new medicines and diagnostics, agricultural products, and powerful forensic tools.

Common Problem Areas

- There are a variety of ways to clone DNA fragments, and trying to remember all the different cloning methods can be challenging. Consider the following as you review the different cloning procedures: size of the cloned DNA, host cell in which the DNA should be cloned, and expression of the cloned DNA in the host cell. Different vectors can be used in different host cells to address each of these considerations.

- Many different experimental questions can be answered using cloning. Ask yourself what cloning strategies could be used to answer the following questions, and review your text for answers. What is the sequence of a gene? What sequences are important for regulation of that gene? What sequences are important for targeting that gene to a particular site in a eukaryotic cell? What is the difference in function between a mutant gene product and a wild-type gene? What kinds of genes are expressed during the development of an organism? How can genes be expressed in plants or in the milk of mammals?

Study Strategies

- Outline the specific steps needed to clone a gene in a bacterial cell. Include how the gene is initially isolated, what kind of vectors can be used, how to introduce that recombinant DNA into the cell, and how to confirm that the recombinant molecule is in the host cell. Now outline the steps needed to clone a gene in a eukaryotic cell.

- Complementary base pairing is important for many aspects of biotechnology. Describe each technique in gene cloning that uses complementary base pairing, detailing specifically how base pairing is involved.

- Review the following animated tutorials and activity on the Companion Website/CD:
Tutorial 16.1 Separating Fragments of DNA by Gel Electrophoresis
Tutorial 16.2 DNA Chip Technology
Activity 16.1 Expression Vectors

Test Yourself

Knowledge and Synthesis Questions

1. Cloning a gene may involve
   a. restriction endonucleases and ligase.
   b. plasmids and bacteriophage λ.
   c. yeast artificial chromosomes and complementary base pairing.
   d. All of the above
   Textbook Reference: *16.2 What Is Recombinant DNA? p. 358*

2. Restriction endonucleases
   a. are enzymes that process pre-mRNAs.
   b. are enzymes that degrade DNA.
   c. protect bacterial cells from viral infections.
   d. All of the above
   Textbook Reference: *16.1 How Are Large DNA Molecules Analyzed? p. 353*

3. DNA fragments are separated using gel electrophoresis
   a. because DNA is pulled through the gel toward the negative end of the field.
   b. because larger DNA fragments move faster through the gel than smaller DNA fragments.
   c. to identify and isolate DNA fragments.
   d. to synthesize DNA for cloning.

Textbook Reference: *16.1 How Are Large DNA Molecules Analyzed? p. 354*

4. Complementary base pairing is important for
   a. ligation reactions with blunt-end DNA molecules.
   b. hybridization between DNA and transcription factors.
   c. restriction endonucleases to cut cell walls.
   d. synthesizing cDNA molecules from mRNA templates.
   Textbook Reference: *16.4 What Are the Sources of DNA Used in Cloning? p. 362*

5. For a prokaryotic vector to be propagated in a host bacterial cell, the vector needs
   a. telomeres.
   b. centromeres.
   c. drug-resistance genes.
   d. an origin of replication.
   Textbook Reference: *16.3 How Are New Genes Inserted into Cells? pp. 359–360*

6. For a eukaryotic vector to be propagated in a host eukaryotic cell, the vector needs
   a. telomeres.
   b. centromeres.
   c. an origin of replication.
   d. All of the above
   Textbook Reference: *16.3 How Are New Genes Inserted into Cells? pp. 359–360*

7. Reporter genes include genes for
   a. drug resistance.
   b. bioluminescence.
   c. DNA origins.
   d. Both a and b
   Textbook Reference: *16.3 How Are New Genes Inserted into Cells? pp. 361–362*

8. Vectors include
   a. bacterial and plant plasmids.
   b. viruses.
   c. artificial chromosomes.
   d. All of the above
   Textbook Reference: *16.3 How Are New Genes Inserted into Cells? pp. 359–360*

9. Which of the following biotechnology products is made by insertion of recombinant DNA into bacteria?
   a. Insulin
   b. Tissue plasminogen activator
   c. Erythropoietin
   d. All of the above
   Textbook Reference: *16.6 What Is Biotechnology? p. 368, Table 16.1*

10. A cDNA clone is
    a. mostly cytosine.
    b. a copy of the DNA identical to the nuclear gene.
    c. a copy of noncoding DNA.
    d. a DNA molecule complementary to an mRNA molecule.

Textbook Reference: 16.4 What Are the Sources of DNA Used in Cloning? p. 362

11. Gene expression can be inhibited by
    a. antisense RNA.
    b. knockout genes.
    c. DNA chips.
    d. Both a and b
    Textbook Reference: 16.5 What Other Tools are Used to Manipulate DNA? p. 365

12. Expression vectors are different from other vectors because they contain
    a. drug-resistance markers.
    b. telomeres.
    c. regulatory regions that permit the cloned DNA to produce a gene product.
    d. DNA origins.
    Textbook Reference: 16.6 What Is Biotechnology? pp. 367–368

13. DNA fingerprinting works because
    a. genes containing the same alleles make it simple to compare different individuals.
    b. PCR allows amplification of proteins from single cells.
    c. there are multiple alleles for some DNA sequences, making it possible to obtain unique patterns for each individual.
    d. DNA in the skin cells is very diverse.
    Textbook Reference: 16.1 How Are Large DNA Molecules Analyzed? pp. 355–356

14. RNAi
    a. is more effective than antisense RNA in inhibiting translation.
    b. inhibits transcription in eukaryotes.
    c. is produced only by viruses.
    d. Both a and c
    Textbook Reference: 16.5 What Other Tools Are Used to Manipulate DNA? p. 365

15. DNA chip technologies can be used to
    a. predict who will get cancer.
    b. show transcriptional patterns in an organism during different times of development.
    c. clone DNA.
    d. make transgenic plants.
    Textbook Reference: 16.5 What Other Tools Are Used to Manipulate DNA? pp. 365–367

Application Questions

1. Describe three useful products that have been produced using biotechnology. Outline two specific dangers that could result from producing organisms that contain foreign genes.
   Textbook Reference: 16.6 What Is Biotechnology? pp. 368, 371–372

2. Your lab assistant has cloned gene *x* into yeast and confirmed that the recombinant DNA molecule is present in the yeast cells. However, the yeast cell is unable to synthesize X protein. Suggest why this part of her experiment is not working and what modifications she needs to make in her cloning procedure.
   Textbook Reference: 16.6 What Is Biotechnology? p. 367

3. You use a radioactively labeled DNA probe to identify a DNA fragment from an *Eco*RI digestion of *Drosophila* DNA. When you complete your hybridization, you notice that the probe has hybridized to several different fragments in the *Drosophila* DNA sample. How do you explain this result?
   Textbook Reference: 16.1 How Are Large DNA Molecules Analyzed? p. 354

4. You are an investigator at a crime scene who is trying to determine if the suspect in custody actually committed the crime. Describe how, using a drop of blood from the crime scene, you could determine whether the suspect was at the crime scene.
   Textbook Reference: 16.1 How Are Large DNA Molecules Analyzed? pp. 355–356

5. Hybridization is a useful technique in biotechnology. Describe three ways hybridization is used experimentally in DNA recombinant technology.
   Textbook Reference: 16.1 How Are Large DNA Molecules Analyzed? pp. 354–355; 16.2 What Is Recombinant DNA? p. 359; 16.4 What Are the Sources of DNA Used to Manipulate DNA? p. 363

Answers

Knowledge and Synthesis Answers

1. **d.** Cloning a gene requires restriction endonucleases to cut the gene of interest and the vector, ligase to covalently join the DNA to the vector, and vectors including plasmids, bacteriophage λ, or yeast artificial chromosomes. Complementary base pairing is important in forming the recombinant DNA molecule if the DNA fragments have sticky ends.

2. **c.** Enzymes that process pre-mRNAs add 5′ caps and 3′ poly A tails to RNAs, and remove introns. Deoxyribonucleases degrade DNA.

3. **c.** DNA fragments migrate toward the positive end of the electric field, with the smallest fragments migrating the fastest. Synthesizing DNA for cloning is done in a test tube, not in a gel.

4. **d.** No complementary base pairing can occur between blunt-end cut DNA molecules. Transcription factors are proteins that bind DNA through interactions with their side chains (which are amino acids) and the nucleotides of the DNA. Restriction endonucleases cut double-stranded DNA, not cell walls.

5. **d.** A prokaryotic vector needs an origin of replication to be propagated in a prokaryotic cell.

6. **d.** A eukaryotic vector requires an origin of replication, telomeres, and a centromere. (It does not require a drug-resistance marker to be propagated in the cell.)

7. **d.** Origins of replication are needed to propagate the vector in the host cell but are not used to detect recombinant molecules in host cells.

8. **d.** All of these molecules can serve as vectors for cloned DNA.

9. **d.** All three of these human proteins can be made by transgenic bacteria.

10. **d.** cDNA clones are not clones that contain mostly cytosine. A cDNA clone is generated by making a DNA copy of a particular messenger RNA using reverse transcriptase. The cDNA clone is not identical to the nuclear gene because in the messenger RNA (which served as a template for the cDNA) the introns have been removed, leaving coding sequence and 5′ and 3′ flanking sequences.

11. **d.** Gene expression can be inhibited by antisense RNA, which complementarily base pairs with the target messenger RNA, making it inaccessible to the translation machinery in the cell. Knockout genes are genes that have been inactivated by the insertion of DNA into their coding sequences. DNA chips are used in hybridization experiments and do not inhibit gene expression.

12. **c.** Expression vectors may contain drug-resistance markers and telomeres and must contain DNA origins, but none of these features distinguish them from other vectors. Expression vectors are unique because they contain regulatory sequences that allow the cloned gene to be expressed in the host cell.

13. **c.** Multiple alleles are needed to distinguish differences in individuals. PCR is a technique that amplifies DNA (not protein).

14. **a.** RNAi is more effective than antisense RNA at inhibiting translation (not transcription). It is produced by viruses and by eukaryotic cells in small amounts.

15. **b.** DNA chips can be used to analyze gene expression at different times in development and to predict which cancer tumors have a better or worse prognosis. It is not used to make transgenic plants or to clone DNA.

Application Answers

1. Useful products include rice grains that produce β-carotene, plants that are resistant to herbicides and insect larvae, sheep that produce human blood-clotting factors and antibodies to colon cancer in their milk, and others (see Tables 16.1 and 16.2). Dangers include creating genetically engineered foods that could adversely affect human nutrition, the transfer of herbicide- and insect-resistant genes from crop plants into noxious weeds, and the introduction into the wild of new organisms that might have unforeseen ecological consequences.

2. The x gene can be cloned into a yeast cell, but unless the vector has the appropriate regulatory signals (promoters, poly A addition sites, translational initiation, and termination signals), no expression of gene x will occur. Recloning gene x in a yeast expression vector will result in the expression of the x gene.

3. The DNA sequence of the probe is not unique in *Drosophila* and may be part of a gene family or a repetitive sequence (see Chapter 14).

4. DNA from the blood drop would be isolated and amplified for DNA fingerprint analysis using PCR amplification. The DNA would be digested with a restriction endonuclease and electrophoresed through a gel. DNA from a blood sample from the suspect would be similarly PCR amplified, digested, and run on the same gel. The gel would be blotted to a nylon filter. A radioactively labeled DNA probe containing a VNTR sequence would be hybridized to that filter to determine if the DNA band pattern was the same from the two samples. This process would be repeated with the two DNA samples using several different restriction endonucleases and several different VNTR probes. A match between DNA band patterns from the crime scene and DNA from the suspect would be needed to confirm the suspect's presence at the crime scene.

5. Hybridization is used to do Southern blots to look for the presence of specific DNA sequences in a DNA gel. Hybridization of complementary base pairs allows ligase to seal DNA fragments with the same sticky ends. Hybridization of newly made DNA to mRNA allows reverse transcriptase to generate cDNAs.

CHAPTER 17 Genome Sequencing, Molecular Biology, and Medicine

Important Concepts

Many diseases in humans are caused by a specific genetic defect.

- Proteins function in cells in many ways—as enzymes, receptors, transporters, and structures. Mutations in genes encoding these proteins can cause many diseases.

- Both phenylketonuria (PKU) and alkaptonuria are caused by a mutation in genes encoding enzymes in the same biochemical pathway (see Figure 17.1).

- In PKU, the enzyme that converts phenylalanine to tyrosine, phenylhydroxylase, is nonfunctional, leading to a buildup of phenylalanine in the blood. As phenylalanine accumulates, it is converted to phenylpyruvic acid. Somehow, high concentrations of phenylpyruvic acid prevent normal brain development in infants, leading to mental retardation. In addition, patients with PKU have lighter skin and hair color because melanin, a skin pigment, is made from tyrosine (see Figure 17.1).

- The molecular alteration in patients with PKU is a change in the 408th amino acid (from arginine to tryptophan) of phenylhydroxylase.

- Many proteins are polymorphic. Protein sequencing has revealed a 30 percent variation in amino acid sequences for the same protein from different individuals. Some mutations in these genes have no effect, producing functional gene products (see Figure 17.2).

- In sickle-cell disease, which is homozygous in 1 in 655 African-Americans, abnormal alleles produce an abnormal protein (β-globin). The resulting hemoglobin protein forms aggregates in the red blood cells, causing the cells to sickle. Sickle cells block narrow capillaries and damage highly vascularized tissues, especially at low blood oxygen concentrations.

- The abnormal allele in sickle-cell disease is caused by a mutation that results in a change at position number 6 (from a glutamic acid to a valine) in the β-globin gene.

- Hemoglobin is easy to isolate and to study. Hundreds of β-globins with alterations in their primary amino acid sequence have been isolated. Some alterations have no effect, whereas others are quite severe. Hemoglobin C disease results from a change at position 6 of β-globin, altering the glutamic acid to a lysine. The result is anemia that is not as severe as sickle-cell disease (see Figure 17.2).

Most common genetic diseases result from altered membrane proteins.

- Familial hypercholesterolemia (FH) is a disease in which elevated levels of cholesterol are present in the blood. The excess accumulation of cholesterol on the inner walls of the blood vessels (a condition known as atherosclerosis) narrows the diameter of the vessel, increasing the probability that clots will lodge in the narrowed vessels, resulting in heart attack and stroke. FH affects about 1 in 500 individuals, resulting in early heart attacks and strokes before the age of 45.

- Cholesterol is taken in through the diet and travels in the blood in lipoproteins. Low-density lipoproteins (LDLs) carry cholesterol to the liver, where they bind LDL receptors and are taken into the interior of the cell via endocytosis. (see Figure 17.3A). Patients with FH lack normal LDL receptors on the surface of their liver cells, causing cholesterol to remain in the blood.

- Cystic fibrosis occurs in 1 in 2,500 individuals and results from a defective version of a membrane protein that is a chloride transporter (see Figure 17.3B). In normal individuals, chloride is transported to the exterior of the cell by this protein, and as a result, water leaves the cell osmotically due to the higher concentration of chloride ions outside the cell. Chloride transport keeps the surface of the lungs (and the mucus) moist. When the chloride transporter is defective, thick, dry mucus accumulates in the lungs and the lining of other organs in the body. The dry mucus prevents the lungs from clearing out bacteria and fungal spores, and the patient is subject to recurrent respiratory infections as well as liver, pancreatic, and digestive failures. Patients with cystic fibrosis usually die in their thirties.

Defects in structural proteins also lead to severe disease.

- Duchenne muscular dystrophy is caused by a defect in dystrophin, the protein in muscle cells that connects the actin filaments to the extracellular matrix. This disease occurs in 1 out of 3,000 individuals and results in weaker muscles, respiratory failure, and eventual death.

- Hemophilia is caused by a mutation in a gene encoding a clotting factor in the blood. Lacking this protein, patients can bleed to death from minor cuts.

Abnormal protein conformations can cause disease.

- Transmissible spongiform encephalopathy (TSE) is a degenerative brain disease in which the brain develops holes and the brain tissue appears spongelike. This disease has been identified in sheep and goats (scrapie), elk and deer (chronic wasting disease), cows ("mad cow disease"), and humans (kuru).

- TSE is transmitted from one animal to another via brain extracts from the diseased animal.

- The infectious agent is a nonfunctional conformation of a membrane protein (PrP^{sc}) termed a prion (proteinaceous infective particles). Once these proteins fold inappropriately, they can induce a conformational change in the normal protein (PrP^c), so that it, too, becomes abnormal (see Figure 17.4). Altered proteins pile up as fibers in the brain tissue, causing cell death.

- The prion (PrP^{sc}) seems to induce a misfolding in (PrP^c) so that it converts to the abnormal misfolded form.

Most diseases are multifactorial.

- Although some diseases are the result of a mutant allele resulting in a single dysfunctional protein, most diseases are multifactorial, caused by many genes and proteins interacting with the environment. Up to 60 percent of people may be affected by diseases that are genetically influenced.

Human genetic diseases have several patterns of inheritance.

- Autosomal recessive diseases include PKU, sickle-cell disease, and cystic fibrosis. Parents who are normal but both heterozygous for these alleles have a 25 percent chance of having a child who expresses the trait (homozygous recessive). Individuals who are homozygous recessive for this trait can synthesize only nonfunctional protein and express the disease. Heterozygotes can synthesize normal proteins 50 percent of the time, which is usually enough for the cell to function.

- Autosomal dominant diseases are diseases in which the presence of one mutant allele is enough to produce the disease. Familial hypercholesterolemia is an autosomal disease in which heterozygotes only produce one-half of the normal LDL receptors, which is not enough to clear the blood of cholesterol. The pattern of inheritance for autosomal dominant alleles is direct transmission from parent to offspring.

- X-linked diseases include hemophilia and Duchenne muscular dystrophy. These diseases are more common in males (because males only have one X chromosome) and are passed from mothers (who are heterozygotes and thus carriers) to their sons. The son who inherits the mutant allele from his mother will express that allele; the daughter who inherits that allele from her mother will be a carrier.

- Chromosome abnormalities include an excess or loss of one or more chromosomes (aneuploidy), a loss of a piece of a chromosome (a deletion), or the transfer of a piece of a chromosome to another chromosome (translocation). Chromosomal abnormalities may be inherited or may be the result of a nondisjunctional event in meiosis (see Chapter 9).

- Fragile-X syndrome (which occurs in 1 in 1,500 males and 1 in 2,000 females) causes mental retardation in some individuals. It is characterized by a constriction in the tip of the X chromosome, which sometimes breaks during preparation for microscopy (see Figure 17.6).

Identifying the genes that cause disease has been accomplished in a variety of ways.

- The β-globin gene was cloned using cDNA generated from mRNA from red blood cells. β-globin mRNA is abundant because it is the major protein in these cells (see Figure 17.7A).

- The dystrophin gene was identified by comparing X chromosomes isolated from several boys with muscular dystrophy to X chromosomes from normal individuals. The X chromosomes from these patients had visible deletions, allowing investigators to position the site of the dystrophin gene on the chromosome (see Figure 17.7B).

- Positional cloning has been used to identify genes that are linked to a particular disease. Genetic markers and family pedigrees are compared to determine if a particular marker is different in affected versus unaffected individuals.

- RFLPs (restriction length fragment polymorphisms, see Figure 17.8) and single nucleotide polymorphisms (SNPs) are genetic markers that can be used to distinguish individual variations in chromosomes. Single nucleotide polymorphisms (SNPs) can be detected using mass spectrometry. If a particular RFLP or SNP is positioned adjacent to a mutated gene, the RFLP or SNP can be used as a landmark to isolate the gene because the marker and the disease gene are inherited together. Comparing RFLP patterns between affected, and unaffected individuals can help determine how close these particular markers are to the gene of interest. DNA fragments from those regions thought to contain a gene can be used to probe cellular mRNA. If a DNA–mRNA hybrid is formed, that DNA fragment contains a gene that is being expressed in the cell and can be sequenced. Comparison of these DNA sequences in affected and unaffected individuals can help determine if this gene is mutated in affected individuals.

Different kinds of mutations can cause human disease.

- Disease-causing mutations may involve a singe base pair, a long stretch of DNA, or even an entire chromosome.

- Hot spots, which are usually sites on the DNA containing methylated cytosines, are more prone to mutation than average. Loss of the amino group on a cytosine converts that base to uracil, which is removed from the DNA by repair enzymes. Loss of an amino group on a methylated cytosine converts that base to thymine, which is retained in the DNA, resulting in a base change at that site. Repair of the GT base pair may result in a correction to the original GC base pair or the introduction of a new AT mutation in the DNA at that site.

- Larger mutations include deletions, which can vary in size. Small deletions are generally less severe than large deletions, which can remove more than one gene.

- Expanding triplet repeats are found in fragile-X syndrome (a repeat of CGG in the *FMR1* gene), in myotonic dystrophy (a repeat of CTG), and in Huntington's disease (a repeat of CAG). Triplet repeats are more extensive in affected individuals and are thought to expand during DNA replication due to slippage of DNA polymerase in those regions. Triplet repeats can be found within or outside the coding sequence of a gene.

- Genomic imprinting is observed in genes or groups of genes whose phenotypic effects differ depending on the parent from whom those genes were inherited. An example of genomic imprinting in humans occurs with the inheritance of a small deletion on chromosome 15. If the chromosome with the deletion is inherited from the mother, the child has Angelman syndrome; if that chromosome is inherited from the father, the child develops Prader-Willi syndrome.

Genetic screening is used to identify people who might be carriers of a particular disease.

- Genetic screening can be done prenatally, on newborns, or on asymptomatic individuals who have close relatives with a genetic disease.

- In PKU screening, blood samples from newborn babies are tested to see if they can support growth of bacteria that are phenylalanine auxotrophs (see Figure 17.11). Individuals who test positive for PKU are fed special diets low in phenylalanine to prevent abnormal brain development.

- Preimplantation screening for mutant alleles for in vitro fertilized human embryos can be done at the eight-cell stage. One of these embryonic cells is removed, and its DNA is analyzed for the disease gene. Once a normal genotype has been confirmed, the seven-celled embryo can be implanted into the mother and will develop normally. Fetuses can be also be screened using chorionic villus sampling or by amniocentesis.

- Newborns can be screened for PKU and other disorders using DNA extracted from their cells and PCR amplification of that DNA that targets the mutant allele. Parents and siblings who are potential carriers for a particular mutant allele (such as sickle cell anemia, cystic fibrosis, or muscular dystrophy) can be screened for that allele using PCR amplification.

- Once the DNA has been isolated, different kinds of genetic screens can be used. Allele-specific cleavage differences can detect changes in restriction endonuclease sites in the DNA containing the gene of interest (see Figure 17.12). Oligonucleotides that are specific for the mutant allele or the normal allele can be used in hybridizations with DNA isolated from different individuals (see Figure 17.13).

Cancer is a genetic disease.

- Cancer is a genetic disease in which there is a loss of normal control over cell division. Cells divide continuously, forming a malignant tumor that metastasizes to other tissues in the body.

- Benign tumors are not cancerous, grow slowly, and do not spread to other parts of the body.

- Malignant tumors have irregular shapes and structures such as irregularly shaped nuclei and can metastasize to other sites in the body.

- Metastasis is characterized by secretion of digestive enzymes as cells begin to relocate, synthesis of cell surface proteins that allow the cancerous cells to invade new host tissue and angiogenesis, which releases chemical signals to stimulate blood vessel growth into the tumor.

- Carcinomas are cancers of surface tissues including skin and epithelial cells that line organs. Carcinomas include lung cancer, breast cancer, colon cancer, and liver cancer. Sarcomas are cancers of tissue such as bone, blood vessels, and muscle. Leukemia and lymphomas affect cells that give rise to blood cells.

- Some cancers (15 percent) are caused by viruses, including hepatitis B and human papillomavirus (see Table 17.1).

- About 85 percent of cancers are caused by mutations, either spontaneous or induced by natural or synthetic carcinogens and various forms of irradiation.

- Two kinds of genes are altered in many cancers: oncogenes (which stimulate cell division when activated) and tumor suppressor genes (which inhibit cell division). Some oncogenes code for growth factors, and some control apoptosis. Point mutations, translocations, and gene amplifications in these genes can lead to uncontrolled cell division. Oncogenes can be activated by one mutational hit. Tumor suppressor genes (such as those involved in retinoblastoma, Wilms' tumor of the kidney, and inherited forms of breast and prostate cancer) require the inactivation of both alleles (the "two-hit" hypothesis). Some tumor suppressor genes code for cell adhesion proteins, for DNA repair enzymes, and for proteins that stop the cell cycle in G1.

- Women who inherit a mutant allele of the *BRCA1* gene (a tumor suppressor) are more prone to develop breast cancer than women with the normal alleles for *BRCA1*.

- Some tumor suppressors control the progress of the cell through the cell cycle. The protein encoded by the *Rb* gene acts during G1 by inhibiting transcription factors that are necessary for progress into DNA synthesis and subsequent cell cycle completion. If a mutation inactivates the *Rb* gene, as occurs in retinoblastoma, cells remain in the cell cycle and continue to divide in the absence of growth factors, forming a tumor.

- *p53* is a gene coding for a protein that stops the cells during the G1 phase of the cell cycle. The protein product of *p53* is a transcription factor for a protein that blocks the interaction of cyclins and protein kinases (see Chapter 10). *p53* is mutant in many types of cancer cells, including lung and colon cancer cells.

- Multiple mutations transform a normal cell into a cancerous cell (see Figure 17.18 and 17.19). These mutations can be screened using oligonucleotide probes for oncogenes and/or tumor suppressor gene alterations.

Treating genetic diseases includes modifying the phenotype and gene therapy.

- There are three ways to treat genetic disease: restrict the substrate of the deficient enzyme, inhibit a harmful metabolic reaction, or supply the normal version of the protein.

- For individuals with PKU, treatment involves restricting the substrate (phenylalanine) for the missing enzyme (phenylalanine hydroxylase), which prevents the buildup of phenylpyruvic acid.

- For individuals with familial hypercholesterolemia, drugs such as statins are administered that block the patient's own cholesterol synthesis.

- Drugs that kill rapidly dividing cells are administered to cancer patients (see Figure 17.20). These drugs also affect noncancerous dividing cells in the body, causing digestive upsets, hair loss, and anemia.

- Patients with hemophilia A are given the missing gene product, which is a clotting factor, in purified form.

- In gene therapy, the new gene is inserted into the affected individual. The gene must have all the important regulatory regions necessary for expression and must be delivered effectively to the individual's cells. This requires effective vectors, efficient uptake, precise insertion into host DNA, appropriate expression and processing of mRNA and protein, and selection of the target cells for the recombinant DNA.

- Ex vivo gene therapy is a technique in which the gene of interest is inserted into the patient's cells (white

blood cells or skin grafts), and then those cells are reintroduced into the patient. This technique has been used to replace the adenosine deaminase gene and the blood clotting factor in hemophilia.

- Another approach to gene therapy involves inserting the gene directly into the cells where the defect exists. For instance, in lung cancer patients, functional alleles of tumor suppressor genes, as well as antisense RNAs against oncogene mRNAs, are provided in aerosols for inhalation.

Sequencing the human genome has been a valuable scientific endeavor.

- To sequence human DNA, it must first be cut into smaller fragments (500 base-pair fragments) and then sequenced. There are two methods to align the different fragments: hierarchical sequencing and shotgun sequencing.

- Hierarchical DNA sequencing involves the careful ordering of DNA sequences using marker sequences along the chromosome.

  - Libraries of human DNA have been created using restriction enzymes that recognize 8–12 base-pair recognition sequences, resulting in much larger DNA fragments. These large DNA fragments have been cloned into a bacterial artificial chromosome (BAC), which can accommodate 250,000 base pairs of inserted DNA.

  - These fragments have been arranged in proper order using markers and comparing those sequences to DNA libraries made with different restriction endonucleases (see Figure 17.22A).

- Shotgun sequencing involves cutting the DNA into random small fragments, which are analyzed for overlapping sequences by computer (see Figure 17.22B) and then aligned. This method was facilitated by powerful computer DNA analyses using bioinformatics. Shotgun sequencing was used to sequence the entire 180 million base pairs of the fruit fly genome.

- The sequence of the human genome has revealed that:

  - Less than 2 percent of the DNA contains coding regions (a total of 54,000 genes).

  - The average gene has 27,000 base pairs.

  - Virtually all human genes have many introns.

  - Over 50 percent of the genome contains highly repetitive sequences.

  - Almost all genes (99.9 percent) are the same in all people.

  - Genes are not evenly distributed over the genome.

  - The functions of many genes are unknown.

  - The ENCODE project (*Encyclopedia of DNA Elements*) has been set up to identify all of the functional sequences in the human genome, including genes for proteins, RNAs, and small RNAs.

- There are many advantages of sequencing the entire human genome:
  - Comparison of genes from different organisms that have similar counterparts in humans.
  - Identifying genes using positional cloning, including disease-related genes.
  - Identifying genes that will react best to particular medications (pharmacogenomics).
  - Constructing DNA chips which can be used to analyze gene expression in different cells in different biochemical states.
  - Gene prospecting, which involves looking for important polymorphisms in specific populations to reveal genes that predispose those populations to certain conditions.
- Now that complete genetic information for each human being is possible, many individuals, including ethicists, psychologists, and geneticists, want to establish guidelines regarding how this information will be used. The search for valuable genes in diverse populations has raised concerns about exploitation and commercialization of a person's DNA sequence.

The proteome is more complex than the genome.

- Because of alternative splicing and posttranslational modifications, the sum total of proteins produced (the proteome) is more complex than the genome.
- Proteome analysis includes two-dimensional gel electrophoresis (see Figure 17.23) and mass spectrophotometry.
- There is little difference in mRNA expression (1.4 percent) in the cortex portion of the brains of humans and chimps. However, proteome analysis has shown that there is a large difference in the kinds (7.4 percent) and amounts (31.4 percent) of proteins expressed in the cortex of humans and chimps.
- Systems biology integrates data from genomics and proteomics to discover emergent properties in the system. This approach helps biologists understand how increasing the amount of a particular molecule might affect the levels of other molecules (see Figure 17.24).

The Big Picture

- Some diseases in humans can be directly related to a change in the genome, causing the production of an abnormal gene product. Other molecular diseases are affected by the individual's environment. Understanding the molecular alterations in a particular gene product and its effect can lead to therapeutic approaches, new diagnostic tools, and preventative measures in health care.

Common Problem Areas

- Sometimes it is difficult to understand how a mutant allele can be dominant to a wild-type allele (as seen in familial hypercholesterolemia). Remind yourself of the function of the wild-type gene product, and then predict how the individual will be affected with only half of those gene products being expressed.
- RFLP mapping, oligonucleotide hybridization, and allele-specific cleavage analyses are similar techniques used in genetic screening to detect mutant alleles. Individual differences in DNA sequences (including simple base changes and DNA deletions, as well as a difference in the number of repeated sequences) can be utilized to distinguish individuals and to map genes. Review Figures 17.6, 17.7, 17.11, and 17.12 to distinguish these screening techniques.

Study Strategies

- Make a list of the molecular diseases discussed in this chapter. Include in this list the specific proteins that, when altered, cause the disease, and describe the treatments that are currently available for each.
- Review the cell cycle and the signals that are important for causing cells to divide. Predict how mutations in those genes can alter the regulation of the cell cycle and lead to unregulated cell division and cancer. Be sure to include a review of tumor suppressors and oncogenes in this analysis.
- Review the following animated tutorials and activities on the Companion Website/CD:
  Tutorial 17.1 DNA Testing by Allele-Specific Cleavage
  Tutorial 17.2 Sequencing the Genome
  Activity 17.1 Allele-Specific Cleavage
  Activity 17.2 Concept Matching

Test Yourself
Knowledge and Synthesis Questions

1. Fragile-X syndrome is
   a. due to a single base change in the DNA.
   b. caused by changing a valine to a glutamic acid.
   c. caused by triplet expansion.
   d. caused by a chromosomal translocation.
   Textbook Reference: *17.2 What Kinds of DNA Changes Lead to Diseases? pp. 382–383*

2. A nonfunctional membrane protein is responsible for
   a. hemophilia.
   b. sickle-cell disease.
   c. cystic fibrosis.
   d. PKU.
   Textbook Reference: *17.1 How Do Defective Proteins Lead to Diseases? p. 377*

3. Triplet repeat expansions
   a. occur during DNA replication due to slippage of DNA polymerase.
   b. are caused by errors in DNA synthesis with reverse transcriptase.
   c. differences can be identified using RFLP mapping.
   d. Both a and c
   Textbook Reference: *17.2 What Kinds of DNA Changes Lead to Diseases? pp. 382–383*

4. In sickle-cell disease the
   a. sixth amino acid is changed from a valine to a glutamic acid.
   b. sixth amino acid is changed to a stop codon.
   c. hemoglobin concentration builds up in the red blood cells.
   d. structure of β-globin is altered, and the hemoglobin protein forms aggregates in the red blood cells.
   Textbook Reference: 17.1 How Do Defective Proteins Lead to Diseases? p. 376

5. Most cancers are
   a. due to the inheritance of mutant genes.
   b. caused by viruses.
   c. caused by mutations in genes whose products help blood clot.
   d. caused by mutations that are spontaneous or induced.
   Textbook Reference: 17.4 What Is Cancer? pp. 387–388

6. Metabolic disorders are
   a. caused by genes that have alterations in their centromeres.
   b. the result of mutations in genes encoding enzymes that are required to synthesize particular compounds in the cell (such as an amino acid).
   c. due to abnormal membrane proteins that transport chloride ions.
   d. caused by prions.
   Textbook Reference: 17.1 How Do Defective Proteins Lead to Diseases? pp. 375–376

7. Prions
   a. cause scrapie in sheep.
   b. are caused by abnormally folded proteins interfering with normal brain cell function.
   c. cause "mad cow disease."
   d. All of the above
   Textbook Reference: 17.1 How Do Defective Proteins Lead to Diseases? p. 378

8. Human genetic disease can be
   a. due to an autosomal recessive trait.
   b. passed from mothers to their sons (if X-linked) 50 percent of the time and also can be caused by translocations.
   c. due to a mutant allele that is dominant to the wild-type allele.
   d. All of the above
   Textbook Reference: 17.1 How Do Defective Proteins Lead to Diseases? p. 379

9. Mutations that result in human disease include
   a. triplet expansions that occur during DNA synthesis.
   b. point mutations that do not change the amino acid sequence of the gene.
   c. prion-like diseases such as kuru.
   d. Both a and b
   Textbook Reference: 17.2 What Kinds of DNA Changes Lead to Diseases? pp. 382–383

10. Genetic screening has been used to identify
    a. embryos carrying a mutant allele.
    b. newborns who have PKU.
    c. fathers who are carriers for X-linked diseases.
    d. Both a and b
    Textbook Reference: 17.3 How Does Genetic Screening Detect Diseases? pp. 384–386

11. Genes that, when mutated cause cancer are
    a. only those genes involved in the regulation of the cell cycle.
    b. transcription factors only.
    c. only those genes inactivated by viruses.
    d. None of the above
    Textbook Reference: 17.4 What Is Cancer? pp. 388–389

12. For gene therapy to be most effective, genes should be inserted in _____ cells.
    a. white blood
    b. red blood
    c. stem
    d. All of the above
    Textbook Reference: 17.5 How Are Genetic Diseases Treated? pp. 391–392

13. The human genome allows scientists to
    a. understand regulatory sequences that are important for gene expression.
    b. locate genes that cause disease.
    c. understand evolutionary relationships by comparing human genes to genes in other organisms.
    d. All of the above
    Textbook Reference: 17.6 What Have We Learned from the Human Genome Project? pp. 394–395

14. Proteomics has been used to compare
    a. DNA sequences between closely related species.
    b. gene expression during embryonic development.
    c. protein expression between closely related species.
    d. shotgun cloned sequences.
    Textbook Reference: 17.6 What Have We Learned from the Human Genome Project? pp. 395–396

15. Allele-specific oligonucleotides
    a. are used to detect mutations in DNA.
    b. can determine if an individual is carrying a disease allele.
    c. can detect a change in a restriction endonuclease site.
    d. All of the above.
    Textbook Reference: 17.3 How Does Genetic Screening Detect Diseases? pp. 385–386

Application Questions

1. Familial hypercholesterolemia (FH) and Duchenne muscular dystrophy (MD) are both molecular diseases that are caused by mutations in particular genes, yet FH is easier to treat than MD. Why?
   Textbook Reference: 17.1 How Do Defective Proteins Lead to Diseases? p. 375

2. Two individuals have a mutation in gene *x* but at different sites. The disease adversely affects the first

individual, and the second individual experiences no effect. Explain this observation

Textbook Reference: *17.1 How Do Defective Proteins Lead to Diseases? p. 375; Figure 17.2*

3. Why are X-linked diseases more common in males than in females?

Textbook Reference: *17.1 How Do Defective Proteins Lead to Diseases? p. 379*

4. The β-globin gene (which when mutated can lead to sickle-cell disease) was much easier to clone than the gene responsible for one of the inherited forms of breast cancer (*BRAC1*). Why?

Textbook Reference: *17.2 What Kinds of DNA Changes Lead to Diseases? pp. 380–382*

5. Explain how mutation in only one of two oncogene alleles can lead to unregulated cell division, whereas mutations in both alleles of a tumor suppressor gene are necessary.

Textbook Reference: *17.4 What Is Cancer? pp. 388–389*

Answers

Knowledge and Synthesis Answers

1. **c.** Fragile-X syndrome is due to a triplet expansion on the X chromosome, not a single base change, a translocation, or an alteration of protein sequence, as seen in sickle-cell disease.

2. **c.** The chloride transporter is a membrane protein that, when nonfunctional, causes cystic fibrosis. Hemophilia is caused by a defect in a clotting factor; sickle-cell disease is caused by a defect in β-globin. PKU is a defect in an enzyme in a biochemical pathway.

3. **d.** Triplet expansion is thought to be due to slippage of DNA polymerase during DNA replication and will result in an increasing number of triplet repeats in those cells. That increase in the DNA sequence can be distinguished from the DNA sequence of a normal individual using RFLP analysis. Reverse transcriptase is an enzyme from HIV that copies RNA into DNA.

4. **d.** In sickle-cell disease, the sixth amino acid is changed from a glutamic acid to a valine (not from a valine to a glutamic acid, and not to a stop codon). The resulting change causes the hemoglobin protein to form aggregates in the red blood cell. Hemoglobin concentration in the red blood cell is not altered in sickle-cell disease.

5. **d.** About 10 percent of cancers are due to the inheritance of a mutant allele, and 15 percent are caused by viruses. Most cancers are caused by mutations that are either spontaneous or induced (note that cancers may require more than one mutation, so any given cancer may have more than one cause). Mutations in genes that encode blood clotting factors result in hemophilia.

6. **b.** Some metabolic disorders are caused by mutations in genes that encode enzymes in a biochemical

pathway (such as the pathway for the synthesis of phenylalanine). Mutations in centromeres and genes for chloride transporters do not result in metabolic disorders. Prions are not the result of a mutation; they are the result of abnormal folding of a protein.

7. **d.** Prions result from abnormally folded proteins that interfere with normal brain cell function. Mad cow disease is caused by prion formation. They are not caused by a mutant allele for this particular protein.

8. **d.** All these mutations and chromosome alterations can cause human disease.

9. **a.** Triplet expansions cause a variety of diseases including fragile-X syndrome, Huntington's chorea, and myotonic dystrophy. Point mutations that do not alter the amino acid sequence of a protein are silent mutations and do not cause any disease (in almost all cases). Prions are not the result of a mutation in a gene but are due to the abnormal folding of a particular protein.

10. **d.** Genetic screening has been used to analyze in vitro fertilized embryos (at the 8-cell stage) and DNA from individuals who might be carrying a mutant allele. Fathers are not carriers for X-linked diseases; they express all their X-linked alleles.

11. **d.** Although all the examples listed are examples of mutations that can result in cancer, none of these mutations are the *only* reason an individual develops cancer.

12. **c.** The most effective approach for genes to reside permanently in the individual's cells is for them to be inserted in stem cells, thus providing a continual supply of differentiated cells under the correct environmental signals. Red blood cells have no nuclei and therefore cannot be used for gene therapy. White blood cells have been used for gene therapy, but because they have a set lifetime in the body, they eventually die, destroying the gene they were carrying.

13. **d.** Human genome sequencing provides all these advantages.

14. **c.** Proteomics used to compare protein sequences between different organisms. Answers a, b, and d all refer to techniques used in genomics.

15. **d.** Allele-specific oligonucleotides can detect specific changes in the DNA, including mutations, disease alleles, and changes in restriction sites.

Application Answers

1. Familial hypercholesterolemia causes an increase of cholesterol in the patient and can be treated with drugs that reduce the patient's own synthesis of cholesterol. Muscular dystrophy is caused by a defect in a structural protein, dystrophin, and results in patients who have dysfunctional muscle tissue. This structural dysfunction cannot be reversed with the application of any drug.

2. The mutation in gene *x* in the first individual must have occurred in an essential region of the gene that is

required for its function. The mutation in gene x in the second individual may be a silent mutation, or it may be in a region that is nonessential for the function of that protein.

3. X-linked diseases are more common in males because males have only one X chromosome, so every allele on that chromosome is expressed in the individual. Mutant alleles can be carried by females but masked by a wild-type allele on the other X chromosome.

4. β-globin is one of the major proteins found in the red blood cell, and thus isolating mRNA from this cell and cloning a cDNA copy of this gene was relatively straightforward. Neither the function nor the location of the *BRAC1* gene was known. Nearby genetic markers had to be identified that were linked to the *BRAC1* gene, and then DNA sequencing of that region had to be done. DNA sequences between affected and unaffected individuals had to be compared to identify the gene.

5. Oncogenes control cell division by responding to an environmental signal and initiating cell division in response to that signal. A mutation in an allele of an oncogene would result in that protein causing the cell to initiate cell division, even in the absence of any environmental signal. Half of the wild-type proteins would respond appropriately to cell division signals, but the dysfunctional proteins that were now unregulated would still signal the cell to divide. A single wild-type copy of a tumor suppressor gene, however, could still provide enough gene product to regulate cell division.

CHAPTER 18 Immunology: Gene Expression and Natural Defense Systems

Important Concepts

Animal defense systems are based on distinguishing self from nonself and include nonspecific defenses and specific defenses.

- Pathogens are the harmful organisms and viruses that cause disease. Nonspecific defenses or innate defenses are inherited mechanisms that act rapidly to protect the body from pathogens. Nonspecific defenses include physical barriers as well as cellular and chemical defenses (see Table 18.1).

- Specific defenses are adaptive mechanisms aimed at specific targets. The specific defenses involve antibodies and are found in the vertebrates.

- Lymphoid tissues, including thymus, bone marrow, spleen, and lymph nodes, are essential to a mammal's defense system. The lymphatic system moves lymph fluid from the tissues to the heart. Both lymph and blood contain white blood cells and platelets. Lymph nodes contain white blood cells and filter the lymph fluid looking for nonself material.

- Granular cells are white blood cells that include the phagocytes involved in engulfing and digesting nonself materials. Macrophages are important phagocytes that are involved in the interaction of nonself materials and T cells.

- Lymphocytes are the most abundant class of white blood cells and include B and T cells. Immature T cells migrate via the blood to the thymus where they mature. T cells participate in specific defenses against foreign or altered cells, including virus-infected cells and tumor cells. B cells circulate through the blood and lymph and make antibodies, which are proteins that bind to nonself or altered self.

- T cell receptors are proteins on the surface of T cells that recognize and bind to nonself substances on the surfaces of other cells.

- Major histocompatibility complex (MHC) proteins stick out from surfaces of most cells in the mammalian body and help in self-identity.

- Cytokines are soluble signal proteins released by T cells, macrophages, and other cells. They can activate or inactivate B cells, macrophages, and T cells or limit tumor growth.

Barriers, local agents, chemicals, and cellular processes act as nonspecific defenses against invaders.

- Nonspecific defenses include skin, mucus-secreting tissues, lysozyme in tears, digestive enzymes, bile salts, and hydrochloric acid in the stomach. Additionally, the bacteria and fungi that compose the normal flora living on the body compete with pathogens and act as a means of defense.

- About 30 complement proteins act as antimicrobial proteins in the vertebrate blood. They function by destroying microbes, activating other immune responses, or lysing microbes.

- Small proteins known as interferons are produced in response to a viral infection. Interferons are glycoproteins that stimulate signaling pathways that inhibit viral reproduction inside infected cells and stimulate lysosome activity.

- Phagocyte cells ingest cells and viruses, which are then killed by defensins inside the phagocyte.

- Natural killer cells are white blood cells that lyse target cells.

- Inflammation defends the body against infectious agents through the release of histamine from mast cells (see Figure 18.4).

- A signal transduction pathway exists in which over 40 genes are activated as an immune response. The receptor that initiates this pathway is known as toll. (see Figure 18.5).

Specific defenses of the immune response occur when immune cells (B and T cells) recognize and destroy nonself substances.

- An antigen is a molecule or organism that is recognized by specific antibodies or T cell receptors to provide specificity. Immune cells recognize antigenic determinants or epitopes (i.e., specific amino acids of a protein) on invading pathogens. There can be many antigenic determinants on a pathogen surface.

- The immune cells can distinguish self from nonself ensuring that they don't attack their own cells.

- Immune cells respond to pathogens by activating lymphocytes of the appropriate specificity. Humans can respond to 10 million different antigenic determinants.

- The immune system "remembers" a pathogen and can respond more rapidly and more effectively the next time it is exposed to it. Immunological memory is the basis of vaccination.

- There are two interactive immune responses against invaders. The humoral immune response involves antibodies from B cells that react with antigenic determinants. The cellular immune response involves T cells binding with antigens.

Clonal selection accounts for the characteristic features of the immune response.

- The body contains millions of different B cells, and each B cell is able to produce only one kind of antibody. The differences among B cells are due to changes in DNA. When an antigen that fits the surface antibody binds to the B cell, that cell is activated and divides to produce clonal B cells, which secrete antibodies. T cells are clonally selected in a similar manner.

Immunologic memory and immunity result from clonal selection.

- In the primary immune response, activated B and T cells produce effector cells and memory cells.

- Effector cells attack a pathogen by producing specific antibodies (in the case of B cells) or cytokines (in the case of T cells). During a first encounter with an antigen, the primary immune response occurs.

- After the primary immune response, long long-lived memory cells remain and divide at a slow rate. The memory B and T cells divide rapidly to produce an effective and more powerful immune response if reexposed to a pathogen. This rapid response, the secondary immune response, provides a natural immunity to those diseases.

- Vaccinations and immunizations are administered to give the recipient artificial immunity and are effective because of this secondary immune response. Safe antigens are produced by treating the antigenic molecule with a chemical (attenuation), producing small peptides through biotechnology, or with DNA vaccines.

- Animals are able to tolerate their own antigens due to clonal deletion, a process that removes any B and T cells that recognize self antigens during their differentiation. An immune cell that recognizes self antigens undergoes programmed cell death (apoptosis).

- Clonal anergy is a suppression of the immune response so those T cells that recognize a self antigen on the body do not send out cytokines. This suppression occurs in those body cells that are missing a co-stimulatory signal, CD28, for the T cell. The co-stimulatory signal for the T cell is only expressed on antigen-presenting cells.

- Immunological tolerance occurs when a foreign antigen is introduced into an animal during fetal development, which results in the animal's recognizing that antigen as self.

B cells and antibodies are the basis of the humoral response.

- Activation of B cells results from the binding of an antibody to a particular antigenic determinant, and the arrival of a signal from a helper T cell that has bound the antigen on an antigen-presenting cell.

- Antibodies (also called immunoglobulins) are proteins that are similar in structure, but can be grouped into five different classes. Antibody molecules consist of two identical heavy polypeptide chains and two identical light polypeptide chains held together by disulfide bonds. Each polypeptide consists of a constant region that is similar in amino acid sequence from one immunoglobulin to another and a variable region that forms the antigen-binding site.

- The variable region produces the specificity of the millions of antibodies. The constant region determines whether the antibody will be secreted or remain on the cell surface of the B cell. It also determines what type of action will be taken to eliminate the antigen.

- The five immunoglobulin classes (IgG, IgM, IgD, IgA and IgE) are based on differences in the constant region of the heavy chain (see Table 18.3).

Monoclonal antibodies are produced by hybridomas.

- Normal immune responses are polyclonal, generated by the activation of many different B cells and resulting in a complex mixture of antibodies.

- Monoclonal antibodies are produced from a clone of B cells that makes only one antibody that binds a unique determinant. Monoclonal antibodies are generated by the fusion of normal B lymphocytes with tumor cells of plasma cells to produce hybridomas.

- Monoclonal antibodies have been used in immunoassays, in immunotherapy to target cancer cells with radioactive ligands or toxins, and for passive immunization.

T cells direct the cellular immune response.

- T cells have specific glycoprotein surface receptors that are made up of two different polypeptide chains, both with a variable and a constant region. The variable region provides the specificity of the T cell receptor.

- T cell receptors bind antigen fragments that are displayed on antigen-presenting cells.

- Cytotoxic T (T_C) cells recognize virus-infected cells and kill them by causing them to lyse.

- Helper T (T_H) cells assist both the cellular and the humoral immune systems. They send out signals to stimulate B cell and T cell proliferation.

The major histocompatibility complex (MHC) encodes proteins that present antigens to the immune system.

- MHC gene products are plasma membrane glycoproteins that present antigens on an antigen-binding site of the MHC protein. These antigens are inspected by T cell receptors to distinguish between self and nonself. MHC proteins can display antigenic peptides of about 10 to 20 amino acids.

- Class I MHC proteins are present on the surface of every nucleated cell in an animal. Degraded cellular proteins are presented to T_C cells on MHC I complexes. T_C cells have a surface protein called CD8 that can recognize MHC I proteins.

- Class II MHC proteins are found on the surfaces of antigen-presenting cells, including B cells and macrophages. T helper cells recognize class II MHC proteins using the CD4 cell surface protein.

- T cell receptors recognize antigenic fragments when they form complexes with the MHC I or MHC II molecules.

- MHC genes, antibody genes, and T cell receptor genes have similar base sequences, suggesting a common ancestral gene.

The humoral response and cellular immune response involve an interaction between T cells and MHC complexes.

- In the humoral response, both macrophages and B cells bind antigens and present antigen fragments on the surface of their cells on MHC II complexes. T_H cells bind both the antigen and the MHC II complex and release cytokines. Those cytokines activate the T_H cells to divide (the activation phase) and activate B cells to produce antibodies (the effector phase).

- In the cellular immune response, cytotoxic T cells destroy cells that are displaying antigens from virus or mutant proteins on their MHC I complexes, eliminating virally infected cells as well as tumor cells. T_C cells must receive a second signal for activation, which involves an interaction between proteins on the T_C cell and CD28 proteins on the antigen-presenting cell. The T_C cell then proliferates and divides, releasing cytokines.

- Later, T_C cells are inhibited by the cell surface protein CTLA4.

MHC molecules play a key role in the tolerance of self.

- T cells are tested in the thymus to see if they bind to MHC displaying one of the body's own antigens, or if they are unable to recognize the body's MHC proteins. If T cells have either of these characteristics, they are destroyed.

- MHC proteins are specific to each individual; tissue transplants are thus recognized as nonself in the recipient. Suppression of rejection of organ transplants by the recipient's immune system is accomplished by treating a patient with drugs such as cyclosporin, which suppress the immune system.

DNA rearrangements and mutations are used to generate all the diverse antibodies within an organism.

- DNA fragments are rearranged and joined during B cell development to generate antibody supergenes.

- The variable region of the light chain originates from two families of genes, and the variable region of the heavy chain originates from three families of genes.

- For the constant and variable regions of the heavy chain in mice there are multiple genes coding for each of four kinds of segments: 100 *V*, 30 *D*, 6 *J*, and 8 *C* (see Figure 18.16). Each B cell randomly selects one gene from each of these multiple genes to make the final coding sequence, *VDJC* for the heavy chain. Light chains are similarly constructed from DNA segments.

- Light and heavy chains are combined to create billions of possible antibodies.

- Increased diversity is achieved by imprecise recombination during DNA rearrangements, addition of extra nucleotides to cut DNA fragments as insertion mutations, and an increased mutation rate in immunoglobulin genes.

- Class switching occurs by rearrangements in the DNA that position the *VDJ* segment (the same variable region on the antibody) adjacent to a constant region farther down the DNA (see Figure 18.18), resulting in the production of an antibody with a different constant region. Class switching in B cells is caused by cytokine release from T_H cells.

Disorders of the immune system.

- Hypersensitivity occurs when the immune system overreacts to a dose of antigen and may cause inflammation as seen in the allergic response.

- Immediate hypersensitivity in allergic reactions is caused by IgE binding the foreign antigen and causing mast cells to release histamines. Pollen allergies can be treated by desensitization in which a small amount of the allergen is injected in the skin to stimulate the production of IgG.

- Delayed hypersensitivity does not happen immediately following exposure to an antigen, but hours later. Antigen-presenting cells process an antigen and initiate a T cell response.

- Autoimmunity occurs when B and T cells direct their response against self antigens. Potential origins of autoimmunity include the failure to delete a clone, which makes antibodies against a self antigen, viral infections, and molecular mimicry. Examples of autoimmune disease include systemic lupus erythematosis, rheumatoid arthritis, Hashimoto's thyroiditis, and insulin-dependent diabetes mellitus.

- AIDS is caused by HIV, a virus that eventually destroys T_H cells. HIV is transmitted through body fluids, including blood and semen.

- HIV initially infects macrophages, T_H cells, and dendrite cells (one kind of antigen-presenting cell).

These cells carry the virus to the lymph nodes, where they preferentially infect activated T_H cells.

- During the first phase of infection, up to 10 billion viruses are made every day, causing mononucleosis-like symptoms. During the second phase, the virus infects T_H cells, and an immune response is mounted, producing HIV antibodies in the blood. Later, the T_H cells are destroyed, and the patient becomes susceptible to a wide variety of infections that would normally be eliminated by T_H cells. At this stage Kaposi's sarcoma, *Pneumocystis carinii*, and virally caused tumors persist in the system of the HIV-infected individual as opportunistic infections.

HIV infects and replicates in T_H cells.

- HIV is an enveloped virus and has a protein coat that surrounds two identical molecules of RNA as well as the enzymes reverse transcriptase, integrase, and a protease.

- The viral envelope proteins gp120 and gp41 attach to T_H cells and macrophages by binding their CD4 proteins. The HIV membrane fuses with the host cell plasma membrane, and the virus enters the host cell.

- A DNA copy of the viral genome is made using reverse transcriptase, which is an error-prone polymerase. Errors made during viral replication create alterations in viral proteins that can now escape the host's immune response.

- Viral cDNA is inserted into the host DNA using the HIV integrase protein.

- Therapeutic treatment for HIV includes drug therapies that interfere with the major steps of the life cycle of the virus. Highly active antiretroviral therapy (HAART) has been used since the late 1990s to combat HIV. However, individuals who take HAART develop resistant strains of HIV. The most promising treatment may be the development of HIV vaccines.

The Big Picture

- Animals have special systems to distinguish self from nonself. Some of these systems are nonspecific defenses (e.g., complement protein, phagocytotic cells). B and T cells can recognize specific antigens on the surface of pathogens and mount an immune response to destroy those pathogens. When the immune system is compromised, as seen in HIV-infected individuals, the individual becomes susceptible to a variety of pathogens.

Common Problem Areas

- The specificity of the immune response may seem complex initially, but there are really only two major components: B cells (which bind antigen) present antigen and secrete antibody, and T cells bind antigen on antigen-presenting cells and help direct the immune response.

- MHC proteins may seem complicated at first, but they have only a few essential functions in the immune system. These proteins are specific for each individual and are essential for presenting antigen on antigen-presenting cells (MHC II) to T_H cells and recognizing one's own cells (MHC I).

- The specificity of the immune response may seem complex, but it can be understood by reviewing cellular communication. Antibodies of B cells bind with antigen on the surface of pathogens; T cell receptors bind antigen presented on MHC I and MHC II proteins on the surface of cells. Once these membrane proteins interact, cytokines are secreted, and cells begin to divide and proliferate. Antibodies are secreted by B cells.

- The rearrangement of DNA in the nucleus of the B cell may seem foreign to students who think the integrity of the genetic information must be preserved. But this change in the genetic makeup of activated B and T cells is essential to provide the organism with a diverse and specific set of antibodies and cell receptors.

Study Strategies

- List the nonspecific responses that a human body can make in response to an invading pathogen. Next list cells that are part of the specific responses. Include proteins that are crucial to generating that specificity. Review how genetic rearrangements generate a diverse set of antibodies for B cells.

- Diagram an antibody and label the important components. Highlight those areas where mutations and alterations occur that change the amino acid sequence of the heavy and light chains.

- Review the expression and function of MHC I and MHC II molecules.

- Make a list of the functions of T_H cells and T_C cells.

- Draw a diagram of a chromosome and label it with 4 *V*s (*V1–V4*), 2 *D*s (*D1* and *D2*), 3 *J*s (*J1–J3*) and 8 *C*s (*C1–C8*). Imagine that a B cell is maturing to produce antibody on its surface. Select one *V*, one *D*, one *J*, and one *C* segment to make an antibody. Repeat the process using different segments. Predict how many different antibodies can be generated from this set of gene segments.

- List some of the disorders of the immune system and their probable causes.

- Review the following animated tutorials and activities on the Companion Website/CD:
  Tutorial 18.1 Cells of the Immune System
  Tutorial 18.2 Pregnancy Test
  Tutorial 18.3 Humoral Immune Response
  Tutorial 18.4 Cellular Immune Response
  Tutorial 18.5 A B Cell Builds an Antibody
  Activity 18.1 The Human Defense System
  Activity 18.2 Inflammation Response
  Activity 18.3 Immunoglobulin Structure

Test Yourself

Knowledge and Synthesis Questions

1. Phagocytes
   a. are T and B cells.
   b. present antigen on MHC II complexes.
   c. digest nonself materials.
   d. Both b and c

   Textbook Reference: 18.1 What Are the Major Defense Systems of Animals? p. 402

2. B cells
   a. secrete antibodies.
   b. present antigen on MHC II.
   c. ingest antigens.
   d. All of the above

   Textbook Reference: 18.1 What Are the Major Defense Systems of Animals? p. 403

3. Nonspecific responses in the immune system include
   a. macrophages.
   b. natural killer cells.
   c. complement proteins.
   d. All of the above

   Textbook Reference: 18.2 What Are the Characteristics of the Nonspecific Defenses? p. 404

4. When the receptor of a T_H cell binds to a pathogen being presented on a macrophage, it
   a. inactivates itself.
   b. secretes cytokines.
   c. inactivates B cells.
   d. All of the above

   Textbook Reference: 18.5 What Is the Cellular Immune Response? pp. 414–415

5. Part of the normal immune response includes
   a. the production of B memory cells.
   b. the production of memory macrophages.
   c. antibody secretion by eosinophils.
   d. the production of B cells that attack the individual's own cells.

   Textbook Reference: 18.3 How Does Specific Immunity Develop? p. 408

6. Antibody molecules are
   a. produced by B cells and have only a constant region.
   b. secreted by B cells, once a signal (a cytokine) is received from a T cell.
   c. produced by T cells and have a variable and a constant region.
   d. Both a and b

   Textbook Reference: 18.4 What Is the Humoral Immune Response? pp. 411–412

7. Cytotoxic T cells
   a. release cytokines that activate B cells.
   b. attack pathogens by binding to cell surface antigens on those pathogens.
   c. destroy pathogens by lysing them.
   d. destroy host cells that are infected with virus.

   Textbook Reference: 18.5 What Is the Cellular Immune Response? p. 414

8. The humoral response
   a. is due to the secretion of antibodies.
   b. occurs when T cells bind antigen-presenting cells.
   c. is due to T cells secreting their receptors.
   d. occurs when natural killer cells engulf cancer cells.

   Textbook Reference: 18.5 What Is the Cellular Immune Response? pp. 414–415

9. Autoimmunity is
   a. active when organ transplantation is successful.
   b. caused by clonal anergy.
   c. the response of the immune cells attacking the body's own tissues.
   d. Both b and c

   Textbook Reference: 18.4 What Is the Humoral Immune Response? p. 411

10. Patients with HIV are susceptible to a variety of infections because
    a. the virus produces cell surface receptors that bind to pathogens, making it easier for those pathogens to be infective.
    b. synthesizing a DNA copy of the viral genome makes a person feel sick.
    c. HIV attacks and destroys the T helper cells, which are central to mounting an effective immune response, making those individuals more susceptible to other infections.
    d. HIV destroys B cells so that antibodies cannot be made in response to invading pathogens.

    Textbook Reference: 18.7 What Happens When the Immune System Malfunctions? pp. 421–422

11. DNA rearrangements in the B cell
    a. are responsible for generating single B cells that can express many different antibodies.
    b. lead to mutations in T cells resulting in the elimination of essential T cell genes.
    c. occur only in B memory cells.
    d. are responsible for generating many different antibodies, with each B cell expressing only one set of identical antibodies.

    Textbook Reference: 18.6 How Do Animals Make So Many Different Antibodies? p. 418

12. Major histocompatibility proteins function in the immune system by
    a. presenting antigen to T cells.
    b. presenting self antigens to T helper cells so those T cells will continue to tolerate self throughout the lifetime of the individual.
    c. generating antibodies to different pathogens.
    d. Both a and b

    Textbook Reference: 18.5 What Is the Cellular Immune Response? p. 415

13. Which of the following is *not* a granular cell?
    a. Basophil
    b. Neutrophil
    c. B cell
    d. Mast cell
    Textbook Reference: *18.1 What Are the Major Defense Systems of Animals? p. 403*

14. Inflammation occurs when _____ release _____.
    a. mast cells; histamine
    b. neutrophils; toxins
    c. B cells; histamine
    d. mast cells; toxins
    Textbook Reference: *18.3 How Does Specific Immunity Develop? p. 406*

15. When an antigen binds to a specific B cell and it begins to divide, this is called
    a. a nonspecific defense.
    b. clonal selection.
    c. normal flora.
    d. None of the above
    Textbook Reference: *18. 3 How Does Specific Immunity Develop? p. 408*

Application Questions

1. Imagine that a particular bacterium has surface molecules that resemble membrane proteins found on the heart valves. What will happen to the individual infected with this bacterium if the infection is not treated?
   Textbook Reference: *18.7 What Happens When the Immune System Malfunctions? p. 421*

2. Organ transplants are more successful when the donor is a relative of the recipient. Why?
   Textbook Reference: *18.7 What Happens When the Immune System Malfunctions? p. 421*

3. Vaccinations are effective against an array of different pathogens. How do they work? Vaccinations against HIV virus have not been effective in preventing the disease. Why not?
   Textbook Reference: *18.3 How Does Specific Immunity Develop? p. 409*

4. If the cytotoxic T cells were eliminated from a person's array of immune defenses, what kinds of disease would he or she be susceptible to?
   Textbook Reference: *18.5 What Is the Cellular Immune Response? p. 414*

Answers

Knowledge and Synthesis Answers

1. **d.** Phagocytes are nonspecific cells (not B and T cells) that digest nonself materials and present protein fragments of those nonself materials on their surface. They do not have antibodies.

2. **d.** B cells are antigen-presenting cells; they bind antigen, digest it, present fragments of that antigen on their MHC II proteins, and secrete antibodies.

3. **d.** Macrophages, natural killer cells, and complement proteins are all part of the nonspecific response of the immune system. B and T cells are part of the specific response of the immune system.

4. **b.** When the T_H cell binds antigen being presented on a macrophage, it secretes cytokines, which activate the T_H cell and B cells.

5. **a.** In a normal immune response, B memory cells are made so that the organism can mount a faster and more effective response to the pathogen if the organism ever encounters it again. Memory macrophages do not exist. Eosinophils do not secrete antibodies. In an abnormal immune response, B cells that attack the individual's own cells can be activated, as seen in autoimmune diseases.

6. **b.** T cells have cell surface receptors with a variable and a constant region and do not produce antibody molecules. Antibodies are not produced by macrophages. Only B cells produce antibodies in response to cytokines released from the T_H cell. Each antibody molecule has two identical heavy chains and two identical light chains, and each of those chains has a variable and a constant region.

7. **d.** Cytotoxic cells bind to virus antigen being presented on MHC I protein and destroy those cells. T_H cells release cytokines to activate B cells. The antibodies of B cells bind cell surface antigens on pathogens. Pathogens are lysed by complement proteins (and not by cytotoxic T cells), which are activated by antibodies bound to those pathogens.

8. **a.** The humoral response refers to that part of the specific response that releases antibodies (B cells) in the lymph and blood (the "humors" of the body). The cellular response is T cell receptors binding antigen on antigen-presenting cells. T cells do not secrete their receptors.

9. **c.** Autoimmunity occurs when the immune cells attack the body's own cells. Transplants can be rejected if the transplanted tissue is recognized by the body as nonself, and is not a result of immune cells attacking the body's own tissue. Clonal anergy suppresses the immune response to self antigens and thus works to prevent autoimmunity.

10. **c.** An HIV-infected individual is more susceptible to a variety of infections because the virus destroys T_H cells, which are essential for mounting an effective immune response. HIV does not bind to pathogens; it binds to T_H cells. HIV does not destroy B cells.

11. **d.** DNA rearrangements occur in B cells and result in each B cell expressing a unique kind of antibody. DNA rearrangements also occur in T cells to generate T cell receptors. This rearrangement does not destroy essential T cell genes. Memory B cells are cells in which the DNA rearrangement has already occurred.

12. **d.** MHC proteins present nonself antigens and self antigens to T cells. T cells that react with self antigens

are destroyed, preventing autoimmunity. The cytokines of T cells (and not MHC proteins) activate B cells. MHC proteins do not generate antibodies.

13. **c.** The B cells are lymphocytes, not granular cells. The basophils, neutrophils, and mast cells are all examples of granular cells (see Figure 18.2).

14. **a.** Inflammation occurs in response to an infection or tissue injury. Once an infection or injury occurs, the local mast cells release histamine. The histamine in turn makes the capillaries leakier, and fluid accumulates in the area to produce an inflammation.

15. **b.** When a B cell attaches to an antigen, it begins to divide and make clone cells. This provides more B cells that recognize the antigen. This process is known as clonal selection.

Application Answers

1. If the bacterial infection is not treated, the individual's immune response to this pathogen will include secreting antibodies that may be able to cross-react with the individual's own membrane proteins on the cells of the heart valve. Such antibodies could create an autoimmune response and heart disease.

2. Organ transplants are successful if the MHC proteins and other cell surface markers are similar, making the immune system more tolerant of the foreign tissue. Close relatives have more genes in common with the patient and may have more surface antigens on their cells that are similar. Organ transplants from unrelated individuals require the administration of immuno-suppressive drugs so that the patient's own immune system does not reject the tissue as foreign.

3. Vaccinations work against foreign pathogens because of B and T memory cells that are made during every immune response. If these memory cells are reexposed at a later time to the same antigen, they will mount a faster and more efficient immune response to that pathogen. One challenge in creating vaccines against HIV is that the virus mutates frequently (HIV uses reverse transcriptase, an enzyme prone to introducing genetic mutations, in its reproductive cycle). A vaccine that creates memory cells specific to a certain protein in one strain of HIV will not be effective against mutant HIV forms in which that protein has been modified if the memory cells do not recognize the modified protein.

4. Cytotoxic T cells target virally infected cells and some cancer cells. If cytotoxic T cells were eliminated, the individual would be much more susceptible to viral infections and cancer.

CHAPTER 19 Differential Gene Expression in Development

Important Concepts

Development includes four processes: growth, determination, differentiation, and morphogenesis.

- During development, an organism changes forms as it progresses through its life cycle. The embryo is the earliest stage of development.

- The developmental fate of a cell occurs during determination and is influenced by internal and external conditions. The fate of a cell is set by differential gene expression and morphogenesis.

- During differentiation, cells become specialized and provide specific structures or functions for the organism.

- Morphogenesis is the creation of form as seen in body shape and organs. It is a result of pattern formation during development, which helps direct differentiated tissues to form specific structures.

- Programmed cell death, known as apoptosis, is an important part of development in both plants and animals.

- Growth is an increase in size due to cell division and cell expansion. The early growth of plant embryos is characterized by cell expansion, whereas early growth of animal embryos is characterized by cell division.

- Transplantation experiments have shown that the environment in which early embryonic cells exist can redirect them to develop down different developmental paths.

- Later embryonic cells become committed to a particular developmental fate, even though they are not yet differentiated. This is because their fate has already been determined. When these cells are transplanted to other locations in the embryo, they continue to develop into the original differentiated tissue, regardless of their environment (see Figure 19.2).

- Cell fate becomes apparent as the cells differentiate.

Cells start out totipotent but lose this ability as they develop.

- Early embryonic cells are totipotent: they can develop into any of the different kinds of cells in the mature organism. The developmental possibilities narrow later in development as the cell fate becomes determined and the cell becomes fully differentiated.

- In plants, differentiation is reversible in some cells. Under the appropriate conditions, plant cells in culture can give rise to a new, genetically identical plant (a clone). They do this by producing a callus or mass of undifferentiated cells. Thus, these differentiated cells contain all the genetic information needed to express genes in the right sequence to generate a whole plant.

- Initially, the fertilized egg or zygote of an animal is totipotent and can give rise to every type of cell in the adult body. Cell descendants lose their totipotency as they become committed to a developmental path. A cell's fate first becomes determined, and then the cell differentiates to form specialized cells.

- The totipotency of early embryonic animal cells has been shown in experiments in which nuclei isolated from frog embryos were fused to enucleated eggs, resulting in the development of tadpoles and mature frogs. Such experiments show that cytoplasm has a large influence on the fate of a cell.

- The principle of genomic equivalence states that no information is lost from the nuclei of cells as they pass through the early stages of development.

- Whole mammals can be cloned by fusing early embryonic cells or somatic cells with enucleated eggs and implanting these eggs in surrogate mothers. The success of cloning mammals from somatic cells has been due to manipulating the donor cells so that they are positioned in G1 of the cell cycle. Upon fusing with the enucleated egg, the donor cells are stimulated by cytoplasmic factors in the egg to enter the S phase. Sheep, mice, and cattle have all been cloned.

Under the right environmental conditions, cells can become differentiated.

- The areas of undifferentiated cells in the growing tips of plant roots are known as meristems. These cells can become any cell type found in the root or stem.

- Undifferentiated dividing cells in animals are known as stem cells. Stem cells in adult animals are specific for the kinds of tissue they replace, typically skin, the lining of the intestines, and blood cells. They are pluripotent, meaning that they can differentiate into a limited number of cell types. However, environmental signals can induce stem cells from one tissue to replace cells that have been depleted from another tissue. For example, stem cells isolated from the mouse brain can replace bone marrow cells in mice that have been depleted of bone marrow.

- Embryonic stem cells are totipotent. They can be grown in culture for multiple generations, injected back into blastocysts, and then develop normally. In culture, the developmental fate of stem cells can be directed by the addition of environmental signals, such as growth factors or vitamins.

- Future medical applications include the use of embryonic stem cells to replace human cells that have been lost to disease or injury. The source of these cells could be from in vitro fertilized embryos. Therapeutic cloning is a method that would produce stem cells that are compatible with the immune system of the recipient.

Differentiated cells retain all their original genetic content, even though they are expressing only a very small subset of genes.

- Scientists use nucleic acid hybridization techniques to look for the presence of genes in cells. These experiments have shown that even though a specific gene is not expressed and not present as mRNA, it is still present in the nuclear DNA.

- Undifferentiated muscle cells (myoblasts) respond to the MyoD protein by fusing to form muscle fibers. This protein is a DNA-binding protein with a helix-loop-helix domain that acts as a transcription factor. When MyoD binds to promoters of muscle determination genes, transcription of those genes increases.

- When MyoD protein is injected into other precursor cells, such as fat cells, those cells are reprogrammed to develop into muscle cells.

Unequal distribution of cytoplasmic factors in eggs and zygotes sets up the initial pathways for precursor cells in developing embryos.

- Cytoplasmic segregation of factors in the egg results in an unequal distribution of maternal elements in each of the cells of the embryo.

- Unequal distribution of cytoplasmic determinants plays an important role in directing embryonic development and controlling the polarity of the organism. For example, if a sea urchin embryo is cut in half at the eight-cell stage and split from left to right by that cut, the resulting larvae are normal, but dwarfed. If the sea urchin is cut so that its upper half is split from its lower half, the cells from the upper half develop into abnormal ciliated larvae, and the cells from the lower half develop into misshapen larvae with an oversized gut (see Figures 19.9 and 19.10).

Tissues in developing embryos can induce other tissues to follow a particular developmental path.

- Cells can induce neighbor cells to differentiate by secreting induction factors.

- The vertebrate eye lens forms when surface cells make contact with a double bulging of the forebrain, known as the optic vesicles. If a barrier is placed between these two tissues in the developing embryo, no lens develops. If the forebrain is cut out before it can contact the lens surface cells, no lens develops. The optic vesicles produce an inducer signal, causing the surface cells to develop into a lens. The lens tissue produces embryonic inducers that control the size of the optic cups and the differentiation of the cornea (see Figure 19.11).

- In *C. elegans,* a common model organism, the egg develops into a larva in 8 hours and an adult in 3.5 days. The animal has a transparent body, making it easy to follow the developmental fate of the 959 somatic cells. The adult is hermaphroditic and has both female and male reproductive organs. Eggs are laid through a pore called the vulva. The vulva is induced to form from a single cell called an anchor cell. If this cell is laser ablated, no vulva develops.

- The anchor cell determines the fates of six cells on the *C. elegans* ventral surface by producing an inducer that diffuses and interacts with adjacent cells. The closest cell becomes the primary vulval precursor cell. It produces an inducer that causes the next two cells to become secondary vulval precursors, and the final three cells (which are the farthest out from the anchor cell) become epidermal cells (see Figure 19.12).

- The primary inducer for the development of the vulva in *C. elegans* is a growth factor, LIN-3, that is homologous to mammalian epidermal growth factor. LIN-3 binds to a receptor on the surface of the vulval precursor cell, which causes a signal transduction cascade in that cell through the Ras protein and MAP kinase proteins. The end result of this cascade is the differentiation of this cell into a vulval cell.

- Molecular switches, such as an inducer, allow a developing cell to proceed down one of two alternative pathways.

Pattern formation is the spatial organization of tissue that results in the appearance of body form (morphogenesis).

- During morphogenesis, some cells are programmed to die by death genes through the process of apoptosis.

- In *C. elegans,* there are a total of 1,090 cells produced. During development, 131 of these cells are programmed to die due to the sequential expression of *ced-4* and *ced-3* genes. *Ced-9* inhibits the actions of *ced-3* and *ced-4* during development.

- In human embryos, apoptosis occurs in the webs of skin that initially form between the fingers and toes so that they disappear by birth. Capsase, a protein homologous to *ced-3*, stimulates this apoptosis. The human homolog to *ced-9* is *bcl-2*.

- The genes controlling apoptosis are crucial for proper development in many organisms. These genes have been preserved in organisms separated by 600 million years of evolution (nematodes and humans).

- As a plant develops, it produces organs including leaves, roots, and flowers. Flowers have four types of organs: sepals, petals, stamens, and carpels. Flower organs occur in whorls organized around a central axis and are derived from meristematic tissue on the plant (see Figure 19.15).

- Plant geneticists have studied the development of flower organs in *Arabidopsis*. There are four whorls of organs in *Arabidopsis*. Three genes expressed in the whorls act as organ identity genes to guide differentiation in each whorl. Gene A is expressed in whorls 1 and 2, which form the sepals and petals; gene B is expressed in whorls 2 and 3, which form petals and stamens; and gene C is expressed in whorls 3 and 4, which form stamens and carpels.

- The three genes, A, B, and C, all encode transcription factors that are active as dimers. Gene regulation is combinatorial, and the combination of different dimers determines which genes are activated. A dimer of the transcription factor encoded by gene A will activate genes that make sepals. A dimer of transcription factor A with transcription factor B will result in petals.

- *Leafy* is a transcription factor that regulates the transcription of the genes A, B, and C. Plants bearing a mutation in *leafy* can produce leaves but no flowers.

- These homeotic genes, A, B, and C, as well as *leafy,* may be able to be genetically modified so that the flower organs can produce more fruit and seeds (which come from the carpels).

- Positional information allows developing cells to determine where they are in the developing organism. Morphogens are signals that establish positional information in the organism. Morphogens act directly on the target cell, and differential concentrations within the embryo cause differential effects.

- In the developing vertebrate limb, concentration gradients of the morphogen BMP2 determine the anterior–posterior axis of the developing limb. The action of such a morphogen works by producing differential gene expression in the target cells.

A cascade of transcription factors controls development in *Drosophila.*

- The body pattern of *Drosophila* is segmented, consisting of a head (which is formed from fused segments), three thoracic segments, and eight abdominal segments.

- The first step of development in *Drosophila* is to establish anterior–posterior and dorsal–ventral polarity. Polarity is based on the cytoplasmic distribution of mRNA and proteins produced in maternal genes. These maternal gene products are distributed to the eggs by the mother in a nonuniform manner to nurture the developing embryo. These maternal effects are not affected by the paternal genotype.

- Mutations in maternal genes (*bicoid* and *nanos*) create larvae lacking anterior structures (*bicoid*) or lacking abdominal segments (*nanos*). Bicoid encodes a transcription factor that positively affects some genes (*hunchback*) and negatively affects others. Nanos protein inhibits the translation of *hunchback.*

- Segmental genes are expressed when the embryo is at the 6,000-nuclei stage and determine the number, boundaries, and polarity of the segments. Three types of segmental genes are involved in regulation of segmental development.

- Gap genes are segmental genes whose products organize large areas along the anterior–posterior axis of the developing embryo. Gap mutants produce larvae that are missing certain larval segments. Pair rule gene products divide the larva into units of two segments each. Mutations in the pair rule genes produce larvae that are missing every other segment. Segment polarity gene products determine the boundary and the anterior–posterior organization of each segment. Mutant larvae lacking segment polarity genes produce larvae in which the posterior structures in the segments have been replaced by anterior structures.

- Gap gene products control the expression of pair rule genes, some of which are transcription factors that control segment polarity.

- Differences between segments are encoded by Hox genes, which give each segment an identity. For instance, thoracic segments produce legs.

- Homeotic mutants include *Antennapedia* (producing legs in place of antennae; see Figure 19.20) and *bithorax* (producing a fly with an extra set of wings). Homeotic genes are clustered on the chromosome in the order of the segments they determine. The first set of genes in the homeotic cluster are thoracic, the next genes are upper abdominal, and so on.

- The cluster of homeotic genes that determine the segment identity in the front of the fly (*Antennapedia*) probably have risen from duplications of an ancestral gene.

- Nucleic acid hybridization experiments confirm that certain clusters of genes contain DNA sequences common to all Hox genes. A DNA sequence, the homeobox, encodes a 60-amino acid sequence, the homeodomain. The homeodomain includes a DNA-binding motif and acts as a transcription factor.

The Big Picture

- The development of a mature differentiated organism from a fertilized egg involves the differential activation of genes in response to environmental signals. Initially in many organisms, positional determinants may be present in different regions of the egg due to maternal factors. The asymmetric distribution of these factors will activate genes differentially, and as cell division proceeds, that asymmetric gene expression continues. As a result, different cells experience different environmental signals based on their position in the developing embryo, and they respond by activating different genes. This differential gene activation determines the fate of the cell.

- Early embryonic cells are totipotent; they can develop into any structure in the adult organism. As development proceeds, the developmental potential of embryonic cells narrows. In the adult organism, differentiated cells express particular genes that give those tissues a particular structure and function, even though all the genes are still present in the nucleus of those cells.

- The concentration of specific genes and their products are involved in the regulation of polarity during development. These genes are classified as maternal effect genes, and their products are supplied by the mother.

Common Problem Areas

- Sometimes it is difficult to visualize how a single cell can divide and grow to produce a mature functional organism with highly differentiated tissues. At some very early point in development (either before fertilization or in one of the first set of divisions), different genes begin to be expressed in different cells. This gene expression can be in response to cytoplasmic factors or signals from other cells. Early gene expression sets up positional determinants that activate another wave of genes that further divides regions of the embryo into different developmental regions. As cell division proceeds in the embryo, these developmental genes continue to be differentially expressed, resulting in differentiation and morphogenesis in the organism.

Study Strategies

- Review the genes that are important in the development of *Drosophila*. Make a list of those genes and what their effects are on the phenotype of the *Drosophila* larva.

- Transplantation experiments help explain totipotency. Describe the transplantation experiments in frogs using early embryonic tissues and those using adult somatic cells to address the totipotency of those tissues.

- Review the experiments that demonstrated how tissue induction causes the lens in the vertebrate eye to form.

- Review some of the important genes whose products direct development in the organisms discussed in this chapter. Identify homologs from different organisms.

- Review the following animated tutorials on the Companion Website/CD:

Tutorial 19.1 Embryonic Stem Cells
Tutorial 19.2 Early Asymmetry in the Embryo
Tutorial 19.3 Pattern Formation in the *Drosophila* Embryo

Test Yourself

Knowledge and Synthesis Questions

1. Cell differentiation is a result of
   a. the loss of particular genes from the nucleus of the differentiated cell.
   b. the differential expression of genes that are responsive to environmental signals.
   c. early embryonic cells remaining totipotent in the mature organism.
   d. mutations in genes that control the synthesis of DNA.

   Textbook Reference: 19.1 What Are the Processes of Development? p. 428

2. A totipotent cell is a cell
   a. whose developmental fate has been decided.
   b. that has differentiated into a specialized tissue.
   c. that is expressing morphogens to form a particular structure.
   d. whose developmental potential is extremely broad.

   Textbook Reference: 19.2 Is Cell Differentiation Irreversible? p. 429

3. Apoptosis is programmed cell death and
   a. only occurs in developing embryos.
   b. only occurs in the mature organism.
   c. is an important process in both development and in mature organisms.
   d. can cause a type of cancer; follicular large-cell lymphoma.

   Textbook Reference: 19.5 How Does Gene Expression Determine Pattern Formation? p. 439

4. The fate of a cell describes
   a. the cell's original type.
   b. its genes.
   c. the type of cell it will differentiate into.
   d. its death.

   Textbook Reference: 19.2 Is Cell Differentiation Irreversible? p. 429

5. Mammals can be cloned using somatic cells fused to enucleated eggs if the
   a. donor cell is in the S phase of the cell cycle.
   b. donor cell is in the G1 phase of the cell cycle.
   c. mammalian embryo can be propagated in culture.
   d. somatic cell is a red blood cell.

   Textbook Reference: 19.2 Is Cell Differentiation Irreversible? p. 431

6. Identify the function of the protein MyoD.
   a. It is a transcription factor that controls the expression of genes that form muscle.
   b. It controls segment identity in mice.
   c. It is a transcription factor that regulates its own transcription.

d. Both a and c

Textbook Reference: 19.3 What Is the Role of Gene Expression in Cell Differentiation? p. 435

7. Tissue induction occurs when
   a. nuclear genes are lost in some tissues.
   b. signals are secreted in one part of the developing embryo and are transported via the circulatory system to another site in the embryo to activate tissue differentiation.
   c. direct contact from one tissue delivers an environmental signal to the tissue that will be induced.
   d. two tissues come into contact with one another, releasing transcription factors.

Textbook Reference: 19.4 How Is Cell Fate Determined? p. 437

8. Which of the following statements about plant development is *not* true?
   a. Both cell division and cell expansion contribute to growth in early plant embryos.
   b. Differentiation in plant cells is irreversible.
   c. Meristems at the tips of the roots and stems consist of undifferentiated cells.
   d. Flower development in a plant involves organ identity genes.

Textbook Reference: 19.1 What Are the Processes of Development? p. 428; 19.2 Is Cell Differentiation Irreversible?, p. 430, 432; 19.5 How Does Gene Expression Determine Pattern Formation? pp. 440–441

9. It may be possible to genetically engineer plants to produce more seeds and fruits by
   a. fertilizing them more in the spring.
   b. manipulating their gap genes.
   c. altering the homeotic genes that control whorl development.
   d. inducing a mutation that eliminates *leafy* gene function.

Textbook Reference: 19.5 How Does Gene Expression Determine Pattern Formation? p. 441

10. Morphogens
    a. diffuse within the embryo to set up a concentration gradient.
    b. are expressed as positional signals in developing embryos.
    c. direct differentiated cells to form organs.
    d. All of the above

Textbook Reference: 19.5 How Does Gene Expression Determine Pattern Formation? p. 441

11. Maternal effect genes
    a. set up positional axes in the egg prior to fertilization.
    b. are mRNA and proteins that are produced by the egg cell.
    c. are masked by the paternal genes.
    d. Both a and b

Textbook Reference: 19.5 How Does Gene Expression Determine Pattern Formation? p. 442

12. Hox genes
    a. encode protein domains that are important in development and that have been highly conserved over evolutionary time.
    b. are found in greatly diverse organisms.
    c. can produce the wrong structure in the wrong place when mutated.
    d. All of the above

Textbook Reference: 19.5 How Does Gene Expression Determine Pattern Formation? pp. 444–445

13. What principle of development states that the nuclei of cells do not lose any information during the early stages of development?
    a. Genomic equivalence
    b. Apoptosis
    c. Totipotent
    d. Transcription

Textbook Reference: 19.2 Is Cell Differentiation Irreversible? p. 430

14. What is the role of cytoplasmic segregation in determining the fate of a cell?
    a. It keeps cells totipotent.
    b. It determines polarity within the organism.
    c. It ensures that gradients do not develop in the developing organism.
    d. All of the above

Textbook Reference: 19.4 How Is Cell Fate Determined? p. 436

15. Organ identity genes are found in what type of organism?
    a. Bacteria
    b. Animals
    c. Plants
    d. All of the above

Textbook Reference: 19.5 How Does Gene Expression Determine Pattern Formation? p. 441

Application Questions

1. Experiments were done that showed fully differentiated cells still contained all their genes, even though they were only expressing a small set of those genes. What kinds of hybridizations demonstrated this?

Textbook Reference: 19.3 What Is the Role of Gene Expression in Cell Differentiation? p. 435

2. Undifferentiated adult stem cells are pluripotent. Describe the experiment that clearly demonstrated this fact.

Textbook Reference: 19.2 Is Cell Differentiation Irreversible? p. 433

3. Transcription factors that control the fate of undifferentiated stem cells can be used to redirect other committed stem cells. Describe the experiment that showed this.

Textbook Reference: 19.3 What Is the Role of Gene Expression in Cell Differentiation? p. 435

4. In cloning whole organisms by fusing nuclei (as seen in frogs) or whole cells (as seen in sheep, cows, and goats) with enucleated eggs, why is the nucleus removed from the egg?

   Textbook Reference: *19.2 Is Cell Differentiation Irreversible? pp. 431–432*

Answers

Knowledge and Synthesis Answers

1. **b.** Genes are not lost from differentiated cells. There are no embryonic cells in a mature organism. Mutations in cells that affect DNA replication would result in cells that were unable to replicate their DNA.

2. **d.** A totipotent cell is a cell that can develop in any number of ways; its fate has not been determined, it has not differentiated, and it is not expressing morphogens.

3. **c.** Apoptosis is important for embryonic development and in adult tissues such as skin and blood cells.

4. **c.** The fate of a cell is the type of cell it will eventually become once it has differentiated.

5. **b.** The donor somatic cell must be in G1. Mammalian embryos must be grown in surrogate mothers. Red blood cells in mammals have no nuclei and could not be used for cloning.

6. **d.** The gene *MyoD* encodes a transcription factor (MyoD) that controls the expression of genes that form muscle in mice and also regulates its own transcription. *MyoD* does not control segment identity in mice. Homeotic genes control segment identity in *Drosophila.*

7. **c.** Induction involves contact between two different tissues, with one tissue providing some kind of environmental signal that is received by the other tissue. Nuclear genes are not lost in developing cells (with the exception of red blood cells and the immune cells). Tissue induction occurs because of tissue contact, not because a signal has been carried from one part of the embryo's body by circulation to another part of its body. Transcription factors are found in the nucleus and are not released to adjacent tissues.

8. **b.** In general, it is easier to reverse differentiation in plant cells than it is in animal cells, as the cloning of a carrot from a differentiated storage cell in its root demonstrates.

9. **c.** Fertilization is not genetic engineering. Gap genes are found in *Drosophila,* not plants. The *leafy* protein transcriptionally activates the homeotic genes that control organ development (and subsequent seed and fruit formation). In its absence, no fruit or seeds would develop.

10. **d.** Morphogens diffuse in the embryo to form concentration gradients that act as positional signals to direct differentiated tissues to form organs.

11. **a.** The paternal genotype does not affect the expression of maternal gene products in the egg. Additionally, the mRNA is placed in the egg when it is made and not by the developing embryo.

12. **d.** Hox genes encode proteins that are highly conserved and important for directing development in many different organisms. Mutations in these genes can produce an abnormal developmental event in the organism.

13. **a.** Genomic equivalence is a fundamental principle of developmental biology and states that none of the genetic information contained in the original zygote is lost during subsequent cell division and growth.

14. **b.** Cytoplasmic segregation is the concept that factors within the developing embryo are unequally distributed. This helps set up, for example, the polarity of the developing organism.

15. **a.** Organ identity genes are found in plants and they help to direct the development of structures such as flowers.

Application Answers

1. Hybridizations were done using a specific DNA probe to nuclear DNA and mRNA. If the DNA probe was complementary to a gene that was not being expressed in the cell, it would not hybridize to the mRNA isolated from that cell, but would still hybridize to the nuclear DNA. This indicated that the gene was still present in the nucleus of the cell, even though it was not being expressed in the cytoplasm.

2. Undifferentiated stem cells from mouse brain were transplanted to the bone marrow of mice whose bone marrow had been depleted. The brain cells differentiated into bone marrow cells, indicating their pluripotency.

3. The MyoD transcription factor from muscle cells directs the transcription of several muscle genes, resulting in the fusion of myoblasts to form muscle cells. When MyoD was applied to precursors of fat cells, it redirected those cells to form muscle cells.

4. The nucleus of the egg contains a full haploid set of chromosomes. Fusing a diploid somatic cell with an egg cell that still contained a nucleus would result in a cell that had three copies of each chromosome and little chance of developing normally.

CHAPTER 20 Development and Evolutionary Change

Important Concepts

Organisms have a molecular tool kit that directs development.

- Many of the genes involved in controlling development are conserved across different organisms. Thus, these genes can be very similar in organisms that are very different in form. The major differences in body form arise because the genes are turned on and off at different times and locations during development.

- The organism's genes and transcription factors, along with the way they interact with extracellular signals, can be thought of as a molecular took kit used to assemble the organism.

- There has been conservation of the genes involved in regulating development in plants and animals. As an example, the development of an anterior–posterior axis in insects and mammals is due to the same types of homeobox-containing genes being expressed at either the anterior or posterior end of the developing embryo.

Changes in gene expression of homologous genes can be localized to single modules.

- The many parts of the developing embryo can change independently for each other because the organism is made up of modules. A module such as the heart can change independently of the other modules in the developing embryo because genes are expressed independently within each module.

- During development, the spatial development in the embryo is controlled by genetic switches. The pattern of development depends on the Hox genes that are expressed within the module. This can be seen in the development of wings in *Drosophila* where Hox genes are differentially expressed in the three thoracic segments of the embryo.

- Heterochrony occurs when the relative timing of developmental processes independently shift in different species and occurs because of the modularity of the developing embryo. The presence or absence of webbed feet in birds is an example of genetic switches acting at different times on a single module.

- Evolutionary changes in morphology have occurred due to the change in the timing or spatial expression of regulatory genes during development. The insect *Ultrabithorax* (*Ubx*) homeotic gene has a mutation that results in the repression of the gene involved in limb formation, the *distal-less* gene. The *Ubx* gene is expressed in the abdomen during development leading to the absence of limbs on the abdomen. Other arthropods, such as millipedes, do not have this mutation and have abdominal legs. Similar patterns of gene expression can be observed in the development of the vertebral column.

The environment can have a large influence on the development of plants and animals.

- Organisms have the ability to exhibit different phenotypes based on a single genotype. The phenotype is dependent on the environment in which the organism develops and lives. This response is known as developmental plasticity.

- Environmental signals can be either accurate predictors of the future environment or poorly correlated with the future environment. Environmental signals such as day length, temperature change, and extent of rainfall are accurate predictors of future conditions. Organisms take advantage of developmental plasticity to respond to environmental signals during development.

- Temperature can influence the coloration of the adult butterfly. The *distal-less* gene is involved in the development of wing eyespots. The temperature at which the pupae develop dictates whether the *distal-less* gene is expressed. Another example involves butterflies that produce two generations during the summer. Each generation must be ready for a different environment. Environmentally controlled development allows organisms to develop according to the environmental conditions that will be present when they become adults.

- Some species respond to specific organisms present in their environment. The body and head sizes of the venomous tiger snake of Australia are dependent on the prey species present as the animals grow.

- The environment of organisms can also change in an uncertain manner. In the presence of predators, many species of *Daphnia* exhibit developmental plasticity by growing large helmets that are thought to increase their survival. The induction of a large helmet has a trade-off in a decreased number of eggs.

- In plants, low light levels result in elongation of cells. This allows a plant to grow faster and potentially reach a sunny patch.

- Some environmental signals are poorly correlated with the future environment that an organism will experience. Plant seeds tend to remain constant in size and not change with the environment because of the unpredictable year-to-year variation in the environment.

- Organisms will most likely be unable to respond to environmental changes to which they have never been exposed. Today the normal development of organisms in the wild is being disturbed by environmental changes brought about by humans.

Evolution is constrained by existing structures and highly conserved genes.

- Evolution work on existing genes and their expression and can occur through changes is Hox gene expression in different modules.

- Similar traits evolve repeatedly because of the highly conserved nature of the genetic code. This can lead to parallel phenotypic evolution of a trait such as the evolutionary loss of body armor in numerous species of freshwater stickleback fish.

The Big Picture

- Evolutionary developmental biology is the study of how organisms evolved by examining the differences in developmental gene expression and regulation between species. Animals have similar conserved genes, such as homeobox genes, that control development. Heterochrony describes how differences in organisms have arisen because of differences in the timing of different developmental processes. The timing of gene expression can differ between organisms because of the modular makeup of animals.

- Plants share similar regulatory genes with animals but also have a large number that are specific to controlling their development. Plants differ from animals in a number of ways that dictate developmental processes.

- Developmental plasticity occurs in response to environmental signals. These signals can either provide an accurate prediction of the future or be poorly correlated with future conditions.

Common Problem Areas

- The concept of heterochrony can be difficult for students to grasp. Try to remember that when we discuss heterochrony, we are talking about differences between species, not within species.

Study Strategies

- This chapter introduces many concepts. Try to learn the concepts involved in evolutionary developmental biology first. Then go back and see how the examples fit with the concepts.

- Review the following animated tutorial and activity on the Companion Website/CD:
  Tutorial 20.1 Modularity
  Activity 20.1 Plant Development

Test Yourself

Knowledge and Synthesis Questions

1. Genes that regulate development are conserved. This means that they
   a. have evolved large differences among multicellular organisms.
   b. have changed very little over the course of evolution.
   c. are always turned on.
   d. None of the above
   Textbook Reference: 20.5 How Do Developmental Genes Constrain Evolution? p. 457

2. _____ genes are involved in the development of the anterior–posterior axis in both humans and insects.
   a. Developmental plasticity
   b. Heterochrony
   c. Hox
   d. All of the above
   Textbook Reference: 20.1 How Does a Molecular Tool Kit Govern Development? p. 450

3. Heterochrony
   a. relates development in at least two species.
   b. involves the relative timing of developmental processes.
   c. has been extensively studied in salamanders.
   d. All of the above
   Textbook Reference: 20.2 How Can Mutations with Large Effects Change Only One Part of the Body? p. 451

4. Module development has allowed for evolutionary changes to occur while still resulting in a viable organism because
   a. all modules must change together.
   b. modules can change independently of one another.
   c. modules are unimportant for development.
   d. None of the above
   Textbook Reference: 20.2 How Can Mutations with Large Effects Change Only One Part of the Body? p. 451

5. Which of the following is *not* considered part of the molecular tool kit that governs development?
   a. Transcription factors
   b. An organism's genes
   c. Environmental signals
   d. All of the above
   Textbook Reference: 20.1 How Does a Molecular Tool Kit Govern Development? p. 449

6. Developmental plasticity is
   a. the ability to express different genotypes under different environmental conditions.
   b. found only in plants.
   c. the ability to express only one phenotype under different environmental conditions.
   d. None of the above

   Textbook Reference: 20.4 How Does the Environment Modulate Development? p. 454

7. Which of the following is *not* a developmental environmental signal?
   a. Temperature
   b. Light
   c. The presence of predators
   d. All of the above are developmental environmental signals.

   Textbook Reference: 20.4 How Does the Environment Modulate Development? pp. 454, 456

8. Genetic switches are involved in
   a. stopping development.
   b. helping various modules develop differently.
   c. ensuring that all modules develop into the same type.
   d. None of the above

   Textbook Reference: 20. 2 How Can Mutations with Large Effects Change Only One Part of the Body? p. 450

9. What triggers the difference in the different appearance of the *Nemoria arizonaria* caterpillars in spring and summer?
   a. Temperature
   b. Length of day
   c. Compounds in their food
   d. None of the above

   Textbook Reference: 20.4 How Does the Environment Modulate Development? p. 455

10. Yearly difference in rainfall is unknown to a plant producing seeds for the next year. Because of this, plants produce
    a. seeds of a constant size.
    b. seeds of varying sizes.
    c. no seeds.
    d. None of the above

    Textbook Reference: 20.4 How Does the Environment Modulate Development? pp. 456–457

Application Questions

1. There are morphological differences between chickens and ducks that are related to the duck's aquatic lifestyle. Why were these adaptations able to occur without influencing the development of the rest of the duck's morphology?

   Textbook Reference: 20.2 How Can Mutations with Large Effects Change Only One Part of the Body? p. 451

2. Day length and temperature are both potential environmental signals that may influence development. What do changes in these two environmental factors tell an organism about future conditions?

   Textbook Reference: 20.4 How Does the Environment Modulate Development? pp. 454–455

3. Why do insects have only six legs, whereas millipedes have legs on each body segment?

   Textbook Reference: 20.2 How Can Mutations with Large Effects Change Only One Part of the Body? p. 451

Answers

Knowledge and Synthesis Answers

1. **b.** Conserved genes are genes that are found in many organisms and have undergone very little change.

2. **c.** Hox genes regulate development along the anterior–posterior axis in animals.

3. **d.** Heterochrony is the process by which the timing of gene expression differs between two or more species. Heterochrony has been studied extensively in salamanders.

4. **b.** Embryos are composed of self-contained units called modules that can change independently of one another. This allows for the evolution of one module without disrupting the other modules of the animal.

5. **d.** An organism's molecular tool kit includes transcription factors, the organism's genes, and environmental signals. All of these contribute to the development of an organism.

6. **a.** Developmental plasticity is found in both plants and animals and allows a developing organism to express different phenotypes, depending on the environment.

7. **d.** Environmental signals are physical or biological properties of the environment that help organisms anticipate changes. Light, predators, and temperature are all environmental signals that some organisms respond to by altering their development.

8. **b.** Genetic switches help each module develop into its appropriate form. They work by turning on and off Hox genes.

9. **c.** The *Nemoria arizonaria* moth population goes through two generations each year. The morphology of the caterpillar is controlled by the presence or absence of a chemical in the oak leaves.

10. **a.** Because current rainfall is an inaccurate predictor for the following season's rainfall, plants will produce seeds each year that are fairly constant in size.

Application Answers

1. Organisms are made up of modules. Change can occur in a module independent of other modules. The webbing found in duck feet is due to changes in the timing of *gremlin* gene expression in the foot module.

2. Both day length and temperature are accurate predictors of the future environment. Changes in day length act as an environmental signal that accurately

predicts when future conditions will change. Shorter days signify that fall and winter are approaching. Similarly, changes in temperature occur with changes in the seasons and can signal to an organism that a change in the season will occur soon.

3. Leg development in arthropods involves the homeotic gene *Ultrabithorax* (*Ubx*) and the gene *distal-less* (*dll*).

The expression of *dll* in a segment results in the development of a pair of legs in that segment. In insects, the *Ubx* gene has a mutation that suppresses the expression of *dll* in the abdomen. As a result, no legs will develop in the abdomen. Millipedes do not have this mutation, therefore legs develop in all body segments.

CHAPTER 21 The History of Life on Earth

Important Concepts

Evolutionary changes may occur over both short and long time frames.

- The development of the science of biology, particularly Darwin's theory of evolution by natural selection, depended on evidence supplied by geologists that Earth is very old.
- Short-term evolutionary changes occur rapidly enough to be studied directly.
- Long-term evolutionary changes involve the appearance of new species and evolutionary lineages. The fossil record provides evidence of such changes.

Several types of evidence have been used to estimate the age of rocks, of fossils, and of Earth itself. All the evidence indicates that both Earth and life on Earth are very old.

- Fossils, which are the remains of ancient organisms, are useful for establishing the relative ages of the sedimentary rocks in which they occur.
- The regular pattern of decay of radioactive isotopes provides a means of estimating the absolute ages of fossils and rocks. The ratio of radioactive ^{14}C to its stable isotope, ^{12}C, is used to date fossils less than 50,000 years old. The decay of ^{40}K to ^{40}Ar is widely used to date ancient evolutionary events.
- Sedimentary and igneous rocks can also be aged because they preserve a record of Earth's magnetic field at the time they were formed.
- Earth's geological history is divided into eras, which are subdivided into periods. The boundaries between these divisions are marked by changes in the types of fossils found in successive layers of sedimentary rock.
- The first forms of life evolved early in the Precambrian era, which lasted for more than three billion years and was marked by stupendous physical changes on Earth.

The history of Earth has been marked by unidirectional and oscillatory changes, as well as by occasional collisions with extraterrestrial objects.

- Earth's crust consists of solid lithospheric plates floating on a fluid layer of magma. Convection currents in the magma result from heat emanating from Earth's core. The currents cause continental drift, which is the gradual shift in the position of the plates and the continents they contain. The drifting of continents has had profound effects on climate, sea level, oceanic circulation, and the distributions of organisms.
- One major unidirectional change has been the increase in concentration of oxygen in Earth's atmosphere. No free oxygen was present in the atmosphere until certain bacteria evolved the ability to use water as a source of hydrogen ions for photosynthesis. The gradual increase in oxygen concentration in the atmosphere resulted in the dominance of organisms using aerobic metabolism and in the evolution of larger eukaryotic cells and of multicellular organisms.
- The climate of Earth has alternated between hot/humid and cold/dry conditions. Though most major climatic changes have been gradual, some have occurred within periods of 5,000 to 10,000 years or less. Rapid climate change is occurring today because of a buildup of atmospheric CO_2, primarily from the burning of fossil fuels.
- The climatic shifts caused by massive volcanic eruptions associated with continental drift are implicated in several mass extinctions. The late Permian mass extinction, which occurred at the end of the Paleozoic era was caused by volcanism triggered by the collision of continents that formed the supercontinent Pangaea.
- The mass extinction that occurred in the late Cretaceous period at the end of the Mesozoic era was probably caused by the collision of Earth with a large meteorite.

The fossil record reveals patterns in life's history.

- The earliest life on Earth appeared about 3.8 billion years ago, but the fossil record of organisms that lived in the Precambrian is fragmentary.
- The first organisms were unicellular prokaryotes. The first eukaryotes evolved about 1.5 billion years ago.
- Because an oxygen-rich environment favors rapid decomposition, organisms that become fossils are likely either to have lived in a poorly oxygenated

environment or to have been transported to such a site soon after death.

- The 300,000 known fossil species represent only a tiny fraction of the species that have ever lived. Some groups, such as hard-shelled marine animals, are much better represented in the fossil record than others.

- The fossil record shows that an organism of any specific type can predictably be found in rocks of a particular age.

- Living organisms resemble recent fossils more closely than they resemble fossils from more ancient periods.

- The best Precambrian fossil deposits, dating from about 600 million years ago, contain diverse soft-bodied invertebrates, some of which may represent lineages with no living descendents.

Beginning with the Cambrian period, the Paleozoic era (542–251 mya) was marked by a dramatic increase in the diversity of life.

- During the Cambrian explosion, most of the major groups of animals that have living representatives appeared. Several continental plates began to come together to form the southern land mass known as Gondwana.

- The Ordovician period was marked by a proliferation of marine filter feeders living on the sea floor. At the end of this period, massive glaciers formed over Gondwana, the sea level and ocean temperatures dropped, and the majority of animal species became extinct.

- The Silurian period witnessed the aggregation of northern continents, the evolution of swimming marine animals, and the appearance of the first terrestrial arthropods and vascular plants.

- During the Devonian period, all major groups of fishes evolved. On land, the first insects and amphibians evolved, and forests of club mosses, horsetails, and tree ferns appeared. A mass extinction of approximately three-quarters of all marine species occurred at the end of this period, possibly caused by the collision of two large meteorites with Earth.

- In the Carboniferous period, swamp forests largely consisting of giant tree ferns and horsetails were widespread. The fossilized remains of these plants formed coal. The first winged insects evolved during this period, while amphibians became better adapted to life on land and gave rise to the lineage leading to the amniotes.

- The Permian period was marked by the formation of a single supercontinent called Pangaea. As the climate cooled drastically, reptiles became the dominant land vertebrates, and the lineage of reptiles from which mammals later evolved became distinct. The occurrence of the most extensive of all mass extinctions brought the Permian period to a close.

During the Mesozoic era (251–65 mya), the continents that formed Pangaea slowly separated to form Laurasia and Gondwana, and distinct assemblages of plants and animals evolved on each continent.

- During the Triassic period, pteridosperms and conifers were the dominant forms of terrestrial vegetation, and reptiles were the dominant vertebrates. A mass extinction occurred at the end of the Triassic.

- The Jurassic period witnessed the radiation of the dinosaurs, the evolution of flying reptiles, and the first appearance of mammals. In the sea, ray-finned fishes also began a great radiation.

- The Cretaceous period was marked by the complete separation of the northern and southern continents. The first flowering plants evolved and began their radiation. The dinosaurs continued as the dominant land vertebrates, despite the presence of many groups of mammals. At the end of this period, the collision of a large meteorite with Earth caused the extinction of the dinosaurs and of many other animal and plant lineages.

Modern groups of plants and animals evolved during the Cenozoic era (65 mya–present).

- During the Tertiary period, the continents drifted toward their present positions. As the climate became cooler and drier, extensive grasslands appeared. Many groups of land vertebrates—especially snakes, lizards, birds, and mammals—radiated extensively.

- The current geological period, the Quaternary, is divided into the Pleistocene and Holocene (recent) epochs. Modern humans evolved during the Pleistocene, which was a time of severe climatic fluctuations, including four major periods of extensive glaciation. As humans spread geographically, many species of large birds and mammals became extinct.

Three major faunas have dominated life on Earth.

- Life on Earth has been dominated by three major animal assemblages resulting from periods of rapid evolutionary diversification. The first was the fauna stemming from the Cambrian explosion; the second fauna dominated the later Paleozoic; and the third originated from the Triassic explosion and resulted in our present-day fauna. Organisms with new body plans appeared in the Cambrian explosion, but not in the others.

The rate of evolutionary change has differed greatly at different times and among different lineages.

- Changes in the physical or biological environment of a species are likely to stimulate evolutionary change.

- Conversely, relatively unchanging environments favor evolutionary stability in the species of organisms ("living fossils") that are adapted to them.

- In most lineages, fossil evidence shows that evolutionary changes have occurred slowly and gradually.

- Though species have become extinct throughout the history of life, extinction rates have varied widely, being relatively low during normal times and very high during periods of mass extinction. Some mass extinctions have affected both terrestrial and marine organisms, whereas others have primarily affected either one group or the other.

- Among animals, large, specialized carnivores are more vulnerable to extinction than smaller species with more generalized diets.

The causes of evolution that have operated in the past continue to operate, but humans are a new and dominant evolutionary agent.

- By changing Earth's vegetation, importing species into regions where they did not previously live, and manipulating the evolution of certain valuable species, humans are having an enormous influence on the future of life on Earth.

The Big Picture

- The evidence provided by nineteenth-century geologists that Earth is very ancient was crucial to Darwin's theory of evolution by natural selection.

- Radiometric dating is an excellent example of how a discovery in one field of science (nuclear physics, in this instance) can lead to advances in other fields (geology and paleontology). Prior to the development of this technique, no absolute dates could be accurately assigned to any of the important geological and evolutionary events described in this chapter.

- The concept of continental drift, first widely accepted in the late 1960s, has revolutionized our understanding of geological processes and of the history of life on Earth.

- The accumulating evidence of the sudden and catastrophic effect of the impact of large meteorites and of other rapid changes (e.g., widespread volcanic eruptions) on the environment of Earth has given paleontologists a new perspective on the causes of mass extinctions and on the dynamics of the evolutionary process in general.

Common Problem Areas

- The concept of the half-life of radioactive isotopes may be confusing to you. Bear in mind that if half of a radioisotope decays in a given time period (its half-life), then at the end of the first half-life only one-half of the original quantity of the isotope is still present, one-half of which will decay in the second half-life. Thus after two half-lives, one-quarter of the original quantity of radioisotope is still present. The same line of reasoning applies to all further half-lives. Study Figure 21.1 to be sure you grasp this point.

Study Strategies

- This chapter contains many names of geological eras and periods and many dates. Instructors vary with respect to the importance they place on memorization of this information. In general, it is best to focus on broad evolutionary patterns and trends in studying this material. Most instructors will also place more importance on your learning phylogenetic sequences (e.g., mammals evolved from reptiles, which evolved from amphibians) than on your memorizing the dates at which these events occurred.

- Making a table or chart to summarize and organize the events that occurred in each geological period is an excellent study strategy.

- Review the following animated tutorial and activity on the Companion Website/CD:
Tutorial 21.1 Evolution of the Continents
Activity 21.1 Concept Matching

Test Yourself

Knowledge and Synthesis Questions

1. The half-life of an isotope is best defined as the
   a. time a fixed fraction of isotope material will take to change from one form to another.
   b. age over which the isotope is useful for dating rocks.
   c. ratio of one isotope species to another in a sample of organic matter.
   d. None of the above
   Textbook Reference: 21.1 How Do Scientists Date Ancient Events? p. 466

2. Mountain ranges are ultimately the result of
   a. plates in Earth's crust that move against one another on top of a fluid layer of molten rock.
   b. climate changes and the movement of glacial ice sheets.
   c. leftover debris from ancient collisions with an asteroid or meteor.
   d. the breakup of Laurasia and Gondwana.
   Textbook Reference: 21.2 How Have Earth's Continents and Climates Changed over Time? p. 468

3. Which of the following would likely be the best setting for fossilization to occur?
   a. The surf zone along a sandy beach
   b. A shallow, cool swamp with good deposition rates of mud sediments
   c. The bottom of a hot, dry cave with no running water
   d. A fast-running mountain stream
   Textbook Reference: 21.3 What Are the Major Events in Life's History? pp. 472–478

4. Despite being incomplete as a whole, the fossil record is rather detailed for
   a. soft-bodied insects.
   b. cnidarians and sponges.
   c. most terrestrial animals.
   d. hard-shelled mollusks.
   Textbook Reference: 21.3 What Are the Major Events in Life's History? p. 473

5. When biologists say that Earth's biota became provincialized, they mean that
   a. mass extinctions took place.
   b. continental drift did not occur.
   c. distinctive assemblages of plants and animals arose on different continents.
   d. reductions in species diversity took place.
   Textbook Reference: 21.3 What Are the Major Events in Life's History? p. 473

6. Which of the following did *not* occur during the Ordovician period?
   a. Marine filter feeders flourished.
   b. The number of classes and orders increased.
   c. Modern mammals appeared.
   d. Many groups became extinct at the end of the period.
   Textbook Reference: 21.3 What Are the Major Events in Life's History? p. 473

7. Most of human evolution has occurred during the
   a. Paleozoic era.
   b. Devonian period.
   c. Quaternary period.
   d. Carboniferous period.
   Textbook Reference: 21.3 What Are the Major Events in Life's History? p. 478

8. For terrestrial animals and plants, the most recent mass extinction event that occurred prior to the evolution of humans took place approximately _____ million years ago.
   a. 10
   b. 65
   c. 220
   d. 400
   Textbook Reference: 21.3 What Are the Major Events in Life's History? p. 477

9. One of the main factors that distinguishes the Cambrian explosion from all others is that
   a. evolutionarily, it was the most recent explosion.
   b. many new major groups of animals appeared at this time, but not during other explosions.
   c. it was the time when the dinosaurs became extinct.
   d. there was a dramatic drop in species diversity, especially among marine organisms.
   Textbook Reference: 21.3 What Are the Major Events in Life's History? pp. 472–473

10. During which of the following geological times did the most new kinds of body plans appear?
    a. Carboniferous
    b. Triassic
    c. Jurassic
    d. Cambrian
    Textbook Reference: 21.3 What Are the Major Events in Life's History? p. 472

11. The sudden disappearance of the dinosaurs some 65 mya may have been the result of
    a. Earth's collision with a large meteorite.
    b. slow climate changes due to planetary cooling.
    c. competition from better-adapted organisms.
    d. the rise of birds and mammals.
    Textbook Reference: 21.3 What Are the Major Events in Life's History? p. 476

12. Which of the following statements about evolution is *not* true?
    a. Some species living in stable environments appear to have changed very little over many millions of years.
    b. Species have become extinct throughout the history of life.
    c. The fossil record of trilobites provides a good example of gradual evolutionary change within a lineage of organisms.
    d. There is no evidence in the fossil record for periods of rapid evolutionary change.
    Textbook Reference: 21.4 Why Do Evolutionary Rates Differ among Groups of Organisms? p. 480

13. Which of the following statements about patterns or processes in the evolution of life is *not* true?
    a. ^{14}C can be used to date the age of dinosaur bones.
    b. The supercontinent Pangaea formed during the Permian period.
    c. Most of the major groups of animals that have species living today appeared in the Cambrian period.
    d. Rates of evolutionary change can be rapid during times of dramatic change in physical environments.
    Textbook Reference: 21.1 How Do Scientists Date Ancient Events? p. 466–467

14. Which of the following pairs of organisms were *not* present on Earth in their living forms at the same time?
    a. Tree ferns and ray-finned fish
    b. Amphibians and birds
    c. Gymnosperms and insects
    d. Humans and dinosaurs
    Textbook Reference: 21.3 What Are the Major Events in Life's History? p. 476–477

15. The most severe mass extinction event, linked to the formation of Pangaea and massive volcanic eruptions, occurred at the end of the _____ period.
    a. Ordovician
    b. Devonian
    c. Permian
    d. Cretaceous
    Textbook Reference: 21.3 What Are the Major Events in Life's History? p. 475

Application Questions

1. ^{14}C decays to ^{12}C with a half-life of about 5,700 years. Suppose you find a fossil in which the amount of ^{14}C is only one-sixteenth of what one would find in a living organism with the same carbon content. Approximately how old is the fossil?
   Textbook Reference: 21.1 How Do Scientists Date Ancient Events? p. 466

2. For each geological period, make a chart summarizing its major geological and evolutionary events.
   Textbook Reference: 21.1 How Do Scientists Date Ancient Events? p. 466–467

3. In what way was the evolution of eukaryotic cells linked to the increase in the oxygen concentration in the atmosphere that occurred during the Precambrian?
   Textbook Reference: 21.2 How Have Earth's Continents and Climates Changed over Time? p. 468

4. In the absence of controls on the emission of CO_2 by the burning of fossil fuels, scientists believe that the concentration of this gas could double by the end of the current century, leading to a significant rise in the average temperature of Earth. What might be the evolutionary effects of this climatic change?
   Textbook Reference: 21.2 How Have Earth's Continents and Climates Changed over Time? p. 470

5. Describe several ways in which the dramatic increase in Earth's human population is affecting the evolutionary process.
   Textbook Reference: 21.4 Why Do Evolutionary Rates Differ among Groups of Organisms? p. 480–481

Answers

Knowledge and Synthesis Answers

1. **a.** Radioactive decay is measured as the time it takes for one-half the amount of a substance to spontaneously convert into another substance.

2. **a.** As the crustal plates are pushed together, one may move underneath the other, pushing up mountain ranges.

3. **b.** Fossilization occurs best in areas of low oxygen concentration and rapid sedimentation, where scavengers cannot destroy the body.

4. **d.** Hard-shelled animals such as mollusks are good candidates for fossilization because their shells can withstand decay for long enough to become buried. Many also tend to live in quiet, shallow waters.

5. **c.** Coming from the same root word as *province,* which is a specific area within a country or state possessing unique characteristics, provincialized biotas are geographically unique assemblages of organisms produced by a variety of physical and biological isolating factors.

6. **c.** The modern mammals did not appear until the Tertiary, more than 350 million years after the Ordovician.

7. **c.** The Pleistocene epoch, which is part of the Quaternary, was the time of most hominid evolution.

8. **b.** The mass extinction at the end of the Cretaceous, 65 mya, was the most recent affecting terrestrial life before the evolution of humans.

9. **b.** The Cambrian explosion produced many novel groups of animals characterized by distinctive body plans. Later explosions caused an increase in diversity only within already existing major lineages.

10. **d.** For the reasons outlined in the previous answer, the Cambrian period is the correct answer.

11. **a.** Paleontologists believe that a large meteorite's collision with Earth so altered conditions that the dinosaurs rapidly succumbed during what we know as the great Cretaceous mass extinction.

12. **d.** In some lineages, evolutionary rates apparently have changed through time. Long periods of little or no change have been interrupted by periods of rapid evolution.

13. **a.** Carbon dating is generally only reliable for fossils less than 50,000 years old.

14. **d.** Dinosaurs disappeared at the end of the Cretaceous, over sixty million years before humans evolved.

15. **c.** Though mass extinctions occurred at the end of all the periods listed as answer choices, the Permian event caused a higher percentage of species to become extinct than any other.

Application Answers

1. If only one-sixteenth of the ^{14}C remains, then four ^{14}C half-lives have passed since the fossil was formed ($\frac{1}{2} \times \frac{1}{2} \times \frac{1}{2} \times \frac{1}{2} = \frac{1}{16}$). Because the half-life of ^{14}C is roughly 5,700 years, the fossil must be about 22,800 years old ($4 \times 5,700 = 22,800$).

2. See Table 21.2 of the textbook.

3. Small prokaryotic cells can obtain enough oxygen by diffusion even when oxygen concentrations are very low. Eukaryotic cells, being larger, have a lower surface area-to-volume ratio and hence need a higher concentration of oxygen in their environment for diffusion to meet their requirement for this gas.

4. The fossil record indicates that rapid evolution often occurs in lineages exposed to a period of environmental instability. The record also shows that major environmental changes occurring over a short time interval, sometimes lead to large-scale, rapid extinctions that appear "instantaneous" in the fossil record. The possible effects of global warming on biodiversity are discussed in more detail in Chapter 57 of the textbook.

5. By dramatically altering Earth's vegetation from forests and grasslands to crops and pastures, humans are causing the extinction of many species, both large and small. By transporting species around the world, humans are reversing the independent evolution of Earth's biota on separate continents. Lastly, by artificial selection and biotechnology (including gene transfer between species), humans are directly determining how certain species are evolving.

CHAPTER 22 The Mechanisms of Evolution

Important Concepts

Darwin's main contribution to biology was to develop the theory of evolution by natural selection.

- The theory of evolution by natural selection rests on two facts:
  - All populations have the capacity to increase in number exponentially. Because populations rarely increase rapidly, high death rates must balance the potential growth rates.
  - All populations show variations among individuals with regard to numerous inherited characteristics, some of which may affect their chances of surviving and reproducing.
- From these two facts, Darwin made the following inference: Those individuals whose heritable traits enable them to survive and reproduce more successfully than others will pass those traits on to greater numbers of offspring.
- In short, natural selection is the differential contribution of offspring to the next generation by various genetic types belonging to the same population.
- An adaptation is a characteristic that helps its bearer survive and reproduce; the term also refers to the evolutionary process that produces such characteristics.
- When selection of individuals with desirable traits is carried out by humans, such as plant and animal breeders, it is referred to as artificial selection.
- The rediscovery of Mendel's laws of inheritance led to the development of population genetics, which applies these laws to entire populations to understand the genetic basis of evolutionary change.

Evolution is any genetic change in a population from generation to generation.

- A genotype is the genetic constitution of a particular trait of an individual. Evolution occurs when individuals with different genotypes survive or reproduce at different rates.
- Agents of evolution actually act on the phenotype, which is the physical expression of the genotype.

- The features of a phenotype are its *characters*. A *trait* is a specific form of a character.
- The gene pool of a population is the sum of all the alleles (alternative versions of a gene) found within it. Although a single diploid individual can have at most two different alleles for any gene, a population may contain many alleles for that gene. The gene pool contains all the variations that result in individuals with differing phenotypes on which agents of evolution act.
- Studies of both domestic and wild species show that virtually all populations contain a certain amount of genetic variation.
- Genotypes do not uniquely determine phenotypes. For example, if one allele is dominant to another, two genotypes (e.g., *AA* and *Aa*) can determine the same phenotype. Conversely, one genotype can produce different phenotypes because of the interaction of genetic and environmental factors during development.
- A Mendelian population is a locally interbreeding group within a geographic population. Such groups are often the subject of evolutionary studies.
- Allele frequencies measure the amount of genetic variation in a population, whereas genotype frequencies show how this variation is distributed among its members. Populations that have the same allele frequencies may nevertheless differ in their genotype frequencies.
- The sum of all allele frequencies at a locus is equal to 1, as is the sum of all genotype frequencies. For a locus with two alleles, p and q are typically used to represent the frequencies of the dominant and recessive alleles, respectively, and thus $p + q = 1$. Biologists estimate allele frequencies by measuring numbers of alleles in a sample of individuals from the population.
- If there is only one allele at a locus, the population is *monomorphic*, and the allele is *fixed*. If two or more alleles exist, the population is *polymorphic* at that locus.
- The genetic structure of a population is described by the frequencies of different alleles at each locus and the frequencies of different genotypes.

A population that is not changing genetically is at Hardy–Weinberg equilibrium.

- To be at Hardy–Weinberg equilibrium, a population must meet five conditions:
  - Mating must be random.
  - The population must be infinite.
  - There must be no migration into or out of the population.
  - There must be no mutation.
  - Natural selection does not affect the survival of particular genotypes.

- If these conditions are met, allele frequencies at a locus remain the same from generation to generation. Moreover, after one generation of random mating, the genotype frequencies will remain in the proportions $p^2 + 2pq + q^2 = 1$, where p^2, $2pq$, and q^2 represent the frequencies of the homozygous dominant, heterozygous, and homozygous recessive genotypes, respectively.

- Though populations in nature never meet the conditions of the Hardy–Weinberg equilibrium, this equation is important because deviations from it may show that evolution is occurring. Moreover, the pattern of deviations is useful in identifying the agents of evolutionary change operating on the population.

Evolutionary agents are forces that change the allele and genotype frequencies in a population.

- The ultimate source of genetic variation in a population is mutation. Because mutations are random changes in genetic material, most are harmful or neutral, but it is the environment that determines whether a particular mutation is disadvantageous or adaptive.

- Though mutations cannot be totally prevented, their frequency is usually so low as to cause little deviation from Hardy–Weinberg expectations.

- Gene flow occurs when individuals migrate from one population to another and breed in their new location. Gene flow can add new alleles to a population's gene pool, change the frequencies of alleles already present, or both.

- Genetic drift is caused by chance events that alter allele frequencies in a population. It has its greatest impact on small populations, in which it may even cause harmful alleles to increase in frequency.

- During a population bottleneck, when a large population is severely reduced in size, allele frequencies may shift drastically, and genetic variation may be reduced as a result of genetic drift.

- The founder effect occurs when a few individuals originate a new population. As in the case of a population bottleneck, some alleles found in the source population will be missing from the founding population, and others will occur with altered frequencies.

- Nonrandom mating occurs when individuals mate preferentially with others either of the same genotype or of a different genotype. The effect of nonrandom mating is to cause either fewer or more heterozygotes to be found in a population than would be expected in a population in Hardy–Weinberg equilibrium. The effect of self-fertilization, another form of nonrandom mating, is to reduce the frequency of heterozygotes. In most types of nonrandom mating, the allele frequencies remain the same despite the changes in genotype frequencies, an important exception being sexual selection.

- Fitness is the contribution of a phenotype to the composition of later generations, relative to the contribution of alternative phenotypes. The fitness of a phenotype is determined by the average rates of survival and reproduction of individuals with that phenotype, as compared to individuals with other phenotypes.

- Differences in fitness among phenotypes (and hence in the controlling genotypes) lead to changes in allele frequencies in the gene pool of later generations. Changes in allele frequencies due to differences in fitness result in adaptation.

- Natural selection, unlike other agents of evolution, adapts organisms to their environment. In cases in which the distribution of a phenotype approximates a bell-shaped curve because it is controlled by many gene loci, selection can produce any one of three results:
  - Stabilizing selection reduces variation in the population by favoring average individuals.
  - Directional selection changes the mean value for a character by favoring individuals that vary in one direction.
  - Disruptive selection, by favoring both extremes, leads to a population with two peaks in the distribution of the character.

- Sexual selection favors traits that benefit their bearers (generally males) in the competition for access to members of the other sex, or make their bearers more attractive to members of the other sex. Sexual selection often results in sexually dimorphic species, in which males and females differ in appearance.

Most populations maintain substantial genetic diversity.

- Genetic drift, stabilizing selection, and directional selection tend to lessen the genetic variation of a population.

- Neutral mutations—that is, mutations that do not affect the fitness of an individual—tend to accumulate and thereby increase the genetic variation in a population.

- By generating new combinations of alleles, sexual reproduction amplifies existing genetic variation.

- Sexual reproduction has short-term disadvantages, including disruption of adaptive combinations of genes and reduction both of the rate at which females pass genes on to their offspring and of the overall reproductive rate. Possible advantages of sexual reproduction that may account for its evolution include facilitation of DNA repair, elimination of harmful mutations, and defense against pathogens and parasites.

- Frequency-dependent selection, in which the less-common genotype or phenotype is favored by natural selection, preserves variation as a polymorphism.

- Geographically distinct subpopulations of a population often vary genetically because they are adapted to different environments.

There are constraints on the evolutionary process.

- Evolutionary changes must be based on modifications of previously existing traits.

- The evolution of adaptations depends on the trade-off between costs and benefits.

- Long-term evolutionary changes are strongly influenced by events that occur so infrequently or so slowly that they are rarely observed during short-term evolutionary studies. To understand the long-term course of evolution, biologists seek evidence of rare events and search for trends in the fossil record.

The Big Picture

- Most biologists rank Darwin's *The Origin of Species* as the most significant book ever written about their subject. Its importance is twofold. First, it presents a vast amount of evidence for the fact of evolution. Second, it describes a mechanism—natural selection—to explain evolutionary change. Evolution through natural selection remains the most important unifying concept in biology.

- Population genetics unites the concepts of Darwin with the insights into the mechanisms of heredity provided by Mendel. It enables biologists to study evolution with quantitative rigor.

- The loss of genetic variation that occurs during a population bottleneck is a major concern for biologists working to preserve endangered species. As the discussion of the greater prairie chicken makes clear, genetic uniformity in a population renders it more vulnerable to extinction.

Common Problem Areas

- A common mistake made by students when attempting to solve Hardy–Weinberg problems is to use the wrong equation. For example, if you are given phenotypic frequencies for a trait that is at genetic equilibrium and shows dominance, remember that the frequency of the recessive phenotype (and genotype) is equal to q^2, not q. By taking the square root of q^2, you can obtain the frequency of the recessive allele and can then determine the frequency of the dominant allele by subtraction ($p = 1 - q$).

- It is important to bear in mind that for any gene locus with two alleles, the allele frequencies must total unity (i.e., it is always the case that $p + q = 1$). The equation that specifies genotype frequencies ($p^2 + 2pq + q^2 = 1$) is only valid for a population at genetic equilibrium.

Study Strategies

- Population genetics is learned best by doing problems, so be sure you can solve problems similar to those presented in the "Test Yourself" section of this chapter of the *Study Guide*. Your instructor may provide a worksheet with additional examples.

- Review the following animated tutorials on the Companion Website/CD:
  Tutorial 22.1 Natural Selection
  Tutorial 22.2 Hardy–Weinberg Equilibrium
  Tutorial 22.3 Assessing the Costs of Adaptations

Test Yourself

Knowledge and Synthesis Questions

1. Evolution occurs at the level of
   a. the individual genotype.
   b. the individual phenotype.
   c. environmentally based phenotypic variation.
   d. the population.
   Textbook Reference: 22.1 What Facts Form the Base of Our Understanding of Evolution? p. 489

2. What does natural selection act on?
   a. The gene pool of the species
   b. The genotype
   c. The phenotype
   d. Multiple gene inheritance systems
   Textbook Reference: 22.3 What Evolutionary Mechanisms Result in Adaptation? p. 497

3. Which of the following about Mendelian populations that is *not* true? A Mendelian population must
   a. consist of members of the same species.
   b. have members that are capable of interbreeding.
   c. show genetic variation.
   d. have a gene pool.
   Textbook Reference: 22.1 What Facts Form the Base of Our Understanding of Evolution? p. 491

4. In comparing several populations of the same species, the population with the greatest genetic variation would have the
   a. greatest number of genes.
   b. greatest number of alleles per gene.

c. greatest number of population members.

d. largest gene pool.

Textbook Reference: *22.4 How Is Genetic Variation Maintained within Populations? pp. 501–503*

5. The ability to taste the chemical PTC (phenylthio-carbamide) is determined in humans by a dominant allele *T*, with tasters having the genotypes *Tt* or *TT* and nontasters having *tt*. If you discover that 36 percent of the members of a population cannot taste PTC, then according to the Hardy–Weinberg rule, the frequency of the *T* allele should be

a. 0.4.

b. 0.6.

c. 0.64.

d. 0.8.

Textbook Reference: *22.1 What Facts Form the Base of Our Understanding of Evolution? pp. 491–493*

6. A gene in humans has two alleles, *M* and *N*, that code for different surface proteins on red blood cells. If you know that the frequency of allele *M* is 0.2, according to the Hardy–Weinberg rule, the frequency of the genotype *MN* in the population should be

a. 0.16.

b. 0.32.

c. 0.64.

d. 0.8.

Textbook Reference: *22.1 What Facts Form the Base of Our Understanding of Evolution? pp. 491–493*

7. If the frequency of allele *b* in a gene pool is 0.2, and the population is in Hardy–Weinberg equilibrium, the expected frequency of the genotype *bbbb* in a tetraploid (4*n*) plant species would be

a. 0.0016.

b. 0.04.

c. 0.08.

d. 0.2.

Textbook Reference: *22.1 What Facts Form the Base of Our Understanding of Evolution? pp. 491–493*

8. Random genetic drift would probably have its greatest effect on which of the following populations?

a. A small, isolated population

b. A large population in which mating is nonrandom

c. A large population in which mating is random

d. A large population with regular immigration from a neighboring population

Textbook Reference: *22.2 What Are the Mechanisms of Evolutionary Change? pp. 494–495*

9. Allele frequencies for a gene locus are *least* likely to be significantly changed by

a. self-fertilization.

b. the founder effect.

c. mutation.

d. gene flow.

Textbook Reference: *22.2 What Are the Mechanisms of Evolutionary Change? pp. 495–496*

10. Select all the following evolutionary agents that produce nonrandom changes in the genetic structure of a population.

a. Self-fertilization

b. Population bottlenecks

c. Mutation

d. Natural selection

Textbook Reference: *22.3 What Evolutionary Mechanisms Result in Adaptation? pp. 494–497*

11. Suppose a particular species of flowering plant that lives only one year can produce red, white, or pink blossoms, depending on its genotype. Biologists studying a population of this species count 300 red-flowering, 500 white-flowering, and 800 pink-flowering plants in a population. When the population is censused the following year, 600 red-flowering, 900 white-flowering, and 1,000 pink-flowering plants are observed. Which color has the highest fitness?

a. Red

b. White

c. Pink

d. All of the above

Textbook Reference: *22.3 What Evolutionary Mechanisms Result in Adaptation? p. 497*

12. The following graph shows the range of variation among population members for a trait determined by multiple genes.

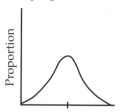

If this population is subject to *stabilizing selection* for several generations, which of the distributions (a–d) is most likely to result?

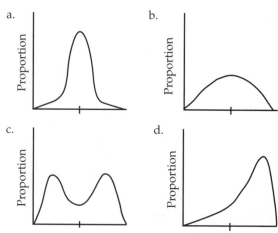

Textbook Reference: *22.3 What Evolutionary Mechanisms Result in Adaptation? p. 497*

13. In areas of Africa in which malaria is prevalent, many human populations exist in which the allele that produces sickle-cell disease and the allele for normal red blood cells occur at constant frequencies, despite the fact that sickle-cell disease frequently causes death at an early age. This is an example of
    a. the founder effect.
    b. a stable polymorphism.
    c. mutation.
    d. nonrandom mating.
    Textbook Reference: 22.4 How Is Genetic Variation Maintained within Populations? p. 502

14. Which of the following choices is *not* a disadvantage of sexual reproduction?
    a. It increases the difficulty of eliminating harmful mutations from the population
    b. It breaks up adaptive combinations of genes
    c. It reduces the rate at which females pass genes on to their offspring
    d. When it involves separate genders, it reduces the overall reproductive rate
    Textbook Reference: 22.4 How Is Genetic Variation Maintained within Populations? p. 501

15. Genetic variation in a population may be maintained by
    a. frequency-dependent selection.
    b. the accumulation of neutral alleles.
    c. sexual recombination.
    d. All of the above
    Textbook Reference: 22.4 How Is Genetic Variation Maintained within Populations? p. 501

16. Which of the following can act as a constraint on the evolutionary process?
    a. There is a trade-off between the cost and benefit of an adaptation
    b. The occurrence of rare catastrophic events, such as meteorite impacts, can disrupt evolutionary patterns
    c. All evolutionary innovations are modifications of previously existing structures
    d. All of the above
    Textbook Reference: 22.5 What Are the Constraints on Evolution? pp. 503–504

Application Questions

1. Discuss the main use of the Hardy–Weinberg rule in evolutionary biology.
   Textbook Reference: 22.1 What Facts Form the Base of Our Understanding of Evolution? p. 493

2. The Hardy–Weinberg expression is an expansion of a binomial or $(p + q)^2 = p^2 + 2pq + q^2$. In this equation, the left-hand side represents _____ frequencies, and the right-hand side represents _____ frequencies.
   Textbook Reference: 22.1 What Facts Form the Base of Our Understanding of Evolution? pp. 491–492

3. In a population with 600 members, the numbers of individuals of three different genotypes are $AA = 350$, $Aa = 100$, $aa = 150$.
   a. What are the genotype frequencies in this population?
      1. $AA =$
      2. $Aa =$
      3. $aa =$
   b. What are the allele frequencies in this population?
      1. $A =$
      2. $a =$
   c. What would be the expected genotype frequencies if this population were in genetic equilibrium?
      1. $AA =$
      2. $Aa =$
      3. $aa =$
   d. Is this population in genetic equilibrium? Explain.
   Textbook Reference: 22.1 What Facts Form the Base of Our Understanding of Evolution? pp. 492–493

4. The following data were collected in a study in which dark and light moths of the peppered moth (*Biston betularia*) were released and recaptured a few days later in several different areas (*a*, *b*, *c*, and *d*). Based on these data, in which area is the light phenotype most advantageous? (Assume that recapture rates reflect the relative survival of the moths, not the ability of the investigators to find and recapture the moths.) What does this tell us about the fitness of light moths in this area?

Area	Moth type	Released	Recaptured
a	dark	100	25
	light	200	50
b	dark	100	25
	light	200	100
c	dark	200	50
	light	400	100
d	dark	200	100
	light	300	75

Textbook Reference: 22.3 What Evolutionary Mechanisms Result in Adaptation? p. 497

5. As peppered moths rest on bark surfaces during the day, they are subject to predation by birds. Assuming that moths that are more camouflaged will be preyed on less, what can you say about the color of bark surfaces in area *d*?
   Textbook Reference: 22.3 What Evolutionary Mechanisms Result in Adaptation? p. 497

6. Describe what is meant by sexual selection and how the sexual selection of tail length in male widowbirds was studied.
   Textbook Reference: 22.3 What Evolutionary Mechanisms Result in Adaptation? pp. 498–500

Answers

Knowledge and Synthesis Answers

1. **d**. Evolution is defined as changes in the genetic structure of a population over time, so evolution occurs at the level of the population.

2. **c**. Natural selection acts on phenotypes, not genotypes. For example, a harmful recessive allele is "invisible" to natural selection when it occurs in a heterozygote, where its harmful effect is masked by the dominant allele.

3. **c**. Although most populations show genetic variation, this condition is not part of the definition of a Mendelian population.

4. **b**. Genetic variation is related to the number of different alleles per gene. Regarding choice **a**, recall that all species members have the same number of genes. Population size per se has little to do with genetic variation, so choices **c** and **d** are also incorrect.

5. **a**. Recall that $q^2 = 0.36$ is the frequency of the *tt* genotype, so q (0.6, the square root of 0.36) is the frequency of the *t* allele. Assuming there are only two alleles for this trait, $T + t = 1$. The frequency of the *T* allele is therefore $1 - t = 1 - 0.6 = 0.4$.

6. **b**. Because $p = 0.2$ and therefore, $q = 1 - 0.2 = 0.8$, the frequency of the *MN* genotype is $2pq$, or $2 \times 0.2 \times 0.8 = 0.32$.

7. **a**. The probability of one allele *b* in a genotype is equal to its frequency, or 0.2, so the probability that all four of the alleles in a tetraploid organism will be *b* would be $(0.2)^4$, or 0.0016.

8. **a**. Genetic drift is most significant in small populations.

9. **a**. Unlike the founder effect, mutation, and gene flow, all of which change allele frequencies in a population, self-fertilization (like other types of nonrandom mating with the exception of sexual selection) only causes a deviation from the frequency of heterozygotes predicted by Hardy–Weinberg equilibrium.

10. **a**. and **d**. Self-fertilization leads to increased numbers of homozygous individuals, and natural selection is also nonrandom.

11. **a**. Fitness measures the relative contribution of a genotype or phenotype to subsequent generations. The red-flowering plants, which doubled in number, had the greatest percentage increase of any of the plants and thus had the highest fitness.

12. **a**. Stabilizing selection results when individuals that are intermediate in phenotype make a larger contribution to future generations than individuals of more extreme phenotype. This leads to reduced variation for the trait and causes the curve to be higher and narrower. Curve *b* shows greater variation, curve *c* would result from disruptive selection, and curve *d* from directional selection.

13. **b**. Polymorphism in a population is the existence of two or more alleles at a particular gene locus. If phenotypes are stable through time, then the underlying alleles will also be constant. In this instance, the polymorphism is stable because malaria is a significant cause of mortality in some parts of Africa, and heterozygotes have greater resistance to this disease than individuals with a "normal" phenotype. Thus the superior fitness of the heterozygotes maintains both alleles in the population.

14. **a**. Sexual recombination produces some individuals in a population who are less fit than others because they carry a greater than average number of deleterious mutations. Because these individuals are selected against, sexual reproduction is able to reduce the number of deleterious mutations in the population over time.

15. **d**. Frequency-dependent selection, the accumulation of neutral alleles, and sexual recombination are the three major forces discussed in Chapter 22 that maintain genetic variation. Another force, hybrid vigor due to superior fitness of the heterozygous condition of some genes, is discussed in Chapter 10. A mating advantage for heterozygous *Colias* butterflies is discussed on p. 502 of Chapter 22 (see Figure 22.19).

16. **d**. Cost-benefit trade-offs, developmental constraints and major environmental disruptions can all limit the freedom of natural selection to produce adaptive traits.

Application Answers

1. No population meets the Hardy–Weinberg conditions for genetic equilibrium. However, by determining how a population deviates from the expectations of Hardy–Weinberg equilibrium, evolutionary biologists can identify what evolutionary agents are affecting the population.

2. allele; genotype

3. a.
   1. $AA = 350/600 = 0.58$
   2. $Aa = 100/600 = 0.1$
   3. $aa = 150/600 = 0.25$

   b.
   1. $A = (700 + 100)/1200 = 0.67$
   2. $a = (300 + 100)/1200 = 0.33$

   c.
   1. $AA = p^2 = (0.67)2 = 0.45$
   2. $Aa = 2pq = 2 \times 0.67 \times 0.33 = 0.44$
   3. $aa = q^2 = (0.33)2 = 0.11$

   d. No. The observed genotypic frequencies differ from those predicted by the Hardy–Weinberg rule using the observed allele frequencies for *p* and *q*. The population is not in genetic equilibrium.

4. Area *b*. In areas *a* and *c*, recaptures were 25% of releases for all moths, but in area *b*, 50% of the light moths were recaptured, indicating that moths with this

phenotype survived best in this area. This suggests that fitness will be higher for light moths in this area, but it won't be possible to determine this for certain until the moths reproduce because fitness, by definition, depends on the relative contribution to subsequent generations (it is possible, for example, that light moths will have a harder time finding mates in this area).

5. Dark moths seem to survive best in area *d*. If the increased survival of the moths is due to cryptic coloration, then the bark surfaces in that area should also be dark.

6. Sexual selection is the spread of a trait that improves the reproductive success of an individual. The trait may improve the ability of its bearer to compete with other members of its sex for access to mates, or it may make its bearer more attractive to members of the opposite sex. By artificially lengthening or shortening the tails of male widowbirds, behavioral ecologists were able to show that increased tail length attracted more females but did not confer an advantage in interactions with other males.

CHAPTER 23 Species and Their Formation

Important Concepts

Species are the fundamental units of evolutionary diversification and of the biological classification system.

- The morphological concept of species used by early biologists grouped organisms into species on the basis of their appearance. For some groups of organisms, morphological similarity is still the major criterion.

- Speciation is the process by which one species splits into two species. Determining whether two populations constitute different species may be difficult because speciation is frequently a gradual process.

- The biological species concept defines a species as a group of actually or potentially interbreeding natural populations that are reproductively isolated from other such groups. Thus speciation depends on the development of reproductive isolation. This definition cannot be applied to organisms that reproduce asexually.

The two principal modes of speciation are allopatric speciation and sympatric speciation.

- Evolutionary change can occur without speciation. A single lineage may change through time without diverging into two species.

- Speciation requires that the gene pool of the original species divide into two isolated gene pools. After separation, the agents of evolution cause sufficient genetic differences to accumulate in the populations for them to be unable to interbreed successfully if they come together again.

- In allopatric speciation, the population is initially divided by a geographic barrier. If the barrier results from a geological or climatic change, the two isolated populations are often large and genetically similar. These populations diverge not only because of genetic drift but especially because the environments in which they live are, or become, different. Alternatively, separation may occur when some members of a population cross a barrier and found a new, isolated population.

- The effectiveness of a barrier in preventing gene flow depends on the size and mobility of the species.

- Evidence suggests that allopatric speciation is the most common mechanism of speciation among animals and most other groups.

- Sympatric speciation occurs without geographic subdivision of the gene pool of the original species. Disruptive selection, in which different genotypes have high fitness on one of two different food resources, may be a widespread mechanism of sympatric speciation among insects.

- Sympatric speciation by polyploidy, the production of duplicate sets of chromosomes within an individual, is common in plants. Polyploidy produces new species because the polyploid organisms cannot interbreed with members of the parent species. Polyploid species that have a single ancestor are called autopolyploids, whereas those that have resulted from the hybridization of two species are referred to as allopolyploids.

- New species arise by polyploidy much more easily among plants than among animals because plants of many species can reproduce by self-fertilization.

Speciation requires that reproductive barriers evolve to prevent previously allopatric populations from exchanging genes if and when they become sympatric.

- Prezygotic barriers prevent members of different species from mating.
  - Species may simply mate in different areas or different parts of a habitat (habitat isolation).
  - Species may not be able to interbreed because they are fertile at different times (temporal isolation).
  - Differences in reproductive organs may prevent interbreeding (mechanical isolation).
  - The sperm and egg may be chemically incompatible (gametic isolation).
  - The two species may not recognize or respond to each other's mating behaviors, or in flowering plants, the behavioral preferences of the pollinating animals may prevent interbreeding (behavioral isolation).

- Postzygotic barriers can prevent effective gene flow between species, even if mating occurs.
  - Hybrid zygotes may not mature normally (low hybrid zygote viability).
  - Hybrids may survive less well than either parent species (low hybrid adult viability).
  - Hybrids may be infertile (hybrid infertility).
- The evolution of more effective prezygotic reproductive barriers is known as reinforcement. It may occur if the hybrid offspring of two species survive poorly.

If two populations reestablish contact before reproductive isolation is complete, several results are possible.

- If hybrid offspring are not at a selective disadvantage, they may spread through both populations with the result that the gene pools of the populations combine. Thus no new species would result from the period of isolation.
- If hybrid offspring are less successful, reinforcement may strengthen prezygotic reproductive barriers.
- If hybrid offspring are at a disadvantage but reinforcement fails to occur, a stable, narrow hybrid zone may form.

Several factors affect the rate of speciation of different evolutionary lineages.

- Characteristics of a species that make it prone to speciation include membership in a large evolutionary lineage and poor dispersal ability.
- Among plants, high rates of speciation are found in groups with specialized animal pollinators.
- Among animals, dietary specialization and sexual selection typically stimulate speciation.

In an evolutionary radiation, many daughter species arise from a single ancestor.

- An *adaptive* radiation results in an array of species that differ in their habitats and resource utilization, whereas a *nonadaptive* radiation is not accompanied by noticeable ecological differences among the daughter species.
- The native biota of the Hawaiian Islands illustrates that populations colonizing environments that have underutilized resources are particularly likely to produce adaptive radiations.

The Big Picture

- Speciation is the process that has produced the millions of life forms—each adapted to a particular environment and way of life—that constitute life on Earth.
- Archipelagos such as the Galápagos and Hawaiian Islands have been called natural laboratories of evolution. Studies of the evolutionary radiations of the Galápagos finches and of several Hawaiian groups,

such as *Drosophila* and the silverswords, have provided crucial insights to evolutionary biologists since the time of Darwin.

- Chapter 57, "Conservation Biology," discusses the reasons for the rapid pace of human-caused species extinctions, including the disappearance of many members of groups that have undergone evolutionary radiations, such as Australian marsupials and Hawaiian honeycreepers. It also describes some strategies for the preservation of Earth's precious biological diversity.

Common Problem Areas

- Some students have difficulty in understanding why hybrids between species that have different numbers of chromosomes are inevitably sterile unless the hybrid is an allopolyploid. Recall that in Meiosis I, homologous chromosomes undergo synapsis. This process cannot occur properly if the haploid sets of chromosomes inherited from the parents contain different numbers of chromosomes because it is then impossible that every chromosome will have a homolog. As a consequence, meiosis does not proceed normally and few if any normal gametes will be produced. Because allopolyploids possess four sets of chromosomes (two from each parent), their chromosomes can synapse normally and they can produce viable gametes.

Study Strategies

- Because most instructors consider the heart of this chapter to be the discussions of allopatric (geographic) speciation and reproductive isolating mechanisms, you should pay particular attention to these topics.
- Review the following animated tutorials and activity on the Companion Website/CD:
  Tutorial 23.1 Speciation Mechanisms
  Tutorial 23.2 Founder Events Lead to Allopatric Speciation
  Activity 23.1 Concept Matching

Test Yourself

Knowledge and Synthesis Questions

1. Select all the correct answers that complete the following sentence: It is difficult to apply the biological species concept to groups of organisms that
   a. are asexual.
   b. produce hybrids only in captivity.
   c. show little morphological diversity.
   d. exist only in the fossil record.
   Textbook Reference: 23.1 What Are Species? p. 510

2. Which of the following statements about allopatric speciation is *not* true?
   a. Allopatric speciation can sometimes involve small populations.
   b. Allopatric speciation only occurs in species that are widely distributed.

c. Allopatric speciation always involves a physical barrier that interrupts gene flow.

d. Allopatric speciation can sometimes involve chance events.

Textbook Reference: *23.2 How Do New Species Arise? p. 511*

3. A long, narrow hybrid zone exists between the ranges of the fire-bellied toad and the yellow-bellied toad in Europe. Which of the following is a factor contributing to the persistence of this zone?

a. Reinforcement strengthens the prezygotic barriers between the two species.

b. Hybrid offspring have the same fitness as nonhybrid offspring.

c. Both species travel long distances over the course of their lives.

d. Individuals from outside the hybrid zone regularly move into the hybrid zone.

Textbook Reference: *23.3 What Happens When Newly Formed Species Come Together? pp. 517–518*

4. Which type of speciation is most common among flowering plants?

a. Geographic

b. Sympatric

c. Allopatric

d. None of the above

Textbook Reference: *23.2 How Do New Species Arise? p. 514*

5. Which of the following would *not* be considered an example of a prezygotic reproductive isolating mechanism?

a. One bird species forages in the tops of trees for flying insects, whereas another forages on the ground for worms and grubs.

b. The males of one species of moth cannot detect and respond to the sex attractant chemicals produced by the females of another species.

c. Sperm of one species of sea urchin are unable to penetrate the egg plasma membrane of another species.

d. Mosquitoes of one species are active in foraging and searching for mates at dusk, whereas those of another species are active at dawn.

Textbook Reference: *23.3 What Happens When Newly Formed Species Come Together? pp. 515–516*

6. Which of the following factors would *not* be expected to increase the rate of speciation in a group of organisms?

a. Fragmentation of populations

b. Poor dispersal ability

c. High birthrates

d. Dietary specialization

Textbook Reference: *23.4 Why Do Rates of Speciation Vary? pp. 518–519*

7. Which of the following is *not* a suggested reason for the adaptive radiation of silverswords on the Hawaiian archipelago?

a. Water is an effective barrier for many organisms.

b. Because islands are small compared with mainland areas, you would expect more species to develop there.

c. Competition is frequently reduced on islands.

d. More ecological opportunities exist on islands that have not been colonized by many species.

Textbook Reference: *23.5 Why Do Adaptive Radiations Occur? pp. 520–521*

8. More than 800 species of *Drosophila* occur in the Hawaiian Islands, representing 30 to 40 percent of all the species in this genus. The occurrence of so many *Drosophila* species in this island chain is

a. the result of many founder events followed by genetic divergence.

b. an example of an evolutionary radiation.

c. largely the result of sympatric speciation.

d. Both a and b are correct

Textbook Reference: *23.2 How Do New Species Arise? pp. 511–512*

9. Which of the following statements about speciation is *not* true?

a. A small founding population can be involved in speciation.

b. Speciation always involves interruption of gene flow between different groups of organisms.

c. The rate of speciation can vary for different groups of organisms.

d. Speciation always requires many generations to be completed.

Textbook Reference: *23.2 How Do New Species Arise? p. 514*

10. Which of the following observations constitutes conclusive evidence that two overlapping populations that had been geographically separated have *not* diverged into distinct species?

a. Matings between members of the two populations produce viable hybrids.

b. A stable hybrid zone exists where their ranges overlap.

c. Interbreeding is common between members of the two populations.

d. None of the above

Textbook Reference: *23.3 What Happens When Newly Formed Species Come Together? pp. 517–518*

11. In allopatric speciation, which process is likely to be *least* important?

a. Founder event

b. Allopolyploidy

c. Behavioral isolation

d. Genetic drift

Textbook Reference: *23.2 How Do New Species Arise? pp. 511–514*

12. A field contains two related species of flowering plants. Species A has a diploid chromosome number of 16, and species B has a diploid number of 18. If a third species arises as a result of hybridization between A and B, how many chromosomes will it have?
    a. 17
    b. 32
    c. 34
    d. 36
    Textbook Reference: 23.2 How Do New Species Arise? p. 514

13. Speciation by polyploidy occurs far more often in plants than in animals because
    a. plants are more likely to be capable of self-fertilization than animals.
    b. plant cells can tolerate extra sets of chromosomes, whereas animal cells cannot.
    c. plants as a rule have higher reproductive rates than animals.
    d. many plants are specialized with respect to their pollinating agent.
    Textbook Reference: 23.2 How Do New Species Arise? p. 514

14. Two species of narrowmouth frogs in the United States have mating calls that differ more in their region of sympatry than in those parts of their ranges that do not overlap. If this difference in their vocalizations has the function of preventing hybridization between the two species, it is an example of
    a. a hybrid zone.
    b. reinforcement.
    c. sympatric speciation.
    d. a postzygotic reproductive barrier.
    Textbook Reference: 23.3 What Happens When Newly Formed Species Come Together? pp. 516–517

15. The following answer choices describe four hypothetical families of birds, each endemic either to a single large island or to an island group (archipelago) far from the nearest continental land mass. In which of the four would you predict the highest rate of speciation?
    a. Family A is endemic to a large island. The species in this family have promiscuous mating systems.
    b. Family B is endemic to an archipelago. The species in this family have monogamous mating systems.
    c. Family C is endemic to a large island. The species in this family have monogamous mating systems.
    d. Family D is endemic to an archipelago. The species in this family have promiscuous mating systems.
    Textbook Reference: 23.2 How Do New Species Arise? p. 511; 23.4 Why Do Rates of Speciation Vary? p. 519

Application Questions

1. Discuss the conditions on the Galápagos Islands that led to the evolution of the birds known as Darwin's finches.
   Textbook Reference: 23.2 How Do New Species Arise? pp. 511–512

2. In autopolyploidy, a new species of plant can arise by the doubling of chromosome numbers in a single individual of one species (provided that the individual is capable of self-fertilization). Why is it virtually impossible for such a tetraploid plant to interbreed successfully with diploid individuals of the "same" species?
   Textbook Reference: 23.2 How Do New Species Arise? p. 514

3. Discuss what is known about the evolutionary radiations on islands, based on studies of Hawaiian silverswords and tarweeds.
   Textbook Reference: 23.5 Why Do Adaptive Radiations Occur? p. 521

4. Suppose that members of two populations are separated by a geographic barrier and begin to diverge genetically. Many generations later, when the barrier is removed, the two populations can interbreed, but the hybrid offspring do not survive and reproduce well. Explain how natural selection might lead to the evolution of more effective prezygotic barriers in these species.
   Textbook Reference: 23.5 Why Do Adaptive Radiations Occur? p. 521

5. The yellow-rumped warbler was formerly split into two species (myrtle and Audubon's warblers) but in 1973 was reclassified as a single species. "Myrtle" and "Audubon's" warblers have largely allopatric ranges but hybridize where they are sympatric in the Canadian Rockies. They are similar in appearance but are readily distinguished by experienced birders. What further data about these two forms should ornithologists collect and analyze to decide whether they should continue to be classified as a single species?
   Textbook Reference: 23.3 What Happens When Newly Formed Species Come Together? p. 517

Answers
Knowledge and Synthesis Answers

1. **a.** and **d.** The key criterion of a biological species is that its members have to be reproductively isolated from other such groups. This criterion is impossible to evaluate in asexual and fossil species.

2. **b.** A wide distribution is not a requisite for allopatric speciation.

3. **d.** These toads provide an example of related species in which reinforcement does not strengthen prezygotic barriers, even though hybrid offspring are only half as fit as nonhybrid offspring. The reason for this is that toads from outside the hybrid zone (not subject to the selective pressure against hybridizing) regularly move into the hybrid zone and mate with members of the other species. The hybrid zone remains narrow because toads do not travel long distances; natural selection removes hybrids from the population before they can disperse very far.

4. **b.** Sympatric speciation is most common among flowering plants. It has been estimated that about 70 percent of all flowering plant species are polyploid.

5. **a.** Provided that the two species are active in the same locality at the same time, a difference in the habitat in which they forage would not in itself be a barrier to interbreeding (though seeking mates in different habitats might well be a barrier to interbreeding). All the other choices describe reproductive barriers that would act prior to fertilization.

6. **c.** Birthrates per se do not seem to affect the rate of speciation in organisms. All other factors have been shown to increase speciation rates in the lineages of some organisms.

7. **b.** Actually, biogeographers have found that larger land masses tend to have more species than smaller land masses, so you might expect the reverse effect.

8. **d.** Island groups are frequently sites of evolutionary radiations through repeated allopatric speciation events that are initiated by individuals (or groups) dispersing from one island to another. The numerous species of *Drosophila* in the Hawaiian Islands are believed to have originated in this way.

9. **d.** New species formed by polyploidy can arise in only two generations.

10. **d.** Interbreeding, production of viable hybrids, and establishment of a hybrid zone do not necessarily mean that speciation is not complete. If, on the other hand, the hybrids were successful, fertile, and bred freely with members of both original populations, their gene pools would merge, and you would conclude that speciation had not taken place.

11. **b.** Genetic drift, founder events, and behavioral isolation may all play a role in allopatric speciation. Allopolyploidy, on the other hand, can only occur as the result of hybridization between individuals of different species; hence the two parent species cannot be allopatric.

12. **c.** If haploid gametes of species A and B joined, the result would be a zygote with 17 chromosomes. A mature plant with 17 chromosomes would be sterile because chromosomes would be unable to pair properly during prophase and metaphase of meiosis I. A fertile allopolyploid would therefore have to have 34 chromosomes so that each chromosome would have a homolog with which to pair. Autopolyploids of species A and B would have 32 and 36 chromosomes respectively.

13. **a.** If a polyploid plant or animal is capable of self-fertilization, then a new species can arise from a single individual. The ability to self-fertilize is far more common among plants than animals.

14. **b.** Reinforcement is defined as the evolutionary strengthening of prezygotic barriers to interbreeding within the zone of sympatry of two closely related species.

15. **d.** Speciation occurs more readily in archipelagos than on isolated islands because the establishment of geographical isolation through founder events involving dispersing individuals occurs much more readily within archipelagos than on a single island. Speciation would be more rapid in a family with promiscuous mating systems because species with this system are frequently characterized by a high degree of sexual dimorphism and the capacity of individuals to make subtle discriminations both between members of their own species and between members of their own species and other species. As a consequence, even slight differences in the appearance or behavior of members of different populations may lead to the evolution of reproductive barriers based on the mating preferences of individuals in the different populations.

Application Answers

1. The relatively great distance between the Galápagos Islands and the South American mainland and also between each of the islands in the archipelago ensured that once immigrants had arrived on an island, they would be genetically isolated for a substantial period of time. Also, because the islands differ greatly in climate and vegetation, the resident birds were subject to different selection pressures. This, in combination with reduced gene flow between the islands, led to a rapid evolutionary radiation of finches.

2. Recall from Chapter 9 that pairs of homologous chromosomes synapse during prophase and metaphase of the first division of meiosis. The homologs then separate so that each cell resulting from meiosis I is haploid, as are the products of meiosis II. In tetraploids as in diploids, meiosis is normal because pairing of homologs can occur. Any offspring of a cross between tetraploid and diploid individuals would be triploid, however, and therefore sterile because correct synapsis of homologs could not occur. Because the tetraploid product of autopolyploidy is reproductively isolated from its diploid relatives, it is a new species.

3. Studies of the silverswords of the Hawaiian archipelago show that taxa which evolve on islands frequently show great morphological diversity because of the reduced competition that immigrants encounter on islands. Thus Hawaiian silverswords have evolved tree- and shrublike species because there were few resident tree and shrub species with which they had to compete, unlike their mainland tarweed relatives.

4. Recall that natural selection tends to remove from a population traits that reduce survival or reproductive success. Individuals that interbreed between populations will have lower fitness (they will contribute fewer offspring to future generations) than those who breed within their own population. If the tendency to avoid interbreeding is heritable (and not just the result of chance), the frequency of alleles that prevent interbreeding will increase in each population. How might such a trait be heritable? Any of the prezygotic

barriers to interbreeding might be heritable traits. For example, if the species in question are frogs, and if their mating calls started to diverge while the populations were separated, the following traits might be heritable: a tendency to make a call that is more distinct from that of the other population, or the ability to distinguish between the existing calls of the two populations (coupled with a preference for the call of one's own population). As alleles for these traits increased in frequency, they would contribute to the behavioral isolation of the two populations and perhaps eventually to more complete speciation.

5. During allopatric speciation, divergence of two populations often occurs gradually. In such cases it is inevitable that intermediate stages of speciation occur, and it may be a matter of opinion whether two forms have diverged sufficiently to be considered separate species. With respect to these two warbler populations, biologists would seek answers to these questions: Are hybrid offspring as fit as those resulting from mating of individuals of the same population? Is there evidence that the zone of hybridization is expanding, indicating that the gene pools of the populations are combining? Is there evidence of reinforcement of prezygotic barriers to interbreeding (e.g., a greater difference in the songs of the two forms in the area of hybridization than in allopatric parts of their ranges)? Lesser fitness of hybrid offspring, a stable, narrow zone of hybridization and evidence of reinforcement would all favor the conclusion that the populations are best regarded as separate species.

CHAPTER 24 The Evolution of Genes and Genomes

Important Concepts

The genome of an organism is its entire set of genes as well as any noncoding regions of the DNA (or, in some viruses, RNA).

- The genome of eukaryotes includes both the nuclear genes located on chromosomes and genes present in mitochondria and chloroplasts.

- A successful gene must be able to function in a wide variety of genetic backgrounds. Among the genes of an individual, there are both divisions of labor and strong interdependencies.

The field of molecular evolution is the study of the evolution of genomes and of particular nucleic acids and proteins.

- Genomes and particular nucleic acids and proteins are investigated to determine how rapidly and why they have changed. The answers to these questions are crucial to an understanding of the evolutionary history of genes and of the organisms that carry them.

- The evolution of nucleic acids and proteins depends on variation introduced by mutation. Nucleotide substitution mutations may result in amino acid replacements on the encoded proteins.

- Evolutionary changes in genes and proteins are detected by comparing the nucleotide and amino acid sequences among different organisms.

The structures of macromolecules can be determined and compared.

- By the use of the sequence alignment technique, biologists identify homologous sequences within nucleic acids or proteins. The concept of homology (similarity that results from common ancestry) extends down to particular positions in nucleotide or amino acid sequences.

- Having identified homologous regions of a nucleic acid or protein, biologists construct a similarity matrix to measure the minimum number of changes that have occurred during the divergence between pairs of organisms. The assumption is that the longer the

molecules have been evolving separately, the more differences they will have.

- For several reasons, the observed number of differences in homologous DNA sequences in two species almost certainly underestimates the number of substitutions that have actually occurred since the sequences diverged from a common ancestor. Molecular evolutionists use mathematical models to correct for this undercounting.

- Because viruses, bacteria, and unicellular eukaryotes have short generation times, biologists use them to study molecular evolution in the laboratory.

Molecular evolution occurs by diverse mechanisms.

- A synonymous (silent) mutation replaces a nucleotide base in a codon but does not change the amino acid specified by the codon. Because synonymous mutations do not affect the functioning of a protein, they are unlikely to be affected by natural selection.

- A nonsynonymous mutation changes the amino acid specified by the codon. Though such mutations are likely to be harmful, they are sometimes selectively neutral, or nearly so, and are occasionally advantageous.

- Within functional genes, nucleotide substitution rates are highest at nucleotide positions that do not change the amino acid being expressed. The rate of substitution is higher in pseudogenes—duplicate copies of genes that are never expressed—than in functional genes.

- The neutral theory of molecular evolution postulates that most evolutionary change in macromolecules, as well as much of the genetic variation within species, is the result of random genetic drift, rather than natural selection. The rate of fixation of neutral mutations is theoretically constant and equal to the mutation rate.

- According to the neutral theory of molecular evolution, it is possible to distinguish among evolutionary processes by comparing the rates of synonymous and nonsynonymous substitutions in a protein-coding gene.

- If an amino acid substitution is neutral in its effect on fitness, then the rates of synonymous and nonsynonymous substitutions in the corresponding DNA sequences are expected to be very similar.

- If an amino acid position is under strong stabilizing selection, then the rate of synonymous substitutions in the corresponding DNA is expected to be much higher that the rate of nonsynonymous substitutions.

- If an amino acid position is under strong selection for change, then the rate of nonsynonymous substitutions in the corresponding DNA is expected to exceed the rate of synonymous substitutions.

- While serving in almost all animals as an important first line of defense against invading bacteria, the enzyme lysozyme also has evolved to take on an essential role in the digestive process of several groups of foregut fermenters. By comparing the lysozyme-coding sequences in foregut fermenters with several of their non-fermenting relatives, molecular evolutionists have discovered that neutral evolution, stabilizing selection, and selection for change have all occurred as lysozyme evolved to take on its new function. The independent evolution in several groups of foregut fermenters of a type of lysozyme adapted to its new environment and function shows that convergent evolution occurs at the molecular level.

- Genome size and organization also evolve. The size of the coding portion of the genome is larger in more complex organisms. Thus, eukaryotes have many times more genes than prokaryotes; vascular plants and invertebrates have more genes than single-celled organisms, and vertebrates have more genes than invertebrates.

- Most of the variation in genome size of various organisms is due not to differences in the number of functional genes, but in the amount of noncoding DNA.

- Although much of the noncoding DNA appears to be nonfunctional, it may alter the expression of surrounding genes. Important categories of noncoding DNA include pseudogenes and parasitic transposable elements.

- Studies of the rate of retrotransposon loss show that species differ greatly in the rate at which they gain or lose apparently functionless DNA. The reason for the differences between species is unclear. It may be related to the rate at which the organism develops or to its population size.

- When a gene is duplicated, four evolutionary outcomes are possible:
  - Both copies retain the gene's original function.
  - Both copies retain the ability to produce the original gene product, but the expression of the genes diverges in different tissues or at different times of development.
  - One copy becomes a functionless pseudogene.
  - One copy retains its original function, whereas the second mutates so extensively that it can perform a different function.

- When an entire genome is duplicated (as in polyploid organisms), there are massive opportunities for new gene functions to evolve. Genome duplication events that occurred in the ancestor of the jawed vertebrates have permitted many individual vertebrate genes to become highly tissue-specific in their expression.

- Successive rounds of gene duplication and mutation can result in a gene family—a group of homologous genes with related functions.

- The globin gene family illustrates that gene diversification can produce molecules with different functions (hemoglobin and myoglobin) as well as functionless pseudogenes.

- Concerted evolution results in similar DNA sequence changes in all copies of highly repeated genes, such as those that code for ribosomal RNA. Two separate mechanisms can cause concerted evolution:
  - If homologous chromosomes align imprecisely during meiosis, *unequal crossing over* may occur with the result that one chromosome will gain extra copies of a highly repeated gene and the other chromosome will be left with fewer copies.
  - If one favored copy of a repeated gene on one homologous chromosome is used as the template for the repair of damage to the copies of the gene on the other homolog, the result is *biased gene conversion*: the rapid spread of the favored sequence across all the copies of the gene.

Molecular evolution has many applications.

- A gene tree depicts the evolutionary history of a particular gene or of the members of a gene family. Orthologs are genes found in different organisms that arose from a single gene in their common ancestor. Paralogs are related genes that have resulted from gene duplication in a single lineage.

- The principles of molecular evolution are used to understand function and diversification of function in many proteins. For example, detection of strong selection for change in a nucleotide sequence can be used to identify molecular changes that have resulted in functional changes.

- Molecular evolutionary principles underlie the field of in vitro evolution, in which new molecules are produced in the laboratory to perform particular desired functions. The basis of in vitro evolution is the creation of random molecular variation followed by selection by the experimenter.

- Biomedical scientists are using principles of molecular evolution to identify and combat human diseases.

The Big Picture

- The growing importance of the study of macro-molecules in the reconstruction of phylogenies is an example of the far-reaching impact that advances in molecular biology have had in the decades since the discovery of the structure of DNA by Watson and Crick in 1953.

- The principles of molecular evolution are critical for our understanding of how new diseases—especially those caused by viruses—originate and evolve. This understanding in turn may lead to more effective strategies for combatting these diseases.

Common Problem Areas

- The concept of concerted evolution and the two mechanisms by which it can occur may be confusing to you. Study Figure 24.11 to understand the difference between unequal crossing over and biased gene conversion.

- The difference between orthologs and paralogs may be troublesome to you. Studying Figure 24.12 should help you to understand the distinction.

Study Strategies

- Study Figure 24.1 carefully to understand how a similarity matrix is constructed. Similarity matrices are essential to reconstructing phylogenies based on molecular data.

- Review the following activities on the Companion Website/CD:
  Activity 24.1 Amino Acid Sequence Alignment
  Activity 24.2 Similarity Matrix Construction
  Activity 24.3 Gene Tree Construction

Test Yourself

Knowledge and Synthesis Questions

1. The *genome* of a eukaryotic organism is best defined as
   a. all of the organism's protein-coding genes.
   b. all of the organism's DNA contained in its nucleus.
   c. all of the organism's genetic material.
   d. a haploid set of all the organism's chromosomes.
   Textbook Reference: 24.1 What Can Genomes Reveal about Evolution? p. 525

2. The sequence alignment technique
   a. permits comparison of sequences of amino acids in proteins or sequences of nucleotides in DNA.
   b. enables the detection of deletions and insertions in sequences that are being compared.
   c. enables the detection of back substitutions and parallel substitutions in sequences that are being compared.
   d. Both a and b
   Textbook Reference: 24.1 What Can Genomes Reveal about Evolution? pp. 526–527

3. Experimental molecular evolutionary studies have shown that
   a. a heterogeneous environment favors adaptive radiation.
   b. a heterogeneous environment induces an increase in the mutation rate.
   c. it requires more than a year to demonstrate substantial molecular evolution even in bacteria.
   d. Both a and b
   Textbook Reference: 24.1 What Can Genomes Reveal about Evolution? pp. 528–529

4. In a eukaryote, where would you expect the rate of nonsynonymous nucleotide substitutions to be the *lowest*?
   a. In an intron of a protein-coding gene
   b. In an exon of a protein-coding gene
   c. In an intron of a pseudogene
   d. In an exon of a pseudogene
   Textbook Reference: 24.2 What Are the Mechanisms of Molecular Evolution? pp. 530–531

5. Which of the following statements about mutations is *false*?
   a. A silent mutation results in no change in the amino acid sequence of a protein.
   b. According to the neutral theory of molecular evolution, most substitution mutations are selectively neutral and accumulate through genetic drift.
   c. A base substitution mutation in the third codon position is more likely to be neutral than a substitution at the first or second codon position.
   d. Nonsynonymous mutations are virtually always deleterious to the organism.
   Textbook Reference: 24.2 What Are the Mechanisms of Molecular Evolution? pp. 531–532

6. Which of the following is *not* a valid conclusion based on studies of the structure of proteins, such as cytochrome *c* and lysozyme, in different species?
   a. The rate of evolution of particular proteins is often relatively constant over time.
   b. Many nucleotide substitutions result either in no change in the amino acid sequence of the protein or in a change to a functionally equivalent amino acid.
   c. Fewer differences were observed in the amino acid sequences of a protein if the organismal sources of the proteins were closely related.
   d. Functionally important regions of a protein can be discovered by identifying the regions with the most amino acid substitutions.
   Textbook Reference: 24.1 What Can Genomes Reveal about Evolution? pp.527–529; 24.2 What Are the Mechanisms of Molecular Evolution? pp. 531–533

7. Which of the following statements about the enzyme lysozyme is true?
   a. A small group of closely related mammals has evolved a special form of lysozyme that functions in digestion.

b. The lysozymes found in the foregut fermenters resulted from convergent evolution.

c. Lysozyme could not have evolved a secondary function if it had been an enzyme with a vital primary function.

d. A higher mutation rate in the foregut fermenters allows their lysozymes to evolve rapidly.

Textbook Reference: *24.2 What Are the Mechanisms of Molecular Evolution? pp. 532–533*

8. Which of the following sequences of organisms ranks them from least to most, in terms of the expected total amount of coding DNA in their genomes?

a. Bacterium, single-celled eukaryote, nematode, bird

b. Bacterium, nematode, bird, single-celled eukaryote

c. Single-celled eukaryote, bacterium, nematode, bird

d. Nematode, single-celled eukaryote, bird, bacterium

Textbook Reference: *24.2 What Are the Mechanisms of Molecular Evolution? p. 533*

9. Which of the following sequences of organisms ranks them from least to most, in terms of the *proportion* of coding DNA to noncoding DNA in their genomes?

a. *E. coli,* yeast, *Drosophila,* human

b. Human, yeast, *E. coli, Drosophila*

c. Human, *Drosophila,* yeast, *E. coli*

d. *Drosophila,* human, yeast, *E. coli*

Textbook Reference: *24.2 What Are the Mechanisms of Molecular Evolution? p. 534*

10. Which of the following is the *least* likely result of gene duplication?

a. The gene produces less of its product than it did before duplication.

b. The two copies of the gene are expressed at different stages in the development of the organism.

c. As a result of evolutionary divergence, one copy retains its original function, and the other copy acquires a different function.

d. One copy remains functional, and the other copy evolves into a functionless pseudogene.

Textbook Reference: *24.2 What Are the Mechanisms of Molecular Evolution? p. 534*

11. Which of the following statements about gene families is *false*?

a. Gene families evolve via gene duplication.

b. Pseudogenes are quickly removed from gene families by deletion.

c. Members of a gene family can include several functional genes.

d. Examples of gene families include the *engrailed* and globin gene families in vertebrates.

Textbook Reference: *24.2 What Are the Mechanisms of Molecular Evolution? p. 534*

12. Gene duplication via the mechanism of polyploidy

a. results in the duplication of the entire genome, apart from extranuclear DNA.

b. has occurred in the evolutionary history of many plants.

c. is believed not to have occurred in the evolutionary history of any animal groups.

d. Both a and b

Textbook Reference: *24.2 What Are the Mechanisms of Molecular Evolution? p. 534*

13. Nonindependent evolution of some repeated genes within a species

a. is called concerted evolution.

b. can be caused by biased gene conversion.

c. can be caused by unequal crossing over.

d. All of the above

Textbook Reference: *24.2 What Are the Mechanisms of Molecular Evolution? pp. 535–536*

14. Orthologous genes are genes that trace back to a common

a. duplication event.

b. substitution event.

c. speciation event.

d. deletion event.

Textbook Reference: *24.3 What Are the Mechanisms of Molecular Evolution? p. 537*

15. In vitro evolution

a. can produce both nucleic acid and protein molecules unknown in living organisms.

b. requires many rounds of production of many variant molecules and the selection of those having (or beginning to have) the desired properties.

c. often involves techniques and molecules employed in recombinant DNA technology, such as PCR and cDNA.

d. All of the above

Textbook Reference: *24.3 What Are Some Applications of Molecular Evolution? p. 538*

Application Questions

1. Determine the amino acid sequence for an eight-residue section of a small protein from five different species (1–5).

Position	1	2	3	4	5	6	7	8
Species 1:	Arg	Cys	Leu	Leu	Ser	Thr	Asn	Met
Species 2:	Arg	Cys	Phe	Leu	Leu	Ser	Thr	Asn
Species 3:	Arg	His	Leu	Leu	Ser	Thr	Asn	Met
Species 4:	Arg	Cys	Leu	Ser	Ser	Thr	Asn	Met
Species 5:	Arg	His	Leu	Leu	Ser	Gln	Asn	Met

Complete the following similarity matrix using these sequences:

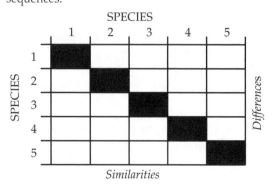

Which species differ from Species 1 because of an amino acid insertion?

Which species differ from Species 1 because of amino acid substitutions?

Textbook Reference: 24.1 What Can Genomes Reveal about Evolution? pp. 526–527

2. How might the proportion of coding to noncoding DNA in the genome of a species be related to the relative importance of natural selection and genetic drift in the evolution of the species?

Textbook Reference: 24.2 What Are the Mechanisms of Molecular Evolution? p. 534

3. How is one of the mechanisms of concerted evolution related to the pairing of homologous chromosomes that occurs during meiosis?

Textbook Reference: 24.2 What Are the Mechanisms of Molecular Evolution? pp. 535–536

4. Compare in vitro evolution with molecular evolution in organisms.

Textbook Reference: 24.3 What Are Some Applications of Molecular Evolution? p. 538

5. How is the study of molecular evolution important in efforts to combat HIV and other viral pathogens that have recently emerged?

Textbook Reference: 24.3 What Are Some Applications of Molecular Evolution? p. 539

Answers

Knowledge and Synthesis Answers

1. **c.** The genome includes not only protein-coding genes but all other DNA in the organism, including noncoding DNA and genes in organelles.

2. **d.** The sequence alignment technique can be employed to compare both amino acid sequences and nucleotide sequences. Although it enables the locations of deletions and insertions to be pinpointed, it is not capable of detecting such phenomena as back substitutions and parallel substitutions.

3. **a.** In the experiments of Rainey and Travisano, heterogeneous environments favored diversification in their bacterial cultures because of natural selection, not because of a difference in either the rates or kinds of mutations that occurred in homogeneous and heterogeneous environments. Experiments of this kind can produce observable evolutionary changes in a matter of months, at most.

4. **b.** Recall from Chapter 14 that eukaryotic genes usually contain both protein-coding regions (exons) and noncoding regions (introns). A substantial proportion of nonsynonymous substitutions occurring in an exon of a gene would most likely be deleterious and therefore be eliminated by selection. Because introns and pseudogenes are not expressed, nonsynonymous substitutions occurring in them cannot be selected against.

5. **d.** A nonsynonymous substitution mutation can be selectively neutral if, as sometimes happens, it results in an amino acid change that has no significant effect on the shape (and hence the functional properties) of the protein.

6. **d.** Functionally important regions of a protein can be discovered by identifying the regions with the fewest amino acid substitutions. Because the sequences in these regions have been optimized by natural selection to accomplish the function of the molecule, random changes in these areas are not tolerated.

7. **b.** Because the animals that evolved a similar lysozyme do not share a recent common ancestor, the mechanism involved is convergent evolution. All other statements are false.

8. **a.** Recall that there is a rough relationship between the amount of coding DNA and organismal complexity.

9. **c.** As shown in Figure 24.9 (p. 534), multicellular eukaryotes with relatively small populations and slow rates of development have the lowest proportion of coding to noncoding DNA, whereas quickly reproducing prokaryotes such as *E. coli* have the highest proportion. It is unclear whether developmental rate or population size (or both) is chiefly responsible for this trend.

10. **a.** If there are two copies of the gene, it is more likely to produce more of its product than less of it.

11. **b.** Pseudogenes (nonfunctional DNA) may be removed via deletion, but numerous pseudogenes persist in many gene families.

12. **d.** Polyploidy is a common event in plant evolution (as discussed in Chapter 23), and it does result in the duplication of the entire nuclear genome. But it is also believed to have occurred at least twice in the lineage that gave rise to the jawed vertebrates.

13. **d.** Concerted evolution is the phenomenon in which all the copies of a highly repeated gene (such as a gene coding for ribosomal RNA) remain highly similar despite experiencing nucleotide substitutions or other mutations that generally cause genes to diverge. Unequal crossing over during meiosis and biased gene conversion during DNA repair are two mechanisms that can produce concerted evolution.

14. **c.** Orthologs are homologous genes whose divergence can be traced to the speciation events that gave rise to the species in which the genes occur.

15. **d.** In vitro evolution can create novel molecules not known to occur in living organisms. The process starts with a very large pool of variant molecules (nucleic acid or protein) and involves selection on the part of the experimenters of those molecules that show some inkling of the desired property. Many rounds of production of variant molecules and selection are typically needed to produce the final product.

Application Answers

1.

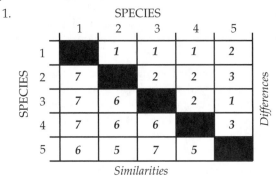

SPECIES 2 shows an amino acid insertion (Phe) at position 3. There are no deletions.

Species 3, 4, and 5 all show amino acid substitutions.

2. According to one hypothesis, the proportion of coding to noncoding DNA in the genome of a species is related to the sizes of the populations typical of the species. As discussed in Chapter 22, genetic drift has more influence on the evolution of small populations than of large ones. If individuals of a species differ in their fitness only slightly as a function of the proportion of noncoding DNA in their genomes, and if the populations of the species are typically small, then genetic drift could lead to retention of the noncoding DNA (or even an increase in its proportion), despite its selective disadvantage. On the other hand, the effect of genetic drift is minimal in very large populations. Therefore, in species characterized by large populations, even a slight disadvantage in fitness for individuals with a greater proportion of noncoding DNA should lead to a decrease in its proportion through the action of natural selection.

3. As discussed in Chapter 9, during the synapsis of homologous pairs of chromosomes in meiosis I, the alignment of the homologs is normally extremely precise. This ensures that the result of crossing over will be the exchange of equivalent sections of genetic material in the two chromatids involved in the crossover event. In the case of highly repeated genes, however, it sometimes happens that the alignment of the homologs is imprecise, as shown in Figure 24.11. The result is that crossing over yields one chromatid with extra copies of the gene, whereas the other has fewer copies. Subsequently, during meiosis II, the chromatids will become independent chromosomes that now possess the altered number of copies of the repeated gene.

4. Variation and selection are involved in both in vitro evolution and in molecular evolution in organisms. With in vitro evolution, a huge number of variant molecules are created in the laboratory, and the human experimenters select those that show any sign of having the desired property. Many repetitions of these two steps—production of variants and selection— eventually produce the targeted molecule. In molecular evolution in organisms, naturally occurring mutations in previously existing DNA sequences are the ultimate source of variation, and the environment determines which mutations are the favorable variants through natural selection. But also recall that according to the neutral theory of molecular evolution, much evolutionary change in DNA (and in the encoded proteins) is not adaptive but rather the result of genetic drift.

5. The principles of molecular evolution are critical for understanding the origin and evolutionary development of such pathogenic viruses as hantaviruses, the SARS virus, and HIV. By studying the evolutionary changes that occur in such viruses, medical researchers gain valuable insights into such questions as how some viruses are able to switch from an animal to a human host and how vaccines can remain effective as viruses evolve. In the future, as more extensive genomic databases and evolutionary trees for viruses are developed, it will be possible to identify and treat a much wider array of human diseases.

CHAPTER 25 Reconstructing and Using Phylogenies

Important Concepts

Phylogenetic trees show the evolutionary patterns of life on Earth.

- A phylogeny is a description of the evolutionary history of relationships among organisms (or their parts). Phylogenetic trees display the order in which lineages are hypothesized to have split. Each split (or node) in a phylogenetic tree represents a point at which lineages diverged in the past. The common ancestor of all the organisms in the tree forms the root of the tree. The timing of separations between lineages of organisms is shown by the positions of nodes on a time or divergence axis.

- A taxon is any named group of species. If a taxon consists of all the evolutionary descendants of a common ancestor, it is called a clade. Just as species that are each other's closest relatives are called sister species, so clades that are each other's closest relatives are called sister clades.

- Systematics is the scientific study of the diversity of life.

- All of life is connected through its evolutionary history. Any claim of an evolutionary association between a trait and a group of organisms is a statement about when during the history of the group the trait first arose and about the maintenance of the trait since its first appearance.

- Homologous traits are keys to reconstructing phylogenetic trees because they are features that are shared by members of a lineage owing to their descent from a common ancestral trait. A trait that differs from its ancestral form is called a derived trait. Conversely, a trait that was present in the ancestor of a group is known as an ancestral trait for that group.

- Synapomorphies are derived traits that are shared among a group of organisms and are viewed as evidence of the common ancestry of the group.

- Homoplasies (homoplastic traits) create confusion when reconstructing the evolutionary history of a lineage because they are features that are similar for some reason other than descent from a common ancestral trait.

- Two processes generate homoplasies: convergent evolution and evolutionary reversals. In convergent evolution, features that were independently evolved become superficially similar. In an evolutionary reversal, a character reverts from a derived state to an ancestral one.

Reconstructing phylogenies accurately involves several steps.

- The first step in reconstructing a phylogeny is to choose the ingroup and the appropriate outgroup. The ingroup is an assemblage of organisms whose phylogeny is to be determined. One method of distinguishing ancestral and derived traits is to compare the ingroup to an outgroup, which can be any species or group of species outside the group of interest. Ancestral traits are expected to be found in both the ingroup and the outgroup, whereas derived traits should occur only in the ingroup. The root of the tree is determined by the relationship of the ingroup to the outgroup.

- In reconstructing phylogenies, systematists use the parsimony principle, which states that the preferred explanation of the observed data is the simplest explanation. In practice, this means that the best phylogenetic reconstruction is the one that minimizes the number of evolutionary changes that need to be assumed over all characters in all groups in the tree. In other words, the best hypothesis is one that requires the fewest homoplasies.

- Phylogenies are constructed from many sources of data.

- Morphology—the presence, size, shape and other attributes of body parts—is an important source of traits for phylogenetic analysis.

- Similarities in developmental pattern may reveal evolutionary relationships. Structures in early developmental stages sometimes reveal evolutionary relationships that are not evident in adults.

- The morphology of fossils is particularly useful in helping to distinguish ancestral and derived traits. The fossil record also reveals when lineages diverged. In groups with few living representatives, information on extinct species may be critical to an understanding of the large divergences between the surviving species.

- Behavior, if genetically determined, is a useful source of information about evolutionary relationships.

- Like morphological characters, molecules are heritable characteristics that may diverge among lineages. The complete genome of an organism contains an enormous set of traits (the individual nucleotide bases of DNA) that can be used to analyze phylogenies. Both nuclear and organelle DNA sequences are used in phylogenetic studies, as are sequences in gene products (such as the amino acid sequences of proteins).

- Because the chloroplast genome has changed slowly over evolutionary time, it is often used for the study of relatively ancient phylogenetic relationships among plants. Animal mitochondrial DNA has changed more rapidly, making it useful for studies of evolutionary relationships among closely related animal species.

- The maximum likelihood method uses computer analysis for the reconstruction of phylogenies. A likelihood score of a tree is based on the probability of the observed data evolving on the specified tree, given an explicit mathematical model of evolution for the characters. The maximum likelihood solution is the most likely tree given the observed data. In comparison with parsimony methods, an advantage of maximum likelihood analyses is that they incorporate more information about evolutionary change.

- Biologists have conducted experiments both in living organisms and with computer simulations that have demonstrated the effectiveness and accuracy of phylogenetic methods.

- Phylogenetic methods are used not only to discover the evolutionary relationships among lineages of living organisms, but also to reconstruct morphological, behavioral and molecular characteristics of ancestral species.

- In a molecular clock analysis, biologists assume that particular DNA sequences evolve at a reasonably constant rate and can be used as a metric to gauge the time of divergence for a particular split in a phylogeny. Molecular clocks must be calibrated using independent data such as the fossil record, known times of divergence, or biogeographic dates.

Phylogenetic trees provide important information to biologists trying to answer many kinds of questions.

- Phylogenetic trees are used to determine how many times a particular trait has evolved within a lineage. They can also be used to determine when, where, and how zoonotic diseases (diseases caused by infectious organisms that have been transferred to humans from another animal host) first entered human populations.

- Phylogenetic analysis is used by biologists to help them make relevant biological comparisons among genes, populations, and species.

- In some cases, biologists can use phylogenies to predict future evolutionary events.

Biological classification systems are designed to express evolutionary relationships among organisms.

- The system of binomial nomenclature developed by Linnaeus in 1758 assigns each species two names, one identifying the species itself and the other the genus to which it belongs. The generic name (always capitalized) is followed by the species name (lowercase), and both are italicized.

- In the Linnaean system of classification, species are grouped into higher-order taxa. The hierarchy of taxa, ranked from most to least inclusive, is kingdom, phylum, class, order, family, genus, species. Thus, a genus includes one or more species, a family includes one or more genera, and so forth.

- Some biologists now de-emphasize the use of ranked classifications, because it is a largely subjective decision whether a taxon should be considered, for example, an order or a class. All biologists, however, believe that evolutionary relationships should be the basis for classifying organisms.

- Taxa in biological classifications are expected to be monophyletic. A monophyletic taxonomic group (also called a clade) contains an ancestor and all descendants of that ancestor and no other organisms.

- A polyphyletic group does not include its common ancestor, whereas a paraphyletic group includes some, but not all, descendants of a particular ancestor. Virtually all taxonomists agree that polyphyletic and paraphyletic groups are inappropriate as taxonomic units.

- Several sets of rules govern the use of scientific names, with the goal of providing unique and universal names for biological taxa. One such rule is that if a species is named more than once, the valid name is the first name proposed.

- In the past, different sets of taxonomic rules were developed by zoologists, botanists, and microbiologists, with the result that there are many duplicated names between groups. Today, taxonomists are developing rules to ensure that every taxon has a unique name.

The Big Picture

- It is fair to say that several decades ago systematics was widely regarded as one of the less dynamic arenas of research within the biological sciences. Several factors have combined in recent years to bring new excitement to the field. First was the development of a new, rigorous approach to systematics, known as cladistics. Later, the invention of methods of sequencing DNA and RNA, coupled with enormous increases in computer power, enabled biologists to apply sophisticated mathematical techniques to phylogenetic studies, such as maximum likelihood analyses. As a result, systematists are now contributing to a wide array of biological studies, including subjects of medical importance such as the origins and types of human immunodeficiency virus (HIV) and other infectious organisms.

- The parsimony principle, which holds that one should prefer the simplest hypothesis capable of explaining the known facts, is fundamental not only to the reconstruction of phylogenies but also to every other field of scientific research.

- The concept of evolution revolutionized taxonomy. Prior to Darwin's time, Linnaeus and others had developed "natural" systems of classification based primarily on similarity of morphology. Modern biologists realize that similarities of organisms (other than homoplasies) are the result of descent from a common ancestor and believe that a truly natural system of classification should reflect evolutionary relationships.

- Knowing that organisms are evolutionarily related enables biologists to make predictions about their characteristics. This knowledge can provide important hints in the search for organisms with valuable properties, such as the ability to produce medically useful drugs.

Common Problem Areas

- The concept of homology can be confusing. You will find Figure 25.2 very helpful for understanding the difference between homologous and homoplastic traits.

- Distinguishing monophyletic, paraphyletic, and polyphyletic groups is difficult for many students. Bear in mind that a monophyletic group is analogous to a branch (or twig) of a tree: a single "cut" can remove it from a phylogenetic tree. If you pursue this analogy, you should be able to work out what kinds of "cuts" would result in paraphyletic and polyphyletic groups.

Study Strategies

- Here is a mnemonic for the hierarchy of taxa (kingdom, phylum, class, order, family, genus, species) in the Linnaean system of classification: "**K**indly **P**rofessors **C**annot **O**ften **F**ail **G**ood **S**tudents." Note that the plural of genus is genera; the words *general* and *generic* come from the same root and can help you remember that the genus is the more general taxon in the genus/species binomial. *Species* and *specific* also come from the same root, and the species is, of course, the most specific taxon.

- Review the following animated tutorial and activities on the Companion Website/CD:
Tutorial 25.1 Using Phylogenetic Analysis to Reconstruct Evolutionary History
Activity 25.1 Constructing a Phylogenetic Tree
Activity 25.2 Types of Taxa

Test Yourself
Knowledge and Synthesis Questions

1. A group that consists of all the evolutionary descendants of a common ancestor is called a(n)
   a. clade.
   b. taxon.
   c. homology.
   d. ingroup.
   Textbook Reference: 25.1 What Is Phylogeny? p. 543

2. A synapomorphy is
   a. the product of convergent evolution.
   b. the result of an evolutionary reversal.
   c. a shared derived characteristic.
   d. a trait that was present in the ancestor of a group.
   Textbook Reference: 25.1 What Is Phylogeny? p. 545

3. Members of genus X, a hypothetical taxon of invertebrates, have antennae with a variable number of segments. Species A and B have 10 segments; species C and D have 9 segments; species E has 8 segments. In all other genera in this family (including genus Y), all species have antennae with 10 segments. Which of the following character states is a synapomorphy useful for determining evolutionary relationships within genus X?
   a. 10-segment antennae in species A and B
   b. 10-segment antennae in genus Y and in two species of genus X
   c. Antennae with fewer than 10 segments in species C, D, and E
   d. 8-segment antennae in species E
   Textbook Reference: 25.1 What Is Phylogeny? p. 545

4. Which of the following would *not* be expected to result in homoplasy?
   a. Convergent evolution
   b. The independent evolution of similar structures in different lineages
   c. Selection for traits that perform similar functions
   d. The inheritance of ancestral traits
   Textbook Reference: 25.1 What Is Phylogeny? p. 545

5. A derived trait is one that
   a. differs from its ancestral form.
   b. is homologous with another trait found in a related species.
   c. is the product of an evolutionary reversal.
   d. has the same function, but not the same evolutionary origin, as a trait found in another species.
   Textbook Reference: 25.1 What Is Phylogeny? p. 545

6. Which of the following statements about reconstructing phylogenies is *false*?
   a. Traits found in the outgroup as well as in the ingroup are likely to be ancestral traits.
   b. Shared traits are generally assumed to be homoplastic until they can be proven to be homologous.

c. Phylogenetic trees do not always provide an explicit time scale by which to date the splits between lineages.

d. In a phylogenetic tree, branches can be rotated around any node without changing the meaning of the tree.

Textbook Reference: *25.2 How Are Phylogenetic Trees Constructed? p. 545*

7. The *most* important limitation of fossils as a source of information about evolutionary history is that
   a. it is sometimes impossible to determine when a fossil organism lived.
   b. the fossil record for many groups is fragmentary or even nonexistent.
   c. most fossils contain no nucleic acids or proteins and therefore are worthless for studies of molecular evolution.
   d. it is impossible to determine if morphologically similar fossils belong to the same species because one cannot know if the fossil species could have interbred.

Textbook Reference: *25.2 How Are Phylogenetic Trees Constructed? p. 548*

8. Which source of molecular data would be most helpful for a study of the evolutionary relationships of closely related animal species?
   a. Chloroplast DNA
   b. Mitochondrial DNA
   c. The amino acid sequences of a protein found in all animals, such as cytochrome *c*
   d. Ribosomal RNA sequences

Textbook Reference: *25.2 How Are Phylogenetic Trees Constructed? p. 548*

9. Which of the following statements about the use of molecular clocks in phylogenetic analyses is true?
   a. A given gene usually evolves at the same rate in two different species independent of differences in generation time of the species.
   b. Molecular clocks must be calibrated using independent data, such as the fossil record.
   c. Even in a group of closely related species, it has been found that different genes evolve at different rates.
   d. Both b and c

Textbook Reference: *25.2 How Are Phylogenetic Trees Constructed? p. 551*

10. Which of the following statements describes a purpose for which biologists use phylogenetic trees?
    a. For human diseases once found only in other animals, phylogenetic trees are helpful in determining when and where the infectious organisms first entered human populations.
    b. Phylogenetic trees are useful for determining how many times a particular trait may have independently evolved within a lineage.
    c. Phylogenetic trees can sometimes be used to make predictions about future evolution.
    d. All of the above

Textbook Reference: *25.3 How Do Biologists Use Phylogenetic Trees? pp. 551–554*

11. The *most* important attribute of a biological classification scheme is that it
    a. avoids the ambiguity created by using common names.
    b. reflects the evolutionary relationships among organisms.
    c. helps us remember organisms and their traits.
    d. improves our ability to make predictions about the morphology and behavior of organisms.

Textbook Reference: *25.4 How Does Phylogeny Relate to Classification? p. 555*

12. Suppose you are writing a scientific paper about a unicellular green alga called *Chlamydomonas reinhardtii*. What would be the proper way to refer to this species after you had used the full binomial earlier in the same paragraph?
    a. *Chlamydomonas reinhardtii*
    b. *Chlamydomonas* spp.
    c. *Chlamydomonas* sp.
    d. *C. reinhardtii*

Textbook Reference: *25.4 How Does Phylogeny Relate to Classification? p. 554*

13. The organisms that make up a class are _____ diverse and _____ numerous than those in a family within that class. The organisms that make up a phylum all diverged from a common ancestor _____ recently than did the organisms in an order within that phylum.
    a. more; more; less
    b. more; more; more
    c. more; less; less
    d. less; less; more

Textbook Reference: *25.4 How Does Phylogeny Relate to Classification? pp. 554–555*

14. Which of the following incomplete lists of taxonomic categories ranks them properly from most inclusive to least inclusive?
    a. Phylum, order, family, genus
    b. Class, phylum, order, species
    c. Order, class, family, genus
    d. Family, order, class, kingdom

Textbook Reference: *25.4 How Does Phylogeny Relate to Classification? p. 554*

15. The ratites are a group of flightless birds comprising the ostrich, emu, cassowaries, rheas, and kiwis. All share certain morphological similarities (such as a breastbone without a keel) not found in other birds, but they live on different continents. In the past, some ornithologists regarded their similarities as homoplasies, but they are now thought to be synapomorphies. Based on this information, you would conclude that the ratites were once regarded as a _____ group but are now believed to be _____.
    a. polyphyletic; paraphyletic
    b. paraphyletic; monophyletic

c. polyphyletic; monophyletic

d. monophyletic; polyphyletic

Textbook Reference: 25.4 *How Does Phylogeny Relate to Classification? p. 555*

Application Questions

1. Based on the following phylogenetic tree showing the evolutionary relationships of five species (A–E) relative to five traits (1–5), fill in the table using 1 to indicate the presence of a derived trait and 0 to indicate the presence of an ancestral trait.

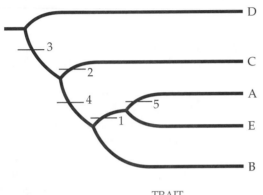

SPECIES	TRAIT				
	1	2	3	4	5
A					
B					
C					
D					
E					

Textbook Reference: 25.2 *How Are Phylogenetic Trees Constructed? p. 546*

2. In the phylogenetic tree shown in Question 1, which species is considered the outgroup? Explain your answer.

Textbook Reference: 25.2 *How Are Phylogenetic Trees Constructed? p. 546*

3. Based on the following table showing the ancestral and derived traits of five species (A–E), construct a phylogenetic tree that represents the evolutionary relationships of this group using conventions presented in the textbook. In this table, the ancestral state of each trait is indicated by 0, and the derived state is indicated by 1.

SPECIES	TRAIT				
	1	2	3	4	5
A	1	1	1	0	0
B	0	0	0	0	0
C	1	1	0	1	0
D	0	1	0	0	0
E	1	1	0	1	1

Textbook Reference: 25.2 *How Are Phylogenetic Trees Constructed? pp. 546–547*

4. Discuss the application of the parsimony principle in the construction of phylogenetic trees.

Textbook Reference: 25.2 *How Are Phylogenetic Trees Constructed? pp. 546–547*

5. Discuss the implications the following statement has for systematics: "DNA is the genetic material for all prokaryotes and eukaryotes."

Textbook Reference: 25.1 *What Is Phylogeny? pp. 544–545*

Answers

Knowledge and Synthesis Answers

1. **a.** A clade can be thought of as a complete branch on the tree of life. It includes the ancestor of a group, all the ancestor's descendants, and no other organisms.

2. **c.** Synapomorphies are traits that are not found in the ancestor of a group (hence they are derived) and that are found in more than one member of a group (hence they are shared).

3. **c.** Because antennae with 10 segments are found in all genera in this family except for genus X, the trait of 10-segment antennae is best regarded as ancestral; hence, having fewer than 10 segments in the antennae is a derived trait. The presence of 8-segment antennae in species E is not a synapomorphy because it is a trait only found in that species.

4. **d.** Homoplasy is the appearance of similar structures in different lineages that were not present in the common ancestor.

5. **a.** Derived traits are those that have undergone a change during evolution from the ancestral (original) character state.

6. **b.** Most shared traits, especially in species with a recent common ancestor, are likely to be homologous, not homoplastic. Therefore it is more consistent with the parsimony principle to assume that traits are homologous until proven homoplastic than to assume the reverse.

7. **b.** The incompleteness of the fossil record is by far the greatest limitation on its usefulness in determining phylogenies.

8. **b.** Mitochondrial DNA in animals changes rapidly over evolutionary time and hence would be most useful in determining evolutionary relationships among species that have only recently diverged from one another.

9. **d.** It is true that molecular clocks must be calibrated by using independent data, and it is true that different genes and other DNA sequences evolve at different rates (as described in Chapter 24). But it is not true that the rate of gene evolution is independent of generation time or several other biological factors.

10. **d.** The textbook describes specific examples of all three of the purposes listed as answers.

11. **b.** Although all of the statements listed are important attributes of biological classification schemes, the most important attribute is that biological classification reflects evolutionary relationships.

12. **d.** After a scientific name is referenced once in a paragraph, typically the genus is abbreviated, but the species name is given in full.

13. **a.** Organisms in a higher taxon are *less* similar, have diverged from a common ancestor *less* recently, and include *more* species than organisms in a lower, included taxon.

14. **a.** The complete hierarchy of taxonomic categories, from most to least inclusive, is: kingdom, phylum, class, order, family, genus, species.

15. **c.** If the shared characteristics of the ratites are homoplasies (meaning that they evolved independently by convergent evolution), then the group did not have a single common ancestor and is polyphyletic. If (as is now accepted) the shared traits are synapomorphies that are shared among all ratites and are not found in other birds, then the group is monophyletic.

Application Answers

1.

	TRAIT				
SPECIES	1	2	3	4	5
A	1	0	1	1	1
B	0	0	1	1	0
C	0	1	1	0	0
D	0	0	0	0	0
E	1	0	1	1	0

2. Species D. The outgroup has the ancestral form for all traits.

3.

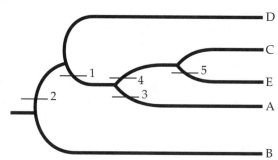

4. In the construction of a phylogenetic tree, you initially assume that derived traits appear only once and never disappear. Given a set of traits for a group of species, sometimes these restrictions must be relaxed to produce a phylogenetic tree for the group. Parsimony involves arranging the species so that you minimize the number of required reversals and multiple origins. Generally, the simplest explanation is most likely to be correct.

5. Some of the implications of this statement are that DNA evolved as the genetic material before eukaryotes had diverged from prokaryotes, that DNA is an ancestral and general homologous trait, and that all surviving eukaryotes use DNA as the genetic material.

CHAPTER 26 Bacteria and Archaea: The Prokaryotic Domains

Important Concepts

Domains attempt to reflect evolutionary relationships among organisms.

- All living things can be placed into one the following three domains: Archaea, Bacteria, and Eukarya. This system reflects evolutionary relationships among these monophyletic groups.

- All organisms have similarities; only derived similarities (synapomorphies) indicate evolutionary relationships. Eukarya have nuclei, membrane-enclosed organelles, and cytoskeletons. All organisms that possess these are descended from the most recent common ancestor of the Eukarya. The absence of these features in Archaea and Bacteria is a retained characteristic inherited from the common ancestor of all life on Earth (a prokaryote). Unique and distinguishing features of Archaea and Bacteria have been discovered through biochemistry and genetic studies.

- All three domains arose from a common ancestor possessing a circular chromosome and machinery for transcription and translation of the DNA code to protein. Divergence from this ancestor occurred billions of years ago. The domain Archaea is more closely related to the Eukarya than to the Bacteria.

Prokaryotes have many features in common.

- Though they are members of two different domains, Archaea and Bacteria have several structural similarities that are indicative of evolutionary adaptation to similar environments.

- Prokaryotes are prolific and are the most numerous of all organisms. Because they are single celled (there are a few exceptions) and very small, they generally escape human notice. They play extremely important roles in the biosphere, for instance in nutrient cycling. They can also metabolize many substances that eukaryotes cannot.

- Prokaryotes generally exist in one of three shapes; spherical (cocci), rod-shaped (bacilli), and spiral-shaped. All of the different shapes may aggregate to form chains or plates of cells. Advanced associations may result in filaments that are encased in a protective sheath.

- Prokaryotes do not have membrane-enclosed organelles; therefore, the organization and replication of DNA proceeds differently in prokaryotes than it does in eukaryotes. Also, the processes of cellular respiration and photosynthesis are carried out differently and rely on infoldings of the cell membrane. Finally, prokaryotes lack a cytoskeleton; therefore, true mitosis does not occur. Prokaryotes go through an asexual process called fission after replicating their DNA.

- Many prokaryotes are mobile and move by gliding, rolling, regulating gas vesicles, or flagella. Bacterial flagella are whiplike and are composed of a single protein called flagellin. In contrast, the flagellum of eukaryotes is complex.

- The cell wall of prokaryotes is thick and rigid and composed of peptidoglycans (bacteria) or protein (archaea). This cell wall is very different from that of plant and fungal cell walls. Cell wall composition is one of many ways of differentiating bacteria and archaea.

- Cell wall structure within the bacteria also serves as a diagnostic characteristic among different bacteria. This difference is used in the Gram staining technique. Whether a cell is Gram-negative or Gram-positive relates to the amount of peptidoglycan in the cell wall.

  - Gram-positive cells stain purple and have a thick layer of peptidoglycan.

  - Gram-negative cells do not retain stain and appear red due to a thin layer of peptidoglycan.

- Some antibiotics work by interfering with the synthesis of peptidoglycan-containing cell walls. Eukaryotic cells are not harmed because their cell walls lack this material.

- Prokaryotes reproduce asexually but may exchange genetic information through transformation, conjugation, or transduction.

- Metabolic processes in prokaryotes are varied and diverse.

- Obligate anaerobes are oxygen-sensitive and can only survive in oxygen-poor conditions. Facultative anaerobes can shift between aerobic and anaerobic modes of respiration. Some facultative anaerobes do not respire aerobically but are not harmed by an oxygen-rich atmosphere as obligate anaerobes are. Obligate aerobes require oxygen and cannot survive in the absence of it.

- Photoautotrophs are photosynthetic and convert light energy to chemical energy. Some photoautotrophs use chlorophyll *a* and give off oxygen as a by-product. Others use bacteriochlorophyll and produce sulfur as a by-product.

- Photoheterotrophs use light for energy but also require outside carbon sources.

- Chemolithotrophs oxidize inorganic substances for energy. Some chemolithotrophs fix carbon dioxide into carbohydrate. Others fix nitrite or ammonia, hydrogen gas, hydrogen sulfide, and other materials.

- Chemoheterotrophs obtain energy and carbon from consuming organic molecules.

- Some bacteria carry out respiratory electron transport using nitrogen or sulfur compounds as electron acceptors. These organisms are extremely important in nutrient cycling.

- Prokaryotes make diverse contributions to their environments.

- The nitrogen cycle depends on nitrification by bacteria. Nitrifying bacteria use nitrogen to make ATP, which is used to fix carbon dioxide into glucose and other food molecules. Through this process, unusable nitrogen sources are converted into biologically available nitrogen sources.

- Cyanobacteria, using photosynthesis, enriched the atmosphere with oxygen. Oxygen levels became too high for many existing organisms; many others evolved to fill those niches, and new organisms arose.

- Prokaryotes live in many mutualistic associations. Many of your digestive processes rely on the bacteria that inhabit your intestine.

- A small minority of bacteria are pathogenic. Bacteria cause disease because they produce toxins that are harmful to host tissues. Endotoxins are released when bacterial cells burst, but are rarely fatal. Exotoxins are secreted by bacteria and may be fatal.

- The German physician Robert Koch came up with criteria for determining if a bacterium was a disease-causing agent:
  - The organism must always be found in the individual with the disease.
  - The organism must be taken from the host and grown in pure culture.
  - A sample of the culture must produce disease if injected into a healthy individual.
  - The newly infected individual must yield a new pure culture of the same organism.

- In order to be pathogenic, an organism must be able to reach a new host, invade the host, evade the host's defenses, multiply, and infect a new host. Relatively few organisms can do this.

- Evolutionary relationships among prokaryotes are determined at the genetic level.

- Physical characteristics are used in classification for identification purposes, but to understand the true evolutionary relationships among very small organisms such as prokaryotes, it is necessary to look at nucleic acid sequences and biochemistry.

- Ribosomal RNA gives us information about relationships among organisms. It is useful for this purpose because it is evolutionarily ancient, is found in all organisms, has the same function in all organisms, and is highly conserved (it evolved slowly). Scientists compare signature sequences to make evolutionary inferences.

- rRNA sequences have not provided clear data. Lateral gene transfer—genetic transfer between species—makes these data hard to understand.

- Mutations are important in prokaryotic variation. Because prokaryotes have single circular DNA pieces, mutations cannot be masked and are passed on through fission. This is in contrast to diploid organisms in which mutation plays a small role in genetic variation.

- Prokaryotes evolve comparatively rapidly because of rapid reproduction, persistent mutation, and selection. Mutations leading to antibiotic resistance are of extreme concern and can occur in a matter of just a few years.

The majority of known prokaryotes are bacteria.

- Bacteria can be subdivided into a series of monophyletic groups. The composition (membership) of these groups is not considered to be the same by all scientists, and this is an active area of research.

- Spirochetes are corkscrew-shaped prokaryotes characterized by the presence of axial filaments that run through their periplasmic space. Many spirochetes are parasites of humans.

- Chlamydias are the smallest bacteria and exist only as parasites. They have a two-stage life cycle and are responsible for several human diseases.

- Actinobacteria or high-GC Gram-positive bacteria develop an elaborately branched system of filaments, some with chains of spores at the tips. This group includes the pathogen *Mycobacterium tuberculosis* (tuberculosis). However, one of our most potent antibiotics comes from this group. Streptomycin is produced by the actinobacteria *Streptomyces*. Most currently used antibiotics come from the actinobacteria.

- Cyanobacteria undergo photosynthesis similar to eukaryotic photosynthesizers. They contribute to the current oxygen-rich environment. These organisms frequently exist in colonies with distinct divisions of labor. Some cells are vegetative and photosynthesize. Others form resting stages called spores that can withstand harsh environmental conditions. Still others, called heterocysts, fix nitrogen.

- Most, but not all, firmicutes are low-GC Gram-positive. They produce endospores that are capable of surviving in extremely harsh conditions. Many medically important bacteria fall within this group, including *Clostridium botulinum* (botulism), *Staphylococcus* (many skin infections), and *Bacillis anthracis*, causative agent of anthrax. Another subgroup, the mycoplasmas, may represent the smallest life form.

- The five groups of proteobacteria represent the largest number of known species. This group is believed to be the origin of the eukaryotic mitochondria. The early ancestor of this group was probably a photoautotroph, but today, the group encompasses multiple metabolic schemes. Included among the proteobacteria are *E.coli*, the nitrogen-fixing *Rhizobium*, and many human pathogens.

Archaea live in some of the harshest known environments.

- Very little is known about the phylogeny of the Archaea, and much research still needs to be done, but there are shared characteristics that distinguish them from the Bacteria. Archaea lack peptidoglycan cell walls and have very distinctive lipids in their cell membranes. These are linked by ether linkages. No other organisms possess this type of lipid.

- Crenarchaeota live in hot, acidic environments and are thermophilic and acidophilic. They require temperatures above 55°C and can survive in an environment with a pH as low as 0.9 while maintaining an internal pH of 7.

- Some members of Euryarchaeota produce methane gas and are obligate anaerobes. These organisms are prevalent in the guts of herbivores. Others live near volcanic vents. Another group of Euryarchaeota live in extremely salty conditions and are referred to as halophiles.

- Korarchaeota are archaea known only from DNA isolated from hot springs.

- Nanoarchaeota, obtained from a deep-sea hydrothermal vent, is minute and lives attached to cells of *Ignicoccus*, a crenarchaeote.

The Big Picture

- All organisms fall within one of three domains. This chapter focuses on the domains Archaea and Bacteria. Subsequent chapters focus on the Eukarya. Archaea and Bacteria are prokaryotic organisms and retain many features of the earliest forms of life on Earth. They have no membrane-enclosed organelles, circular chromosomes, or cell walls of peptidoglycans or proteins. Their structures are simple, and they exist as single cells. Some prokaryotes are stationary, others move by means of flagella. Their metabolic processes

are highly varied, as are their modes for procuring energy. Many prokaryotic organisms make valuable contributions to their environments through nitrogen fixation, photosynthesis, and other metabolic processes. The majority of prokaryotes are free-living, but some live in mutualistic relationships, and yet others are pathogenic to other organisms.

- The evolutionary relationships among prokaryotes are still highly disputed and best understood at the gene level. The most ubiquitous extant prokaryotes are the bacteria. They are grouped according to physical and biochemical characteristics, and these groupings do not necessarily reflect their evolutionary relationships. The Archaea are more closely related to the Eukarya, and they exist in some of the harshest known environments.

Common Problem Areas

- Because this chapter covers a large number of diverse, microscopic organisms included in two of the three domains of life it is easy to get overwhelmed by the details and exceptions. Many of the names of these groups are long, complex, and unfamiliar. Because many of the distinguishing features of the different groups of bacteria and archaea are biochemical, you may need to review information from Part II of your textbook.

Study Strategies

- This chapter is an overview of two domains of organisms. A single chapter can only cover a few points of prokaryotic diversity. Focus your study on what makes each domain distinct (its synapomorphies) and the unique features that indicate a closer evolutionary relationship between archaea and eukaryotes than between archaea and bacteria. Move on to learning about sample organisms from each domain and groups within each domain.

- The temptation is to focus on the pathogenic organisms, but remember that only a very small percentage of prokaryotes are pathogenic.

- Review the following animated tutorial and activity on the Companion Website/CD:
  Tutorial 26.1 The Evolution of the Three Domains
  Activity 26.1 Gram Stain and Bacteria

Test Yourself

Knowledge and Synthesis Questions

1. Which of the following characteristics is unique to prokaryotes?
   a. Lack of membrane-enclosed organelles
   b Presence of cell walls
   c. Presence of plasma membranes
   d. All of the above
   Textbook Reference: *26.1 How Did the Living World Begin to Diversify? p. 562*

2. A soil sample contains large numbers of rod-shaped bacteria. These bacteria are referred to as
   a. bacilli.
   b. cocci.
   c. spiral.
   d. helici.
   Textbook Reference: 26.2 Where are Prokaryotes Found? p. 563

3. Which of the following statements concerning prokaryotes is true?
   a. Because prokaryotes do not contain organelles, they cannot photosynthesize or carry out cellular respiration.
   b. Prokaryotes have no chromosomes and therefore lack DNA.
   c. Prokaryote flagella are similar in structure to eukaryote flagella.
   d. None of the above
   Textbook Reference: 26.1 How Did the Living World Begin to Diversify? pp. 561–562

4. Gram-negative bacteria stain pink because
   a. they have specialized lipids in their cell walls.
   b. their peptidoglycan layer is thin.
   c. their peptidoglycan layer is thick.
   d. they are receptive to antibiotics.
   Textbook Reference: What are Some Keys to the Success of Prokaryotes? p. 565

5. Archaea are more closely related to _____ than to any other monophyletic group.
   a. bacteria
   b. eukaryotes
   c. bacteria and eukaryotes
   d. fungi
   Textbook Reference: 26.1 How Did the Living World Begin to Diversify? p. 562

6. The dense films laid down by many prokaryotes are
   a. endotoxins.
   b. denitrifiers.
   c. biofilms.
   d. pathogens.
   Textbook Reference: 26.2 Where are Prokaryotes Found? pp. 563–564

7. Which of the following groups of bacteria has the highest proportion of pathogens?
   a. Proteobacteria
   b. Cyanobacteria
   c. Chlamydias
   d. None of the above
   Textbook Reference: 26.5 What Are the Major Known Groups of Prokaryotes? p. 575

8. The mitochondria of eukaryotes were derived from _____ by endosymbiosis.
   a. proteobacteria
   b. chemoheterotrophs
   c. eukaryotes
   d. archaea
   Textbook Reference: 26.5 What Are the Major Known Groups of Prokaryotes? p. 575

9. Autoclaves must pass a "spore test" in many states to demonstrate that they are able to sterilize equipment. The spore-producing bacteria they use for this test are most likely from which of the following groups of bacteria?
   a. Low-GC Gram-positive bacteria (Firmicutes)
   b. Proteobacteria
   c. Cyanobacteria
   d. Chlamydias
   Textbook Reference: 26.5 What Are the Major Known Groups of Prokaryotes? p. 573

10. Archaea often live in harsh environments. Archaea that live in extremely salty conditions are referred to as
    a. thermophiles.
    b. halophiles.
    c. salinophiles.
    d. None of the above
    Textbook Reference: 26.5 What Are the Major Known Groups of Prokaryotes? p. 577

11. Methane gas contributes to the greenhouse effect that is raising atmospheric temperatures. A large portion of all methane emission is from grazing cattle. This is because cows harbor methane-producing archaea from which of the following groups?
    a. Crenarchaeota
    b. Euryarchaeota
    c. Anarchaeota
    d. Proteobacteria
    Textbook Reference: 26.5 What Are the Major Known Groups of Prokaryotes? p. 577

12. One of the major diagnostic characteristics that distinguishes archaea from bacteria is the lack of
    a. peptidoglycan cell walls in archaea.
    b. peptidoglycan cell walls in bacteria.
    c. ribosomes in archaea.
    d. chemoautotrophic bacteria.
    Textbook Reference: 26.5 What Are the Major Known Groups of Prokaryotes? p. 575

13. A bacterium that requires a carbon source other than carbon dioxide, yet can convert light energy to chemical energy, is called a
    a. photoautotroph.
    b. photoheterotroph.
    c. chemoautotroph.
    d. chemoheterotroph.
    Textbook Reference: 26.3 What Are Some Keys to the Success of Prokaryotes? p. 568

14. A bacterium that *cannot* live in the presence of oxygen is called a(n)
    a. obligate aerobe.
    b. facultative aerobe.
    c. obligate anaerobe.
    d. facultative anaerobe.
    Textbook Reference: 26.3 What Are Some Keys to the Success of Prokaryotes? p. 568

15. Another name for the high-GC Gram-positive bacteria is
    a. proteobacteria.
    b. actinobacteria.
    c. firmicutes.
    d. thermophiles.
    Textbook Reference: *26.5 What Are the Major Known Groups of Prokaryotes? p. 572*

Application Questions

1. You work for a wastewater treatment plant. An organism is getting past your treatment process, and fish are dying in a stream that the runoff enters. You are able to isolate the organism. Using a microscope, how can you tell if the organism is a prokaryote or a eukaryote?
   Textbook Reference: *26.1 How Did the Living World Begin to Diversify? p. 562*

2. You work with the Centers for Disease Control in Atlanta. You have been called to a remote area of Uganda to study a mysterious disease that is causing respiratory ailments in a small village. You isolate a bacterium from several patients that seems like a good candidate for the pathogen. How can you determine if this bacterium is causing the illness?
   Textbook Reference: *26.6 How do Prokaryotes Affect Their Environments? p. 579*

3. As a U.S. Department of Agriculture field representative, you counsel a young farmer to plant alfalfa in fields with soils that have low nitrogen levels. You know alfalfa roots are hosts for nitrogen-fixing bacteria. How will the alfalfa and its associated bacteria help "fertilize" this farmer's soil?
   Textbook Reference: *26.6 How do Prokaryotes Affect Their Environments? p. 578*

4. A patient comes to your medical practice with a bacterial infection. You are terribly concerned because the bacteria she is harboring produce an exotoxin. Why are exotoxins often more dangerous than endotoxins?
   Textbook Reference: *26.6 How do Prokaryotes Affect Their Environments? p. 579*

5. Describe how Gram staining depends on bacterial cell wall structure. Would Gram staining work for archaea?
   Textbook Reference: *26.3 What Are Some Keys to the Success of Prokaryotes? p. 565*

6. Differentiate between the following terms:
   a. obligate anaerobe/obligate aerobe
   b. photoautotroph/photoheterotroph
   c. chemolithotroph/chemoheterotroph
   Textbook Reference: *26.3 What Are Some Keys to the Success of Prokaryotes? p. 568*

Answers

Knowledge and Synthesis Answers

1. **a.** Prokaryotes are differentiated from eukaryotes by lacking membrane-enclosed organelles. They also have a single circular piece of genomic DNA and lack a cytoskeleton.

2. **a.** A rod-shaped bacterium is known as a bacillus (plural, bacilli). Cocci are roughly spherical.

3. **d.** Bacteria do carry out both photosynthesis and multiple forms of cellular respiration. All life forms have DNA. The flagella of bacteria consist of a single protein called flagellin rather than multiple proteins as found in eukaryotes.

4. **b.** Gram-negative bacteria have thin peptidoglycan layers that do not retain crystal violet stain.

5. **b.** The Archaea and the Eukarya share a more recent common ancestor with each other than either does with the Bacteria. The evidence for this is, for example, a signature sequence from rRNA that has been found in all archaea and eukaryotes tested so far, but in none of the bacteria.

6. **c.** Biofilms are gel-like polysaccharide matrices that are laid down by prokaryotes and trap other bacteria.

7. **c.** Chlamydias are all parasitic and cause several human diseases.

8. **a.** Endosymbiosis of proteobacteria gave rise to the mitochondria of eukaryotes.

9. **a.** Among the groups listed, only low-GC Gram positive bacteria (firmicutes) and cyanobacteria produce spores. Firmicute spores can withstand extreme environmental conditions, so they would make good spores for testing an autoclave's ability to sterilize equipment.

10. **b.** The term "halophile" means "salt loving."

11. **b.** Methanogenic bacteria that reside in the guts of cows belong to the Euryarchaeota.

12. **a.** Archaea lack peptidoglycans. Instead they have a unique lipid in their cell walls.

13. **b.** Photoheterotrophs require an outside carbon source other than carbon dioxide, yet are able to harvest light energy.

14. **c.** Oxygen gas is toxic to obligate anaerobes.

15. **b.** The actinobacteria are the filament-forming high-GC Gram positives that resemble the mycelia produced by fungi.

Application Answers

1. The most immediately apparent difference between a prokaryote and a eukaryote would be the absence of membrane-enclosed organelles in prokaryotes.

2. You can use Koch's postulates. These postulates stipulate that the organism must always be found in individuals with the disease, the organism must be taken from the host and grown in pure culture, a sample of the culture must produce disease if injected into a healthy individual, and the newly infected individual must yield a new pure culture of the same organism.

3. Nitrogen-fixing bacteria can convert atmospheric nitrogen into ammonia. Other bacteria can then convert the ammonia into nitrates that plants can use.

4. Exotoxins are continuously produced by the infecting bacteria. Endotoxins are produced only at cell death. There are a limited number of endotoxin molecules released. Exotoxins can be released until the host dies.

5. Gram staining results depend on the arrangement and amounts of peptidoglycans in the bacterial cell wall. This technique does not work with archaea because they do not have peptidoglycan in their cell walls.

6. Obligate anaerobes die in the presence of oxygen gas. Obligate aerobes die without oxygen gas for cellular respiration. Photoautotrophs convert light energy to chemical energy using carbon dioxide as their carbon source. Photoheterotrophs rely on an outside carbon source other than carbon dioxide but still can convert light energy to chemical energy. Chemolithotrophs can produce what they need from inorganic molecules. Chemoheterotrophs depend on nutrients from external sources.

CHAPTER 27 The Origin and Diversification of the Eukaryotes

Important Concepts

Protists are a paraphyletic group of very diverse organisms.

- To avoid the notion that the term "protist" has any real taxonomic meaning, your textbook instead uses the term "microbial eukaryotes" for all eukaryotes that are neither land plants, fungi, nor animals.

- Because of the "catch-all" nature of this grouping, it is extremely diverse, containing many different organisms that are not necessarily closely related to each other. Most are unicellular, but many are colonial and some are multicellular.

The origin of the eukaryotic cell was pivotal in evolution.

- The origin of the eukaryotic cell is still being debated. The evolution of the eukaryotic cell was revolutionary and occurred at a time of great environmental change.

- The eukaryotic cell contains a flexible cell surface, a cytoskeleton, a nuclear envelope, digestive vesicles, and organelles.

- The first step in the path toward a eukaryotic cell may have involved loss of the rigid cell wall, which allowed infoldings of the cell surfaces to increase the surface area-to-volume ratio. This allows cells to become larger. It also allows vesicles to form from bits of infolded membrane that pinch off. Such infolding may have been the origin of the nuclear envelope (see Figure 27.6).

- New structures evolved. Ribosomes became associated with internal membranes to form the endoplasmic reticulum, a cytoskeleton of actin and microtubules formed, and digestive vesicles developed. The cytoskeleton opened up modes of locomotion with actin/myosin-based flagella. How these structures formed is currently conjecture.

- It is thought that these early motile cells may have been phagocytes that engulfed prokaryotic cells.

- Subsequent endosymbiotic events may have led to the evolution of mitochondria and chloroplasts.

- See Figure 27.7 for a diagram of events in the evolution of the eukaryotic cell.

- An understanding of the evolution of eukaryotes is complicated by lateral gene transfer. The prokaryotic genes in many eukaryotes are so numerous that lateral transfer cannot explain them all, nor can endosymbiosis. Additional hypotheses are being tested.

Microbial eukaryotes have diverse ways to accomplish the same functions.

- Most microbial eukaryotes are aquatic. Even those that seem terrestrial live in soil water or other extremely moist environments.

- Mechanisms of metabolism are relatively uniform among microbial eukaryotes, but nutritional acquisition varies. Some microbial eukaryotes are autotrophic, some are heterotrophic, and some have the ability to switch between the two modes. Many microbial eukaryotes are involved in symbioses from mutualism to parasitism.

- Most microbial eukaryotes are motile and move via pseudopodia, flagella, or cilia. Cilia and eukaryotic flagella are identical in cross section; they differ only in length.

- Many microbial eukaryotes are microscopic, though some algae literally form aquatic forests of very large, multicellular organisms. Single-celled microbial eukaryotes are able to increase their cell size with the presence of vesicles, which effectively increase surface area. Vesicles assist with water regulation (contractile vacuoles) and nutrient procurement (food vacuoles).

- Many microbial eukaryotes have cell walls or "shells" to protect their cell membranes.

- Many microbial eukaryotes contain endosymbionts. Some harbor other microbial eukaryotes that photosynthesize or carry out other metabolic processes.

- Most microbial eukaryotes undergo both sexual and asexual reproduction, but some lack the ability to reproduce sexually. Sexual and asexual reproductive practices are as diverse as the microbial eukaryotes themselves.

The excavates are a monophyletic group.

- The excavates consists of four major subgroups: the euglenoids, the kinetoplastids, the diplomonads, and the parabasalids.

- The excavates are single-celled flagellates that reproduce asexually by binary fission.

- Euglenids have anterior flagella. Many are photosynthetic and are characterized by a complex cell structure (see Figure 27.28). They are flexible in nutritional requirements and may switch between autotrophism and heterotrophism. Euglenid chloroplasts have triple membranes.

- Kinetoplastids are parasitic and are characterized by having a single large mitochondrion. The mitochondrion is unique in having a kinetoplast housing multiple circular DNA molecules and associated proteins. Many tropical human diseases are caused by kinetoplastids, including African sleeping sickness.

- Diplomonads and parabasalids lack mitochondria. *Giardia lamblia* is a well-known diplomonad causing the human intestinal disorder giardiasis. *Trichomonas vaginalis* is a parabasalid responsible for a sexually transmitted disease in humans.

The alveolates are a monophyletic group.

- The synapomorphy that defines the Alveolata is the possession of cavities called alveoli just below their cell surfaces.

- This group consists of dinoflagellates, apicomplexans, and ciliates.

- Dinoflagellates are mostly marine organisms with two flagella. They are yellowish in color due to photosynthetic and accessory pigments and are primary producers in marine environments. Many live in symbiosis with other organisms. They are common endosymbionts in corals, for example. Dinoflagellates are responsible for "red tides" in warm marine waters and can damage fish. Toxins produced by these organisms can accumulate in shellfish and be fatal to people eating the shellfish. Dinoflagellates are also bioluminescent and produce light when disturbed.

- Apicomplexans are parasitic microbial eukaryotes. Their name comes from the cluster of organelles at the apex of their spores, which assist with invasion of their host organism. The life cycles of these parasites are complex. Apicomplexans in the genus *Plasmodium* cause malaria, a disease that claims more than a million lives annually (see Figure 27.3).

- Ciliates move via cilia and are characterized by multiple macronuclei and micronuclei. Most ciliates are heterotrophic. Paramecia are common and well-studied ciliates.

  - Paramecia move by cilia, their membranes are protected by a pellicle, and they protect themselves with trichocysts.

  - Paramecia reproduce asexually by binary fission. Genetic recombination is accomplished through conjugation, which involves the exchange of equal amounts of genetic material (see Figure 27.13). The resulting recombined paramecia then go through binary fission.

Stramenopiles, most likely a polyphyletic group, are referred to as the brown algae.

- All members are either flagellated or have lost their flagella over the course of evolution.

- This group consists of diatoms, brown algae, oomycetes, and a few smaller groups.

- Diatoms are yellowish brown and store chrysolaminarin and oils as photosynthetic products. They are most noted for their silica-containing cell walls. The cell walls occur in two halves and fit together like a petri plate. All diatoms are symmetrical. Diatoms reproduce both asexually and sexually (see Figure 27.21). Diatom cell walls have significant human uses from filters to reflective paint.

- Brown algae are multicellular and are the largest of the microbial eukaryotes.

  - Brown algae are almost exclusively marine. They produce leaflike growths called thalli. The giant kelp that grow in large oceanic "forests" are brown algae.

  - Brown algae are brown because of chlorophylls *a* and *c* and fucoxanthin. They have specialized regions called holdfasts, which anchor them to a substrate and allow them to resist water movement.

  - Brown algae are commercially important for the presence of alginic acid, used as a binder in many food and cosmetic products.

  - Like many other microbial eukaryotes and all plants, brown algae go through alternation of generations in which a diploid sporophyte gives rise to haploid spores through meiosis to form the haploid gametophyte, which, in turn, produces haploid gametes (see Figure 27.23). Heteromorphic alternation of generations occurs when the two generations differ morphologically, and isomorphic alternation of generations occurs when the two generations are morphologically similar.

- Oomycetes are a nonphotosynthetic group of stramenopiles. These are commonly known as water molds and downy mildews, but they are not fungi. Oomycetes are coenocytic, having multiple nuclei per cell. Many are saprobic and feed on dead material. Others are infectious to plants. The Irish potato famine was caused by an oomycete.

The red algae are mostly multicellular and are part of the Plantae.

- The red algae are characterized by the pigment phycoerythrin, which gives them their characteristic red color. Depending on light intensity, red algae alter the ratio of pigments in the chloroplast. In high light conditions, red algae may appear very green.

- Most red algae are large marine organisms and have specialized structures.

- Red algae store the products of photosynthesis as floridean starch and have no motile cells at any point in their life cycles.

- Red algae are the source of agar, which is important in food products as well as in the study of biology.

The green algae are not monophyletic.

- The green algae are not a monophyletic group. This is because plants are also descended from the most recent common ancestor of all green algae, but plants are not included among the green algae. Chlorophyta, the largest lineage of green algae, is a monophyletic group.

  - Chlorophytes contain chlorophylls *a* and *c*, store photosynthetic products as starch in plastids, and are mainly aquatic.

  - Morphology of the chlorophytes is diverse and ranges from simple single-celled organisms to very complex multicellular organisms. Colonial and aggregate forms exist. Colonial forms show the first indications of cell specialization without multi-cellularity. Individual cells range from single-nucleated to coenocytic.

  - Life cycles within the chlorophytes are diverse. Figures 27.15 and 27.16 illustrate the life cycle of *Ulva* and *Ulothrix*, respectively. Some chlorophytes are isogamous, indicating that their gametes are identical. Others are anisogamous, with female gametes much larger than the male gametes. Some chlorophytes have isomorphic life cycles, and others are heteromorphic. In one variation of the hetero-morphic life cycle—the haplontic life cycle—a multicellular haploid individual (gametophyte) produces gametes that fuse to form a zygote. Some chlorophytes exhibit diplontic life cycles much like animals, in which meiosis of sporophytes produces gametes directly.

Choanoflagellates are thought to be the closest relatives of the animals.

- Choanoflagellates are flagellated, and the majority are colonial.

- Choanoflagellates are structurally similar to sponges, the most basal animal lineage.

Endosymbiosis is the origin of chloroplasts.

- Some chloroplasts are enclosed in double membranes, and others are enclosed in triple membranes. This phenomenon can be explained by endosymbiosis.

- All chloroplasts can be traced to the engulfment of an ancestral cyanobacterium with a single membrane. Its engulfment in a vesicle of the phagocyte's membrane resulted in a double membrane. This initial event is called primary endosymbiosis. Primary endosymbiosis gave rise to the green and red algae.

- Photosynthetic euglenoids, with their triple membranes, arose from secondary endosymbiosis. In this case, a chlorophyte was ingested with its housed chloroplast, resulting in three membranes. Ultimately all of the constituents of the chlorophyte except the chloroplast were lost. Another line is thought to have arisen by means of a red algal endosymbiont.

Some body plans recur in several microbial eukaryote lineages.

- The amoeboid body plan is seen in several groups of microbial eukaryotes. It is characterized by pseudo-pods, cytoplasmic extensions that are used in locomotion. This body plan is highly effective in nutrient-rich environments. Some organisms have an amoeboid body plan at different stages of their life cycle.

- Actinopods have very thin pseudopods reinforced by microtubules. The pseudopods increase surface area, help the organisms float, provide locomotion, and are used in feeding. Radiolarians, a specific type of actinopod, secrete glassy endoskeletons. Other types, called, heliozoans, lack an endoskeleton.

- Foraminiferans secrete shells of calcium carbonate and live either as plankton or on the sea bottom. Their pseudopods are used to trap food. Limestone deposits are often the result of discarded foraminiferan skeletons.

- Slime molds occur in three distinct groups. All groups are motile, feed by endocytosis, form spores on erect fruiting bodies, and undergo drastic changes during their life cycle.

  - Plasmodial slime molds form multinucleate masses. During the vegetative stage, an acellular slime mold is a wall-less mass of cytoplasm with multiple diploid nuclei. It oozes over a network of strands called plasmodium. This phenomenon is called cytoplasmic streaming and is used to move and engulf food. If exposed to adverse environmental conditions, the slime mold will form a dormant sclerotium from which it can turn back into a plasmodium when environmental conditions are right. It may also transform into a fruiting structure called a sporangiophore.

  - Cellular slime molds are made of large numbers of cells called myxamoebas with single haploid nuclei. This life cycle can continue as long as conditions are favorable. Once conditions become unfavorable, myxamoebas secrete cAMP, aggregate, and form a motile slug that eventually produces fruiting bodies. Spores from the fruiting bodies germinate to form more myxamoebas. Sexual reproduction occurs with the fusion of two myxamoebas, which then undergo meiosis to release haploid myxamoebas.

The Big Picture

- The many lineages of microbial eukaryotes as well as all fungi, animals, and plants are eukaryotic. Eukaryotic cells are characterized by membrane-enclosed organelles, the presence of a cytoskeleton, and a nuclear envelope. The evolution of the eukaryotic cell is a major evolutionary milestone. The eukaryotic cell allows for compartmentalization of processes and increased adaptability. Though the evolutionary lineage is still debated, it is thought that eukaryotes evolved via multiple symbiotic events.

- Microbial eukaryotes as a group are not monophyletic. This is because the most recent common ancestor of microbial eukaryotes gave rise to the plants, animals, and fungi, and these groups are not included in the microbial eukaryotes. Microbial eukaryotes are diverse, ranging from microscopic single-celled organisms to complex multicellular organisms. Several subgroups of microbial eukaryotes are covered in this chapter. Some groups are monophyletic; others are not.

- Certain body plans and modes of nutritional acquisition are repeated throughout the microbial eukaryote groups and are good examples of form following function. Microbial eukaryotes as a group play a vast role in the environment and medicine and are economically valuable.

Common Problem Areas

- There is a tendency for students to become over-whelmed with organism names and characteristics. Focus on the trends outlined in the chapter and use lab opportunities to understand the organism groupings. For a start, see Table 27.1.

Study Strategies

- This chapter has a lot of information in it. It is best not to try to learn it all at once, but to break it up into smaller pieces and study it over several sessions.

- To learn the organism groups, create a chart with key characteristics. This will help you to compare and contrast the groups.

- Review the following animated tutorials and activities on the Companion Website/CD:
  Tutorial 27.1 Family Tree of Chloroplasts
  Tutorial 27.2 Food Vacuoles Handle Digestion and Excretion
  Activity 27.1 An Isomorphic Life Cycle
  Activity 27.2 A Haplontic Life Cycle
  Activity 27.3 Anatomy of a *Paramecium*

Test Yourself

Knowledge and Synthesis Questions

1. Which of the following modes of reproduction can be found in at least some microbial eukaryotes?
   a. Binary fission
   b. Sexual reproduction
   c. Spore formation
   d. All of the above
   Textbook Reference: *27.4 How Do Microbial Eukaryotes Reproduce? p. 593*

2. During the evolution of eukaryotes from prokaryotes, which of the following did *not* occur?
   a. Infolding of the flexible cell membrane
   b. Loss of the cell wall
   c. A switch from aerobic to anaerobic metabolism
   d. Endosymbiosis of once free-living prokaryotes
   Textbook Reference: *27.2 How Did the Eukaryotic Cell Arise? p. 588*

3. Which of the following statements about microbial eukaryotes is *false*?
   a. Apicomplexans are the only microbial eukaryote group without parasitic representatives.
   b. Foraminiferans and radiolarians are shelled microbial eukaryotes.
   c. Ciliates have great control over the direction their cilia beat.
   d. Although they appear structurally simple, amoebas are not primitive organisms.
   Textbook Reference: *27.1 How Do Microbial Eukaryotes Affect the World Around Them? p. 585*

4. Ciliates, as represented by paramecia, have defensive organelles called _____ in their pellicles.
   a. trichonympha
   b. trichocysts
   c. trichomes
   d. trochlea
   Textbook Reference: *27.5 What Are the Major Groups of Eukaryotes? p. 598*

5. A major difference between the vegetative states of cellular and plasmodial slime molds is that plasmodial slime molds
   a. have haploid nuclei, and cellular slime molds have diploid nuclei.
   b. produce fruiting bodies, and cellular slime molds do not.
   c. undergo aggregation when conditions become adverse, and cellular slime molds do not.
   d. exist as a coenocytic mass, and cellular slime molds exist as individual myxamoebas.
   Textbook Reference: *27.5 What Are the Major Groups of Eukaryotes? pp. 606–607*

6. Organisms that exhibit isogamy have
   a. similar male and female gametes.
   b. female gametes that are significantly larger than male gametes.
   c. male gametes that are significantly larger than female gametes.
   d. a diplontic mode of reproduction.
   Textbook Reference: *27.4 How Do Microbial Eukaryotes Reproduce? p. 594*

7. Red tides often cause massive fish kills and human illness in those eating shellfish. Which group of microbial eukaryotes is responsible for red tides?
   a. Parabasalids
   b. Red algae
   c. Euglenozoans
   d. Dinoflagellates
   Textbook Reference: *27.1 How Do Microbial Eukaryotes Affect the World Around Them? p. 586*

8. Holdfasts and alginic acid are characteristic of which group of microbial eukaryotes?
   a. Parabasalids
   b. Red algae
   c. Brown algae
   d. Stramenopiles
   Textbook Reference: *27.5 What Are the Major Groups of Eukaryotes? p. 600*

9. A water mold is discovered growing on the surface of a freshwater green alga. Which two groups of microbial eukaryotes are involved?
   a. Choanoflagellates and red algae
   b. Oomycetes and chlorophytes
   c. Apicomplexans and chlorophytes
   d. Apicomplexans and ciliophora
   Textbook Reference: *27.5 What Are the Major Groups of Eukaryotes? pp. 600–602*

10. Which of the following statements regarding conjugation in *Paramecium* is *false*?
    a. Conjugation results in genetic recombination.
    b. Conjugation results in clones.
    c. Conjugation results in offspring.
    d. Both b and c
    Textbook Reference: *27.4 How Do Microbial Eukaryotes Reproduce? p. 593*

11. Which of the following statements regarding microbial eukaryotes in general is *false*?
    a. Microbial eukaryotes are always parasitic.
    b. Microbial eukaryotes are all single celled.
    c. Microbial eukaryotes are all heterotrophic.
    d. All of the above
    Textbook Reference: *27.1 How Do Microbial Eukaryotes Affect the World Around Them? p. 583*

12. According to Figure 27.17, which of the following statements is true?
    a. The alveolates are more closely related to the stramenopiles than to any other group.
    b. The most recent common ancestor of the parabasalids and euglenoids also gave rise to the animals.
    c. The red algae are more closely related to the brown algae than they are to any other taxon.
    d. The opisthokonts are not monophyletic.
    Textbook Reference: *27.5 What Are the Major Groups of Eukaryotes? p. 597*

Application Questions

1. Explain alternation of generations. Why are the microbial eukaryotes the first group in which this process could be seen?
   Textbook Reference: *27.4 How Do Microbial Eukaryotes Reproduce? pp. 593–594*

2. Differentiate between asexual and sexual reproduction. Select one mode of microbial eukaryote sexual reproduction and explain it fully.
   Textbook Reference: *27.4 How Do Microbial Eukaryotes Reproduce? pp. 593–594*

3. You have identified a new organism. What would the criteria be for placing it with the microbial eukaryotes?
   Textbook Reference: *27.1 How Do Microbial Eukaryotes Affect the World Around Them? p. 583*

4. Explain one line of thought regarding the origin of the eukaryotic cell. Why are scientists unsure of the evolutionary relationships among the microbial eukaryotes?
   Textbook Reference: *27.2 How Did the Eukaryotic Cell Arise? pp. 588–589*

5. Most microbial eukaryotes are motile. Describe some of the means of motility.
   Textbook Reference: *27.3 How Did the Microbial Eukaryotes Diversify? p. 591*

6. Describe the origin of double-membrane-enclosed and triple-membrane-enclosed chloroplasts.
   Textbook Reference: *27.2 How Did the Eukaryotic Cell Arise? pp. 589–590*

7. Many microbial eukaryotes are significant human pathogens. Select one pathogenic microbial eukaryote, describe its life cycle, identify the microbial eukaryote group to which it belongs, and discuss the disease it causes.
   Textbook Reference: *27.1 How Do Microbial Eukaryotes Affect the World Around Them? p. 586*

8. Draw the phylogeny of the three domains of life and add the various lineages of microbial eukaryotes to it. Now add the plants, animals, and fungi.
   Textbook Reference: *27.5 What Are the Major Groups of Eukaryotes? p. 597*

Answers

Knowledge and Synthesis Answers

1. **d.** Methods of reproduction are quite varied among the microbial eukaryotes.

2. **c.** At the time the first eukaryotes evolved, the environment was becoming oxygen-rich. There was a switch from anaerobic to aerobic metabolism, not vice versa.

3. **a.** Malaria is caused by *Plasmodium*, which is an apicomplexan. In fact, all apicomplexans are parasitic.

4. **b.** Trichocysts are defensive barbs ejected from ciliates when they are disturbed.

5. **d.** The vegetative (feeding) state of an acellular slime mold is called a plasmodium; it consists of multiple diploid nuclei enclosed in a single membrane. The vegetative state of a cellular slime mold is a myxamoeba with a single haploid nucleus. Although the myxamoebas of cellular slime molds do aggregate to form fruiting structures; individual myxamoebas never fuse into multinucleated structures.

6. **a.** *Isogamy* means *same gametes*. The gametes appear identical.

7. **d.** Dinoflagellates are the cause of red tides.

8. **c.** The brown algae are multicellular microbial eukaryotes notable for their organ and tissue differentiation and the presence of alginic acid in their cell walls.

9. **b.** Water molds are oomycetes, and green algae are chlorophytes.

10. **d.** Conjugation does result in genetic recombination but does not result in the production of clones or offspring.

11. **d.** Microbial eukaryotes do not have a unifying characteristic other than being eukaryotic organisms that do not fit into the kingdoms Plantae, Animalia, or Fungi.

12. **a.** The only choice that is supported by the cladogram in Figure 27.17 is that the alveolates are more closely related to the stramenopiles than to any other group.

Application Answers

1. In alternation of generations an organism exists in a haploid gamete-producing form and a diploid spore-producing form. Prokaryotes have a single chromosome, so only in eukaryotes, with multiple copies of chromosomes, is a diploid stage of a life cycle possible.

2. Asexual reproduction results in clones of the original organism, and there is no genetic recombination or variation associated with the creation of offspring.

Sexual reproduction allows for new genetic combinations. Microbial eukaryotes undergo varied types of sexual reproduction, from fusing haploid myxamoebas in cellular slime molds, to haplontic alternation of generations, to a diplontic type of reproduction. Conjugation between paramecia is an example of sexual recombination without reproduction.

3. The organism must be a eukaryote and not fit the criteria for land plants, animals, or fungi.

4. See Figure 27.7. The evolutionary relationships are difficult to understand due to limited fossilization, lateral gene transfers, and the contributions of symbiotic prokaryotes.

5. Microbial eukaryotes move by flagella, cilia, and pseudopodia. Because many are aquatic, additional modifications allow for floating and movement along currents.

6. Double-membrane-enclosed chloroplasts most likely arose from the endosymbiosis of a cyanobacteria in a process called primary endosymbiosis; triple-membrane-enclosed chloroplasts most likely arose from the endosymbiosis of a chlorophyte or a rhodophyte (see Figure 27.8).

7. You may select from a variety of pathogenic microbial eukaryotes: malaria caused by one of the apicomplexans, African sleeping sickness caused by a kinetoplastid, or others. See Figure 27.3 for the life cycle of *Plasmodium*, the microbial eukaryote that causes malaria.

8. This takes practice! Begin by drawing just the Bacteria, Archaea, and Eukarya. Now "add" the branches representing groups of microbial eukaryotes, animals, plants, and fungi. Check your final tree against Figure 27.17.

CHAPTER 28 Plants without Seeds: From Sea to Land

Important Concepts

The plant kingdom is monophyletic.

- Plants are photosynthetic eukaryotes that utilize chlorophylls *a* and *b* to produce and store carbohydrates as starches.

- All land plants (embryophytes) develop from an embryo that is protected by tissues of the parent plant.

- Land plants undergo alternation of generations.

  - Haploid gametophytes produce haploid gametes through mitosis, and diploid sporophytes produce haploid spores through meiosis.

  - The diploid sporophyte arises from the fusion of gametes to produce a diploid zygote that develops into the mature sporophyte.

- The extent of the development of the two life stages is characteristic of the different phyla.

  - Nonvascular plants typically have a dominant gametophyte. The small sporophyte is nutritionally dependent on the gametophyte.

  - In the nonseed vascular plants, the sporophyte is dominant, but the gametophyte, though small, is nutritionally independent of the sporophyte.

  - In the most recently evolved vascular plants, the sporophyte is dominant, while the gametophyte is greatly reduced and nutritionally dependent on the sporophyte.

- There are ten extant (surviving) groups of land plants (see Table 28.1). Land plants as a whole are monophyletic, but the nonvascular plants, as a group, are paraphyletic.

- Molecular and fossil evidence indicates that land plants arose from a green algae ancestor. Land plants most likely arose from a group of green algae called the Charales that lived at edges of marshes and ponds between 400 and 500 mya.

Survival on land poses a series of challenges.

- To survive on land, an aquatic organism must evolve ways to reduce its dependence on water.

- Aquatic organisms can float in their environment. Terrestrial plants need to support themselves against gravity if they are to grow upward.

- Terrestrial plants need a mechanism to move water to all their cells.

- Terrestrial plants need a mechanism to prevent loss of water to evaporation.

- The sperm of aquatic organisms can swim in search of an egg. Terrestrial plants need some other mechanism to move sperm from one plant to another.

- Terrestrial plants need a mechanism to prevent desiccation of the developing embryo.

Many characteristics that distinguish land plants from green algae are adaptations to a terrestrial environment.

- The earliest land plants developed the following modifications for terrestrial life: a waxy cuticle, gametangia, embryos, protective pigments, thick spore walls, and mutualistic associations with fungi.

  - Waxy cuticles, though minimal in early land plants, prevented water loss from tissues as they were exposed to dry air. The extent of the cuticle varies among plant groups and is most well developed in the more recently evolved gymnosperms and angiosperms.

  - Multicellular gametangia mark the transition from algae to plant. Gametangia enclose the gamete-producing tissue and protect the developing gametes.

  - Plant embryos are significant in that they contain the developing sporophyte and are housed within protective tissues. Recently evolved angiosperms have a complex protective mechanism.

  - Water protects aquatic organisms from damaging ultraviolet radiation. To compensate for the loss of this protection, land plants have evolved pigments that screen out UV radiation.

  - Spore walls protect spores from desiccation during dispersal. This is particularly important in the nonseed land plants.

- Mutualistic associates with fungi help with the absorption of nutrients from the soil that might otherwise be inaccessible to land plants.

The development of vascular tissue was key to terrestrial existence.

- Vascular tissue provides for the transport of water and food through a plant. The fundamental unit of a vascular system is the tracheid, a strawlike cell that conducts water in the plant. Some mosses have cells that are precursors to tracheid development, but these are not true tracheids.

- Land plants without vascular tissue depend on abundant water sources, thin tissues that can absorb water easily, mechanisms for capturing water vapor in the air, and capillary action to move water through the plant. They lack the support of lignin and closely hug the ground.

- Not all land plants have vascular tissue; therefore, it is not an absolute requirement for terrestrial life. Land plants that lack vascular tissue, however, cannot grow very large or live far from a water source. The evolution of vascular tissue has allowed land plants to occupy many niches in the biosphere that would otherwise be inaccessible.

Nonvascular plants were the first plant colonizers of land.

- Most nonvascular plants live in dense mats that hug the ground in moist environments.

- The life cycle of the nontracheophyte is dominated by the gametophyte generation. The sporophyte is very tiny and is completely dependent on the gametophyte. Figure 28.5 illustrates the life cycle of a moss as an example of the nontracheophyte life cycle. You should be able to follow this life cycle and understand which stages are haploid, which are diploid, and where meiosis takes place.

- Liverworts (*Hepatophyta*) may be the oldest of all plant lineages.

  - The liverwort gametophyte is usually a flat plate of cells (see Figure 28.13). Gametophytes produce antheridia and archegonia on their upper surfaces (sometimes on stalks) and rhizoids (rootlike structures for anchoring and water absorption) on their lower surfaces.

  - The liverwort sporophyte has a stalk that, in many species, is elongated to aid in spore dispersal. Some sporophytes also have structures that forcefully eject spores.

  - Liverworts can reproduce asexually by shedding small clumps of cells called gemmae.

- Hornworts (*Anthocerophyta*) evolved simple stomata, pores that help with the exchange of gas and water vapor. Unlike stomata in other land plants, however, hornwort stomata are unable to close.

- Hornworts are distinguished from other nonvascular plants by two characteristics: the presence of a single, platelike chloroplast in each cell, and sporophytes that are capable of indeterminate growth. The hornwort sporophyte resembles a long, slender horn (see Figure 28.14).

- Hornworts often have symbiotic relationships with cyanobacteria that are able to fix atmospheric nitrogen and make it available to the hornwort.

- Mosses (*Bryophyta*) are the most familiar nonvascular plants.

  - Mosses have specialized cells called hydroids. These hydroids are thought to be precursors to the tracheid seen in vascular plants, but they lack lignin for structural support. Like tracheids, the hydroid cell dies and leaves a tiny channel that can transport water through the plant.

  - Mosses also differ from other nonvascular plants in that they exhibit apical cell division. This characteristic is seen in "higher" land plants and allows for regular, upright growth.

  - The moss life cycle is presented in Figure 28.5.

The vascular plants have specialized vascular tissues.

- All vascular plants are characterized by the presence of water-conducting tracheids. Tracheids were a significant advance in the evolution of land plants, providing a means of water transport and support for growth.

- Vascular plants fossilize much better than nonvascular plants. Thus, our morphological knowledge of early vascular plants is much more extensive than our knowledge of nonvascular plants.

- Vascular plants are also characterized by branching, independent sporophytes.

- Figure 28.7 illustrates the relationships between the different groups of vascular plants.

- The first vascular plants were members of a now-extinct group of land plants called the rhinophyta. *Rhynia* had characteristics of extant vascular plants and showed the earliest features of this group, including an independent, branched sporophyte and tracheids, but they lacked more advanced characteristics such as true leaves and roots (see Figure 28.9).

Significant new features arose in early vascular plants.

- Roots are believed to have evolved over time from branches that grew underground. Because branches above ground and below ground would have been exposed to different selective pressures, underground branches might have given rise to what we know today as roots.

- True leaves are flattened photosynthetic structures with vascular tissue. Vascular plants have evolved two different kinds of leaves: microphylls and megaphylls (see Figure 28.10).

- Microphylls are present in the club mosses. They probably evolved from sterile sporangia.
- Most familiar leaves are megaphylls. These are thought to have evolved when photosynthetic tissue developed between the ends of small, lateral branches.
- Heterospory appears to have evolved several times in the vascular plants (see Figure 28.12).
  - Most early vascular plants exhibit homospory. In homospory, spores (and the gametophytes that grow from them) are all of the same type; gametophytes produce both archegonia and antheridia. In heterospory, one spore type called a megaspore gives rise to the female, egg-producing megagametophyte; another spore type called a microspore develops into a male, sperm-producing microgametophyte.
  - Land plants typically produce many more microspores than megaspores.

Extant nonseed vascular plants are varied and abundant.

- Club mosses (*Lycophyta*) diverged from the tracheid lineage relatively early and are the most "primitive" of the extant nonseed vascular plants. They have simple leaves arranged spirally on the stem. Their sporangia are held in apical strobili, which are simple branching structures of fertile sporangia and sterile leaves (see Figure 28.16).
- The horsetails, whisk ferns, and ferns were once treated as separate phyla, but recent evidence suggests that they form a clade, the Pteridophyta.
  - Horsetails are represented by few extant species. They have true roots, their sporophytes are large and independent, and their gametophytes are highly reduced but also independent. The leaves of horsetails are simple, forming whorls around the stem (see Figure 28.17). This group is one of the few plant groups that exhibit basal growth.
  - Whisk ferns once were thought to be the "missing link" to *Rhynia,* but molecular evidence indicates that they evolved much later from more highly complex land plants. Their relatively simple body plan works well and demonstrates that evolution does not necessarily move toward greater complexity.
  - Ferns are the largest surviving group of nonseed vascular plants. Ferns are characterized by large complex leaves with branching vascular strands. Like all other nonseed vascular plants and nonvascular plants, ferns continue to be dependent on water to carry motile sperm. Ferns have advanced vascular structures and may reach great heights, but they do not produce true wood. Their root systems are poorly developed.

The fern life cycle has independent sporophytes and gametophytes.

- The fern life cycle is dominated by the sporophyte, but the gametophyte is an independent photosynthetic structure (see Figure 28.20).
- Most ferns are homosporous, but some fern groups are heterosporous. A few genera of ferns produce gametophytes that depend on a mutualistic fungus for nutrition.

The Big Picture

- Land plants are photosynthetic eukaryotes that utilize chlorophylls *a* and *b,* undergo alternation of generations, and develop from multicellular embryos that are protected by the parent plant. The ten extant groups can be classified as nonvascular plants (those without highly developed vascular tissue) and vascular plants (those with highly developed vascular tissue). This chapter focuses on the nonvascular plants and the nonseed vascular plants.
- In order to colonize land, plants had to evolve strategies for dealing with desiccation and gravity. This involved mechanisms for extracting water from soil, means of transporting water throughout the plant, methods of ensuring fertilization, and modes of protecting developing embryos.
- This chapter introduces and describes the liverworts, hornworts, and mosses, which are all considered nonvascular plants, and the club mosses, horsetails, whisk ferns, and ferns, which are nonseed vascular plants.

Common Problem Areas

- It is very tempting to memorize plant types and characteristics, but it is more important to focus on the trends and relationships among the land plants. Your goal is to understand the evolutionary relationships among the land plants.

Study Strategies

- Focus on relationships among the land plants and how to differentiate one group from another. Many of the important distinctions between groups are related to adaptations for terrestrial life.
- When studying life cycles, be sure to note whether the sporophyte or the gametophyte is dominant and make note of where mitosis and meiosis take place.
- Review the following animated tutorial and activities on the Companion Website/CD:
  Tutorial 28.1 Life Cycle of a Moss
  Activity 28.1 Homospory
  Activity 28.2 Heterospory
  Activity 28.3 Fern Life Cycle

Test Yourself

Knowledge and Synthesis Questions

1. A land plant may be reliably distinguished from green algae by which of the following characteristics?
   a. Chlorophyll type
   b. The presence of an embryo protected by parent tissue
   c. The presence of roots
   d. Swimming sperm
   Textbook Reference: *28.1 How Did the Land Plants Arise? p. 611*

2. Which of the following characteristics was (were) necessary for plants to colonize land?
   a. Vascular tissue for moving water throughout the plant
   b. A mechanism to prevent desiccation of tissues
   c. The ability to screen ultraviolet radiation
   d. Both b and c
   Textbook Reference: *28.1 How Did the Land Plants Arise? p. 612*

3. The main difference between nonvascular plants and vascular plants is
   a. lack of gametophytes.
   b. spore production.
   c. the presence of tracheids.
   d. All of the above
   Textbook Reference: *28.3 What Features Distinguish the Vascular Plants? p. 617*

4. In alternation of generations, the sporophyte generation is _____ and the gametophyte generation is _____.
   a. haploid; diploid
   b. diploid; haploid
   c. haploid; haploid
   d. diploid; diploid
   Textbook Reference: *28.2 How Did Plants Colonize and Thrive on Land? pp. 614–615*

5. Which of the following characteristics helps to differentiate between liverworts and mosses?
   a. The presence of hydroids
   b. Gametophyte dominance
   c. Sporophyte dominance
   d. Swimming sperm
   Textbook Reference: *28.4 What Are the Major Clades of Seedless Plants? p. 622*

6. During a plant's life cycle, meiosis takes place in the _____ to produce haploid _____.
   a. gametophyte; gametes
   b. sporophyte; gametes
   c. sporophyte; spores
   d. gametophyte; spores
   Textbook Reference: *28.2 How Did Plants Colonize and Thrive on Land? pp. 614–615*

7. Asexual reproduction in liverworts is accomplished by _____.
   a. gametophytes

b. spores
c. gemmae
d. tracheids
Textbook Reference: *28.4 What Are the Major Clades of Seedless Plants? p. 622*

8. Which of the following limits the size of a hornwort's sporophyte?
   a. The sporophyte's ability to distribute water to all its cells
   b. The gametophyte's ability to produce enough nutrients for the sporophyte
   c. Developmental genes that prevent growth of the sporophyte beyond a certain size
   d. All of the above
   Textbook Reference: *28.4 What Are the Major Clades of Seedless Plants? p. 623*

9. You are walking along a roadside and find a plant with the following characteristics: a very thin, waxy cuticle, stomata, simple leaves in whorls around a central stem, independent sporophytes and gametophytes, and sporangia in strobili. This plant is most likely a member of which of the following phyla?
   a. Bryophyta
   b. Pteridophytes
   c. Anthocerophyta
   d. Lycophyta
   Textbook Reference: *28.4 What Are the Major Clades of Seedless Plants? p. 625*

10. Match the phyla with their diagnostic characteristics.
    ____ Mosses   a. No stomata
    ____ Liverworts   b. Descended from a more complex ancestor
    ____ Hornworts   c. Complex branching leaves
    ____ Club mosses   d. Unique chloroplasts
    ____ Horsetails   e. First strobili of sterile leaves and sporangia
    ____ Whisk ferns   f. Hydroids
    ____ Ferns   g. Basal growth
    Textbook Reference: *28.4 What Are the Major Clades of Seedless Plants? pp. 622–627*

Application Questions

1. Explain why the largest mosses are less than a meter tall.
   Textbook Reference: *28.2 How Did Plants Colonize and Thrive on Land? p. 614*

2. Diagram the evolutionary relationships among the nonvascular plants and nonseed vascular plants. Label the major differentiating characteristics at each branch of your tree.
   Textbook Reference: *28.3 What Features Distinguish the Vascular Plants? p. 617*

3. Discuss how the relationship between sporophytes and gametophytes changes as you move from the first nonvascular plants to later nonseed vascular plants.
   Textbook Reference: *28.2 How Did Plants Colonize and Thrive on Land? pp. 614–616*

4. Compare and contrast the moss life cycle and the fern life cycle.
   Textbook Reference: *28.2 How Did Plants Colonize and Thrive on Land? p. 615; 28.4 What Are the Major Clades of Seedless Plants? p. 627*

5. Discuss the challenges of terrestrial life that land plants needed to address for survival.
   Textbook Reference: *28.2 How Did Plants Colonize and Thrive on Land? p. 613*

6. Compare and contrast homospory and heterospory. What reproductive structures result from meiosis in each type of life cycle, and what structures result from mitosis?
   Textbook Reference: *28.3 What Features Distinguish the Vascular Plants? p. 621–622*

Answers

Knowledge and Synthesis Answers

1. **b.** By the definition your textbook uses, all land plants produce embryos that are protected by tissue of the parent plant. Green algae and land plants use the same types of chlorophyll. Not all land plants have roots, so a plantlike organism lacking roots will not necessarily be a green algae.

2. **d.** There are several successful groups of terrestrial plants that lack vascular tissue.

3. **c.** Tracheids are found only in the vascular plants.

4. **b.** Meiosis occurs in the diploid sporophyte to produce haploid spores that develop into the haploid gametophyte.

5. **a.** Mosses are the only group with hydroids, the precursors to true vascular tissue.

6. **c.** The outcome of meiosis is four cells, each of which has half the genetic material of the parent cell. A haploid cell already has only half the normal number of chromosomes of most eukaryotes, so a cell must be diploid (or have higher ploidy) to undergo meiosis. In all plant life cycles, the sporophyte is diploid and the gametophyte is haploid; therefore, only the sporophyte can undergo meiosis. Spores are the products of sporophyte meiosis.

7. **c.** Liverworts have specialized asexual reproductive structures called gemmae.

8. **a.** The sporophyte of a hornwort is nutritionally dependent on the gametophyte, but its growth is indeterminate—the sporophyte will continue to grow as long as water is able to reach all its cells.

9. **b.** You have come across a horsetail. These plants are very common along roadsides in damp ditches, particularly in the Midwest.

10. **f.** Mosses
    a. Liverworts
    d. Hornworts
    e. Club mosses
    g. Horsetails
    b. Whisk ferns
    c. Ferns

Application Answers

1. Mosses have very rudimentary water transport cells called hydroids. Hydroids lack the waterproofing and support molecule lignin. Because of this, they can carry water only short distances and cannot support tall growth.

2. See Figure 28.7 and Table 28.1.

3. Early nonvascular plants have reduced sporophytes that are highly dependent on the gametophyte for nutrition. Nonseed vascular plants, which have evolved more recently, have independent sporophytes and gametophytes, and the gametophyte is highly reduced.

4. See Figures 28.5 and 28.20. Pay particular attention to the relative dominance of the sporophyte versus the gametophyte.

5. Terrestrial life poses problems of support, water conduction, UV radiation, water loss, and embryo protection. The tracheid was an important evolutionary innovation that helped with both water conduction and support. Special pigments protect against UV radiation. A waxy cuticle prevents water loss from cells. Plant embryos are protected in early development by parental tissue.

6. See Figure 28.12. In homospory, meiosis results in a single type of spore, but in heterospory, meiosis results in megaspores and microspores. Mitosis occurs throughout both types of life cycle as cells divide and plants grow, but the reproductive cells that result from mitosis in both homospory and heterospory are sperm and eggs. In homospory, the antheridium produces sperm by mitosis, and the archegonium produces eggs; in heterospory, the microgametophyte produces sperm and the megagametophyte produces eggs.

CHAPTER 29 The Evolution of Seed Plants

Important Concepts

Seed plants have characteristics that set them apart from nonseed plants.

- Seed plants consist of two groups of tracheophytes: the gymnosperms and the angiosperms. Gymnosperms include the following phyla: Cycadophyta, Ginkgophyta, Coniferophyta, and Gnetophyta.

- Seed plants have highly reduced gametophyte generations that are dependent on the sporophyte for survival. When you look at a seed plant, what you see is the sporophyte.

- Very few seed plants have retained swimming sperm. Most have evolved other means for dispersing male gametes.

- All seed plants are heterosporous. Microspores and megaspores develop in specialized cones or in flowers.

- In the megasporangium, meiosis produces four megaspores, but in seed plants only one of the four megaspores is retained. This megaspore divides by mitosis to form the multicellular (yet tiny) megagametophyte, which produces the eggs.

- When the eggs are ready to be fertilized, they are surrounded by megagametophyte cells, which in turn are still within the megasporangium. The megasporangium is surrounded by sterile sporophyte tissues (the integument).

- Microspores develop into pollen grains, the male gametophytes. Pollen grains consist of sperm and supporting cells; they are dispersed by wind and animals.

- Because the female gametophyte is retained within sporophyte tissue, pollen grains do not have direct access to gametophytes and their eggs. The sporophyte housing a gametophyte creates tissue for receiving pollen grains; upon reaching this tissue, pollen produces pollen tubes that deliver the sperm through the sporophyte tissue to the eggs for fertilization.

- The embryo resulting from fertilization grows to a certain size and then becomes dormant within the surrounding tissues. This dormant, protected embryo and its surrounding tissues constitute the seed.

Seeds are complex structures that may contain tissue from three plant generations.

- Recall that alternation of generations involves a multicellular sporophyte generation that alternates with a multicellular gametophyte generation. The embryo within a seed represents the beginning of a new sporophyte generation. It is still surrounded by its gametophyte, which will provide the embryo with nutrients to begin growth (particularly if it is a gymnosperm). The tough coat that surrounds the seed consists of tissue provided by the embryo's sporophyte parent.

- Seeds protect the embryo until conditions are right for germination. Many seeds have adaptations that allow dispersal by wind or another vector.

Gymnosperms are naked-seed plants.

- Gymnosperms are seed plants but they do not form flowers or true fruits. They are diverse and live in a variety of habitats, from sparse deserts to vast forests. The Coniferophyta are the most abundant of the gymnosperms.

- Gymnosperms have significant vascular tissue and are capable of woody secondary growth.

- The gymnosperm life cycle can be illustrated by that of the pine, a conifer (see Figure 29.8). Conifers differ from other gymnosperms in that they produce cones and strobili, which are specialized structures for reproduction. Cones house the megasporangia and produce megaspores, megagametophytes, and eggs; strobili house the microsporangia and produce microspores, microgametophytes (pollen), and sperm.

- Pines do not have swimming sperm, and their pollen is modified for wind dispersal. Pollen lands on the female cone and lodges in the pollen chamber. The pollen tube grows through the maternal sporophyte tissue to the female gametophyte, where it releases two sperm. Only one of the two sperm will fertilize an egg. The resulting zygote develops into an embryo that remains encased in the tissues of the megasporangium and gametophyte. The seed is also protected by the scale of the cone (sporophyte tissue) until it is mature and ready for dispersal. Dispersal is aided by modifications of the seed coat.

Angiosperms are characterized by flower formation.

- Angiosperms are a diverse group of plants that all produce flowers and are thus called flowering plants. They are presently the dominant plant form on Earth.

- Angiosperms have the most reduced gametophyte generation and the most highly developed sporophyte generation of all plant lineages.

- Angiosperms differ from all other plants in that they produce triploid endosperm, they have double fertilization, their ovules are enclosed in a carpel, they produce flowers and fruits, and their xylem and phloem have multiple modified cell types, including vessel elements, fibers, and companion phloem cells.

Flower structure is varied among plants but consists of similar structural elements.

- All flower parts are modified leaf structures. Although flowers are very diverse, they are constructed from the same small set of structures (see Figure 29.9).

- The male structures in a flower are stamens. Each stamen consists of a filament and an anther; the anther contains microsporangia, which produce pollen. Flowers often have several stamens.

- The female structure in a flower is the pistil, which consists of a stigma, a style, and an ovary. The stigma is modified to receive pollen. The style is a stalk that separates the stigma from the ovary. The ovary contains one or more ovules, each of which houses a megasporangium.

- Nonreproductive floral structures include petals, sepals, and the receptacle. Petals and sepals are often modified to attract animal pollinators. Sepals also serve to protect the developing flower bud. The receptacle is the attachment site for the sepals, petals, stamens, and carpels.

- Flowers that have both megasporangia and microsporangia are called perfect. If either structure is missing, the flower is called imperfect. Species that have both megasporangiate flowers and microsporangiate flowers on the same plant are called monoecious. If they are on separate plants, they are called dioecious.

- Many plants and animals have coevolved in such a way that the nutrition of the animal and pollination of the plant are interdependent. Some plants have evolved methods of limiting pollination to a single species of insect, but most can be pollinated by a range of species.

The angiosperm life cycle is dominated by the sporophyte.

- The angiosperm life cycle differs from that of all other plants in that double fertilization occurs (see Figure 29.14). In double fertilization, one sperm unites with the egg to produce a diploid zygote; the other sperm unites with two other haploid cells of the female gametophyte to produce a triploid cell. This triploid cell divides mitotically to create the (triploid) endosperm tissue. The endosperm provides nutrition for the developing embryo.

- Angiosperm embryos have one or two seed leaves called cotyledons. The number of cotyledons distinguishes the two main lineages of flowering plants. Monocots have one cotyledon; eudicots have two. A few other relatively small groups of angiosperms do not fit into these two lineages (see Figures 29.16 and 29.17).

- Fruits develop from the ovary and its supporting tissues. Fruit types depend on the number of carpels associated with the fruit and the extent of support tissue incorporated into the fruit structure.

The evolutionary relationships among angiosperms have not yet been resolved.

- Different phylogenetic methods have led to different conclusions regarding angiosperm phylogeny. Investigators are still working on the question of how angiosperms first arose and whether they are sister to one gymnosperm phylum.

- Evidence now points to *Amborella* as the living species most similar to the first angiosperms. *Amborella* has a variable number of carpels and stamens, and it lacks vessel elements.

- Botanists recently discovered fossils of two species (both in the genus *Archaefructus*) that may belong to the sister taxon of all angiosperms. The flowers of *Archaefructus* had ovules enclosed in carpels, but no petals or sepals.

The Big Picture

- Seed plants can be characterized into two main groups: the gymnosperms and the angiosperms. Seed plants have developed complex ways of protecting embryos, dispersing gametes, and moving water, nutrients, and food throughout the plant. Both gymnosperms and angiosperms are characterized by the presence of seeds for protecting the embryo until conditions favor germination. The main difference between gymnosperms and angiosperms is the presence of flowers and ovaries in angiosperms. Both groups have highly developed vascular tissue, may exhibit secondary woody growth, and show significant diversity.

- There is great variation among flower types, but the basic flower consists of the structures shown in Figure 29.9. Many plants and animals have coevolved in such a way that the flower structure and the animal are integrally linked for nutrition of the animal and pollination of the plant. The angiosperm life cycle is different from that of all other plants in that double fertilization occurs.

- Both the angiosperm and gymnosperm life cycles are dominated by the sporophyte. Both groups are heterosporous. The roles of megasporangia and microsporangia are similar in both groups.

Common Problem Areas

- Not all gymnosperms are pines. The examples used in this chapter focus on the pine, but gymnosperms are a varied group, and cones are not a diagnostic characteristic of this group. Cones define the conifers only.

- Angiosperms are extremely varied as well. Not all flower parts are found on all plants, and some are highly modified. Also note that cultivated flowers are often sterile, have aberrant parts, and may be altered through selective breeding and treatment.

Study Strategies

- Much of the terminology in this chapter was introduced in Chapter 28, and it is worth reviewing that chapter in light of the material presented here. The details of heterospory can be particularly confusing; reviewing Figure 28.12 can help you to sort them out.

- Be sure that you truly understand what sets angiosperms and gymnosperms apart. Use Figures 29.8 and 29.14 to compare their life cycles.

- Review the following animated tutorials and activities on the Companion Website/CD:
  Tutorial 29.1 Life Cycle of a Conifer
  Tutorial 29.2 Life Cycle of an Angiosperm
  Activity 29.1 Life Cycle of a Pine Tree
  Activity 29.2 Flower Morphology

Test Yourself

Knowledge and Synthesis Questions

1. You are enjoying a stroll in the botanical gardens. You notice a plant with a beautiful flower. Upon closer inspection you find that the flower has a pistil but no stamens. You look at several more flowers on the same plant, but you are unable to find any that have stamens. This plant is _____ and its flowers are _____.
   a. monoecious; perfect
   b. monoecious; imperfect
   c. dioecious; perfect
   d. dioecious; imperfect
   Textbook Reference: 29.3 What Features Distinguish the Angiosperms? p. 639

2. Why are gymnosperms referred to as naked-seed plants?
   a. They lack ovules.
   b. They lack ovaries.
   c. They do not protect their embryos.
   d. They do not have seed coats.
   Textbook Reference: 29.2 What Are the Major Groups of Gymnosperms? p. 634

3. Which of the following functions make seeds useful to plants?
   a. They provide a mechanism for dispersal.
   b. They protect the embryo.
   c. An embryo may remain dormant in a seed until optimum growth conditions are available.
   d. All of the above
   Textbook Reference: 29.1 How Did Seed Plants Become Today's Dominant Vegetation? pp. 633–634

4. Many angiosperms and animals have coevolved. What roles do animals play in the life cycle of plants?
   a. They act as pollinators.
   b. They assist in dispersal of seeds.
   c. They promote fertilization.
   d. All of the above
   Textbook Reference: 29.3 What Features Distinguish the Angiosperms? p. 641

5. In gymnosperms, fertilization results in a _____. In angiosperms, fertilization results in a(n) _____.
   a. zygote; zygote
   b. zygote; endosperm nucleus
   c. zygote; zygote and an endosperm nucleus
   d. zygote and an endosperm nucleus; zygote
   Textbook Reference: 29.3 What Features Distinguish the Angiosperms? p. 638

6. The diagnostic characteristic of an angiosperm is the presence of
   a. multiple carpels.
   b. woody growth.
   c. a flower.
   d. All of the above
   Textbook Reference: 29.3 What Features Distinguish the Angiosperms? p. 638

7. Which of the following statements regarding gymnosperms is true?
   a. All gymnosperms produce cones.
   b. Gymnosperms are heterosporous.
   c. Gymnosperm seeds have no protection.
   d. All gymnosperms are woody.
   Textbook Reference: 29.1 How Did Seed Plants Become Today's Dominant Vegetation? p. 632

8. Match the following flower structures with their function.

   _____ Ovule   a. Assists in attracting pollinators
   _____ Anther   b. Secretes sticky material to help pollen adhere
   _____ Stigma
   _____ Style   c. Site of ovule development
   _____ Ovary   d. Protects the immature flower bud
   _____ Petal   e. Houses the megasporangium
   _____ Sepal   f. Holds the stigma in position
               g. Houses microsporangia
   Textbook Reference: 29.3 What Features Distinguish the Angiosperms? pp. 638–639

9. What is the purpose of a fruit?
   a. It aids in dispersal of seeds.
   b. It protects seeds until they are mature.
   c. It attracts pollinators.
   d. Both a and b
   Textbook Reference: 29.3 What Features Distinguish the Angiosperms? p. 643

10. Vascular tissue in angiosperms is highly developed. The purpose of this vascular tissue is to move
    a. water.
    b. food.
    c. nutrients.
    d. All of the above
    Textbook Reference: 29.3 What Features Distinguish the Angiosperms? p. 638

Application Questions

1. Discuss the significance of double fertilization in angiosperms.
   Textbook Reference: 29.3 What Features Distinguish the Angiosperms? p. 642

2. Refer back to the obstacles encountered by plants when they began to colonize terrestrial environments. How have gymnosperms surmounted those obstacles?
   Textbook Reference: 29.1 How Did Seed Plants Become Today's Dominant Vegetation? pp. 631–634

3. Pollen is found only in seed tracheophytes. What is the significance of pollen in tracheophyte evolution? Is pollen a sporophyte or gametophyte?
   Textbook Reference: 29.1 How Did Seed Plants Become Today's Dominant Vegetation? pp. 632–633

4. Where would you find a female gametophyte in an angiosperm? Explain your answer.
   Textbook Reference: 29.3 What Features Distinguish the Angiosperms? p. 638

5. What is a cotyledon? How is it used to classify angiosperms into two monophyletic groups?
   Textbook Reference: 29.3 What Features Distinguish the Angiosperms? p. 643; 29.4 How Did the Angiosperms Originate and Diversify? p. 645

6. What is the significance of the angiosperm fruit? What types of tissues does the fruit arise from?
   Textbook Reference: 29.3 What Features Distinguish the Angiosperms? p. 643

Answers

Knowledge and Synthesis Answers

1. **d.** If, upon inspection, you find only female flowers, then you are most likely looking at a dioecious plant in which male and female flowers appear on separate plants. The absence of stamens makes these flowers imperfect.

2. **b.** Gymnosperms lack ovaries and thus lack the ability to produce fruit. Their embryos are protected, and they have ovules and seed coats.

3. **d.** Seeds protect the embryo, provide a mechanism for dispersal, and allow an embryo to remain dormant.

4. **d.** By acting as pollination vectors, animals promote fertilization of the plant. Animals also assist with dispersal of seeds.

5. **c.** Angiosperms differ from gymnosperms by having double fertilization that results in a triploid endosperm nucleus in addition to the zygote.

6. **c.** Only angiosperms have flowers. It is true that only angiosperms can have multiple carpels, but multiple carpels are not a necessary characteristic of an angiosperm.

7. **b.** All gymnosperms are heterosporous. Only conifers are cone producers.

8. **e.** Ovule
   g. Anther
   b. Stigma
   f. Style
   c. Ovary
   a. Petal
   d. Sepal

9. **d.** By the time fruit forms, a flower has already been pollinated, and there is no need to attract additional pollinators.

10. **d.** Vascular tissues in angiosperms move food, water, and nutrients throughout the plant.

Application Answers

1. Double fertilization allows for a diploid embryo and a triploid endosperm. The endosperm provides nutrition to the developing embryo at the time of seed germination.

2. Gymnosperms have a sophisticated vascular system that allows movement of nutrients and also support for large plants competing for sunlight. Wind-dispersed pollen removes the dependence on water for fertilization and increases the range of gymnosperms. Seeds protect embryos from desiccation and provide a mechanism for dispersal.

3. Pollen is the male gametophyte of seed tracheophytes; in other words, it is the male multicellular haploid structure of the seed tracheophyte's life cycle (see Figure 29.14). As a gametophyte, pollen's most fundamental function is to create gametes (sperm, in this case), but it also plays a role in helping those gametes reach and fertilize female gametes. Pollen is significant in that it provides a means of dispersal that eliminates the need for water in fertilization.

4. The female gametophyte is highly reduced and is part of the ovule of the angiosperm.

5. A cotyledon is the first "seed leaf" produced in an angiosperm embryo. Monocots have only one seed leaf, whereas eudicots have two.

6. Angiosperm fruit arises from tissues of the sporophyte and the gametophyte. How that fruit comes to develop depends on fruit type. Different fruits encompass more or less of the parent sporophyte tissues.

Important Concepts

Fungi are absorptive heterotrophs characterized by hyphae.

- Most fungi belong to one of five major groups: chytrids, zygomycetes, glomeromycetes, ascomycetes, and basidiomycetes. Based on evidence from DNA analysis, the chytrids and zygomycetes appear to be paraphyletic, while the other three groups are clades (i.e., monophyletic; see Figure 30.2 and Table 30.1). The chytrids are aquatic; the other four groups are terrestrial.

- All of these groups except the glomeromycetes contain unicellular as well as multicellular species. Except for the chytrids, unicellular fungi are commonly referred to as yeasts. Yeasts may reproduce by budding (see Figure 30.3), by fission, or sexually.

- The body of a multicellular fungus is called a mycelium. A mycelium is made up of individual tubular filaments called hyphae, Hyphae grow rapidly into a substrate; the hyphae in a single mycelium may collectively grow as much as 1 kilometer per day. Hyphal cell walls are strengthened by the polysaccharide chitin. The ability to produce chitin is a synapomorphy that distinguishes the fungi from choanoflagellates and animals.

- Hyphae may be septate (divided into compartments by chitinous walls) or coenocytic (continuous and multinucleate) (see Figure 30.4).

- Most fungi are multicellular, but multicellularity in fungi is significantly different from the familiar multicellularity of plants and animals. In plants and animals, each cell is usually enclosed in its own membrane and has only one nucleus. In multicellular fungi, cells are separated by porous septa that do not completely block the movement of organelles (see Figure 30.4). In many species, each cell has two nuclei; in some, even nuclei can pass through the septa.

- Hyphae provide a large surface area-to-volume ratio, thus a large surface area for absorption of nutrients from the substrate.

- Fungi are tolerant of hypertonic environments and temperature extremes.

Fungi are saprobes, pathogens, predators, or symbionts.

- Fungi absorb the nutrition needed for their survival from dead matter (saprobes), from living hosts (parasites), and from mutually beneficial symbiotic relationships with other organisms (mutualists).

- Adaptations of hyphae allow fungi to exploit different sources of nutrition. Among fungi that are plant parasites, for example, hyphae enter a leaf from a spore on the surface and form a mycelium within the leaf (see Figure 30.5). Rhizoids are hyphae that anchor chytrids and some other fungi to the substrate.

- Parasitic fungi are either facultative (possessing the ability to grow independently) or obligate (dependent on their living host for growth).

- Some parasitic fungi not only derive nutrition from their hosts, but also sicken or kill the host. Such fungi are called pathogens.

- Some fungi are active predators that trap microscopic protists or animals in a sticky substance; others form a constricting ring around their prey (see Figure 30.6).

Many fungi are beneficial to other organisms.

- Saprobic fungi secrete enzymes into the environment that help in absorption of dead matter and the decomposition and recycling of elements—especially carbon— used by living organisms. As decomposers, saprobic fungi are essential life on Earth.

- Fungi form two crucial types of symbiotic, mutualistic relationships, lichens and mycorrhizae.

- Lichens are associations of fungi with photosynthetic protists (green algae) or cyanobacteria. The fungus provides the photosynthetic partner with minerals and water, and the photosynthesizing organism provides the fungus with organic compounds. Lichens are among Earth's hardiest organisms, and can thrive in barren and extreme environments such as Antarctica. Lichens are characterized by their appearance as either crustose (crusty), foliose (leafy), or fruticose (shrubby) (see Figure 30.8).

- Mycorrhizae are associations between fungi and the roots of plants in which a fungus obtains the products of photosynthesis from the plant and provides minerals and water to the plant. The mycorrhizal symbiosis is essential to the survival of most plants, and the evolution of this relationship may have been the most important step in allowing plants to colonize land.

- Ectomycorrhizal fungi wrap their hyphae around a plant root. The hyphae of arbuscular mycorrhizal fungi penetrate the root cells and form treelike (arbuscular) structures that provide the plant with nutrients (see Figure 30.10).

- Endophytic fungi are symbionts living within the aboveground plant parts. They help certain plants, especially grasses, resist pathogens and herbivores; their role in other plants is not well understood.

Fungi have many different life cycles can reproduce both sexually and asexually.

- The different fungal groups display different life cycles. Most display alternation of generations. The glomeromycetes reproduce only asexually.

- Fungi have several means of asexual reproduction: They can produce spores enclosed within sporangia or naked spores known as conidia. Unicellular fungi can reproduce by fission or budding. Just about any part of a mycelium is capable of living independently of the rest; therefore, simply breaking a mycelium into two or more parts is a method of reproduction.

- Mechanisms of sexual reproduction are used to distinguish members of the different fungal groups from one another. Sexual reproduction involves two or more mating types. Self-fertilization is prevented because individuals of the same mating type cannot mate with one another (see Figure 30.13).

- The first step in sexual reproduction is fusion of two hyphae of different mating types. Eventually, two nuclei (one from each mating type) will fuse to create a zygote nucleus. The zygote nucleus may be the only diploid nucleus in the life cycle of a fungus.

- In the zygomycetes, ascomycetes, and basidiomycetes, a unique dikaryon condition is the first stage of sexual reproduction. The cytoplasms of two individuals fuse, but the nuclei do not fuse right away, and the resulting hyphae contain genetically different haploid nuclei. A fruiting structure is eventually formed, and the two nuclei fuse to form a zygote. This form of reproduction has no gamete cells (only gamete nuclei). The hyphae are not truly diploid ($2n$), but dikaryotic ($n + n$).

- Many parasitic fungi have complex life cycles that require a number of different hosts (see Figure 30.14).

- Sexual stages have not yet been identified in about 25,000 species of fungi, so these have not yet been assigned to any of the five major groups. Such fungi are pooled in a polyphyletic group known as the imperfect fungi, or deuteromycetes.

Chytrids are aquatic fungi with flagellated gametes.

- Chytrids probably resemble the common ancestor of all fungi more closely than any other fungal group (see Figure 30.15).

- Chytrids are aquatic, have chitinous cell walls, and are the only fungi with flagellated gametes. They can be either parasitic or saprobic.

- The genus *Allomyces* displays alternation of generations. A haploid zoospore becomes a small organism (the gametophyte) with both female and male gametangia that produce gametes (see Figure 30.13A). When the gametes fuse, they form a diploid zygote and a diploid organism (the sporophyte). The sporophyte produces diploid zoospores that grow into other diploid sporophytes. Eventually, a sporophyte produces sporangia that give rise to haploid zoospores via meiosis.

Zygomycetes have coenocytic hyphae and (usually) no fleshy fruiting body.

- Over 700 species of zygomycetes have been described. The hyphae of a zygomycete spread randomly over a substrate, periodically producing stalked sporangiophores bearing sporangia (see Figure 30.16).

- The zygomycetes include black bread mold and many other saprobic and parasitic species.

- Sexual reproduction in zygomycetes occurs between adjoining individuals with different mating types (see Figure 30.13B). Branches from each individual grow toward each other until they fuse to produce gametangia. A thick-walled zygosporangium with a zygospore forms at the fusion site. A sporangium sprouts from the zygospore.

Glomeromycetes form arbuscular mycorrhizae.

- Glomeromycetes are entirely asexual and coenocytic. They are also entirely terrestrial; about half the fungi found in soils are glomeromycetes. Along with ascomycetes and basidiomycetes, they form a clade known as the "crown fungi" (see Figure 30.2).

- The 200 species of glomeromycetes form arbuscular mycorrhizae with the roots of 80 to 90 percent of all plants (see Figure 30.10B).

Ascomycetes include baker's yeast, truffles, and *Penicillium*.

- The ascomycetes are distinguished by their ascus, a sexual reproductive structure (see Figure 30.13C). There are approximately 30,000 species of ascomycetes, divided into two groups: The euascomycetes have an ascocarp, whereas the hemiascomycetes lack this structure.

- The hyphae of ascomycetes have regularly spaced septa that nonetheless do not prevent movement of nuclei and other organelles.

- Most hemiascomycetes are microscopic.

- Unicellular hemiascomycetes can reproduce asexually by fusion or budding.

- Sexual reproduction in hemiascomycetes occurs between cells of different mating types and involves the forming of a diploid cell population in some species or direct meiosis of the zygote in others. Meiosis results in the whole cell, becoming an ascus with either eight or four ascospores.

- The species of yeast used to make bread and alcoholic beverages, *Saccharomyces cerevisiae*, is a hemiascomycete.

- The euascomycetes produce a dikaryon during sexual reproduction.

  - Euascomycetes include molds and the cup fungi and have a number of uses for humans (see Figure 30.17). Mold from the genus *Penicillium* produces the antibiotic penicillin; other euascomycete molds are important in making cheese. Brown molds are used in brewing sake (a Japanese alcoholic beverage) and soy sauce.

  - Asexual reproduction in euascomycetes involves the production of conidia at the tips of hyphae (see Figure 30.18).

  - A dikaryon is produced during sexual reproduction by the fusing of two mating structures (see Figure 30.13C). From the dikaryon hyphae ($n + n$), dikaryotic asci form where the nuclei fuse. The zygote ($2n$) undergoes meiosis and then produces ascospores (n). The ascospores germinate and grow into new mycelia.

The fruiting structures of basidiomycetes are familiar as mushrooms.

- Puffballs and mushrooms are among the 25,000 species of basidiomycete fungi, which produce the group's most spectacular fruiting structures.

- In basidiomycetes, meiosis and nuclear fusion occur in the basidium, found at the tips of hyphae (see Figure 30.13D). After meiosis, four basidiospores are formed on tiny stalks. The basidiospores are released into the environment and germinate to form haploid hyphae. The hyphae grow and fuse with different mating types to form the dikaryotic mycelium. The fruiting structure, or basidiocarp, grows from the dikaryotic mycelium. The cap of a mushroom is the fruiting structure of a basidiomycete (see Figure 30.19). Basidia form on the underside of the cap along gills and discharge their spores, which will form the new developing basidium.

The Big Picture

- Fungi are heterotrophic organisms that absorb nutrients from the environment. Fungi can be unicellular or multicellular. The mycelium is the body of a multicellular fungus and is composed of many tubular filaments known as hyphae. Many fungi form symbiotic relationships with photosynthetic organisms, creating mycorrhizae and lichens.

- Reproduction in the fungi occurs both sexually and asexually. Asexual reproduction involves the production of spores, breakage, fission, or budding. Sexual reproduction in multicellular fungi requires the union of hyphae with two different mating types.

- The five major phyla of Fungi are the chytrids, zygomycetes, glomeromycetes, ascomycetes, and basidiomycetes. Fungi that do not fall into these groups because they lack known sexual reproductive structures are pooled into the paraphyletic deuteromycetes, or "imperfect fungi."

  - Chytrids are aquatic fungi with flagellated gametes. Many chytrids display alternation of generations.

  - Zygomycetes have coenocytic hyphae. Sexual reproduction occurs when hyphae of two mating types join to form a zygosporangium from which a sporangium will eventually grow.

  - Glomeromycetes are terrestrial, coenocytic, and asexual. They are the predominant arbuscular fungi and as such are essential to plant life.

  - Ascomycetes produce an ascus as their reproductive structure. The ascomycetes are separated into the euascomycetes and the hemiascomycetes. The hemiascomycetes reproduce by budding, fission, or the production of an ascus. The euascomycetes have a dikaryotic ($n + n$) stage during reproduction.

  - The basidium is the reproductive structure of the basidiomycetes. The fusion of haploid hyphae forms a dikaryotic mycelium that produces the fruiting body, a basidiocarp.

Common Problem Areas

- Students are often confused by the numerous ways that fungi reproduce. Create flowcharts to help you understand the ploidy level at each stage of the life cycle in the different phyla of fungi. Make sure to include both the sexual and asexual stages in your charts. Comparisons of your flowcharts for the different phyla will help you learn the differences between the groups.

Study Strategies

- Organisms are placed in particular systematic groupings because they share unique characteristics. Look for patterns that distinguish the five major fungal groups from each other. Most fungi are grouped according to the reproductive structures that they produce.

- Review the following animated tutorial and activities on the Companion Website/CD:
  Tutorial 30.1 Life Cycle of a Zygomycete
  Activity 30.1 Fungal Phylogeny
  Activity 30.2 Ascomycete Life Cycle

Test Yourself

Knowledge and Synthesis Questions

1. Many species of fungi can be placed into one of five main groups based on
   a. methods of sexual reproduction.
   b. nucleic acid sequences.
   c. the presence or absence of septa in the hyphae.
   d. All of the above
   Textbook Reference: *30.1 How Do Fungi Thrive in Virtually Every Environment? pp. 651–652; p. 663*

2. Which of the following would you *not* expect to find in any of the typical fungal life cycles?
   a. Haploid nuclei
   b. Diploid nuclei
   c. Spores
   d. Chloroplasts
   Textbook Reference: *30.3 How Do Fungal Life Cycles Differ from One Another? p. 659*

3. Fungi are absorptive heterotrophs. Which of the following is an adaptation that greatly aids this mode of nutrient procurement?
   a. Dikaryosis
   b. A large surface area-to-volume ratio
   c. Conjugation
   d. A complex life cycle
   Textbook Reference: *30.1 How Do Fungi Thrive in Virtually Every Environment? p. 652*

4. Assume that two normal hyphae of different fungal mating types meet. After a period of time, the cell walls between these hyphae will dissolve, producing a
   a. mycelium.
   b. fruiting body.
   c. zygote.
   d. dikaryotic cell.
   Textbook Reference: *30.3 How Do Fungal Life Cycles Differ from One Another? p. 659*

5. Which of the following is the best way to represent the ploidy of dikaryotic hyphae?
   a. 2*n*
   b. *n*
   c. *n* + *n*
   d. *n/n*
   Textbook Reference: *30.3 How Do Fungal Life Cycles Differ from One Another? p. 662*

6. Which of the following statements about sexual reproduction in fungi is *false*?
   a. Motile gametes are present in all fungal species.
   b. Water is not required for fertilization to occur in most fungi.
   c. There is no true diploid tissue in the life cycle of most sexually reproducing fungi.
   d. Sexual reproduction often begins with contact between hyphae of different mating types.
   Textbook Reference: *30.4 How Do We Tell the Fungal Groups Apart? p. 663*

7. A mycorrhiza is
   a. an imperfect fungus.
   b. the fruiting structure of a basidiomycete.
   c. a symbiotic association between a fungus and cyanobacterium or green algae.
   d. a symbiotic association between a fungus and a plant.
   Textbook Reference: *30.2 How Are Fungi Beneficial to Other Organisms? p. 657*

8. Suppose a scientist investigating the classification of an imperfect fungus (a fungus that has never been observed to reproduce sexually) discovers that it has DNA sequences characteristic of basidiomycetes. If this scientist could coax fungi of this species to reproduce sexually, which of the following would he or she most likely observe?
   a. Dikaryotic hyphae segmented by septa
   b. Dikaryotic hyphae without septa
   c. Asymmetrical cell division
   d. A dikaryotic ascus
   Textbook Reference: *30.4 How Do We Tell the Fungal Groups Apart? p. 663*

9. Which of the following fungi have coenocytic hyphae and stalked sporangiophores?
   a. Chytrids
   b. Zygomycetes
   c. Ascomycetes
   d. Basidiomycetes
   Textbook Reference: *30.4 How Do We Tell the Fungal Groups Apart? p. 664*

10. Which of the following fungi display alternation of generations?
    a. Chytrids
    b. Glomeromycetes
    c. Ascomycetes
    d. Basidiomycetes
    Textbook Reference: *30.4 How Do We Tell the Fungal Groups Apart? pp. 663–664*

Application Questions

1. Early taxonomists considered fungi to be members of the plant kingdom. What evidence indicates that they are in fact more closely related to animals?
Textbook Reference: *30.1 How Do Fungi Thrive in Virtually Every Environment? p. 651*

2. Complete the following diagram, which shows the generalized life cycle of a basidiomycete, by adding the following terms in their appropriate locations: spores, dikaryotic hyphae, meiosis, mitosis, fertilization. Also, circle and label the portions of the life cycle corresponding to the haploid (*n*), diploid (*2n*), and the dikaryotic (*n + n*) stages.

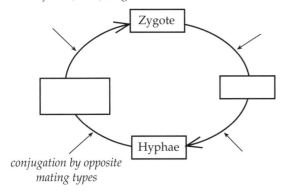

conjugation by opposite mating types

Textbook Reference: *30.4 How Do We Tell the Fungal Groups Apart? p. 662*

Answers

Knowledge and Synthesis Answers

1. **d.** Traditionally, a fungus's method of sexual reproduction was the primary criterion in its classification, but all the traits listed are useful in classifying fungi.

2. **d.** Fungi have haploid and diploid nuclei at different stages of their life cycle. They also produce spores. The chloroplasts are not found in the fungi, but in plants.

3. **b.** The large surface area-to-volume ratio of the hyphae increases the ability of a fungus to absorb nutrients.

4. **d.** When two hyphae of different mating types fuse, they form a dikaryotic hyphae.

5. **c.** Dikaryotic hyphae are neither truly diploid (*2n*) nor haploid (*n*). Because dikaryotic hyphae include genetic material from two haploid nuclei that remain separate, the best way to represent their ploidy is as *n + n*.

6. **a.** No fungi have motile gametes. Only the gametes of chytrids are motile.

7. **d.** Mycorrhizae are associations between fungi and the roots of plants. Lichens are symbiotic relationships between fungi and cyanobacteria or green algae.

8. **a.** If this fungus is indeed a basidiomycete, when hyphae of different mating types fuse, they will most likely create dikaryotic hyphae that are segmented by septa.

9. **b.** Coenocytic hyphae are characteristic of both the chytrids and the zygomycetes, but only the zygomycetes regularly produce sporangiophores.

10. **a.** Among the fungi, only the chytrids have a multicellular haploid stage and a multicellular true diploid stage.

Application Answers

1. At the cellular level, fungi bear very little resemblance to plants. Fungi don't even particularly resemble the surviving green algae that are the most likely common ancestor of all plants. They do not produce any sort of chlorophyll, nor do they have plastids for storing reserves of photosynthetic products (see Section 28.1). Fungi and plants both have cell walls, but the cell walls of fungi contain chitin, a polysaccharide molecule (see Chapter 3) that plants do not produce. Chitin *is* found in some animals (particularly among the ecdysozoans; see Chapter 32) and in choanoflagellates, the protist group most closely related to the animals. It is unlikely that this complex molecule evolved more than once, so fungi and animals probably shared a chitin-producing common ancestor more recently than either group did with the plants.

2.

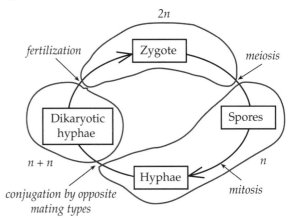

CHAPTER 31 Animal Origins and the Evolution of Body Plans

Important Concepts

Animals are descendants of a common ancestor.

- The animals, or metazoans, are a monophyletic group of motile, multicellular, heterotrophic organisms. Many gene sequences support their monophyly, and animals display similarities in the organization and function of Hox genes. In addition, animals share certain characteristics important for multicellularity: special types of cell–cell junctions (tight junctions, desmosomes, and gap junctions) and extracellular matrix molecules (collagen and others).

- The common ancestor of modern animals may have been a colonial flagellated protist such as a choanoflagellate (see Figure 27.28).

- Differences in patterns of embryonic development, including cleavage patterns, gastrulation patterns, and the number of cell layers present, show the evolutionary relationships among animals. On these bases, animals can be grouped into diploblastic (two cell layers, the endoderm and ectoderm) and triploblastic (three layers, with the addition of a mesoderm) animals, and into protostomes ("mouth first"; the blastopore becomes the individual's mouth) and deuterostomes ("mouth second"; the blastopore becomes the anus).

Every animal has a body plan.

- The overall organization of an animal's body is known as its body plan. An animal's body plan includes symmetry, body cavity structure, support (skeletal) structure, segmentation, and the presence or absence of appendages.

- An animal that can be divided along at least one plane into similar halves is said to be symmetrical. Asymmetrical animals such as sponges have no plane of symmetry. The sea anemone is radially symmetrical; any plane running along a sea anemone's main axis will divide it into roughly equal halves (see Figure 31.3).

- Animals exhibiting bilateral symmetry can be divided into two mirror images by only one plane. Bilateral symmetry is common among animals that are able to move quickly.

- Bilaterally symmetrical animals often have sense organs and nervous tissue concentrated at the anterior end; this type of organization is known as cephalization, from the Greek word for "head."

- Animals have three different types of body cavities (see Figure 31.3). Acoelomates have no enclosed body cavity, pseudocoelomates have a liquid-filled space known as the pseudocoel in which many of the internal organs are located, and the coelomates have a true body cavity, a coelom, developing within the mesoderm. In coelomates, the internal organs are in pouches of the peritoneum.

The structure of the body plan influences movement.

- Fluid-filled body cavities act as hydrostatic skeletons for many animals. Other animals evolved rigid supportive skeletons that can be internal (bones or cartilage) or external (a shell or cuticle). Muscles attached to hard skeletons allow the animal to move.

- Segmentation allows specialization of the different body regions and can improve control of movement.

- Appendages, especially jointed limbs, enhance locomotion. Jointed limbs in the arthropods and vertebrates are a major factor in their evolutionary success.

- Animals can lose as well as gain body plan attributes over the course of evolution; snakes, for example, have no limbs but are descended from a four-limbed (tetrapod) vertebrate ancestor; snakes are close relatives of the lizards, which retain the four-limbed state.

All animals are ingestive heterotrophs.

- To obtain food, animals either move through the environment to the food or move the environment and the food to them.

- Most animals are motile, but some are sessile (nonmoving).

- As heterotrophs, animal feeding strategies fall into a few broad categories. Animals may be filter feeders, herbivores, predators, parasites, or detritivores. Some animals change their feeding strategies at different developmental stages.

There are a variety of animal life cycles.

- The life cycle of an animal encompasses embryonic development, birth, growth to maturity, reproduction, and death.

- In direct development, newborns look very similar to adults. In many species, however, newborns differ strikingly from adults and pass through distinct larva, pupa, and adult stage. Metamorphosis refers to the dramatic changes that can occur between the larval and adult stages, when the individual is a pupa.

- All life cycles have a least one dispersal stage. The larva is the dispersal stage for most sessile organisms. Common larval types seen in animals are trochophore and nauplius.

- Every life cycle necessarily involves evolutionary trade-offs. Common trade-offs involve reproduction, such as courtship behavior to obtain a mate and producing altricial or precocial young.

Animals are classified into broad groups.

- The simplest animals are the three groups of sponges, which have no body symmetry and no distinct cell layers. All animals other than the sponges comprise the eumetazoans.

- The bilaterians are a monophyletic group that includes all of the eumetazoans except the ctenophores and the cnidarians. Bilaterian synapomorphies include bilateral symmetry, three cell layers, and the presence of at least seven Hox genes. The Bilateria has two major subgroups, the protostomes (see Chapter 32) and the deuterostomes (see Chapter 33).

Sponges have differentiated cells but no true organs.

- Although the sponge body plan is relatively simple, the three sponge groups all have cells that are differentiated for specific functions.

- Most of the 8,000 species of sponges are marine filter feeders that remove small organisms and nutrient particles from seawater as it flows through pores in the walls of their inner cavity. Choanocyte cells use flagella to divert water through the pores and filter out food particles. The water exits the sponge through large openings called oscula (see Figure 31.7).

- Sponges have a supporting skeleton composed of branching spines (spicules). The spicules of glass sponges and demosponges (the largest sponge group) are made of silicon. The calcareous sponges take their name from their calcium carbonate spicules and are the sponge group most closely related to the eumetazoans.

- The huge variety of sponge body sizes and shapes is a response to the different movement patterns of water, specifically tides and currents.

- Sponges reproduce sexually by producing both egg and sperm, and asexually by budding and fragmentation.

Ctenophores have two cell layers separated by mesoglea.

- The 100 known species of ctenophores—the comb jellies—are marine, diploblastic animals found primarily in the open ocean, where they feed on planktonic organisms that are filtered by sticky filaments on the tentacles. Ctenophores have a complete gut (i.e., a gut with an entrance and exit, or mouth and anus).

- Ctenophores get their name from the eight rows of comb-like plates of cilia, known as ctenes. The cilia are used to propel the animal through the water. Prey is caught on sticky filaments on the tentacles or body (see Figure 31.17).

- The ectoderm and endoderm are separated by a thick, gelatinous layer, the mesoglea.

- Ctenophores have a simple life cycle and reproduce sexually. In most species, the externally fertilized egg hatches into a miniature ctenophore (i.e., direct development).

The cnidarian life cycle has two stages: the polyp and the medusa.

- All cnidarians are diploblastic and have radial symmetry.

- The cnidarian gastrovascular cavity is a blind sac. There is only one opening, which serves as both mouth and anus. The gastrovascular cavity functions in food digestion, respiratory gas exchange, and circulation. It also lends support as a hydrostatic skeleton.

- Like ctenophores, a large amount of mesoglea is found between the two cell layers of cnidarians. Because of the inert nature of the mesoglea, cnidarians have low metabolic rates. Many species, however, are able to capture large prey; this combination of characteristics makes it possible for them to live in environments with little prey.

- Cnidarian tentacles have specialized cells, cnidocytes, which inject toxins into their prey with the help of stingers called nematocysts (see Figure 31.10).

- The life cycle of most cnidarians includes a sessile polyp stage and a motile medusa stage (see Figure 31.18). The polyp stage usually reproduces asexually. The medusa stage reproduces sexually, with the fertilized egg becoming a planula larva and eventually a polyp.

- Cnidarians possess muscle fibers that enable them to move and simple nerve nets that integrate their activities. They also possess the structural molecules collagen, actin, and myosin, which, along with their Hox genes, link them to the bilatarians.

- All but a few of the 11,000 or so species of cnidarians are marine. The smallest individuals are almost microscopic, while individual jellyfish can be quite large. Many cnidarians are colonial.

- Scyphozoans include the marine jellyfish, which have medusae with thick mesoglea. These species spend most of their lives in the medusa stage. Polyps are produced sexually by adult medusae. The polyps produce young medusae by budding (see Figure 31.18).

- Anthozoans (about 6,000 species of sea anemones, sea pens, and corals) lack a medusa stage altogether. Their polyps can reproduce both sexually and asexually. Sexual reproduction produces a planula that will develop into a polyp. Sea anemones are solitary and occasionally motile; corals are sessile and colonial. The coral skeleton is composed of calcium carbonate and varies greatly between species. Corals live in symbiosis with photosynthetic protists that provide nutrients for the coral colony through photosynthesis, contributing to the success of corals in clear tropical waters (see Figure 31.20).

- Hydrozoans are typically dominated by the polyp stage of the life cycle. The polyps tend to be colonial, with many polyps sharing a common gastrovascular cavity (see Figure 31.21).

The Big Picture

- The animals are a monophyletic group sharing a number of morphological and genetic traits. Animals are motile, multicellular organisms that must ingest nutrients.

- Animals are distinguished by the number of cell layers found in their embryos, their type of body cavity, and their body symmetry.

- The importance of acquiring nutrition (food) has led to the evolution of a variety of feeding strategies that maximize the available sources of nutrition. Life cycles also vary considerably among the animals, and involve a considerable amount of trade-off.

- The eumetazoans encompass all animals except the three groups of sponges. Ctenophores and cnidarians are diploblastic eumetazoans that are not bilaterally symmetrical.

- Sponges can be classified in three different groups based on molecular evidence and the morphology of their spicules.

- Both ctenophores and cnidarians are characterized by the presence of a largely inert layer of gelatinous mesoglea. Ctenophores have a complete gut with a mouth and an anus, whereas cnidarians have a blind gastrovascular cavity in which a single opening serves to ingest food and expel wastes. Cnidarians have simple nerve nets and muscle fibers, which allow them a level of control over their movements not found among the ctenophores.

Common Problem Areas

- Many of the same evolutionary "themes" can be found in widely divergent species. Venomous or venom-producing structures, for example, are found in many different animal groups, from the cnidarians to spiders to snakes. As the different animal groups are described, create table of the various characteristics or mechanisms that consistently appear. Use this table as a tool to help you organize your thinking about evolution and the creation of diversity.

Study Strategies

- The phylogenetic trees in your textbook are good skeletons on which you can add more information. Doing so will provide you with a framework and a point of reference when you study the different groups. Remember that animals have a set of traits they inherit from and share with their ancestors, as well as distinctive derived traits that distinguish them.

- Review the following animated tutorial and activities on the Companion Website/CD:
  Tutorial 31.1 Life Cycle of a Cnidarian
  Activity 31.1 Animal Classification
  Activity 31.2 Animal Features

Test Yourself

Knowledge and Synthesis Questions

1. Which of the following is *not* a derived trait that is shared by all animals?
   a. Hox genes
   b. The extracellular matrix molecule collagen
   c. Tight junctions, desmosomes, and gap junctions
   d. Bilateral symmetry
   Textbook Reference: *31.1 What Evidence Indicates the Animals Are Monophyletic? p. 672*

2. An animal is known to be a deuterostome. Which of the following does *not* apply to this animal?
   a. Three distinct layers of tissue were present during development.
   b. If a coelom is present, it formed within the embryonic mesoderm.
   c. Its early embryonic cleavage pattern was radial.
   d. It is diploblastic.
   Textbook Reference: *31.1 What Evidence Indicates the Animals Are Monophyletic? p. 673*

3. An important factor contributing to the evolution of diversity among animals is the considerable variation in its
   a. multicellularity.
   b. cell junctions.
   c. methods of food acquisition.
   d. symmetry.
   Textbook Reference: *31.3 How Do Animals Get Their Food? p. 676*

4. Which of the following statements about the body cavity of animals is true?
   a. The body cavity of coelomates develops from the embryonic ectoderm.

b. The acoelomate's body cavity is filled with liquid.

c. The pseudocoel of the pseudocoelomates has a peritoneum.

d. The acoelomates do not have an enclosed body cavity.

Textbook Reference: *31.2 What Are the Features of Animal Body Plans? p. 674*

5. Which one of the following items is *not* involved with some mechanism of movement in animals?

a. Hydrostatic skeleton

b. Jointed appendages

c. Being sessile

d. Segmentation

Textbook Reference: *31.3 How Do Animals Get Their Food? p. 676*

6. Which one of the following feeding strategies requires a life cycle in which much time and energy are devoted to dispersal?

a. Herbivory

b. Parasitism

c. Predation

d. Filter feeding

Textbook Reference: *31.4 How Do Animal Life Cycles Differ? p. 681*

7. Which of the following would likely be true of a species of bird with precocial young?

a. Their eggs would have a small amount of yolk.

b. They would have to care for the young for a long time.

c. They would have a long incubation period.

d. They would produce many small eggs.

Textbook Reference: *31.4 How Do Animal Life Cycles Differ? p. 681*

8. Which of the following statements about sponge structure or function is *false*?

a. Choanocytes are flagellated cells that play a role in feeding.

b. Large species are found in areas of heavy wave action, where food is most abundant.

c. Individual sponges are both male and female.

d. Water enters a sponge through pores and exits via one or more oscula.

Textbook Reference: *31.5 What Are the Major Groups of Animals? p. 684*

9. Which of the following statements is *not* true about ctenophores and cnidarians?

a. Most of them are marine organisms.

b. Both have radial symmetry.

c. Both have complete guts.

d. Both have feeding tentacles.

Textbook Reference: *31.5 What Are the Major Groups of Animals? pp. 684–685*

10. Why does the mesogleal layer found in both ctenophores and cnidarians help these organisms survive even when prey is scarce?

a. Prey adheres to the sticky surface of the mesoglea.

b. The mesoglea is biologically inert.

c. The mesoglea allows the animal to capture large prey items.

d. Mesogleal molecules provide nutrients to the host animal.

Textbook Reference: *31.5 What Are the Major Groups of Animals? pp. 684–686*

11. Which of the following traits is *not* shared by sea anemones and jellyfishes?

a. A medusa is the dominant stage in the life cycle

b. Possession of a gastrovascular cavity

c. Sexual reproduction

d. Nematocysts on the tentacles

12. Which cnidarian group is dominated by species that live symbiotically with photosynthetic protists?

a. Hydrozoans

b. Ctenophores

c. Scyphozoans

d. Anthozoans

Textbook Reference: *31.5 What Are the Major Groups of Animals? p. 686*

Application Questions

1. Discuss two evolutionary trade-offs confronting animals that involve reproduction.

Textbook Reference: *31.4 How Do Animal Life Cycles Differ? pp. 680–681*

2. As ingestive heterotrophs, discuss the two major strategies animals use to get food, and indicate which of these would most likely be used by sessile organisms.

Textbook Reference: *31.3 How Do Animals Get Their Food? p. 676*

Answers

Knowledge and Synthesis Answers

1. **d.** Not all animals have bilateral symmetry. Sponges, cnidarians, and ctenophores do not exhibit bilateral symmetry in their body plan.

2. **d.** Deuterostome embryos have three layers, the ectoderm, the mesoderm, and the endoderm, making deuterostomes triploblastic, not diploblastic.

3. **c.** Animals eat foods from all kingdoms of life. Thus, they have evolved a wide variety of methods for acquiring diverse types of food.

4. **d.** The body cavity of coelomates develops from the mesoderm and contains a peritoneum. The acoelomates lack a body cavity.

5. **c.** Being sessile means that you are stationary and do not move. All other items are involved with movement in animals.

6. **b.** Because many parasites die when the host dies, parasites must have a way to disperse their progeny to new hosts.

7. **c.** Because precocial young hatch as well-developed individuals who can forage for themselves, they undergo more of their development in the egg and tend to be incubated for longer periods of time.

8. **b.** Because they are not structurally robust, heavy wave action would destroy large, upright sponges.

9. **c.** Ctenophores and cnidarians are mostly marine, have radial symmetry, and are largely composed of mesoglea. Ctenophores have a complete gut with a mouth and two anal pores; cnidarians have a blind gut with only one opening.

10. **b.** The biologically inert mesoglea does not require energy; thus these animals have a very low metabolic rate and can survive for an extended time with only minimal amounts of food.

11. **a.** Although the jellyfish have both a medusa and polyp stage in their life cycle, the sea anemones have lost the medusa stage and spend their life cycle as polyps.

12. **d.** Anthozoans include the corals and sea anemones, both of which contain many species that live symbiotically with photosynthetic protists.

Application Answers

1. One trade-off would be to produce many small eggs or just several large eggs. Another trade-off would be to produce altricial young that hatch early but require much parental care or to produce precocial young that require a longer incubation period, but little parental care after they hatch.

2. Animals can either move through the environment to where the food is located or can move the environment and the food to themselves. Sessile organisms would more likely have adaptations to move the environment and food to themselves.

CHAPTER 32 Protostome Animals

Important Concepts

Protostomes are diverse and exhibit many body plans.

- Among many (but not all) protostomes, the blastopore becomes the mouth. Other traits that protostomes share are a ventral nervous system and an anterior brain that surrounds the opening to the digestive tract. Beyond these aspects, protostome body organization varies extensively from group to group.

- Protostomes includes some groups that are coelomate and some that are pseudocoelomate. One important group, the flatworms, is acoelomate. Two prominent protostome groups; the arthropods and the mollusks, have experienced secondary evolutionary modifications of the coelom (into a hemocoel in the arthropods; the mollusks have secondarily returned to a virtually open circulatory system).

- With the exception of the arrow worms, protostome animals are classified as either lophotrochozoans or ecdysozoans. A number of groups in both of these divisions possess a wormlike body plan.

- Several lophotrochozoans groups are characterized by a complex U-shaped structure, the lophophore, which is used both as a feeding apparatus and in gas exchange. Many groups have a type of free-living larva known as a trochophore. (Hence, "lophotrochozoans.") Most species have internal skeletons and many use cilia for locomotion.

- Several different lophotrochozoan lineages exhibit a derived form of early development known as spiral cleavage. These are sometimes grouped into the spiralians, but gene sequence analysis indicates that the different spiralian groups are not monophyletic.

- Ecdysozoans are characterized by a rigid exoskeleton that provides both protection and support. These animals grow by molting their exoskeletons ("ecdysis" is the Greek word for "shedding"). Increasing molecular evidence, including a set of Hox genes shared by all ecdysozoans, supports the monophyly of these animals and suggests that exoskeletal molting is a lifestyle that may have evolved only once.

- The need for rapid locomotion is met in the arthropod ecdysozoans by jointed appendages controlled by muscles. Over evolutionary time, different arthropod groups have experienced extensive modification of the appendages to suit life in a multitude of different environments. Arthropods are the dominant animals on Earth today, in both number of species and number of individuals.

- The arrow worms are an enigmatic group. Because of their early developmental morphology, they were once classed with the deuterostomes, but molecular evidence now clearly identifies them as protostomes. It is unclear whether the arrow worms are a sister group of the entire protostome clade, or whether they are most closely related to the lophotrochozoans. Arrow worms have no circulatory system and no larval stage. They are major predators of small organisms in the open ocean (see Figure 32.5).

Many lophotrochozoans have unsegmented bodies.

- The 4,500 species of ectoprocts are colonial: strands of tissue connect individuals in each colony, and in some species individuals are specialized for feeding, reproduction, defense, and support. Individuals have a great deal of control in manipulating their lophophores to increase contact with prey (see Figure 32.3). Colonies grow via asexual reproduction of the founding members. Sexual reproduction also occurs, and larvae emerge to seek suitable sites to form new colonies. Ectoprocts can cover large areas of coastal rock and can even form small reefs in shallow seas.

- Flatworms lack respiratory organs and have only simple cells for waste removal. The flat shape of the animal helps in oxygen transport and waste removal (see Figure 32.7). The digestive tract has only one opening, but it has many branches that aid in nutrient absorption. Although there are some dramatic-appearing free-living flatworm species, most of the 25,000 species of flatworms are tapeworms or flukes, both of which are parasites of other animals. Parasitic flatworms feed on the nutrient-rich body tissues of their host animal and disperse their eggs in the host's feces.

- Most of the 1,800 species of rotifers are very small (some smaller than single-celled protists), but they have specialized internal organs (see Figure 32.8). Most species live in freshwater habitats and feed using a ciliated organ called a corona. This is the only group of animals known to have existed for millions of years without the benefits of sexual reproduction.

- Nemerteans, or ribbon worms, have a complete digestive tract with two openings. Small ribbon worms use their cilia for movement, and large ribbon worms use muscle contractions to move. The feeding organ of the ribbon worms is a proboscis that is kept in a rhynchocoel, or fluid-filled cavity (see Figure 32.9). The proboscis has a sharp stylet and can be forcefully ejected from the body to catch prey. Most of the 1,000 or so species are marine, although a few species are found in freshwater or on land. Most are small, but some species can be up to 20 meters long.

- The phoronids include 20 species of tiny, sedentary worms that live in chitinous tubes and extract food from the water with their lophophores (see Figure 32.10).

- Brachiopods are solitary marine animals that live attached to the substratum. Their divided shell with two halves makes them superficially resemble clams (see Figure 32.11). The lophophore is located in the shell, and cilia help draw water and food into the shell. More than 26,000 fossil brachiopod species have been described, but only about 335 species are known to exist today.

Annelids undergo spiral cleavage and have segmented bodies.

- The annelids consist of approximately 16,500 species of segmented worms living in marine, freshwater, and moist terrestrial environments. The segmentation that evolved in this group gives these worms extremely good control of their movement. A separate nerve center called a ganglion controls the movement of each segment, and in most cases each segment also contains an isolated coelom (see Figure 32.12). Gas exchange occurs across the thin surface of the animal; the thinness of this external covering restricts most annelids to aquatic or very moist terrestrial environments, because they lose moisture rapidly when exposed to air.

- More than half of all annelids are polychaetes ("many hairs"). Outgrowths used in gas exchange, the parapodia, extend from segments over much of the body. Setae extending from the parapodia help to attach the animal to the substrate and aid in movement. The joining of male and female gametes produces trochophore larvae. A relatively recently discovered polycheate group, the pogonophorans, have secondarily lost their digestive tract. Pogonophorans are found in the deep ocean near hydrothermal vents. They harbor a number of endosymbionts that contribute to their nutrition.

- There are two major clades of clitellate annelids. The oligochaetes ("few hairs") include the most familiar annelids, called the earthworms. Oligochaetes live mainly in freshwater and terrestrial environments and are hermaphroditic, containing both male and female reproductive organs in the same individual. The second clitellate group, the leeches, are also hermaphroditic species that live either in freshwater or on land. The coelom of these parasitic annelids is not segmented but is composed of undifferentiated tissue. Clusters of segments at the leeches' anterior and posterior ends are modified into suckers, which the leech attaches to the substratum for movement, or to a host mammal from which it sucks blood (its nutritional source).

- The evolutionary relationship of the clitellates compared to the different polychaete groups is unclear, and the latter may be a paraphyletic group.

The three-part molluscan body plan has undergone dramatic evolutionary radiation.

- Mollusks are another group that undergoes spiral cleavage. The mollusks have evolved into a morphologically diverse group based on a distinctive three-part body plan. All species have a large muscular *foot*, and a tissue fold known as the *mantle* covers a *visceral mass* of internal organs. The mantle is extended in some species to create a mantle cavity holding the gills used in respiration; in many species the mantle secretes a hard, calcareous shell. Mollusks have a secondarily reduced coelom (see Figure 32.14).

- The molluscan body plan has resulted in a diverse array of some 95,000 species that fall into four major modern groups: the chitons, bivalves, gastropods, and cephalopods. A fifth group, the monoplacophorans, was abundant 500 million years ago, but only a few species survive today.

  - Chitons are marine mollusks that feed on algae, which they scrape off rocks using a razor-like body structure called the radula. A chiton's shell consists of iterated plates. They have simple internal organs, multiple gills, and bilateral symmetry.

  - The bivalve mollusks include the familiar clams, oysters, scallops, and mussels. They are found in both salt and fresh water, but they are all aquatic. Bivalves use a siphon to bring water into their two-part hinged shells; they are filter-feeders who extract foodstuffs from these water currents. Large gills inside the shell extract food and also function as respiratory organs. Among the clams, the molluscan foot has been modified into a digging device that allows the animal to burrow into the mud or sand.

  - Gastropods are the most species-rich and widely distributed of the mollusks, and are found in all environments. There are shelled and unshelled gastropod species, including the snails and slugs (the only terrestrial mollusks), as well as the marine nudibrachs (sea slugs), whelks, limpets, and abalones. Gastropods use their foot either to crawl or swim.

- Cephalopods include the octopuses, squids, and nautiluses. They have evolved the ability to control water movement into the mantle, allowing them to use ejected water as a jet for propulsion. They capture prey with their tentacles, and their greatly enhanced mobility makes them dominant ocean predators. They are also able to control gas movement in the mantle, which helps in buoyancy control. As rapidly moving, active predators, cephalopods have a head with complex sensory organs, most notably the eyes, which are in many ways comparable to those of vertebrates.

An external skeleton that does not grow distinguishes ecdysozoans.

- Ecdysozoans have an exoskeleton that ranges from thin and flexible to hard and rigid. Exoskeletons provide protection and support, but present obstacles to growth, locomotion, and gas exchange.

- Ecdysozoans have overcome the obstacles of the exoskeleton to become the most numerous and diverse of all animal groups.

- Some marine ecdysozoans have a thin exoskeleton called a cuticle that is molted periodically. The cuticle allows for gas exchange with the environment, but offers little protection. The priapulids, kinorhynchs, and loriciferans are groups of tiny, wormlike marine animals that live burrowed in ocean sediments (see Figure 32.16).

- The 230 species of horsehair worms are, as their name suggests, long and very thin (see Figure 32.17). They live in fresh water or very damp soil, near the edges of ponds and streams and feed in the larval stage as parasites of terrestrial and aquatic insects and crabs. Adults have reduced guts and in some species the adults may not feed; it is also possible that adult worms may absorb nutrients from the environment during those times between molting the old cuticle and the hardening of the new one.

- Nematodes, or roundworms, range from microscopic to up to 9 meters in length. About 25,000 species of this diverse group have been described, and this may represent less than a quarter of extant nematode species. They live in soil, on the bottoms of lakes and streams, in marine sediments, and as parasites of plants and animals.

- Nematodes use their gut for both gas exchange and nutrient uptake (see Figure 32.18). Many species prey upon protists and other microscopic organisms, but of most significance to humans is the large number of parasitic species; several of these species, including *Trichinella spiralis*, are dangerous to humans. The free-living nematode *Caenorhabditis elegans* is a widely used model organism for geneticists and developmental biologists.

Arthropods are Earth's dominant animals.

- Arthropods have a hard exoskeleton of chitin that provides protection, waterproofing, and muscle attachment sites. Their bodies are segmented, with individual muscles attached to the exoskeleton that operate each segment. The jointed appendages that give the group its name ("arthros" = joint, "poda" = limb) allow for a greater range of movement and the specialization of different appendages for different purposes.

- Two groups that are closely related to the arthropods, the onychophorans and the tardigrades, have unjointed legs and thin cuticles composed of chitin (see Figure 32.20). Members of both groups use their fluid-filled body cavities as hydrostatic skeletons. Onychophorans have probably changed relatively little from their common ancestor with the arthropods.

- The jointed appendages that characterize the arthropods appeared during the Cambrian, among a once-widespread group known as the trilobites ("three sections"). The trilobites died out during the great Permian mass extinction, but their heavy exoskeletons were readily fossilized and left an abundant record of their existence (see Figure 32.21). At least 10,000 species have been described based on the fossil record.

- Four major arthropod groups survive: the crustaceans (crabs, lobsters, scallops, barnacles, etc.), the hexapods (insects and their relatives), the myriapods (centipedes and millipedes), and the chelicerates (including the arachnids—spiders, mites, ticks, etc.).

Most marine arthropods are crustaceans.

- Crustaceans include many familiar animals (shrimp, lobsters, crabs, barnacles) as well as the less-familiar isopods, copepods, krill, and others. About 50,000 species have been described so far (see Figure 32.22).

- The crustacean body is divided into the head, thorax, and abdomen. In many species, a carapace extends dorsally from the head to protect and cover the body. The thorax and abdomen have one pair of appendages each; crustacean appendages are specialized for walking, swimming, feeding, sensation, and gas exchange. In some species, complex branched appendages have evolved.

- The typical crustacean larva is called a nauplius. It has three pairs of appendages and one eye.

Insects dominate the terrestrial environment.

- More than a million species of insects have been described so far, and many biologists believe this is only a small fraction of the species that actually exist.

- Three groups of wingless hexapods—the springtails, two-pronged bristletails, and proturans—are related to the insects and probably most closely resemble the insect ancestral form. Members of these three groups differ from insects in having internal rather than external mouthparts. In addition, insects carry paired antennae with sensory receptors called Johnston's organs.

- Like the crustaceans, the insect body has three parts: the head with a pair of antennae, the thorax with three pairs of legs, and the abdomen. Unlike the crustaceans, the abdominal segments do not bear appendages (see Figure 32.24). Gas exchange occurs in a system composed of a series of air sacs and tubular channels called tracheae.

- There are two classes of insects: the wingless apterygotes and the winged pterygotes (some species of which have secondarily become wingless). The apterygotes include the springtails and silverfish.

- Pterygotes typically have two pairs of wings attached to the thorax. They are divided three major groups: those that cannot fold their wings over their bodies, those that can fold their wings and undergo incomplete metamorphosis, and those that can fold their wings and undergo complete metamorphosis (see Table 32.2 and Figure 32.27).

  - Insects that cannot fold their wings over their bodies include dragonflies, damselflies, and mayflies.

  - Insects that undergo incomplete metamorphosis include grasshoppers and crickets, termites, stone flies, earwigs, thrips, true bugs, aphids, cicadas, and many others. In these groups, the hatchlings resemble small, usually wingless adults; as they molt from one stage to the next, they gradually acquire more adult characteristics.

  - Complete metamorphosis occurs in lacewings, beetles, caddisflies, butterflies, moths, flies, wasps, bees, and ants, among others. In these groups, the larvae and adult forms are substantially different; the young pass through at least two stages (larva and pupa) before becoming adults.

- The evolution of wings probably occurred only once during insect evolution. Studies of homologous genes in insects and crustaceans suggest that insect wings are modified from a respiratory structure on the leg of an ancestral crustacean (see Figure 32.28).

Myriapods and chelicerates have only two body regions.

- The myriapods include 3,000 described species of centipedes and 10,000 described species of millipedes. Members of both groups have similar body plans, consisting of a well-formed head and a long, segmented trunk. Centipedes have one pair of legs on each trunk segment; millipedes have two pairs on each segment (see Figure 32.29).

- The chelicerates include the pycnogonids, the horseshoe crabs, and the arachnids. All chelicerates have two body parts, and most have eight legs.

  - The pycnogonids, or sea spiders, are a group of about 1,000 exclusively marine species. Most are very small and are rarely seen. There are only four living species of horseshoe crabs. These animals have changed so little in morphology over evolutionary time that they are often referred to as "living fossils" (see Figure 32.30).

  - The arachnids are the most prominent chelicerates and include the spiders, scorpions, mites, and ticks. They have a simple life cycle, with young resembling small adults. Spiders, the most familiar arachnids, build webs of protein threads that they use to capture prey. Spider webs are strikingly varied, often species-specific, and increase the predatory ability of spiders in many different environments.

The Big Picture

- The protostomes include two major groups: the lophotrochozoans and the ecdysozoans. The arrow worms are not placed in either group; they may be sister to the protostomes as a whole, or may be more closely related to the lophotrochozoans. Protostomes have an anterior brain and a ventral nervous system. Many species in both groups have a wormlike appearance.

- The lophotrochozoans get their name from two structures: the lophophore; a feeding and gas-exchange structure found in a number of groups in this clade, and the trochophore larvae; also shared by a number of lophotrochozoan groups.

- A number of lophotrochozoan groups, including the annelids (segmented worms) and mollusks, undergo spiral cleavage during early development.

- The segmentation found among the annelid worms helps improve their locomotion. The unique molluscan body plan is based on a muscular foot structure, a visceral mass containing the organs, and a mantle that covers and protects the visceral mass. Diverse adaptations of this plan are seen in the major molluscan groups.

- The ecdysozoans include more species than all other lineages combined. They are characterized by a rigid external covering (cuticle or exoskeleton) that must periodically be molted in order to allow the animal to grow.

- Wormlike, unsegmented ecdysozoans with a thin cuticle include the priapulids and the kinorhynchs. Horsehair worms and nematodes are also unsegmented, with a tougher cuticle.

- A segmented body with a hard exoskeleton and jointed appendages characterize the arthropods, which are Earth's dominant animals in both number of species and number of individuals. The trilobites are extinct arthropods. Today four arthropod groups—crustaceans, hexapods, myriapods, and chelicerates—are found in all environments. The hexapods (which include the numerous and diverse insects), chelicerates, and myriapods are found in terrestrial and aquatic environments. The crustaceans are the dominant arthropods in marine environments.

Common Problem Areas

- With the advent of molecular phylogenetic studies (see Chapter 24), we have had to rethink many of our previous views of where different organisms are classified. The arrow worms, for example, used to be considered deuterostomes, but have recently been placed among the protostomes, based primarily on molecular (i.e., gene sequence) studies. When compared with the more stable deuterostome phylogeny that you will study in Chapter 33, protostome phylogeny is in a vast state of flux and there is a great deal of disagreement among systematists about the ancestry and monophyly—and thus the classification—of many protostome groups. You should review the overview of evolutionary trends summarized on page 714 of the textbook, and note how these same themes will appear in Chapter 33.

Study Strategies

- Try to use the phylogeny presented in Figure 32.1 as a base for learning about different groups. Understanding that certain characteristics are shared between related groups can take you a long way toward remembering which characteristics define each group.

- Sometimes learning the Latin root of a group you are studying can help you organize your thinking. Although the names are unfamiliar at first, they all "make sense" if you understand what they mean.

- Review the following activities on the Companion Website/CD:
  Activity 32.1 Animal Classification
  Activity 32.2 Animal Features

Test Yourself

Knowledge and Synthesis Questions

1. The rhynchocoel is a body plan feature found
   a. only in trematodes.
   b. in rotifers and nematodes.
   c. only in rotifers.
   d. only in nemerteans.
   Textbook Reference: 32.2 What Are the Major Groups of Lophotrochozoans? p. 697

2. All of the following possess lophophores *except*
   a. ectoprocts.
   b. phoronids.
   c. annelids.
   d. brachiopods.
   Textbook Reference: 32.1 What Is a Protostome? pp. 692–693

3. Which of the following statements about rotifers is *false*?
   a. They have a complete gut with an anterior mouth and posterior anus.
   b. They are coelomates.
   c. The corona is a ciliated organ used in acquiring food.

d. They have a hydrostatic skeleton.
Textbook Reference: 32.2 What Are the Major Groups of Lophotrochozoans? p. 696

4. An appetizer "sampler" of escargot (snails), clams on the half shell, and calamari (octopus) would contain _____ mollusks.
   a. bivalve
   b. cephalopod
   c. gastropod
   d. All of the above
   Textbook Reference: 32.2 What Are the Major Groups of Lophotrochozoans? p. 702

5. The combination of a true coelom and repeating body segmentation allowed the annelids to do which of the following?
   a. Evolve complex body shapes and control movement more precisely
   b. Move through loose marine sediments
   c. Become hermaphroditic
   d. Inject paralytic poisons into their prey
   Textbook Reference: 32.2 What Are the Major Groups of Lophotrochozoans? p. 698

6. Lobsters, millipedes, and butterflies share which of the following traits?
   a. Parapodia
   b. Setae
   c. Gas exchange across the skin
   d. Jointed appendages
   Textbook Reference: 32.4 Why Do Arthropods Dominate Earth's Fauna? p. 705

7. Which of the following has a specialized internal gas exchange system?
   a. Priapulids
   b. Arrow worms
   c. Hexapods
   d. Horsehair worms
   Textbook Reference: 32.4 Why Do Arthropods Dominate Earth's Fauna? p. 708

8. Which insect group is made up of species that undergo *complete* metamorphosis and *can* fold their wings back over their bodies?
   a. Coleoptera (beetles)
   b. Apterygota (silverfish and springtails)
   c. Orthoptera (grasshoppers and their kin)
   d. Odonata (dragonflies and damselflies)
   Textbook Reference: 32.4 Why Do Arthropods Dominate Earth's Fauna? pp. 709–710

9. Which of the following insect groups consists of winged species that undergo *incomplete* metamorphosis and *can* fold their wings back over their bodies?
   a. Coleoptera (beetles)
   b. Apterygota (silverfish and springtails)
   c. Orthoptera (grasshoppers and their kin)
   d. Odonata (dragonflies and damselflies)
   Textbook Reference: 32.4 Why Do Arthropods Dominate Earth's Fauna? pp. 709–710

10. Which of the following insect groups consists of winged species that *cannot* fold their wings back over their bodies?
    a. Coleoptera (beetles)
    b. Apterygota (silverfish and springtails)
    c. Orthoptera (grasshoppers and their kin)
    d. Odonata (dragonflies and damselflies)
    Textbook Reference: 32.4 Why Do Arthropods Dominate Earth's Fauna? p. 710

11. The majority of terrestrial arthropod species are
    a. myriapods.
    b. hexapods.
    c. cirripeds.
    d. tardigrades.
    Textbook Reference: 32.4 Why Do Arthropods Dominate Earth's Fauna? p. 708

12. The typical crustacean larva is called a(n)
    a. nauplius.
    b. pupa.
    c. instar.
    d. cyst.
    Textbook Reference: 32.4 Why Do Arthropods Dominate Earth's Fauna? p. 707

13. Which of the following chelicerate groups has a long fossil history showing little morphological change?
    a. Pycnogonids (sea spiders)
    b. Arachnids (mites, ticks, and spiders)
    c. Horseshoe crabs
    d. None of the above
    Textbook Reference: 2.4 Why Do Arthropods Dominate Earth's Fauna? p. 712

14. Which of the following ecdysozoan groups is extinct?
    a. Chelicerates
    b. Crustaceans
    c. Tardigrades
    d. Trilobites
    Textbook Reference: 32.4 Why Do Arthropods Dominate Earth's Fauna? p. 706

15. Which of the following attributes is *not* seen in the arthropod body plan?
    a. Segmentation
    b. Jointed appendages
    c. A closed circulatory system
    d. A hard exoskeleton
    Textbook Reference: 32.1 What Is a Protostome? p. 691

Application Questions

1. The mollusks are a very diverse group morphologically. However, they all share three common morphological traits. Describe these traits and their functions.
   Textbook Reference: 32.2 What Are the Major Groups of Lophotrochozoans? p. 700

2. The following are all characteristics of earthworms: They have segmented bodies, a spiral cleavage pattern, a terrestrial lifestyle, a complete gut, a blastopore that develops into a mouth, tight junctions between cells, and bilateral symmetry. In what order did these characteristics evolve?
   Textbook Reference: 32.1 What Is a Protostome? pp. 691–693

3. Scientists have not found a fossilized body of the species that is the common ancestor of all bilaterally symmetrical animals. Is there much hope that they will? How have scientists been able to learn about this common ancestor?
   Textbook Reference: 32.1 What Is a Protostome? p. 693

4. A thick cuticle provides an animal with protection, but it also can seal it off from the world, obstructing the exchange of gases. How would you design a simple, hypothetical animal with a covering that provided for both protection and gas exchange?
   Textbook Reference: 32.4 Why Do Arthropods Dominate Earth's Fauna? p. 708

5. The letters *pter* occurs in the names of many insect groups. Can you discover the Greek word that is the root of this nomenclature? (Hint: the same word is the root name for a well-known dinosaur group, the pterodactyls.) Most good dictionaries give the meanings of word prefixes and roots; using such a dictionary, determine the meaning of the following names of insect groups:
   Apterygota
   Ephemeroptera
   Orthoptera
   Isoptera
   Hemiptera
   Homoptera
   Coleoptera
   Trichoptera
   Lepidoptera
   Diptera
   Hymenoptera
   Siphonaptera

Answers

Knowledge and Synthesis Answers

1. **d.** The rhynchocoel is a fluid-filled cavity containing a proboscis, a feeding organ found in the taxon Nemertea.

2. **c.** Annelids belong to the spiralian lineage, not the lophophorates.

3. **b.** Rotifers are pseudocoelomates and have a pseudocoel.

4. **d.** Snails are in the Gastropoda, clams are in the Bivalvia, and octopuses are in the Cephalopoda.

5. **a.** The segmentation of the annelids allows for more complex, coordinated movement.

6. **d.** All of these animals have jointed appendages.

7. **c.** The Hexapoda have an internal system of tracheae for gas exchange.

8. **a.** The Coleoptera, or beetles, undergo complete metamorphosis and are able to fold their wings over their bodies.

9. **c.** The Orthoptera (grasshoppers, crickets, roaches, etc.) undergo incomplete metamorphosis and are able to fold their wings over their bodies.

10. **d.** The Odonata (dragonflies and damselflies), along with the Ephemeroptera (mayflies), have wings that they cannot fold over their bodies. Apterygotes (silverfish and springtails) are wingless.

11. **b.** The 1.4 million species of insects that have been described so far are only a small fraction of all insect species.

12. **a.** The larvae of crustaceans are known as nauplius larvae.

13. **c.** Horseshoe crabs have changed little during their fossil history; they are sometimes referred to as living fossils.

14. **d.** The Chelicerata include spiders, sea spiders, and horseshoe crabs; the Crustacea include decapods (crabs, shrimp, and others), barnacles, isopods, and copepods; the Tardigrada are a small (but not extinct) group also known as water bears. Of the groups listed, only the Trilobita are extinct.

15. **c.** Arthropods have a hard exoskeleton, a segmented body, jointed appendages, and an open circulatory system (see Table 32.1).

Application Answers

1. The mollusk body plan has three morphological components shared by all molluscan groups. The first trait is a muscular foot used for locomotion. In cephalopods the foot has been modified into arms and tentacles. The second trait is a mantle, which covers the third trait, the visceral mass. The mantle secretes a calcareous shell in many species and often houses the gills. The visceral mass consists of the internal organs.

2. The first of these characteristics to evolve was the tight junction. This type of cell junction occurs in all animals and was probably present in the common ancestor of all animals. Figure 32.1 shows that the evolution of a complete gut came next, and then bilateral symmetry. After the evolution of bilateral symmetry, the phylogenetic tree in Figure 32.1 splits into the deuterostomes and protostomes; as protostomes, the annelids have a mouth that develops from their blastopore. The next characteristic to evolve was a segmented body. The move from aquatic to terrestrial lifestyle is not covered in this figure, but given that the annelids are overwhelmingly aquatic, and even terrestrial annelids are confined to moist habitats, it seems likely that the terrestrial lifestyle was the most recent of the listed characteristics to evolve.

3. The vast majority of species that have lived on Earth have gone extinct without leaving fossils (see Chapter 21), so it is very unlikely that scientists will ever discover a fossilized body of the common ancestor of all bilaterally symmetrical animals. Some fossilized animal tracks from the Precambrian appear to have been made by a bilaterally symmetrical animal, but these offer limited information. The best source of information about the common ancestor of any group of animals is the set of characteristics that are common to all members of that group. All bilaterally symmetrical animals share a triploblastic organization and certain Hox genes, and it is probable that these were also present in the common ancestor of this group.

4. The animal would need special mechanisms for gas exchange with the environment. The insects have overcome this problem with their system of tracheae. This network of tubes is found throughout the body and opens to the environment. The network is in close contact with all the cells of the animal. In an aquatic environment, the animal could have an internal chamber with gills used for respiration.

5. As a root or prefix, *pter* comes from the Greek word *pteris* and refers to wings.

   Apterygota: The prefix *a-* in this case means *without*. Apterygote insects have no wings.

   Ephemeroptera: *Ephemeral* means *short-lived*. Adult mayflies typically live only a few hours or days.

   Orthoptera: *Ortho-* means *straight*. Orthopterans tend to have very straight wings (see Figure 32.27A).

   Isoptera: *Iso-* means *equal* or *uniform*. All four wings on termites tend to be very similar.

   Hemiptera: *Hemi-* means *half*. In many true bugs, the front half of the forewing is hardened, and they appear to have half a shell (like a beetle's) and half wings.

   Homoptera: *Homo-* means *same*. Unlike their close relatives the hemipterans, homopterans have forewings and hind wings that are similar to each other.

   Coleoptera: *Coleo-* means *sheath*. The forewings of beetles are modified into a protective sheath that usually covers the hind wings and abdomen.

   Trichoptera: *Tricho-* means *hair*. The bodies and wings of adult caddisflies are covered in small hairs.

   Lepidoptera: *Lepido-* means *scale*. Butterfly and moth wings are covered with tiny scales.

   Diptera: *Di-* means *two*. Unlike most winged insects, flies have only two wings.

   Hymenoptera: *Hymen-* means *membrane*. Bees, wasps, and the reproductive castes of ants have membranous wings.

   Siphonaptera: *Siphon-* means *tube* or *pipe*. This is a reference to the flea's feeding method. Note that this name ends in *-aptera*, not *-optera*: Fleas do not have wings (*a-* once again meaning *without*), though as pterygotes they are descended from winged ancestors, having secondarily lost their wings over the course of evolution.

CHAPTER 33 Deuterostome Animals

Important Concepts

The deuterostomes include the echinoderms, hemichordates, and chordates.

- There are many fewer species of deuterostomes than of protostomes, but the deuterostomes are of special interest because mammals—a group that includes the largest living animals as well as the human lineage—are deuterostomes.

- Deuterostomes fall into three major clades: the echinoderms, the hemichordates, and the chordates (see Figure 33.1). All are triploblastic, coelomate animals with internal skeletons. The chordates are characterized by a structure called the notochord and a dorsal nerve cord.

- Fossil beds, most of them from a 520-year-old site in China, have revealed insights into deuterostomate ancestry. Fossils and phylogenetic analysis of living species indicate that the earliest deuterostomes were bilaterally symmetrical, segmented animals with a slitted pharynx and external gills (see Figure 33.2).

Echinoderms have pentaradial symmetry and a water vascular system.

- Although echinoderm larvae have bilateral symmetry, a trait found in the larvae and adults of all other deuterostomes, the echinoderm adults display a unique pentaradial symmetry (radial symmetry in fives or multiples of five). They have no heads. Echinoderms move slowly but can move equally well in any direction. They do not have anterior–posterior (head–tail) body organization, but have an oral (mouth) and aboral (anus) surface (see Figure 33.3).

- In addition to their pentaradial symmetry, echinoderms are unique in possessing a system of water-filled canals called a water vascular system. The canals end in external structures called tube feet. The water vascular system functions in feeding, gas exchange, and locomotion. Echinoderms also have an internal skeleton made up of calcified plates.

- About 7,000 echinoderm species exist today; many more species are known only from the fossil record. They diverged into two major clades. The first of these,

the crinoids, includes about 80 species of sea lilies and feather stars (see Figure 33.4A and B). The body form is of a cup-shaped structure and has anywhere from five to several hundred arms. The arms have tube feet that catch food particles. Cilia on the tube feet move the food to a groove that leads to the mouth. The tube feet also provide a surface for respiratory gas exchange and nitrogenous waste excretion.

- The remaining echinoderms include the sand dollars and sea urchins, the sea cucumbers, sea stars, brittle stars, and sea daisies (see Figure 33.4C–F).

  - Sand dollars and sea urchins feed on algae and organic debris. Neither has arms, but the sea urchin has spines that are attached to the skeleton.

  - Sea cucumbers have tube feet used for anchoring the animal to the substrate rather than for moving. The anterior tube feet have evolved into tentacles that catch food particles.

  - The sea stars or starfish are major predators, feeding on marine animals including polychaetes, mollusks, crustaceans, and fish. Some species use their tube feet to pry open the shells of bivalve mollusks. Once the bivalve is open, the sea star inserts its stomach into the shell and digests the mollusk's internal organs.

  - Brittle stars have five flexible arms and feed on the surface sediments and the accompanying organic material through their single opening to the digestive tract.

  - Sea daisies are a recent discovery, having only been known since 1986. Little is known about the sea daisies other than that they probably feed on prokaryotes found in decaying oceanic driftwood. Recently molecular data indicates that they are highly modified sea stars.

Hemichordates have pharyngeal slits.

- The hemichordates include the acorn worms and the pterobranchs. Both groups have a body plan composed of a trunk, a collar, and a proboscis (see Figure 33.5). The sticky proboscis is used for catching prey and digging. Cilia move the food from the proboscis to the mouth. Acorn worms burrow in sand or mud and extract food items from it; pterobranchs are sedentary marine animals and may be either solitary or colonial.

- A pharynx that opens to the outside through pharyngeal slits is found behind the mouth. The pharyngeal slits are used as respiratory surfaces; oxygen is extracted from the water that is moved through the slits.

All chordates have a notochord at some point in their development.

- The features that reveal the evolutionary relationship between chordates and echinoderms, and among the chordates, are primarily seen in the larvae.

- Chordates share five anatomical features: At some point in their lives, they all have pharyngeal slits, a hollow dorsal nerve cord, a ventral heart, a tail beyond the anus, and a notochord. In ascidians, the notochord is lost during its metamorphosis from larval to adult forms. In vertebrates, the notochord is replaced by skeletal structures called the vertebrae (spinal column).

- The three major urochordate groups are the ascidians (sea squirts, also known as tunicates), thaliaceans, and larvaceans. All members of these three groups are marine, and more than 90 percent of urochordate species are ascidians.

- Ascidian larvae have pharyngeal slits, a nerve cord, and a notochord; it is the larva, not the adult form, that suggests their chordate ancestry. The adults lose the notochord and nerve cord but have an enlarged pharynx—the baglike pharyngeal basket—for catching food. Ascidians sometimes reproduce asexually, by budding, thus forming colonies (see Figure 33.7).

- The 30 species of cephalochordates (lancelets) are small, fishlike animals that retain the notochord for their entire life and use their pharyngeal baskets to catch prey items (see Figure 33.6). Cephalochordates are the sister group of the vertebrates (see Figure 33.1).

Vertebrates evolved an internal skeleton.

- The vertebrate body plan can be categorized by the following traits: a vertebral column that replaces the notochord and supports an internal skeleton with two pairs of appendages; an anterior head with a large brain and protective skull; a closed circulatory system; and a large coelom in which the internal organs are suspended. This body plan is capable of supporting large-sized animals (see Figure 33.10).

- The elongate, eel-like hagfishes are the sister group of the remaining vertebrates. These jawless fishes look superficially similar to the lampreys, which are also jawless, but hagfishes lack true vertebrae and have only a partial cranium (skull). Lampreys have true vertebrae.

- Jawless fishes were dominant during the Devonian, but during that time jaws evolved via modifications in the skeletal arches supporting the gills (see Figure 33.11). This led to greatly enhanced feeding efficiency, and today the 100 or so species of hagfishes and lampreys are the only surviving jawless vertebrates. The remaining vertebrates fall into the category of gnathastomes, or "jaw mouths."

- The evolution of jaws provided access to new food sources, and the evolution of teeth made for extremely effective predation. Jawed fishes were the major predators of the Devonian. The evolution of unjointed appendages called fins helped control movement through the water. In some early fishes, lung-like, gas-filled sacs called swim bladders evolved. With the aid of the swim bladder and fins, many fish can control their position in the water column with minimal energy expenditure.

- The chondrichthyan fishes, including sharks, skates, rays, and chimaeras, have a skeleton composed of cartilage (see Figure 33.12). The ray-finned fishes and virtually all other vertebrates have a bony skeleton. The approximately 24,000 species of ray-fins encompass a remarkable variety of shapes, sizes, and lifestyles. They exploit every kind of food source in the aquatic environment (see Figure 13.13).

The evolution of jointed fins aided the colonization of land by vertebrates.

- The evolution of the lung-like swim bladders in some ray-finned fishes set the stage for the move to the land. Some fishes may have used these sacs to supplement their oxygen supply in low-oxygen freshwater and they may have allowed some fishes to survive out of water.

- Jointed fins evolved in the ancestors of the coelocanths and the lungfishes, which along with the tetrapods are known as sarcopterygians. The change in fin structure allowed these fish to support themselves in shallow water and, eventually, to move onto land and exploit food sources there. These early land-dwellers gave rise to the tetrapods—the four-legged vertebrates.

- Amphibians evolved from a common ancestor shared with sarcopterygian fishes. Their walking legs allowed them to exploit the new dry territory. However, most spend at least part of their life cycle in the water (see Figure 33.15). Many species come back to the water to lay their eggs, producing aquatic larvae. The skin surface is a common site of respiratory gas exchange in amphibians, but they also have lungs.

- There are approximately 6,000 species of extant amphibians, falling into three groups:
  - The caecilians are wormlike, legless, burrowing or aquatic amphibians found only in moist tropical regions.
  - Anurans include the tail-less frogs and toads, and account for the vast majority of amphibian species. Anurans undergo metamorphosis from an aquatic to a terrestrial life form. The adults have a very short vertebral column and a pelvic region modifed for hopping on the hind legs. Some species have adapted

to life in very dry, even desert environments, while others have returned to an entirely aquatic lifestyle.

- Salamanders are tailed amphibians that exchange respiratory gases through both lungs and gills. Paedomorphic evolution (retention of the juvenile form in the adult) has led to several entirely aquatic salamander species.

The amniote egg allowed tetrapods to invade dry environments.

- Amphibians are largely confined to moist environments by their need to reproduce in water. In one ancestral lineage, the amniote egg evolved. The egg has a protective calcium-based shell that inhibits dehydration while allowing oxygen and carbon dioxide to pass through. Within the shell, extraembryonic membranes further protect the embryo and large quantities of yolk provide nutrition. The amniote animals—reptiles (including birds) and mammals—were thus freed from reliance on water to reproduce.

- Other evolutionary adaptations to life on land appeared among the amniotes, including a tough, impermeable skin covered with scales or modified scales (i.e., hair or feathers). The excretory systems of amniotes also evolved adaptations that allowed these animals to excrete nitrogenous wastes in the form of concentrated urine with a minimal loss of valuable water.

- During the Carboniferous period, about 250 mya, the amniote animals diverged into two groups, the reptiles and the mammals. Birds would evolve from one of the reptilian groups.

Reptiles and birds belong to a monophyletic group.

- One relatively small reptilian group, the turtles and tortoises, has a unique body plan that has changed very little over the millenia since it evolved. These animals are characterized by dorsal and ventral bony plates that evolved from the ribs to form a protective shell. Their relationship to other reptiles is not clear.

- The lepidosaurs include the squamates (lizards, snakes, and amphisbaenians—another group of legless, wormlike burrowers) and the tuataras. The limbless condition of snakes is a secondary evolutionary condition. Most lizards and all snakes are carnivores. Adaptations to the jaws of snakes allow them to swallow prey much larger than themselves. Tuataras are represented by only two living species.

- The archosaurs include the extant crocodilians (crocodiles and alligators), the extinct dinosaurs, and the "living dinosaurs" we know as birds. The crocodilians—crocodiles, alligators, caimans, and ghalials—live only in tropical and warm temperate regions. They spend most of their lives in water, but build nests on land. They are all carnivores, feeding on other vertebrates, including large mammals.

- Dinosaurs arose among the reptiles around 215 mya and survived for about 150 million years. Both the fossil record and molecular evidence support the position of birds as a sister group of the saurischian dinosaurs, and as such birds can be viewed as living dinosaurs.

Birds evolved from an ancestral dinosaur.

- Birds probably evolved from an ancestral theropod, a bipedal, predatory dinosaur that appears to have had hollow bones, a wishbone, limbs with three digits, and a backward-thrusting pelvis. Recent fossil discoveries in China have demonstrated that some dinosaurs had feathers.

- *Archaeopteryx* lived 150 mya and represents the oldest known fossil of a bird. It was covered in feathers and had well-developed wings and a wishbone like present-day birds (see Figure 33.21). It also had teeth, which were lost in later members of this lineage.

- Both the lightweight bones and feathers of birds have been modified for flight. About 80–90 mya, the birds diverged into two major groups. The few modern descendants of one group, the palaeognaths have secondarily lost the ability to fly, or can fly only weakly. These birds are all endemic to the continents (Africa, Australia, and South America) and include the tinamous, rheas, kiwis, emus, and the largest bird, the ostrich. The rest of the 9,600 bird species fall into the group known as the neognaths, most of whose members retained the ability to fly.

- Along with the ability to fly came a high metabolic rate needed to fuel flight. Their metabolic rate means birds generate a great amount of heat, and their feathers are adapted for insulation to control body heat. The lungs of birds also function differently from those of other vertebrates, an adaptation to the needs of flight. Different bird groups feed on many different types of animal and plant material, including carrion. Birds that feed on fruits and seeds are major agents of plant dispersal.

Mammals radiated after the dinosaurs went extinct.

- The earliest mammals lived side by side with reptiles from the first split of the two lineages early in the Mesozoic era. However, only after the large dinosaurs became extinct did mammals begin to flourish in size and number.

- A number of unique traits distinguish the mammals. Sweat glands in the skin produce secretions that are crucial to cooling the body. Adaptations to the sweat glands may have led to the mammary glands that give the group its name and provide nutrient fluid for their young. Modifications of scales became the external body hair that protects and insulates most mammals.

- A four-chambered heart that completely separates oxygenated from deoxygenated blood evolved in the mammals. It also evolved convergently (i.e., separately) in the crocodilians and birds.

- In mammals, the amniote egg became modified for growth of the embryo within the mother's uterus. Three species of egg-laying mammals, the prototherians, exist today but are found only in Australia and New Guinea.

- The remaining mammals are classed as therians. Marsupial therians give birth to tiny, underdeveloped young which they then nurture externally, usually in a pouch on the mother's belly (see Figure 33.25). Marsupials were once widespread, but today the approximately 330 marsupial species are largely found only in Australia and South America, with minor representatives in North America.

- The 4,500 species of eutherian mammals develop in the mother's uterus (see Figure 33.26). This group is widely known by the name placental mammals, from the placenta that provides nourishment for the growing embryo. However, some marsupials also have placentas.

- Eutherians exist in virtually all Earth's environments. Several groups, most notably the cetaceans (whales and dolphins) returned to a marine lifestyle. Flight evolved in the bats (which represent the second largest number of eutherian species, after the rodents). The 235 species of primates are the best studied eutherians because human beings belong to this group.

Primate evolution is well understood.

- The primate ancestor was a small, arboreal, insectivorous mammal. They are identifiable by the presence of opposable digits (i.e., thumbs). Early in their evolutionary history—about 65 mya—the primates split into two clades, the prosimians (lemurs, bush babies, and lorises) and the anthropoids (tarsiers, monkeys, apes, and humans). Prosimian species were once found on all continents, but today they are restricted to Africa, especially the island of Madagascar.

- Soon after the prosimian–anthropoid split, the anthropoids further split into the New World and Old World monkeys. The break up of the African and South American continents meant that the two groups evolved in isolation. About 35 mya, an Old World lineage broke off that would lead to modern apes and humans. Another split occurred around 22 mya. The Asian apes (gibbons and orangutans) descended from two groups in this lineage (see Figure 33.28).

- The ancestor of modern African apes is believed to belong to the extinct genus *Dryopithecus*. The African apes include the modern gorillas, chimpanzees, and humans.

Early human ancestors evolved bipedalism.

- Ardipithecines (the earliest protohominids) and their descendants, the australopithecines, were adapted for bipedalism, freeing up their hands for other tasks and elevating the eyes. *Australopithecus afarensis* is currently regarded as the ancestor of the modern humans (genus *Homo*; see Figure 33.31).

- The oldest known member of the genus *Homo*, *Homo habilis*, lived about 2 mya and fossil tools have been found with the bones of *H. habilis*. A number of species of *Homo* existed simultaneously over evolutionary time, but by 200,000 years ago *H. sapiens* was predominant, and by about 20,000 years ago, all other species had been supplanted by *Homo sapiens* (modern humans), the only currently existing human species.

- Brains of *H. sapiens* had reached modern size by about 160,000 years ago. Larger brain size appears to have evolved concurrently with a smaller, less muscular jaw structure, suggesting that the two traits may be functionally correlated. The rapid increase in brain size may have been favored by an increasingly complex social life that thrived on ever-more sophisticated communication; any trait that allowed more effective communication would have been favored in a society based on cooperative hunting and other complex social interactions.

- The expansion of language and other behavioral abilities allowed humans to develop culture and pass knowledge and traditions from generation to generation. Human societies were transformed from communities of hunters and gatherers to those of farmers and eventually urban dwellers.

The Big Picture

- The deuterostomes are not as diverse or numerous as the protostomes. Most fit into one of two groups; the echinoderms and the chordates.

- Echinoderms have an internal skeleton of calcified plates and a water vascular system used in feeding and gas exchange. They include the sea stars (starfish), sea lilies, sea cucumbers, and sea daisies, all of which exhibit pentaradial symmetry as adults.

- The chordates have several shared traits: pharyngeal slits, a nerve cord, a ventral heart, a tail beyond the anus, and a notochord. The chordates consist of the ascidians, lancelets, and vertebrates. Vertebrates have a dorsal vertebral column that supports a rigid endoskeleton.

- The most numerous vertebrates, both in terms of number of species and in number of individuals, are the fishes. The amphibians adapted to life on land but their life cycle requires that they return to the water to reproduce.

- The evolution of the amniote egg allowed reptiles freedom from water-based reproduction. During the Carboniferous, amniotes diverged into two major clades, the reptiles and the mammals. Reptiles were the dominant vertebrates for many millenia. One reptilian group gave rise to both the now-extinct dinosaurs and to the birds. Among the birds, the reptilian scales evolved into feathers, allowing the rise of flight in this group.

- The extinction of the dinosaurs opened the door for the radiation of the mammals. Most mammals belong to one of 20 major groups of eutherians. The largest eutherian group is the rodents; the primate eutherians include the lemurs, monkeys, apes, and humans.

- Humans (genus *Homo*) are primates that evolved from bipedal australopithecine ancestors on the African continent. Several species of *Homo* arose and went extinct over evolutionary time. In the lineage leading to *Homo sapiens*, the only surviving species, brain size expanded greatly relative to body size, allowing humans to develop culture and learn language.

Common Problem Areas

- As you work your way through the phylogenies presented in the textbook, try to recall the representative organisms that you know from each group and what their special traits are. If you base your learning on information you already know, you will have an easier time learning the facts.

Study Strategies

- Many students become confused with the evolution of the chordates and vertebrates and the traits that separate the vertebrates from other chordates. Create a time line with all the major chordate classes. Insert the specific traits that distinguish one class from another. This will help you organize the evolution of the chordates so that memorizing it becomes a manageable task.

- Review the following animated tutorial and activities on the Companion Website/CD:
  Tutorial 33.1 Life Cycle of a Frog
  Activity 33.1 Deuterostome Phylogeny
  Activity 33.2 Amniote Egg
  Activity 33.3 Deuterostome Classification

Test Yourself

Knowledge and Synthesis Questions

1. Which of the following is *not* a characteristic of echinoderms?
   a. They have a water vascular system.
   b. They have an internal skeleton.
   c. They are protostomes.
   d. They have bilateral symmetry as larvae.
   Textbook Reference: *33.1 What Is a Deuterostome? p. 718*

2. Which one of the following echinoderm groups has more extinct than living members?
   a. Sea stars
   b. Crinoids
   c. Sea urchins
   d. Sea daisies
   Textbook Reference: *33.2 What Are the Major Groups of Echinoderms and Hemichordates? p. 719*

3. Which of the following chordate groups evolved prior to the appearance of cartilaginous fishes?
   a. Ray-finned fishes and sea squirts
   b. Ascidians, lancelets, and hagfishes
   c. Ascidians, lancelets, and ray-finned fishes

d. Lampreys and ray-finned fishes
   Textbook Reference: *33.3 What New Features Evolved in the Chordates? pp. 722–723*

4. The vertebrate body plan includes all of the following traits *except*
   a. a ventral spinal cord.
   b. an internal skeleton.
   c. a well-developed circulatory system.
   d. organs suspended in the coelom.
   Textbook Reference: *33.3 What New Features Evolved in the Chordates? pp. 724–725*

5. The swim bladder of many fishes evolved from a lunglike sac. What important function does the swim bladder provide?
   a. It aids in prey capture.
   b. It controls swimming speed.
   c. It controls buoyancy.
   d. It aids in reproduction.
   Textbook Reference: *33.3 What New Features Evolved in the Chordates? p. 726*

6. Which of the following traits do the chondrichthyans and the ray-finned fishes share?
   a. The gills are the major site of gas exchange.
   b. The skeleton is composed of cartilage.
   c. Their outer surface is covered with bony plates.
   d. They have a swim bladder.
   Textbook Reference: *33.3 What New Features Evolved in the Chordates? p. 726*

7. The transition from aquatic to terrestrial lifestyles required many adaptations in the vertebrate lineage. Which of the following is *not* one of those adaptations?
   a. Switch from gill respiration to air-breathing lungs
   b. Improvements in water resistance of skin
   c. Alteration in mode of locomotion
   d. Development of feathers for insulation
   Textbook Reference: *33.4 How Did Vertebrates Colonize the Land? pp. 728–729*

8. The amniotes evolved the ability to reproduce by laying eggs with shells. What is the major advantage of shelled eggs?
   a. The embryo needs only a small amount of yolk for development.
   b. It does not have to be laid in a moist environment.
   c. Evaporation from the egg is increased.
   d. Nitrogenous wastes can be excreted across the shell.
   Textbook Reference: *33.4 How Did Vertebrates Colonize the Land? p. 730*

9. How do birds differ from reptiles?
   a. Birds have a lower metabolic rate.
   b. Reptiles lay eggs and birds give birth to live young.
   c. Birds can breathe and run at the same time, whereas reptiles cannot.
   d. They do not share a common ancestor.
   Textbook Reference: *33.4 How Did Vertebrates Colonize the Land? pp. 732–733*

10. Which of the following is *not* a trait that you could use to identify an animal as a mammal?
    a. Mammary glands
    b. Hair
    c. Sweat glands
    d. Kidneys
    Textbook Reference: 33.4 How Did Vertebrates Colonize the Land? p. 735

11. Which of the following vertebrate groups is a living representative of the dinosaur lineage?
    a. Crocodilians
    b. Birds
    c. Lobe-finned fishes
    d. Snakes
    Textbook Reference: 33.4 How Did Vertebrates Colonize the Land? p. 731

12. Which of the following statements about human evolution is *false*?
    a. Bipedalism was a hominid adaptation for life on the ground.
    b. Increases in the size of hominid brains preceded the appearance of language and culture.
    c. The extinction of the Neanderthal people was caused by the emergence of *Homo habilis*.
    d. Humans are not the direct descendants of modern-day chimpanzees.
    Textbook Reference: 33.5 What Traits Characterize the Primates? p. 740

Application Questions

1. Ascidians, or sea squirts, are armless and legless organisms that spend their adult lives attached to a substrate under water. They feed by pulling water into one tube, filtering out planktonic organisms, and pushing the water out another tube. Grasshoppers have legs, move around freely on land, and feed by ingesting food through a mouth. What evidence has led biologists to believe that humans are more closely related to sea squirts than to grasshoppers?
   Textbook Reference: 33.3 What New Features Evolved in the Chordates? p. 722

2. The evolution of a hinged jaw resulted in an extensive radiation of the fishes into the many modern jawed forms. What was the main advantage and significance of hinged jaws in this radiation?
   Textbook Reference: 33.3 What New Features Evolved in the Chordates? pp. 724–725

Answers

Knowledge and Synthesis Answers

1. **c.** Species from the echinoderm clade are deuterostomes.

2. **b.** The crinoids are a relict group with many more extinct than extant species.

3. **b.** The ascidians, lancelets, and hagfishes all evolved before the cartilaginous fishes.

4. **a.** Along with an internal skeleton, well-developed circulatory system, and organs suspended in a coelom, the vertebrates have a dorsal spinal cord.

5. **c.** The swim bladder of modern-day fishes is involved in controlling buoyancy.

6. **a.** In both the ray-finned fishes and cartilaginous fishes the major site of gas exchange is the gills. The ray-finned fishes have a skeleton of bone and have a swim bladder. These traits are not shared with the cartilaginous fishes. Neither group has bony plates in their outer surface.

7. **d.** In the move onto land, the development of feathers for insulation was not required. Amphibians and reptiles do not have an insulation layer, and many mammals have hair for insulation.

8. **b.** The shelled egg of the birds and reptiles allowed them to occupy dry terrestrial habitats because the shell decreases water loss from the egg.

9. **c.** Birds and reptiles differ in the morphology of the muscles that control movement and breathing. Birds can run and breathe at the same time, whereas reptiles must stop running to take a breath.

10. **d.** Kidneys are present in reptiles, birds, fishes, and mammals. The presence of kidneys is not a trait that could be used to identify a mammal.

11. **b.** Birds are now thought to be direct descendants of a group of dinosaurs.

12. **c.** The extinction of Neanderthals is thought to have been due in some part to the presence of Cro-Magnon. *Homo habilis* was extinct for more than a million years by the time the Neanderthals emerged.

Application Answers

1. Although adult ascidians and adult humans seem like very different animals, our embryonic stages share certain important characteristics, including the presence of a notochord (like all deuterostomes) and a blastopore that develops into the anus (see Chapter 31). A look at embryonic grasshoppers, however, shows them to be fundamentally different. The embryonic grasshopper's blastopore develops into the mouth, putting grasshoppers squarely among the protostomes. Grasshoppers have an external rather than an internal skeleton.

2. The hinged jaw of the fishes evolved during the Devonian period. The evolution of the jaw opened up a new untapped food source for these animals. With a jaw, fish could now grasp and kill larger living prey and chew and tear the body parts of prey.

CHAPTER 34 The Plant Body

Important Concepts

Vegetative structures serve a variety of roles, including nutrient procurement, support, water uptake, and transport.

- The angiosperms are the flowering plants and consist of two clades with distinct characteristics, the monocots and the eudicots. The narrow-leaved monocots, such as the grasses, lilies, orchids, and palms, differ from the eudicots, broad-leaved flowering plants, such as soybeans, roses, and maples, in a number of key anatomical characteristics.
  - Monocots have one cotyledon, parallel leaf veins, flower parts in multiples of three, and scattered primary vascular bundles.
  - Eudicots have two cotyledons, a netlike leaf venation, flower parts in fours or fives, and primary vascular bundles in a ring.
- Plants are composed of root systems and shoot systems (see Figure 34.2). Root systems are responsible for mineral and water uptake and support. Shoot systems consist of leaves (and leaf derivatives) involved in photosynthesis and stems for support.
- Root systems are generally one of two types.
  - Taproots of many eudicots are large single roots that grow deep within the ground and serve as storage tissue. Smaller lateral roots absorb water as the taproot enlarges.
  - Fibrous roots are numerous thin roots that spread far out from the plant, hold soil in place, rarely grow very deep, and are found in both monocots and eudicots.
- In some plants, adventitious roots occur along the stem and assist with rerooting or support.
- Stem systems produce buds, leaves, and flowers. Buds are embryonic shoots. At each node where leaves meet stem, an axillary bud is produced that may generate a new branch. At each stem tip is an apical bud responsible for elongation of that stem.
- Stems may be modified into tubers (such as potatoes) or runners. Stems may remain photosynthetic or become nonphotosynthetic woody material.

- Leaves are the site of photosynthesis but may have modifications to allow for specific jobs. The blade of a leaf is attached to the stem by a petiole. The leaf blade, with help from the petiole, is able to maintain a constant orientation toward sunlight for maximum photosynthesis. Leaves may be simple and consist of a single blade or may be compound with leaflets arranged along a central axis (see Figure 34.5). Vein patterns may be netted as in eudicots or parallel as in monocots.
- Some leaves are modified for storage of nutrients or water. Many have color modifications to attract pollinators. Some are modified into structures like tendrils and suckers to help the plant hold onto a support. Humans have taken advantage of these characteristics by breeding for these modifications.

Plant cells are organized into tissues.

- Simple tissues are composed of only one cell type that functions together; complex tissues are composed of multiple cell types. Xylem and phloem are examples of complex tissues that perform multiple functions by means of multiple cell types.
- Three tissue systems extend through the plant: vascular, dermal, and ground.
  - The vascular tissue system is made of xylem and phloem, and is the conductive tissue of the plant.
  - The dermal tissue system provides the outer covering of the plant. This includes the epidermis and the layer of cuticle it secretes.
  - The ground tissue system composes the rest of the plant. This tissue functions in storage, support, and photosynthesis.

Plant cells are specifically suited for their functions.

- Plant cells have the characteristics common to all eukaryotic cells but differ in that some have plastids such as chloroplasts, many have vacuoles, and every plant cell is bounded by cellulose-containing cell walls. Some plant cells are only functional after they die.
- Cell walls play important roles in plants, including providing turgor pressure, being chemically active, allowing for transport, and becoming modified for specific functions. Cell walls are formed after

cytokinesis. The daughter cells secrete the middle lamella and then begin laying down the primary cell wall. Once the cell is full size, it may begin secreting other substances to form a secondary wall.

- Cell walls do not block communication between cells. Small cytoplasmic strands called plasmodesmata extend through the walls allowing movement of substances from cell to cell.

- Plant cells are often specialized to carry out specific functions.

  - Parenchyma cells are frequently photosynthetic or used for storage. They are thin-walled and have large central vacuoles. They may also continue dividing and proliferate in a wounded area.

  - Collenchyma cells are support cells with special thickenings at the cell wall corners. They are elongated and allow for support without rigidity, which is an important characteristic for plants in windy areas.

  - Sclerenchyma cells with highly thickened cell walls provide rigid support after they die. They may be elongated fibers or variously shaped sclereids. Fibers strengthen bark and woody stems. Sclereids produce the gritty texture of pears.

  - Xylem transports water from the roots to the rest of the plant. The tracheary elements involved in the water movement are functional after they die. In the gymnosperms, tracheids make up the xylem transport. Vessel elements are found in angiosperms.

  - Living phloem moves carbohydrates and nutrients. Individual phloem cells are called sieve tube elements. Plasmodesmata enlarge where sieve tube elements join, making sieve plates. Sieve tube elements may lose nuclei and other organelles to prevent clogging of the sieve plates. These cells are filled with sieve tube sap that is high in carbohydrates. Adjacent companion cells regulate the function of the sieve tube elements.

Plants develop in specific patterns from embryonic tissue.

- In the plant body plan, the plant is laid out along a main axis from root to shoot and in a concentric pattern that establishes the layout of the tissue systems.

- Plants frequently lose parts during growth. To counter this loss, plants are laid out in modules or units. A module consists of a node and its attached leaf, the internode (a section of stem) below the node, and the axillary buds at the base of the internode. New units are formed as the plant grows. Leaves in some respects can be thought of as units themselves. The root system contains lateral roots, which are semi-independent units.

- The primary plant body is made of all the non-woody parts of the plant. The secondary plant body is found in plants that are composed of wood and bark.

- Plants differ from animals in having indeterminate growth. They continue to grow throughout their life span. This is possible because plants have regions of continual cell division called meristems.

- Apical meristems give rise to the primary plant body and are located at the tips of roots and stems and in buds. They are responsible for elongating the plant body that occurs as primary growth. Apical meristems give rise to primary meristems that produce the primary plant body. These are called the protoderm (dermal tissue system), the ground meristem (ground tissue system), and the procambium (vascular tissue system). All plant parts arise from division of the apical meristems.

- In some plants, lateral meristems lay down wood and bark. The vascular cambium arises from the lateral meristem and forms new secondary xylem and secondary phloem. The cork cambium produces new dermal tissues to accommodate increasing diameter and inhibits water loss with the production of periderm cells. The action of the vascular and cork cambiums is called secondary growth. Wood results from secondary xylem, and bark is everything outside the cork cambium.

- Root apical meristems (protoderm, ground meristem, and procambium) produce root tissues. At the tip of a root, the root apical meristem forms a root cap and quiescent center. The root has three primary tissue systems. The zone of cell division includes the apical and primary meristems. The zone of elongation is found above this and is the site of new cell formation. The upper layer is the zone of maturation where the cells differentiate and take on special functions.

- The protoderm gives rise to the epidermis and root hairs. Ground meristem gives rise to the cortex and endodermis. The endodermis is specialized with a waxy coating of suberin to assist with water movement. The procambium gives rise to the stele, which houses three tissues: the pericycle, the xylem, and the phloem. In eudicots, the very center of the root is xylem, but in monocots the center is pith tissue (see Figures 34.16 and 34.17).

- The shoot primary meristem also gives rise to three primary meristems that produce the three tissue systems. Leaves arise from leaf primordia with bud primordia forming at each base. Shoot vascular tissues are arranged in vascular bundles. In eudicots, the vascular bundles are arranged in a cylinder allowing for woody growth. In monocots the vascular bundles are scattered.

- Stems and roots may undergo secondary growth from the lateral meristems. The orientation of the vascular cambium gives rise to cylinders of secondary xylem (wood) and secondary phloem. Connections of living cells to secondary phloem occur via vascular rays. Vascular rays are perpendicular to the vessel elements for conduction. Only eudicots have secondary growth and the associated vascular cambium and cork cambium.

- Annual rings seen in wood are a result of climate shifts in temperate zones, particularly water availability. You do not see rings in tropical trees. Heartwood in trees appears darker because the xylem has been clogged with water insoluble substances. Sapwood is still conducting at the time of cutting and appears lighter and less dense. Knots are the result of branches becoming buried in the expanding trunk.

- Stretching, breaking, and flaking off of epidermis and cortex produce bark. This leaves the secondary phloem at risk. Cells at the surface of the phloem produce cork that is thickened and reinforced with suberin. New cork is produced as secondary growth proceeds. The areas that allow gas exchange through the bark are known as lenticels.

Leaves are designed to maximize photosynthesis.

- Leaves have two zones of photosynthetic cells called mesophyll. The upper level of cylindrical mesophyll is called palisade mesophyll, and the lower level is called spongy mesophyll. Air space around mesophyll cells is necessary for carbon dioxide to reach the photosynthesizing cells.

- Vascular tissue extends throughout leaves in veins. Veins carry water to cells and transport carbohydrates to sink tissues.

- The entire leaf is covered by a protective epidermis and is waterproofed by a waxy cuticle. Gas exchange occurs through guarded stomata. Guard cells open and close stomata to limit water loss.

The Big Picture

- A plant can be thought of as having vegetative structures that carry out the major functions of day-to-day life and reproductive structures that are responsible for reproducing the plant. You will see that there are cases in which the vegetative portions of the plant are quite adept at reproducing asexually. The vegetative plant body consists of the root system, which anchors the plant and absorbs water and nutrients, and the shoot system, which photosynthesizes and supports the plant against gravity. Modifications of the root and shoot systems lead to specialization of the plant.

- Plant cells are uniquely suited for support, transport, and the carrying out of cellular functions. Groups of cells form tissues that have specific roles within a plant. Patterns of development are under the control of hormones and regulatory genes. Indeterminate growth and the modular organization of plants allow for regeneration of parts lost to damage and disease.

- Plant growth occurs from meristems. Apical meristems allow for elongation of the plant, whereas lateral meristems allow for secondary or woody growth. Not all plants exhibit secondary growth. All tissue types

arise from the meristems and go through a process of elongation, then differentiation. Review stem, root, and leaf anatomy using the figures in the textbook.

Common Problem Areas

- This chapter covers plant anatomy. Basic plant anatomy is not difficult, but it is new to most students. The more time you spend looking at diagrams and live specimens, the easier it will be to understand the anatomy. As you study, think about the function of each structure. You will find that "form follows function."

- This chapter exposes you to many new vocabulary words. Resist the temptation to merely memorize. By looking at the words and understanding their roots, you will be better able to understand the terms.

Study Strategies

- The best study strategy for this material is to use the figures, pictures, and live material. Plant anatomy is the study of structure. Structures are three dimensional and best understood visually.

- Make a vocabulary list. Though you don't want to memorize the terms, a list will help you organize your study.

- If your school has three-dimensional models, make use of them. There are 3-D models available now on the Internet. Using these will make it easier to understand the orientation of the structures.

- Review the following animated tutorial and activities on the Companion Website/CD:
  Tutorial 34.1 Secondary Growth: The Vascular Cambium
  Activity 34.1 Eudicot Root
  Activity 34.2 Monocot Root
  Activity 34.3 Eudicot Stem
  Activity 34.4 Monocot Stem
  Activity 34.5 Eudicot Leaf

Test Yourself

Knowledge and Synthesis Questions

1. You are studying tropical plants in a Costa Rican cloud forest. You have identified a plant that has not been studied before. It has compound leaves with netted veins, vascular bundles arranged in a cylinder, and five petals on its very large and showy flower. This plant is most likely a
   a. monocot.
   b. eudicot.
   c. gymnosperm.
   d. None of the above
   Textbook Reference: *34.1 How Is the Plant Body Organized? p. 746*

2. The plant described in Question 1 grows to approximately 40 m at maturity and is in a forest ecosystem. This plant most likely has which of the following root types?

a. Fibrous
b. Taproot
c. Adventitious
d. Rhizoids

Textbook Reference: *34.1 How Is the Plant Body Organized? p. 746*

3. You planted sweet peas along your back garden fence. You note that they are very effective at climbing the fence. Upon closer inspection, you notice that the plants are attached to the fence by tendrils. These tendrils are modifications of
   a. stems.
   b. roots.
   c. branches.
   d. leaves.

Textbook Reference: *34.1 How Is the Plant Body Organized? p. 748*

4. Plant cells are easily distinguished from animal cells because they have
   a. rigid cell walls.
   b. plastids.
   c. large vacuoles.
   d. All of the above

Textbook Reference: *34.2 How Are Plant Cells Unique? p. 749*

5. Plant cells that are photosynthetically active are found in the _____ layer of the leaf and are _____ cells.
   a. mesophyll; parenchyma
   b epidermis; parenchyma
   c. mesophyll; sclerenchyma
   d. epidermis; sclerenchyma

Textbook Reference: *34.4 How Does Leaf Anatomy Support Photosynthesis? p. 761*

6. Water is conducted in _____ tissue, and carbohydrates and nutrients are transported in _____ tissue.
   a. xylem; phloem
   b. phloem; xylem
   c. parenchyma; phloem
   d. parenchyma; xylem

Textbook Reference: *34.2 How Are Plant Cells Unique? p. 748*

7. Plants are capable of indeterminate growth because of
   a. meristem tissues.
   b. regions of continually dividing cells.
   c. their modular nature.
   d. All of the above

Textbook Reference: *34.3 How Do Meristems Build the Plant Body? p. 754*

8. Which of the following best describes the origin of wood?
   a. Xylem cells enlarge and deposit large amounts of lignin.
   b. Primary meristems increase the amount of xylem deposited.

c. Lateral meristems contribute to continuous increases in vascular tissue.
d. None of the above

Textbook Reference: *34.3 How Do Meristems Build the Plant Body? p. 759*

9. Which of the following statements concerning wood is true?
   a. All woody plants show annual growth rings.
   b. Patterns of both cylindrical secondary growth and lateral vascular rays are visible in wood.
   c. Sapwood and hardwood result from patterns in primary xylem.
   d. All of the above

Textbook Reference: *34.3 How Do Meristems Build the Plant Body? p. 759*

10. Which of the following best describes the function of the cork cambium?
    a. It lays down a protective cork covering over exposed phloem tissue.
    b. It inhibits the sloughing of epidermal tissue.
    c. It allows for diameter shrinking in stems and roots.
    d. All of the above

Textbook Reference: *34.3 How Do Meristems Build the Plant Body? pp. 754–755*

11. Sieve tube members have sieve plates where they join other sieve tube members. Which of the following best describes the sieve plates?
    a. Sieve plate pores are enlargements of meristems.
    b. Sieve plates are necessary to allow conduction between sieve tube cells.
    c. Sieve plates allow joining of cytoplasm between adjacent stomata.
    d. All of the above

Textbook Reference: *34.3 How Do Meristems Build the Plant Body? p. 752*

12. Plants regulate gas exchange and water loss via
    a. the cuticle.
    b. guarded stomata.
    c. coated pits.
    d. sieve plates.

Textbook Reference: *34.4 How Does Leaf Anatomy Support Photosynthesis? p. 761*

13. The protoderm becomes what type of tissue system?
    a. Dermal tissue system
    b. Ground tissue system
    c. Vascular tissue system
    d. All of the above

Textbook Reference: *34.3 How Do Meristems Build the Plant Body? p. 754*

14. Primary growth occurs at the
    a. lateral meristems.
    b. fruit.
    c. quiescent center.
    d. apical meristems.

Textbook Reference: *34.3 How Do Meristems Build the Plant Body? p. 754*

15. Vascular bundles are composed of _____ and
    _____.
    a. root hairs; xylem
    b. cork; phloem
    c. xylem; phloem
    d. wood; cork
    Textbook Reference: *34.3 How Do Meristems Build the Plant Body? p. 757*

Application Questions

1. Draw a typical eudicot plant. Label the root system and shoot systems. Indicate on your drawing where you would find axillary buds and where you would find apical buds. Label the following structures: internode, petiole, leaf blade, taproot, lateral roots.
   Textbook Reference: *34.1 How Is the Plant Body Organized? p. 746*

2. Differentiate between apical and lateral meristems. Do all plants have apical meristems? Do all plants have lateral meristems?
   Textbook Reference: *34.3 How Do Meristems Build the Plant Body? p. 754*

3. Describe the development of a xylem cell in a monocot from its origin in the apical meristem.
   Textbook Reference: *34.2 How Are Plant Cells Unique? p. 750*

4. Draw a growing root. Label the primary meristems, root cap, cortex, stele, and epidermis. Discuss the function of each of these structures.
   Textbook Reference: *34.3 How Do Meristems Build the Plant Body? p. 755*

5. Compare and contrast monocots and eudicots. Identify the following plants as monocot or eudicot: corn, oak tree, day lily, rose, dandelion.
   Textbook Reference: *34.1 How Is the Plant Body Organized? p. 746*

6. Look at the leaf in Figure 34.23 in the textbook. Which surface of that leaf faces the sun? How do you know? Why are the stomata opposite the palisade layer?
   Textbook Reference: *34.4 How Does Leaf Anatomy Support Photosynthesis? p. 761*

Answers

Knowledge and Synthesis Answers

1. **b.** See Figure 34.1 for a comparison of eudicots and monocots.

2. **b.** A taproot would be necessary to anchor a plant of that size. Most forest trees have taproots.

3. **d.** Tendrils are modified leaves. Some climbing plants produce tendrils, and others produce suckers.

4. **d.** Plant cell shape is maintained by their cell walls. No animal cells have cell walls. Plants have plastids and vacuoles as well. Though these characteristics are seen in some protists, no animal cells have them.

5. **a.** Palisade and spongy mesophyll cells are photosynthetically active and derived from parenchyma cells.

6. **a.** Xylem tissue transports water from the roots throughout the plant. Phloem tissue transports carbohydrates and nutrients from source tissue to sink tissue.

7. **d.** The apical and (when present) lateral meristems are regions of continually dividing cells that can contribute to the growth of a plant throughout its life. A modular body plan also contributes to the ability of plants to grow continually.

8. **c.** Lateral meristems are responsible for the growth of new xylem and phloem. The secondary xylem gives rise to wood.

9. **b.** Wood grain is a result of patterns in the vascular cylinders. Wood porosity results from the lateral phloem rays.

10. **a.** The cork cambium is a meristematic region outside the secondary phloem. As girth increases and splits the epidermis causing loss of those protective layers, the cork cambium produces new cells to cover the expanding vascular tissue.

11. **b.** The sieve plates allow for the conduction of sap from one sieve tube cell to another sieve tube cell. The pores are due to enlargements of the plasmodesmata. These cells may also lose nuclei and organelles so that carbohydrates can easily pass through the sieve tubes.

12. **b.** Guard cells at the edges of stomata respond to changes in osmotic pressure. These changes lead to the opening and closing of the stomata to regulate gas exchange and water loss.

13. **a.** The protoderm becomes the dermal tissue system of the growing plant.

14. **d.** The apical meristems are the site of primary growth. Apical meristems are found at the tips of the roots, in stems, and in buds.

15. **c.** Vascular bundles are composed of the vascular tissue of the plant, which includes the xylem and phloem.

Application Answers

1. See Figure 34.2 for a labeled diagram.

2. Apical meristems are responsible for elongation of the plant body. Lateral meristems are responsible for an increase in girth. Lateral meristems are found only in woody eudicots and are responsible for creating wood. All plants have apical meristems.

3. The apical meristem gives rise to the protoderm, the ground meristem, and the procambium. The procambium gives rise to the vascular tissue system, including xylem cells. Monocots do not have secondary xylem, so only primary xylem is possible in this plant.

4. See Figure 34.14 for a diagram of root growth. The protoderm gives rise to the epidermis for protection. The ground meristem gives rise to the cortex for storage. The procambium gives rise to the stele for

transport. The root cap protects the meristem as it pushes through the soil.

5. See Figure 34.1 for a comparison of monocots and eudicots. Corn is a monocot. The easiest way to be sure of this is to look at the parallel veins in its leaves. An oak tree is a eudicot because only eudicots have woody secondary growth. Day lilies are monocots. They have the characteristic parallel veins and classically three-petaled flowers. Roses are eudicots. Many roses have woody secondary growth, petals in multiples of five, and netted veins. Dandelions have netted veins, but the clearest indication that they are eudicots is their well-developed taproot (see Figure 34.14).

6. The "top" of the diagram in the textbook would be the surface that faces the sun. This is the surface where photosynthetic cells, which need maximum sun exposure, are located. The stomata are on the opposite side to reduce water loss due to evaporation during photosynthesis.

CHAPTER 35 Transport in Plants

Important Concepts

Water and nutrients are taken up in the roots of plants.

- The movement of water across a semipermeable membrane is a special type of diffusion known as osmosis (see Figure 35.2).

  - For osmosis to occur across a semipermeable membrane, there must be a solute potential (or difference in solute concentrations) great enough to initiate movement and a pressure potential (turgor pressure in plants) small enough to allow for movement.

  - The overall tendency of a solution to take up water across a membrane is called water potential and is the sum of the negative solute potential and the positive pressure potential. All three parameters can be measured in megapascals (MPa). Water always moves to a region of more negative water potential.

- The structure of plants is maintained by osmotic phenomena. If a plant loses turgor pressure by a decrease in pressure potential, it wilts. Movement of water from cell to cell depends on water potential. Long-distance movement depends on pressure potential and is referred to as bulk flow.

- Specialized membrane channel proteins in plant cells called aquaporins can increase the rate of water movement by allowing water to cross the plasma membrane without interacting with the hydrophobic bilayer that slows water flow. Though aquaporins can increase the rate of osmosis, they cannot influence the direction of flow.

- Mineral uptake from the soil solution requires active transport via proteins. When mineral concentrations are greater in the soil solution than in the plant, they are taken up by facilitated diffusion. If minerals are in smaller concentrations outside the plant than inside the plant, or if they must be moved against an electro-chemical gradient, then the plant must rely on active transport.

- Plants rely on a proton pump for active transport of minerals into cells. Plants actively pump protons out of cells, causing the area outside the cell to become more positive. This assists facilitated diffusion of positive ions through protein channels. It also drives the movement of negatively charged ions into the cell by active transport (see Figure 35.3). The result of this pumping action is to make the internal environment of the plant cell highly negative compared to its environment. This difference in charge is called membrane potential. The proton gradient that develops across the membrane can also facilitate secondary active transport of ions such as Cl^-.

- Water moves into a root because the root has a more negative water potential than the soil solution it is bathed in. Movement inside the root takes place because the stele (vascular tissue) has a more negative water potential than the cortex. Minerals dissolved in water are moved via bulk flow of water once it is in the vascular system.

- Minerals follow two paths for reaching the vascular tissue: the apoplast or the symplast.

  - The apoplast is formed by cell walls and intercellular spaces. Water and minerals may move unregulated through this space without ever having to cross a membrane.

  - The symplast is the living portion of the plant and is enclosed in plasma membranes. Movement of water and minerals in the symplast is highly regulated.

  - Water and minerals can travel through the apoplast as far as the endodermis. At the endodermis, water and minerals are stopped by the Casparian strips, which are waxy structures surrounding the endodermal cells.

  - Because of this, water can reach the stele only via the symplast. The transport proteins in the endodermal cells determine which minerals enter the stele.

  - Once past the endodermal barrier, water and minerals can again leave the symplast and move back to the apoplast with the aid of parenchyma cells.

- Ultimately, water and minerals from the soil solution end up in xylem cells and are referred to as xylem sap.

Water and nutrients are moved through the plant in the xylem.

- It was originally thought that a pumping mechanism for the movement of fluids might be active in plants. This was shown to be false by experiments in 1893 by Edward Strasburger. Tree trunks immersed in poison showed progressive death of all living cells as the poison progressed through the plant. This experiment led to three important conclusions:

  - Because movement continued even as cells were killed, no "pumping" cells were active.

  - Leaves were critical to transport because transport continued until the leaves died.

  - Transport did not depend on the roots because it occurred in the absence of roots.

- Root pressure, as shown by guttation (the forcing of water out of openings in leaves), was another theory for water movement. It was thought that the pressure exerted by root tissue might be sufficient to force water up the xylem. In actuality, xylem sap is under negative pressure as it is ascending.

- Pulling forces due to transpiration at the leaf surface are responsible for the movement of water through the xylem. Water evaporates from mesophyll cells during transpiration, creating tension on the water associated with the mesophyll cell wall. Transpiration generates tension on the water molecules in the xylem water column, which pulls them up from the roots and through the apoplast of the leaves. Water molecules are cohesive enough to resist the tension and pull other water molecules along because of their hydrogen–bonding resulting in bulk flow (see Figure 35.6).

- This mechanism, which pulls water from the roots up through the plant is known as the transpiration–cohesion–tension mechanism. This is a passive process requiring no energy input by the plant. Minerals are drawn passively along with the water column. Transpiration also assists with temperature regulation through evaporative cooling of the leaves.

- Per Scholander measured the tension in the xylem sap with a pressure chamber (see Figure 35.7). To conduct the experiments, Scholander cut the stem of the plant and placed the stem and leaves into a pressure chamber, leaving the cut portion of the stem out of the chamber. By placing the leaves under pressure and measuring the pressure needed to push the sap back to the surface of the cut end, the tension in the xylem was measured.

- The flow rate of xylem sap depends on a number of environmental factors such as temperature, light, and wind. Changes in K^+ concentration also influence xylem flow.

Water loss in a plant must be controlled.

- Leaf surfaces are covered with a waxy cuticle to prevent excessive water loss. However, the leaf must take up CO_2 for photosynthesis. Any time a plant surface is open enough to allow gas exchange, water is lost to the environment.

- Stomata with guard cells are pores that regulate gas exchange and water loss from a leaf. Guard cells, in response to osmotic differences, shrink and swell to open and close the stomata. Guard cells open when light is sufficient to maintain photosynthesis and carbon dioxide levels are low. Guard cells are also regulated by water potential. If the water potential in mesophyll cells is low, mesophyll cells release abscisic acid, which causes the guard cells to close.

- Blue light stimulates a proton pump that helps regulate guard cell activity. Guard cells open when potassium ions diffuse into the cell as a result of the electrical gradient set up by the proton pump. High potassium levels cause water to move in by osmosis. Pressure potential builds in the guard cells, and they are pulled apart to reveal the stoma. For guard cells to shut, the proton pumps stop, and potassium ions move back across the membrane. Water follows, and the cells go limp and seal off the stoma (see Figure 35.9).

- Scientists are looking for ways to reduce water loss by transpiration without reducing carbon dioxide availability to plants. This has been attempted through foliar sprays of antitranspirants, like abscisic acid, and more recently through genetic modification of plants to make them more sensitive to abscisic acid.

Phloem moves materials from sources to sinks by translocation.

- Sources are organs that produce more sugars than are used by metabolism, storage, and growth. Sinks are organs that do not make enough sugar for their own growth or storage needs. Sugars, amino acids, minerals, and other substances are translocated between sources and sinks in the phloem.

- Translocation proceeds in both directions along the stem. Translocation stops if phloem tissue is killed and is inhibited whenever respiration and the availability of ATP are limited.

- Scientists can sample the sieve tube sap of a single sieve tube member using aphids. An aphid drills into a single cell, and sap is forced out. Once the aphid begins eating, it is frozen and its feeding organ is used as a tap to collect the sap from a single sieve tube.

- The pressure flow model explains how materials move through the phloem. Once in the sieve tubes, sieve tube sap moves via bulk flow, which requires no energy input by the plant. Energy is required for the loading of the sieve tubes at the sources and the unloading of solutes when the sink is reached. Sucrose is actively transported into sieve tubes at the sources. This causes water to move into sieve tubes by osmosis, thereby increasing the pressure potential at the source end and pushing the contents toward the sink end of the tube, resulting in bulk flow.

- For the pressure flow model to work, the sieve plates must be open to allow uninterrupted flow of sieve tube sap from one sieve tube member to another. Microscopic analysis indicates that the sieve tube plates are open in undamaged cells.

- There must also be a mechanism for loading and unloading sucrose and amino acids. Neighboring cells assist with the loading and unloading of sucrose at sources and sinks.

- Plasmodesmata assist in moving substances from cell to cell and help with the loading and unloading of sieve tube elements. Plasmodesmata in source areas differ from those in sink areas. They are larger and more abundant in sink tissues for unloading. Scientists are exploring the regulation of plasmodesmata in crops to enhance production.

The Big Picture

- In terrestrial plants, water must be acquired from the soil, transported through the plant, and used in the leaves for photosynthesis. At the same time, the nutritional products of photosynthesis must be transported throughout the plant to nonphotosynthetic tissues. This two-way transport is achieved through specialized cells that make up the vascular tissue of the plant. Water with dissolved mineral nutrients is absorbed through the roots and transported to cells in the xylem of the plant, where it is pulled up the stem to the leaves via the transpiration–cohesion–tension mechanism. Sugars and solutes are moved out of the leaves and to the rest of the plant through cells in the phloem.

- Water uptake is regulated by osmotic and water potentials in the root cells, and the rate of transport is controlled by the rate of evaporation at the leaf surface. Guard cell activity in the leaves regulates the opening and closing of stomata to match water availability, light, and drying conditions. Sucrose movement is regulated by active transport and facilitated diffusion in the phloem tissue. The rate of sucrose transport depends on the rates of loading and unloading at source and sink tissues.

Common Problem Areas

- Students often fail to look at the anatomy of the structures that are used in transport. Each structure is uniquely suited to its function. If you couple form and function, the process is much easier to understand.

Study Strategies

- Water and sucrose transport are pathways that can be understood by visually tracing a molecule of water or sucrose through the plant. Use the figures in your textbook to follow these pathways.

- In order to understand water transport, you should understand osmosis and the properties of water molecules. If you are struggling with this chapter, review these concepts from earlier chapters.

- The basis of nutritional transport in plants is cell-to-cell transport. Review the sections in your book on diffusion, osmosis, and active transport.

- Review the following animated tutorial and activity on the Companion Website/CD:
  Tutorial 36.1 The Pressure Flow Model
  Activity 36.1 Apoplast and Symplast of the Root

Test Yourself

Knowledge and Synthesis Questions

1. The function of the Casparian strips is to
   a. divert water and minerals through the membranes of endodermal cells.
   b. prevent water and minerals from entering the stele through the apoplast.
   c. provide regulation for water and mineral movement in the plant.
   d. All of the above
   Textbook Reference: 35.1 How Do Plants Take Up Water and Solutes? p. 768

2. The primary difference between the apoplast and the symplast is that the
   a. apoplast consists of nonliving spaces and cell walls.
   b. apoplast relies on active transport.
   c. symplast consists of nonliving spaces and cell walls.
   d. apoplast prevents passive diffusion.
   Textbook Reference: 35.1 How Do Plants Take Up Water and Solutes? p. 768

3. Which of the following regarding water transport is true?
   a. Root pressure is sufficient to drive xylem sap movement.
   b. Bulk flow is not a mechanism by which water and minerals are transported.
   c. The cohesive nature of water is central to water movement in a plant.
   d. None of the above
   Textbook Reference: 35.2 How Are Water and Minerals Transported in the Xylem? p. 770

4. Tension is a result of which of the following?
   a. Transpiration at the leaf surface
   b. The cohesive nature of water
   c. The narrowness of the xylem tube
   d. All of the above
   Textbook Reference: 35.2 How Are Water and Minerals Transported in the Xylem? p. 771

5. The fact that water transport continues as long as leaves are alive and active indicates that
   a. leaves pump water.
   b. leaves are necessary for transport of water.
   c. roots are active.
   d. water is not needed for leaves to remain alive.
   Textbook Reference: 35.2 How Are Water and Minerals Transported in the Xylem? pp. 769–771

6. Which of the following is true regarding transport in phloem?
   a. Transport in phloem is always in the direction of leaves to roots.
   b. Transport in phloem is from source tissue to sink tissue.
   c. Transport in phloem cells requires no energy inputs from the plant.
   d. None of the above
   Textbook Reference: *35.4 How Are Substances Transported in the Phloem? p. 775*

7. If the pressure potential is 0.16 megapascals (MPa) and the solute potential is –0.24 MPa, then the water potential would be
   a. 0.04 MPa.
   b. 0.08 MPa.
   c. –0.08 MPa.
   d. –0.24 MPa.
   Textbook Reference: *35.1 How Do Plants Take Up Water and Solutes? p. 766*

8. If you were to order the water potential of the following root cells or regions from least to most negative, which cell or region would be third?
   a. Xylem
   b. Soil next to root
   c. Cortex apoplast
   d. Stele apoplast
   Textbook Reference: *35.1 How Do Plants Take Up Water and Solutes? p. 765–767*

9. The movement of water up the stems of tall plants is least dependent on which of the following factors?
   a. Guttation
   b. Transpiration
   c. Cohesion of water molecules
   d. Tension within water columns
   Textbook Reference: *35.2 How Are Water and Minerals Transported in the Xylem? p. 771*

10. Which of the following is true of both xylem transport and phloem transport?
    a. Both are passive processes that do not require energy from the plant.
    b. Both involve only living cells.
    c. Both rely on a water potential gradient.
    d. The direction of flow can reverse in both.
    Textbook Reference: *35.2 How Are Water and Minerals Transported in the Xylem? pp. 769–770; 35.4 How Are Substances Translocated in the Phloem? p. 775*

11. Which of the following regulates stomatal opening and closing?
    a. Abscisic acid levels
    b. Light levels
    c. Carbon dioxide concentrations
    d. All of the above
    Textbook Reference: *35.3 How Do Stomata Control the Loss of Water and the Uptake of CO_2? pp. 770–771*

12. The opening and closing of the stomata are accomplished by the
    a. sieve tube.
    b. guard cells.
    c. translocation.
    d. aquaporins.
    Textbook Reference: *35.3 How Do Stomata Control the Loss of Water and the Uptake of CO_2? p. 770*

13. Regulators of stomatal opening and closing work by activating the
    a. proton pump in guard cells.
    b. proton pump in stomata.
    c. sodium–potassium pump.
    d. All of the above
    Textbook Reference: *35.3 How Do Stomata Control the Loss of Water and the Uptake of CO_2? p. 772*

14. Mineral ions enter the cell due to the force of an electrochemical gradient set up by pumping _____ out of the cells.
    a. K^+
    b. Ca^{2+}
    c. Na^+
    d. H^+
    Textbook Reference: *35.1 How Do Plants Take Up Water and Solutes? p. 766*

15. In the pressure flow model describing translocation, the movement of water by osmosis occurs from the
    a. xylem to the phloem at the sink.
    b. phloem to the xylem at the source.
    c. xylem to the phloem at the source.
    d. All of the above
    Textbook Reference: *35.4 How Are Substances Translocated in the Phloem? p. 775*

Application Questions

1. Differentiate between source and sink tissues. What happens relative to phloem in each?
   Textbook Reference: *35.4 How Are Substances Translocated in the Phloem? p. 774*

2. Under what conditions does transpiration occur most rapidly? What effect will increased transpiration have on water flow in a plant? What happens if adequate water for the plant is not available?
   Textbook Reference: *35.2 How Are Water and Minerals Transported in the Xylem? p. 769*

3. Trace a water molecule as it moves from the soil solution to the stele of a plant. Identify where the molecule is traveling through the apoplast and where the molecule is traveling through the symplast.
   Textbook Reference: *35.1 How Do Plants Take Up Water and Solutes? p. 767*

4. Describe the role of the proton pump in moving minerals into the root.
   Textbook Reference: *35.1 How Do Plants Take Up Water and Solutes? p. 765*

5. Explain how transpiration, cohesion, and tension work together to move water in a large plant.
   Textbook Reference: 35.2 How Are Water and Minerals Transported in the Xylem? p. 770

6. Describe the pressure flow model of phloem transport.
   Textbook Reference: 35.4 How Are Substances Translocated in the Phloem? p. 776

Answers

Knowledge and Synthesis Answers

1. **d.** Not all minerals that enter the apoplast of a plant's root are beneficial to the plant. The Casparian strips prevent water and minerals from reaching the stele through the apoplast, diverting them instead through the plasma membranes of the endodermal cells. Channel proteins in these plasma membranes determine which minerals can enter the symplast, and from there, the stele. The Casparian strips thus contribute to the regulation of water and mineral movement in the plant.

2. **a.** The intercellular spaces and cell walls of the plant constitute the apoplast.

3. **c.** Water movement depends on the cohesive nature of water to withstand the tension placed on the water column by transpiration.

4. **a.** Transpiration causes tension.

5. **b.** Leaves are necessary for transpiration to take place.

6. **b.** Transport in phloem does not always go from leaf to root, but it is always from source tissue to sink tissue. The plant must contribute energy to create the water pressure gradient by pumping solutes into the phloem at the source and out of the phloem at the sink.

7. **c.** Water potential is equal to pressure potential plus solute potential.

8. **c.** The xylem would be the most negative, followed by the stele, then cortex, then the area outside the root.

9. **a.** Guttation occurs under extremely humid conditions when water is plentiful.

10. **c.** Both xylem transport and phloem transport depend on water potential.

11. **d.** Abscisic acid, light, and carbon dioxide levels all regulate stomatal opening and closing.

12. **b.** Guard cells are specialized epidermal cells that regulate the opening and closing the stomata. They do this by covering the stomata opening.

13. **a.** Stomatal regulators work by activating and deactivating the proton pump in guard cells.

14. **d.** Cells pump H^+ ions out into the soil with the help of proton pumps.

15. **c.** In the pressure flow model, water moves by osmosis from the xylem into the phloem at the source and from the phloem to the xylem at the sink. This movement is all driven by the solute concentrations at each location.

Application Answers

1. Source tissues produce sugars in excess of what can be used and stored. Phloem loading occurs in source tissues and creates a pressure potential that results in bulk flow of sieve tube sap toward sink tissues. Sink tissues produce fewer sugars than can be stored or used and unload phloem through active transport.

2. Transpiration will occur most rapidly in high light conditions when stomata are open, along with high wind conditions and low humidity when evaporation is greatest. This will result in faster bulk flow through the xylem and increased water demands by the plant. If water is not available, plant cells will lose turgor and the plant will wilt.

3. The water molecule enters the epidermis due to water potential. If it crosses the plasma membrane into an epidermis or cortex cell, it enters the symplast and can travel via plasmodesmata to the stele. If it does not enter the symplast, it can travel through the apoplast until it reaches the endodermis and a Casparian strip. There, it crosses the plasma membrane of the endodermis and enters the symplast. Once past the endodermis, it may again enter the apoplast (see Figure 35.4).

4. Plants rely on a proton pump for active transport of minerals into cells. Plants pump protons out of cells, which causes the area outside the cell to be more positive. This assists facilitated diffusion of positive ions through protein channels. It also drives the movement of negatively charged ions into the cell by active transport (see Figure 35.3).

5. The transpiration–cohesion–tension mechanism pulls water from the roots up through the plant. Water evaporates from mesophyll cells during transpiration. This puts tension on the film of water associated with the mesophyll cell wall. The tension at the mesophyll cell draws water from the xylem of the nearest vein. This creates tension in the entire xylem column, and the column is drawn upward from the roots.

6. The difference in solute concentration between sources and sinks creates a pressure potential along sieve tubes, resulting in bulk flow. For this to occur, sugars must be loaded at the source tissue and unloaded at the sink tissue through active transport, and the sieve plates must remain open and unclogged along the phloem column.

CHAPTER 36 Plant Nutrition

Important Concepts

All organisms must acquire nutrients through the cycling of other compounds or by uptake from their environment.

- The basic nutrient requirements for all living things are carbon, hydrogen, oxygen, and nitrogen. These elements are the fundamental building blocks of all macromolecules.

- Carbon enters organisms through photosynthesis. Oxygen and hydrogen enter plants as water.

- Nitrogen's entry into biological organisms is dependent on nitrogen-fixing bacteria in the soil.

- Mineral nutrients are essential to life. Sulfur, phosphorous, magnesium, and iron are all essential components of many macromolecules.

- Mineral nutrients enter biological organisms through soil solutions, which plants take up through their roots.

- Plants form the basis of the terrestrial food chain because they make organic compounds that are available to other organisms. Organisms that have the ability to make organic compounds from inorganic compounds are called autotrophs. Plants rely on photosynthesis to produce the organic molecules that allow them to be autotrophs. Chemolithotrophs are autotrophs that derive their energy from hydrogen sulfide. Heterotrophs require preformed organic compounds.

- Plants are sessile, meaning they cannot move around and therefore cannot "search" for nutrients. They overcome this problem by growing toward new resources. A plant grows taller to procure more sunlight for itself and to outcompete nearby plants, and its roots spread to acquire mineral nutrients in the soil.

Mineral nutrients are vital to proper plant growth.

- Every plant requires specific essential nutrients necessary for growth and development. These nutrients cannot be replaced by another element and must be directly obtained for the day-to-day functioning of the plant. A deficiency in any essential nutrient leads to an unhealthy plant.

- Macronutrients are essential elements required at a rate of 1 g per 1 kg of dry plant matter, and micronutrients are essential elements required at a rate of 100 mg per 1 kg of dry plant matter.

- See Table 36.1 for a list of macro- and micronutrients, their sources, and their functions.

- Though many deficiencies ultimately lead to plant death, specific deficiency symptoms are displayed in a plant before it dies. Nitrogen deficiency is the most common deficiency and can be corrected by the addition of fertilizers to supplement nitrogen in soils. Plant growth will slow to match the availability of nutrients until the nutrient loss becomes severe.

- Yellowing of leaves is the most common symptom of nitrogen deficiency. This occurs when the green pigment chlorophyll cannot be synthesized because of inadequate nitrogen. Yellowing of the leaves is also evident in iron deficiencies for the same reason and commonly occurs in the youngest leaves. Iron cannot be moved through a plant easily, and the growing young leaves show iron deficiency before the older leaves.

- Many essential elements have multiple roles in plants. Phosphorus is used in ATP, nucleic acids, and intermediates of photosynthesis and glycolysis. The role of essential elements was identified by growing plants hydroponically, depriving them of specific nutrients, and observing the plants' response. Missing nutrients that prevented a plant from completing its life cycle were deemed essential.

Soils are nature's nutrient sink.

- Plants and soils interact in a complex fashion. Plants change soils, and soils influence the growth of plants.

- Soils are composed of living and nonliving matter. The living portion of soil contains roots, protists, bacteria, fungi, and many small animals. The nonliving portion consists of rock fragments, clay, water, air spaces, and dead organic matter. The air spaces provide the O_2 needed for a plant to survive.

- All soils have a soil profile consisting of two or more horizons (layers). Water-soluble nutrients are leached to deeper horizons through rainfall. Three major horizons (also called zones) can be identified in soils. A horizon is topsoil that is organically rich and very biologically active and is the most agriculturally important layer. Loam soil with sand, silt, and clay, high nutrient content, plentiful water, and air spaces is ideal topsoil for agriculture. Sands typically do not hold nutrients or water well, and clays are too dense for the trapping of air. B horizon is the subsoil, and it holds many leached nutrients. C horizon is parent rock, which roots cannot penetrate.

- Soils form from mechanical and chemical weathering that breaks down rocks. Mechanical weathering occurs through freeze/thaw cycles, rain, and drying. Chemical weathering occurs by oxidation, hydrolysis, or breakdown by acids, and can change the composition of rock. Chemical weathering is very important in clay formation.

- Mineral nutrients are tied to clay particles in the soil. Because many nutrients are positively charged cations, clays with a negative charge can hold these nutrients and make them available to plants. Roots release protons into the soil that bind to clay particles, and the cation minerals are released. This process is called ion exchange. Nutrients that are negatively charged and therefore do not participate in ion exchange are rapidly leached from soil. Thus, nitrate and sulfate are often not available to plants.

- Agricultural fertilizers add nitrogen, phosphorous, and potassium to soils and are rated by their "N-P-K" percentages. A 10-10-10 fertilizer contains 10 percent nitrogen, 10 percent phosphate, and 10 percent potash (potassium source). These nutrients need to be replenished in soils because they are negatively charged and leach from soils rapidly.

- Fertilizers may be organic in nature (manures, compost) or inorganic (chemical fertilizer). Organics release nutrients much more slowly and do not leach as quickly as inorganics. Inorganic fertilizers provide a much more rapid release of nutrients.

- The pH of soils affects nutrient availability. Specific plants have specific pH needs. Soils tend to become slightly acidic from leaching and rain. The practice of liming raises soil pH. This has the secondary effect of making calcium available to plants. Adding sulfur reduces soil pH.

- Copper, iron, and manganese can be added by foliar spraying of nutrients.

- Plants affect the pH of soil, adding decaying matter in the form of humus, and altering soil temperature. The plant roots also can alter the soil pH by excreting H^+ or OH^- ions.

Nitrogen is vital to plant growth.

- Nitrogen gas is readily available in the atmosphere, but plants cannot use it directly. Only a few bacteria species can convert nitrogen gas into biologically usable ammonia through nitrogen fixation.

- The most biologically important nitrogen fixers are associated with plant roots and release up to 90 percent of the nitrogen they fix to the soil. *Rhizobium* bacteria have a close association or mutualistic symbiosis with the roots of plants in the legume family; both species benefit from the association. These bacteria and the associated legumes are important agriculturally and are the basis of crop rotation.

- The fixation of nitrogen requires the enzyme nitrogenase to catalyze the reaction, lots of energy (ATP), and a strong reducing agent. Nitrogenase is inhibited by oxygen and therefore is active only under anaerobic conditions. The protein leghemoglobin binds with oxygen to keep oxygen levels low.

- The anaerobic *Rhizobium* bacterium resides in a root nodule of the legume where it is protected from oxygen. The bacteria are provided with a low-oxygen growing environment, and the plant benefits from the released ammonia. To establish the symbiosis, the plant releases flavonoids to attract the bacteria. In response to the flavonoids, the bacteria turn on *nod* genes. The resulting gene products cause formation of the nodule by the plant. Once housed in the nodule, the bacteria form swollen bacteroids capable of nitrogen fixation. The plant surrounds the bacteroids with leghemoglobin to support respiration (see Figure 36.9).

- Bacterial nitrogen fixation is not adequate to supply all the nitrogen needed for agriculture. Currently, nitrogen fertilizers are produced via industrial fixation through the Haber process, an energy-expensive process. Research is looking into the possibility of engineering plants that have their own nitrogenase.

- In high levels, ammonia is toxic to plants, but at low levels it is used to produce amino acids. Therefore, soil bacteria that convert ammonia to nitrate are necessary. These bacteria are called nitrifiers. Plants, through their metabolism, reduce nitrate to ammonia. This is accomplished by the plant's own enzymes.

- The entire nitrogen cycle is completed by denitrifiers, which convert the nitrogen from waste and dead organic matter back to nitrogen gas. See Figure 36.10 for a review of the nitrogen cycle.

Some plants are heterotrophic.

- Some plants that live in soils that are nitrogen- or phosphorous-deficient are carnivorous. Carnivorous plants acquire nitrogen from trapped decaying animals. Examples of carnivorous plants are Venus flytraps, pitcher plants, and sundews. Although these plants can survive and grow without consuming insects, they do best when they have a continuous supply of them.

- Some plants have lost the ability to photosynthesize. These plants must acquire their nutrients from other sources. Some are parasitic and acquire some or all of their nutrients from a host plant at the host plant's expense.

The Big Picture

- Plants require specific macro- and micronutrients. Deficiencies in any of these produce consequences for the health of the plant. Essential nutrients must be present, and no substitutions will sustain the plant. Nutrients are procured from the soil solution that bathes the roots of a plant. The availability of nutrients depends on quantity, solubility, and structure of soil. Many agricultural practices deplete soils of nutrients, which must be added back through fertilization.

- Nitrogen availability is essential to plant growth. Bacteria and plants are intrinsically linked in the nitrogen cycle (see Figure. 36.10). Biological nitrogen fixation is utilized in agriculture by rotating crops with those that harbor nitrogen-fixing bacteria in their root nodules. Commercial nitrogen fixation for chemical fertilizers is extremely energy demanding, and alternatives through genetic engineering are being sought.

- A small number of plants do not photosynthesize. These heterotrophic plants are often parasites of other plants and acquire their nutrients solely through their host plant.

Common Problem Areas

- The interactions between nitrogen-fixing bacteria and plants can become confusing. Remember that it is a mutualistic relationship and that the bacteria have the enzymes necessary to fix atmospheric nitrogen.

- The most difficult concept of this chapter is the nitrogen cycle. Use the figures in your textbook to visualize this process and understand how the organisms involved interact with the environment.

Study Strategies

- Be sure you understand the consequences of nutrient shortages in plants.

- Review the following animated tutorial and activity on the Companion Website/CD:
  Tutorial 36.1 Nitrogen and Iron Deficiencies
  Activity 36.1 Nitrogen Cycle

Test Yourself

Knowledge and Synthesis Questions

1. Which of the following nutrients is *not* considered essential for plant growth?
   a. Cadmium
   b. Nitrogen
   c. Manganese
   d. Potassium
   Textbook Reference: 36.1 How Do Plants Acquire Nutrients? p. 781

2. Macronutrients are _____ than micronutrients.
   a. larger molecules
   b. needed in greater quantities
   c. more essential
   d. more important for growth
   Textbook Reference: 36.2 What Mineral Nutrients Do Plants Require? p. 783

3. Nitrogen and potassium are acquired from
   a. soil solution.
   b. heterotrophs.
   c. air.
   d. All of the above
   Textbook Reference: 36.1 How Do Plants Acquire Nutrients? pp. 781–782

4. You notice that the young leaves of your tomato plants are very yellow, whereas the older leaves are still green. What type of deficiency does this suggest?
   a. Nitrogen
   b. Carbon
   c. Water
   d. Iron
   Textbook Reference: 36.2 What Mineral Nutrients Do Plants Require? p. 783

5. Years of cotton farming in the South has stripped away much of the A horizon of the soils. Subsequent agriculture has been difficult for farmers because the
   a. A horizon contains the most available nutrients.
   b. B horizon contains significantly more available nutrients.
   c. C horizon is most conducive to root growth.
   d. All of the above
   Textbook Reference: 36.3 What Are the Roles of Soil? p. 786

6. Clay particles in soils are important for
   a. holding soil together.
   b. ion exchange.
   c. holding water.
   d. All of the above
   Textbook Reference: 36.3 What Are the Roles of Soil? p. 786

7. Most clays form from the _____ of rock.
   a. mechanical weathering
   b. chemical weathering
   c. heaving
   d. All of the above
   Textbook Reference: 36.3 What Are the Roles of Soil? p. 786

8. You purchase a commercial fertilizer at your local garden center. The label says that it is 10-20-10. This label refers to the
   a. percentages of nitrogen, phosphate, and potassium.
   b. percentages of nitrogen, carbon, and oxygen.
   c. rate at which nitrogen is released from the fertilizer.
   d. ratio of organic to inorganic matter in the fertilizer.
   Textbook Reference: 36.3 What Are the Roles of Soil? p. 787

9. The relationship between *Rhizobium* and the roots of legumes can best be described by which of the following terms?
   a. Parasitic
   b. One-sided
   c. Mutualistic
   d. Carnivorous
   Textbook Reference: *36.4 How Does Nitrogen Get from Air to Plant Cells? p. 788*

10. Nitrogen gas is reduced to ammonia by which of the following enzymes?
    a. *Rhizobium*
    b. Nitrogenase
    c. Nitrification
    d. Denitrification
    Textbook Reference: *36.4 How Does Nitrogen Get from Air to Plant Cells? p. 789*

11. Plants are able to take up and use nitrogen in the form of _____ and _____.
    a. ammonia; nitrate
    b. ammonia; nitrogen gas
    c. nitrogen gas; nitrate
    d. All of the above
    Textbook Reference: *36.4 How Does Nitrogen Get from Air to Plant Cells? p. 791*

12. Carnivorous plants are often found in acidic and nutrient-poor environments. The main selective pressure for carnivory is
    a. lack of nitrogen and phosphorous sources.
    b. lack of iron and calcium sources.
    c. incomplete ion exchange.
    d. All of the above
    Textbook Reference: *36.5 Do Soil, Air, and Sunlight Meet the Needs of All Plants? p. 792*

13. Why do nitrate and sulfate leach from the soil?
    a. Because they bind with ions such as K^+ and Mg^{2+}.
    b. The H^+ ions released by the roots push them out.
    c. Because they are unable to bind with the negatively charged clay particles.
    d. All of the above
    Textbook Reference: *36.3 What Are the Roles of Soil? p. 786*

14. What process is used to lower the pH of soil?
    a. Ion exchange.
    b. Leaching.
    c. 5-10-10 fertilizing.
    d. Liming.
    Textbook Reference: *36.3 What Are the Roles of Soil? p. 787*

15. The role of leghemoglobin is to maintain _____ levels in the root nodule.
    a. high O_2
    b. high CO_2
    c. low O_2
    d. low CO_2
    Textbook Reference: *36.5 Do Soil, Air, and Sunlight Meet the Needs of All Plants? p. 792*

Application Questions

1. Explain how it was determined whether a plant nutrient was essential.
   Textbook Reference: *36.2 What Mineral Nutrients Do Plants Require? p. 784*

2. Sketch the nitrogen cycle. Which components of the cycle are impacted by plants and how?
   Textbook Reference: *36.4 How Does Nitrogen Get from Air to Plant Cells? p. 791*

3. Describe how plants and bacteria interact to form nitrogen-fixing root nodules. Why do many farmers plant crops such as alfalfa and soybeans without harvesting them?
   Textbook Reference: *36.4 How Does Nitrogen Get from Air to Plant Cells? p. 789*

4. Most carnivorous plants are found in boggy, wet, acidic environments. What effect would an environment such as this have on nutrient availability?
   Textbook Reference: *36.5 Do Soil, Air, and Sunlight Meet the Needs of All Plants? p. 792*

5. Explain how plants "grow into their nutrients."
   Textbook Reference: *36.1 How Do Plants Acquire Nutrients? p. 782*

6. Differentiate between micro- and macronutrients. Where are most of these nutrients acquired?
   Textbook Reference: *36.2 What Mineral Nutrients Do Plants Require? pp. 782–783*

Answers

Knowledge and Synthesis Answers

1. **a.** Cadmium is not one of the 14 essential micro- and macronutrients.

2. **b.** The main difference between micronutrients and macronutrients is in how much of them a plant needs to survive.

3. **a.** Nitrogen and mineral nutrients are acquired in soil solution (dissolved in water).

4. **d.** If the younger leaves are yellow, an iron deficiency would be suspected because iron is easily relocated in the plant whereas iron is not.

5. **a.** Topsoil, or the A horizon, is most conducive to root growth. Ample nutrients are available, as are air spaces and water for ease of root growth.

6. **d.** Clay particles are critical for ion exchange. They are also important for retaining water and for the integrity of soil.

7. **b.** Chemical weathering leads to clay formation.

8. **a.** Your fertilizer is 10 percent nitrogen, 20 percent phosphorous, and 10 percent potash (a potassium source).

9. **c.** The relationship is mutualistic in that both the plant and the bacteria benefit from the association.

10. **b.** Nitrogenase catalyzes the reduction of nitrogen gas to ammonia. This is an energy-expensive process.

11. **a.** Plants can take up and use nitrogen that is in the form of ammonia or nitrate.

12. **a.** Carnivory supplements insufficient nitrogen and phosphorous availability.

13. **c.** Because nitrate and sulfate are anions, they do not have any positively charged molecules in the soil to interact with and they leach from the soil.

14. **d.** If a soil is too basic, it can be acidified by adding lime.

15. **c.** Leghemoglobin is a protein produced by the root nodule of plants to help maintain low levels of O_2 so that the nitrogen-fixing bacteria can have an anaerobic environment.

Application Answers

1. Plants were grown in cultures that lacked specific nutrients. If a plant could not complete its life cycle, the missing nutrient was said to be essential. These experiments are well controlled hydroponic studies.

2. See Figure 36.10. Plants are actively involved in nitrogen fixation through the symbiosis with *Rhizobium*. *Rhizobium* fix nitrogen only in close association with the roots of plants in the legume family.

3. See Figure 36.9. By rotating crops (and even more by rotating and plowing legume crops under) farmers can organically add nitrogen to depleted soils. This results in significantly larger yields to offset the years when the land is not utilized for harvested crops.

4. Acidic environments limit decomposition and thus nitrogen sources. They also limit ion exchange. Both conditions result in reduced nutrient availability that can be offset by carnivory.

5. Plants cannot physically move to areas with greater nutrients. They are limited to extending their roots into soils that contain a larger nutrient reserve. When nutrients are limited, plant growth matches the availability of resources.

6. Macronutrients are needed at a rate of 1 g/1 kg of dry plant tissue. Micronutrients are needed at a rate of 100 mg/1 kg of dry plant tissue. The majority of these nutrients are in soil solution and are taken up as water is drawn into roots.

CHAPTER 37 Regulation of Plant Growth

Important Concepts

Plant development is influenced by multiple regulatory factors.

- Plant development depends on the interplay of environmental cues, receptors that sense these cues and hormones that mediate the effects of the environment. Enzymes influence development at all stages.

- Hormones are regulatory compounds that are produced in one area of a plant and translocated throughout it. The effects of the hormones result from their relative concentrations, and they play multiple regulatory roles in plants.

- Photoreceptors are pigment proteins that sense light and are altered by light quality to induce changes within a plant.

- A plant's genome ultimately regulates its growth and development, under the influence of hormones and photoreceptors.

- Signal transduction mediates hormone and photoreceptor action. A receptor receives an environmental cue, a biochemical signal transduction pathway is initiated, and ultimately a cellular response is generated through a protein kinase.

Plant development depends on cell division, cell expansion, and cell differentiation.

- Plant seeds are dormant and remain so until seed germination. Germination is triggered by one or more mechanical or environmental cues. To germinate, a plant must imbibe water and draw polysaccharides, fats, and protein nutrients from the endosperm or cotyledons. A plant is considered a seedling when its radicle breaks through the seed coat to end germination.

- Because most seeds germinate underground, there must be rapid elongation of the stem to reach light. This rapid growth is directed by signals from a photoreceptor. Subsequent plant growth is under hormonal control.

- Flowering is initiated by plant age, size, or season. Seasonal flowering is controlled by light duration and is sensed by photoreceptors. Flower formation is most likely induced by hormones, as is subsequent flower development.

- Perennials, which live year after year, have buds that enter dormancy during winter. Hormones such as abscisic acid maintain dormancy. In plants that lose their leaves, senescence is controlled by hormone interactions of ethylene and auxin. Plant death may follow senescent activity.

Defined events regulate the breaking of seed dormancy and germination.

- Dormancy, lasting for different amounts of time depending on the plant, involves exclusion of water or oxygen from the embryo by the seed coat, mechanical restraint of the embryo by the seed coat, and chemical inhibition of the embryo.

- Dormancy can be broken by mechanical abrasion or fire. Prolonged exposure to water may leach chemical inhibitors away from the seed and induce germination.

- Dormancy ensures survival through adverse conditions. Some seeds rely on other environmental cues before germination, such as cold temperatures, light, or the passage of a specific amount of time. These cues help ensure that the seed will germinate in the correct location and time.

- Germination begins with a seed's imbibing water. The seed has a negative water potential allowing for water to be taken up by the seed. Water results in metabolic changes. DNA synthesis is halted until the radicle emerges from the seed coat. Breakdown of starch and protein reserves in the cotyledons and endosperm begins to provide nutrients for the developing embryo. In cereal seeds, the embryo secretes gibberellins, which promote enzymes that break down starch and proteins.

Gibberellins regulate growth from germination to fruiting.

- Gibberellins are produced in both plants and fungi. The first gibberellin was isolated from a fungus that infects rice plants and causes them to grow tall and spindly. Initial experiments indicated that the medium that the fungus was grown in was sufficient to cause rice plants to be spindly; therefore, the agent that caused the phenomenon was a chemical produced by the fungus.

- Experiments using dwarf and normal corn plants showed that plants have innate gibberellins. Normal corn plants exposed to gibberellins showed no alteration in appearance, but dwarf plants showed shoot elongation to near normal lengths when exposed to gibberellins. These experiments proved that the chemical was present in plants and that its addition did nothing to alter the plant when it existed in sufficient quantities. In plants lacking the chemical, however, application of gibberellin induced stem elongation.

- Gibberellins also have roles in fruit development. Developing seeds produce gibberellins that enhance development of fruit tissue. It is common agricultural practice to spray seedless fruits (especially grapes) with gibberellins to enhance fruit growth. Gibberellins also induce bolting and seed set in biennials under appropriate environmental cues.

- Additional roles of gibberellins include fruit development from unfertilized flowers, germination of some seeds, and breaking of winter dormancy.

Auxin affects plant growth and form.

- The discovery of auxin (indoleacetic acid) was the result of work of Charles and Francis Darwin in the 1880s. They were interested in how plants grew toward light by phototropism and which part of an emerging plant coleoptile was responsible for sensing light. They found that the tip was the light-receptive portion, but that the growing region (responsible for the bending to or away from light) was some distance below the receptor region. From this they reasoned that some chemical must be transmitted from the tip to the growing region. Additional experiments with removed tips and tips on gelatin blocks indicated that a chemical was indeed moving from the tip to the growing region. Subsequent experiments by other researchers showed that gelatin exposed to the tips was sufficient to cause altered growth. The chemical was later isolated and determined to be auxin.

- Auxin movement in plant tissues is unidirectional and polar from apex to shoot. Auxin anion efflux carriers are carrier proteins responsible for export of auxin anions from cells and contribute to the unidirectional movement of auxin. Auxin enters the cell by passive diffusion and by active transport along with H^+. Because a gradient is established via transport, auxin acts as a morphogen and directs how the cells differentiate along the auxin gradient.

- Redistribution of auxin laterally is responsible for phototropism and gravitropism. Figure 37.10 illustrates how the lateral distribution of auxin affects the curvature of coleoptiles in response to light and gravity. Higher auxin concentrations correspond to increased rates of growth.

- Auxin affects vegetative growth by initiating root growth in cuttings, causing abscission or dropping of leaves, maintaining apical dominance, and promoting stem elongation and inhibiting root elongation. Auxin also can stimulate unfertilized fruit to form (parthenocarpy). Synthetic auxins are used as selective herbicides.

- The effects of auxin on growth are mediated at the cell walls. The cell wall determines the rate and direction of cell growth. Cells grow by taking up more water. The amount of water that can be taken up is restricted by a rigid cell wall. Cell walls must loosen, stretch, and add polysaccharides and cellulose to maintain structure. Cell walls exhibit elasticity when their shape is altered but later return to normal. Plasticity means the shape of the cell is altered and does not return to normal. Plasticity is necessary for cell growth. It is thought that auxin affects plasticity and "loosens" cell walls so that a plant can grow. Auxin stimulates a cell-wall-loosening factor that results in a pH change. The pH change allows proteins called expansins to alter polysaccharide bonding so that they slide past one another during expansion.

- In 2005, it was shown that there is a receptor that binds with auxin and promotes gene expression. A similar pathway has been observed for the action of gibberellin.

Cytokinins have multiple roles within plants.

- Cytokinins are powerful stimulators of cell division and bud formation, aid in seed germination, inhibit stem elongation, and delay leaf senescence. They are synthesized primarily in the roots and are translocated throughout the plant.

- Auxins and cytokinins regulate organ development based on the relative concentrations of the two. High auxin levels favor root formation, and high cytokinin levels favor bud formation.

Ethylene promotes senescence.

- Ethylene is a gaseous hormone that promotes senescence of leaves and ripening of fruit. In many instances it is given off by rotting fruit. Commercial use of ethylene spray to promote ripening is common for a variety of fruits. Ethylene scrubbers are used in fruit storage to prevent ethylene from accumulating and causing fruit to spoil. Other chemicals are used in the flower industry to inhibit ethylene's affects on flower senescence.

- Ethylene plays a role in maintaining the apical hook on emerging eudicots by inhibiting cells on the inner portion of the hook. Ethylene inhibits stem elongation, promotes lateral swelling of stems, and inhibits sensitivity to gravitropic stimulation. These three responses are known as the triple response.

- Figure 37.18 illustrates the signal transduction pathway initiated by ethylene. The process begins when ethylene binds to its receptor on the endoplasmic reticulum. This binding initiates a cascade starting with the activation of endoplasmic reticulum channels and ultimately the expression of genes that results in the physiological changes stimulated by ethylene.

Abscisic acid is responsible for maintaining dormancy.

- Abscisic acid is a terpene that inhibits seed germination and promotes protein storage during embryo formation. It accumulates during times of stress and is responsible for maintaining winter dormancy. It also inhibits stem elongation.

- Plants deficient in abscisic acid often show vivipary, in which seeds sprout while still attached to the plant.

- Abscisic acid regulates guard cell activity by opening calcium channels. Increased calcium levels eventually lead to the opening of potassium channels and the closing of the stoma as guard cells sag together.

Brassinosteroids are involved in the response to light.

- Brassinosteroids were originally isolated from a member of the Brassicaceae. They have been shown to stimulate cell elongation, pollen tube elongation, and vascular tissue differentiation and to inhibit root elongation, all of which are similar to the effects of auxin. The receptor for brassinosteriod is found on the cell wall and initiates a transduction pathway that influences gene expression.

Light and photoreceptors interact to stimulate a variety of plant events.

- Photoreceptors interpret light, its duration, and its wavelength. Light regulates a wide variety of plant processes, including germination, flower production, and the onset of winter dormancy.

- The effects of red and dim blue light are mediated by five phytochromes. Blue and red light promote germination. The wavelength of far-red light is centered at 730 nm, whereas red light is centered 660 nm. Exposure to red and far-red light has different effects on plants. Far-red light reverses the effect of prior exposure to red light. The "switching" effects occur because phytochrome can be shifted from one form (red P_r) to the other (far-red P_{fr}) upon absorption of light. When exposed to red light, P_r is converted to P_{fr}. When exposed to far-red light, P_{fr} is converted to P_r.

- Phytochromes regulate the initiation of chlorophyll production to produce etiolated seedlings. Etiolation functions to conserve resources until plants can begin photosynthesizing. In seedlings that have not been exposed to light, all the phytochrome is the P_r type. Once exposed to light, it shifts to P_{fr} and begins a series of events that reverse etiolation and is known as photomorphogenesis.

Cryptochromes and phototropin are blue-light receptors.

- Cryptochromes absorb blue and ultraviolet light. These receptors are found both in plants and animals. In plants, they carry out many of the same regulatory functions as phytochrome, but they are active in the nucleus rather than in the cytoplasm.

- Phototropin, under blue light, initiates the signal responsible for phototropic curvature. Zeaxanthin plays a role in light-stimulated opening of the stomata.

The Big Picture

- Plant growth is controlled by interactions between a plant's environment, hormones, and genetic makeup. Changes in plant growth are dependent on hormone receptors receiving information from hormones, the subsequent signal transduction pathways that are turned on, and the alteration of gene expression. Receptors are often highly specific and respond to select hormones. Hormones are produced in a specific region of the plant body and translocated throughout the plant; therefore, their effects are often concentration dependent. All growth in a plant is a result of changes in cell division, cell expansion, and cell differentiation.

- Seed germination, growth of the vegetative structures, reproduction, and senescence all proceed in defined patterns. Seed dormancy is broken, and development begins when the seed coat is abraded, inhibitory chemicals are diluted, and the seed imbibes water. This begins a series of events that mobilize nutrients and induce growth. Once a seedling emerges from the soil, light begins to influence subsequent development under the control of hormones.

- Hormones exist in several classes, each with its own effects on growth and development. Some of these effects are antagonistic in nature, and therefore control is maintained via relative concentration of several hormones. Hormones influence every step of development, from breaking of seed dormancy to senescence.

- Light regulates plant processes through photoreceptors. Photoreceptors respond to very specific wavelengths of light and induce cascades that lead to changes in a plant.

Common Problem Areas

- Understanding how phytochrome shifts from the red to far-red forms frequently gives students difficulty. This is understandable, because phytochromes have given researchers a difficult time for many years.

- The tendency when studying hormones is simply to memorize functions. You need to understand what relative ratios of hormones will do in a plant, not what one individual hormone does.

- It is easy to get overwhelmed with information in this chapter. Chances are the material is completely new to you. Take your time in learning how growth is regulated and the effects of the various hormones.

Study Strategies

- Avoid focusing too much on details. Try to assimilate the big picture of how the environment, receptors, hormones, and genome interact. From there, begin to work toward the details. A big mistake is to jump in and memorize functions of hormones or sequences of development without understanding the broad picture.

- Review the following animated tutorials and activities on the Companion Website/CD:
  Tutorial 37.1 Tropisms
  Tutorial 37.2 Went's Experiment
  Tutorial 37.3 Auxin Affects Cell Walls
  Activity 37.1 Monocot Shoot Development
  Activity 37.2 Eudicot Shoot Development
  Activity 37.3 Events of Seed Germination

Test Yourself

Knowledge and Synthesis Questions

1. Plant growth is ultimately regulated by which of the following?
   a. Environmental cues
   b. Hormones
   c. Signal transduction pathways
   d. A plant's genome
   Textbook Reference: *37.1 How Does Plant Development Proceed? p. 797*

2. Which of the following schemes best represents how environmental cues are transduced to changes in a plant?
   a. Receptors receive environmental cues, a signal transduction pathway is initiated, there is an alteration in the particular genes that are transcribed and translated, and a cellular response is generated.
   b. Receptors are triggered, hormones are released, signal transduction pathways are initiated, there is an alteration in expression of genes, and a cellular response is generated.
   c. None of the above
   d. Both a and b
   Textbook Reference: *37.1 How Does Plant Development Proceed? p. 797*

3. Which of the following triggers germination of a seed?
   a. It imbibes water.
   b. It is released from fruit.
   c. It undergoes chemical changes.
   d. All of the above
   Textbook Reference: *37.1 How Does Plant Development Proceed? p. 800*

4. Which of the following may function to break dormancy in seeds?
   a. Penetration of the seed coat
   b. Leaching of inhibitory compounds by water
   c. Exposure to fire
   d. All of the above
   Textbook Reference: *37.1 How Does Plant Development Proceed? pp. 799–800*

5. Which of the following hormones is responsible for bud break in the spring in deciduous trees?
   a. Auxins
   b. Cytokinins
   c. Gibberellins
   d. Ethylene
   Textbook Reference: *37.2 What Do Gibberellins Do? p. 802*

6. Auxin regulates cell growth by which of the following mechanisms?
   a. Altering the elasticity of cell walls
   b. Altering the plasticity of cell walls
   c. Synthesizing new cell walls
   d. Breaking down cell walls in growing cells
   Textbook Reference: *37.3 What Does Auxin Do? p. 807*

7. Which of the following hormones is responsible for maintaining bud dormancy in deciduous trees?
   a. Auxins
   b. Cytokinins
   c. Gibberellins
   d. Abscisic acid
   Textbook Reference: *37.4 What Do Cytokinins, Ethylene, Abscisic Acid, and Brassinosteroids Do? p. 811*

8. You have installed an outdoor gas-burning grill on your back patio next to your favorite camellia bush. After the first few chilly nights of using your grill, you notice that your camellia, which does not normally lose its leaves, is beginning to do so. Which of the following is the best explanation for what is happening?
   a. The bush is getting too warm next to your grill.
   b. Ethylene is a by-product of the gas you are burning and is causing senescence in your plant.
   c. Abscisic acid is a by-product of the gas you are burning and is causing senescence in your plant.
   d. The plant is a biennial and is bolting.
   Textbook Reference: *37.4 What Do Cytokinins, Ethylene, Abscisic Acid, and Brassinosteroids Do? p. 810*

9. You are slicing a green pepper for a pizza you are making at home. As you slice into it, you notice lots of tiny pepper plants emerging from the seeds of the pepper. This pepper is exhibiting _____ and may be lacking in _____.
   a. parthenocarpy; gibberellins
   b. parthenocarpy; abscisic acid
   c. vivipary; gibberellins
   d. vivipary; abscisic acid
   Textbook Reference: *37.4 What Do Cytokinins, Ethylene, Abscisic Acid, and Brassinosteroids Do? p. 811*

10. Which of the following light receptors is responsible for absorbing blue and ultraviolet light?
    a. Phytochrome P_r
    b. Phytochrome P_{fr}
    c. Cryptochrome
    d. Phototropin
    Textbook Reference: *37.5 How Do Photoreceptors Participate in Plant Growth Regulation? p. 814*

11. Etiolated seedlings are produced by germinating seeds and kept in total darkness. Under which of the following conditions will plants kept in the dark begin to synthesize chlorophyll?
    a. After being given a pulse of blue light
    b. After being given a pulse of red light
    c. After being given a pulse of red light followed by a pulse of far-red light
    d. After being given a pulse of far-red light followed by a pulse of red light
    Textbook Reference: 37.5 How Do Photoreceptors Participate in Plant Growth Regulation? p. 813

12. Ethylene is produced by what part of a plant?
    a. The seedling
    b. The leaves
    c. The fruit
    d. All of the above
    Textbook Reference: 37.4 What Do Cytokinins, Ethylene, Abscisic Acid, and Brassinosteroids Do? p. 810

13. Auxin transport within a plant is said to be _____ and dependent on the action of _____ pumps.
    a. polar; proton
    b. nonpolar; potassium
    c. polar; potassium
    d. bidirectional; proton
    Textbook Reference: 37.3 What Does Auxin Do? p. 805

14. Auxin initiates all the following *except*
    a. stimulation of root initiation.
    b. inhibition of leaf abscission.
    c. stimulation of leaf abscission.
    d. maintenance of apical dominance.
    Textbook Reference: 37.3 What Does Auxin Do? pp. 805–806

15. Red light activates the phytochorome into the P_{fr} state, which leads to which of the following events?
    a. Inhibition of chlorophyll
    b. Leaf expansion
    c. Hook folding
    d. None of the above
    Textbook Reference: 37.5 How Do Photoreceptors Participate in Plant Growth Regulation? p. 813

Application Questions

1. You have planted pea seeds in your garden. Trace the steps that occur between your planting of a seed and the emergence of the pea plant from your garden.
   Textbook Reference: 37.1 How Does Plant Development Proceed? pp. 799–800

2. What hormonal influences are affecting a pea plant from the moment you put it in your garden until its emergence?
   Textbook Reference: 37.1 How Does Plant Development Proceed? pp. 799–800; 37.2 What Do Gibberellins Do? p. 801; 37.4 What Do Cytokinins, Ethylene, Abscisic Acid, and Brassinosteroids Do? pp. 809–811

3. Discuss how gibberellins were discovered. How did researchers determine they were chemical in nature?
   Textbook Reference: 37.2 What Do Gibberellins Do? pp. 802–803

4. Explain how auxin distribution regulates phototropism and gravitropism.
   Textbook Reference: 37.3 What Does Auxin Do? p. 805

5. Auxins and cytokinins appear to cancel out the effects of each other. Why would a hormone that affects bud growth be produced in the roots and one that affects root growth be produced in the shoots? How does this relate to polar distribution of hormones?
   Textbook Reference: 37.3 What Does Auxin Do? p. 804; 37.4 What Do Cytokinins, Ethylene, Abscisic Acid, and Brassinosteroids Do? p. 809

6. Genetic engineering has produced fruits that are deficient in the ability to produce ethylene. How is this deficiency useful in the storage and marketing of fruits?
   Textbook Reference: 37.4 What Do Cytokinins, Ethylene, Abscisic Acid, and Brassinosteroids Do? p. 810

7. You are experimenting in the laboratory with tissue culture methods. You have isolated pith tissue and grown an undifferentiated mass of cells (called a callus). You divide up the tissue and place it in the following culture media.
   a. Necessary nutrients plus indoleacetic acid
   b. Necessary nutrients plus zeatin (a cytokinin)
   c. Necessary nutrients plus equivalent concentrations of indoleacetic acid and zeatin
   d. Necessary nutrients plus an excess of zeatin and minimal indoleacetic acid

   Explain what happens to the mass of tissue under each condition.
   Textbook Reference: 37.4 What Do Cytokinins, Ethylene, Abscisic Acid, and Brassinosteroids Do? p. 809

Answers

Knowledge and Synthesis Answers

1. **d.** Though the environment influences development, final regulation is at the genomic level.

2. **d.** Receptors receive environmental cues and trigger hormone release or a signal transduction pathway that involves hormones. The signal alters gene expression by altering which genes are transcribed and translated. Cellular response depends on expression of those genes.

3. **a.** The uptake of water by a seed begins the processes that lead to seed germination.

4. **d.** Mechanical abrasion, leaching of inhibitors by water, and exposure to fire may all trigger germination. Actual germination cannot begin until a seed imbibes water.

5. **c.** Gibberellins are responsible for bud break in deciduous trees.

6. **b.** Altering plasticity allows for permanent changes in cell wall shape. The cell wall must increase in size for cell growth to occur.

7. **d.** Abscisic acid is responsible for maintaining bud dormancy in winter in deciduous plants.

8. **b.** Ethylene gas promotes senescence and is one of the by-products of burning your gas grill. You should move your grill or your camellia bush.

9. **d.** Vivipary is the germination of seeds before they leave the parent plant and is caused by a deficit of abscisic acid.

10. **c.** Cryptochromes respond to blue and ultraviolet light wavelengths.

11. **c.** The pulse of red light converts P_r to P_{fr}. Subsequent pulses of far-red light stimulate changes that lead to chlorophyll synthesis.

12. **d.** Ethylene can be produced by the fruit, seed, or leaves in most plants.

13. **a.** Auxin transport is a polar process, meaning it moves in only one direction with the help of proton pumps and auxin anion efflux carriers. See Figure 37.9 for further details.

14. **c.** Auxin is involved in all of the processes. Auxin inhibits leaf abscission rather than stimulates it.

15. **b.** The P_{fr} phytochrome stimulates chlorophyll synthesis, hook unfolding, and leaf expansion.

Application Answers

1. Upon planting a pea seed, you watered it, promoting the leaching of germination inhibitors. At this point the seed also began to imbibe water, resulting in metabolic changes. DNA synthesis is halted until the radicle emerges from the seed coat. Breakdown of starch and protein reserves in the cotyledons and the endosperm begins to provide nutrients for the developing embryo. The apical hook begins to push through the soil. Upon its emergence and exposure to light, chlorophyll synthesis begins.

2. Dormancy of the seed is maintained by abscisic acid until it is leached from the seed by water. Once water is imbibed, cytokinins begin to influence germination. Gibberellins assist with the mobilization of storage products to the growing embryo. Ratios of auxins and cytokinins balance root and bud formation. The apical hook is maintained by ethylene.

3. The first gibberellin was isolated from fungus that infects rice plants and causes them to grow tall and spindly. Initial experiments indicated that the medium that the fungus was grown in was sufficient to cause rice plants to be spindly; therefore, the agent that caused the phenomenon was a chemical produced by the fungus.

4. Lateral distribution of auxin controls phototropism and gravitropism. Auxin accumulates in the shaded portions of a stem and stimulates cell growth. This uneven cell growth results in a bending of the stem toward the light. The same mechanism works in response to gravity. An accumulation of auxin occurs where the gravitational pull is the strongest. Cells grow in response to auxin, and stems bend upward, away from the gravitational force.

5. Roots need shoots and vice versa. Increases in roots require increases in shoots for photosynthesis. Regulation of the development of one by the other keeps growth in tandem. Because the distribution of the molecules is polar, a concentration gradient can be established. It is this relative gradient that controls development.

6. Fruits can be kept from ripening until they reach their destination. Once in markets they can be sprayed with ethylene to stimulate ripening. The result is fruit that can be shipped more easily yet can be ripe at any time in the market.

7. a. Roots will develop from the callus.
   b. Buds will develop from the callus.
   c. Both roots and buds will develop from the callus.
   d. Buds will develop from the callus.

CHAPTER 38 Reproduction in Flowering Plants

Important Concepts

Plants may reproduce asexually or sexually.

- Asexual reproduction results in plants with the same genetic makeup as the parent plant. In agriculture, this may be desirable.

- Sexual reproduction is necessary for genetic recombination. This promotes genetic variability which provides the plants with the ability to adapt to their environment.

Sexual reproduction provides for genetic variability.

- Sexual reproduction is advantageous because it provides for greater adaptability within a population; however, it is energy consuming and can break up useful alleles.

- The flower is the basis of sexual reproduction in angiosperms. The basic flower structures include carpels, stamens, petals, and sepals, all of which are modified leaves. The carpel and stamen are the male and female parts of the flower. Review the description of the flower in Chapter 29.

- Plant reproduction involves the concept of alternation of diploid and haploid generations. The flower is produced in the diploid sporophyte generation. The flower produces spores that develop into highly reduced gametophytes and produce gametes. The female gametophytes, called embryo sacs, develop in the megasporangia. Male gametophytes, called pollen grains, develop in microsporangia.

- Within the ovule, a megasporocyte produces four haploid megaspores through meiosis. Only one of these megaspores survives and it divides mitotically to produce eight nuclei within a single large cell. The nuclei migrate to either end of the cell, but two remain in the middle. Cell walls form, isolating the three nuclei at either end into individual cells, whereas the two nuclei in the middle remain together in one cell. At one end, the three cells become two synergid cells and one egg cell; at the other are three antipodal cells, which degenerate. In the large central cell are the two polar nuclei. This seven-celled embryo sac is the gametophyte (see Figure 38.1).

- Pollen grains develop when a microsporocyte undergoes meiosis. All products of meiosis are retained and undergo mitosis to form a two-celled pollen grain composed of the tube cell and the generative cell. Further development is halted until after pollination.

- Pollination of gymnosperms and angiosperms allows for fertilization without water. Pollen grains are carried to female flowers via wind, animals, or other vectors. Some plants are able to self-fertilize within the same flower. Pollen grains must attach to the stigma for fertilization to occur.

- The stigma allows plants to practice "mate selection." Pollen of the right species sticks readily to the stigma via cell-to-cell signaling; however, pollen of other species does not adhere, and germination cannot occur. A single gene, the *S* gene, regulates self-incompatibility and prevents self-fertilization in many plants. Other plants self-fertilize without a problem.

- When a pollen grain germinates, a pollen tube grows down the style toward the embryo sac. Chemical signals in the form of small proteins from synergids within the ovule direct the growth of the pollen tube.

- The pollen grain consists of two cells at the time of pollination—the tube cell and the generative cell. The tube cell controls the growth of the pollen tube. As the tube is growing, the generative cell undergoes meiosis and produces two haploid sperm cells. Once the pollen tube enters the embryo sac, the two sperm cells are released into a synergid that disintegrates and releases the sperm nuclei. One nucleus fuses with the egg cell, the other with the polar nuclei. This results in the zygote ($2n$) and the endosperm ($3n$). All other cells disintegrate.

- Double fertilization resulting in a zygote and the nutritive endosperm is a characteristic feature of angiosperms (see Figure 38.5).

Embryos develop within seeds.

- The success of the embryo depends on its own development, the development of the endosperm, the integuments, and the carpel. Ultimately a seed coat is produced that protects the dormant embryo.

- The zygote divides mitotically to produce the embryo itself and the suspensor. The suspensor pushes the embryo to the endosperm and becomes a route for nutrients. The suspensor continues to divide and produces a filament, whereas the embryo produces a globular structure. This establishes polarity and symmetry in the embryo.

- In eudicots, the embryo progresses through the heart stage and torpedo stage as cotyledons and other organs are developed. The hypocotyl, a shoot apex, and a root apex is also established (see Figure 38.6).

- The endosperm develops and holds nutrient reserves. The cotyledons may absorb the nutrients from the endosperm and increase in size.

- Once development is nearly complete, the seed loses water, and the embryo ceases development. It remains in this state until germination.

- As the embryo and seed are developing, the ovary begins to form the fruit. Other parts of the flower and plant may be included in the fruit, but to be considered a fruit, only the ovary wall and the seed need be involved. The fruit disperses by various means, including attaching to the coats of animals or being consumed and later deposited.

Specific environmental factors signal the onset of reproduction through flower development.

- Flowering may halt, interrupt, or have little effect on plant growth. Seed production is ultimately the purpose of the plant.

- The first transition from vegetative growth to floral production is the alteration of the apical meristem to an inflorescence meristem. The inflorescence meristem can produce bracts and floral meristems. The floral meristem differs from the apical meristem in that growth is determined. The floral meristem is programmed to produce four consecutive whorls of flower organs.

- A gene cascade leads to flower formation. Floral formation begins with the activation of a set of meristem identity genes. The products of the transcription and translation of these genes influence the pattern formation known as cadastral genes, which direct the spatial organization of the floral organs. The expressions of floral organ identity genes specify successive whorls.

- Plants fall within three different life cycle patterns. Annual plants go from seed to seed set and die within one growing season. Biennial plants require one vegetative growing season before reproducing. Perennial plants repeatedly flower and live for many years.

- The gene cascade leading to flower development is controlled by environmental cues. Seasonal flowering is a result of changes in photoperiod, specifically the duration of continuous darkness. Experiments have shown that plants actually respond to the length of the night, rather than amount of daylight. In experiments in which daylight was interrupted, there was little effect on flowering, but in experiments in which dark was interrupted, there were significant effects on flowering. The mechanism controlling this phenomenon is far from understood, however. Each plant type has a critical day length corresponding to light availability that induces flowering.

- Short-day plants flower when the amount of light available is shorter than the critical maximum. Long-day plants flower only when the day is longer (more light available) than the critical minimum. Some plants require complex combinations of day lengths. Some are day-neutral and depend on other factors, such as temperature, to flower.

- Plants, like all other organisms, have an internal mechanism for measuring the length of continuous dark periods. The duration of dark periods appears to be detected by special phytochromes and blue-light receptors.

- Manifestations of an internal biological clock are referred to as circadian rhythms. Circadian rhythms have a duration referred to as a period and an amplitude, or magnitude of change over time. Circadian rhythms have specific characteristics in all organisms. They are not influenced by temperature, they are highly persistent, they can be entrained to coincide with different light–dark cycles, and light exposure causes a phase shift in the rhythm. Plants provide numerous examples of 24-hour circadian rhythms.

- Plants use phytochromes and blue-light receptors to control the biological clock. The biological clock in *Arabidopsis* controls the expression of the *CONSTANS* gene and the build up of the gene product: CO protein. When CO protein levels are high, light absorbed by phytochrome A and the blue-light receptor cryptochrome 2 leads to flowering (see Figure 38.15).

- Florigen is thought to be the hormone responsible for flowering, even though it has yet to be isolated. A series of experiments provided evidence that once a leaf is light induced, it sends a hormonal signal to the rest of the plant to initiate flowering.

- Vernalization is the induction of flowering by low temperatures. The discovery that plants could be induced to flower by exposure to cold has been enormously useful to agriculture.

In certain conditions, asexual reproduction is advantageous to plants' success.

- Asexual reproduction occurs without genetic recombination. During the process of vegetative reproduction, asexual reproduction occurs through the modification of a vegetative organ. Stolons, tip layers, tubers, rhizomes, bulbs, corms, and suckers are all modifications of stems or roots that allow vegetative reproduction.

- Some plants such as dandelions reproduce asexually through seeds in a process called apomixis. Apomictic plants skip over meiosis and fertilization and produce seeds with the identical genetic makeup of the maternal plant. Sometimes apomixis does require pollination and the union of sperm nuclei with polar nuclei to form endosperm, yet no genetic recombination occurs.
- Asexual reproduction is used in agriculture. Cuttings are commonly used in horticulture. Grafting is widely used in fruit crops where the root containing stock is grafted to a scion (shoot system). This allows for a hardy root stock to be combined with a good but often-fragile fruit producer. Tissue culture (production of entire new plants from a small tissue sample) and recombinant DNA technology are leading to daily advances in agriculture.

The Big Picture

- Though plants can reproduce both asexually and sexually, maintenance of genetic variability depends on sexual reproduction. The flower is the basis of sexual reproduction in plants. The flower not only produces the necessary gametes, but it is also integrally involved in ensuring pollination. Eggs are produced through megasporogenesis, and pollen is produced through microsporogenesis. Angiosperms exhibit double fertilization resulting in a diploid embryo and a triploid endosperm.
- Embryo development takes place in the seeds of angiosperms and progresses until cotyledon(s) and the hypocotyl with its apical meristems are developed. At that point, the embryo halts development, the seed desiccates, and dormancy begins until conditions are right for germination. The seed protects the embryo, and in many cases a fruit is formed for protection until maturation and to aid in dispersal of the mature seed.
- Flowering and seed set is triggered by the length of continuous dark exposure. Some plants flower in response to short days, others to long days, and yet others to complex combinations of the two. Day-to-day functions of plants are triggered by photoperiod and circadian rhythms. Plants may flower, set seed, and die in one growing season or two growing seasons, or they may continue to do so for even longer periods of time.

Common Problem Areas

- Megasporogenesis is probably the most difficult concept in this chapter, but it is relatively simple if you look at the source of each cell.
- Recall that flowering, like all other plant processes, is a result of environmental signals, receptors, enzyme cascades mediated by hormones, and alterations in gene expression.

Study Strategies

- Refer to the figures in the textbook to understand flower structure, megasporogenesis, microsporogenesis, and fertilization.
- Spend some time thinking about the advantages and disadvantages of sexual and asexual reproduction. Related questions often show up on exams.
- Be sure you understand the particular features of double fertilization.
- Refer to the following animated tutorials and activity on the Companion Website/CD:
  Tutorial 38.1 Double Fertilization
  Tutorial 38.2 The Effect of Interrupted Days and Nights
  Activity 38.1 Early Development of a Eudicot

Test Yourself

Knowledge and Synthesis Questions

1. You manage a greenhouse that produces roses for Valentine's Day. Roses normally bloom in June. Which of the following will most likely be the best lighting schedule for your roses?
   a. 16 hours of light, 8 hours of interrupted dark
   b. 16 hours of light, 8 hours of uninterrupted dark
   c. 10 hours of light, 14 hours of dark
   d. None of the above
   Textbook Reference: 38.2 What Determines the Transition from the Vegetative to the Flowering State? p. 827

2. After setting the correct photoperiod, you still don't have roses. Which of the following has most likely contributed to the problem?
   a. The heating system is allowing fluctuations in temperature between 20°C and 25°C.
   b. The furnace mechanic accidentally turned off the lights for an hour two days in a row.
   c. The cleaning crew turned the lights on for an hour three nights in a row.
   d. None of the above
   Textbook Reference: 38.2 What Determines the Transition from the Vegetative to the Flowering State? p. 827

3. You have moved into a new house. During the first summer you notice lots of plants that do not bloom. During the second summer your yard is a sea of blooms. It is now spring of the third year, and there are no plants. This can best be explained by which of the following?
   a. Your plants are annuals.
   b. Your plants are biennials.
   c. Your plants are perennials.
   d. Your plants are being affected by drought.
   Textbook Reference: 38.2 What Determines the Transition from the Vegetative to the Flowering State? p. 826

4. You mother gives you a houseplant. You notice it sending out long stems with what look like "little plants" attached. You allow one of these to rest in a cup of water and note that roots form. This is an example of
   a. asexual reproduction.
   b. apomixis.
   c. heterospory.
   d. parthenogenesis.
   Textbook Reference: *38.3 How Do Angiosperms Reproduce Asexually? p. 832*

5. If plants produce pollen and mature eggs at the same time, how are self-pollination and fertilization prevented?
   a. Some plants have self-incompatibility genes.
   b. Some plants rely on physical barriers.
   c. Both a and b
   d. None of the above
   Textbook Reference: *38.1 How Do Angiosperms Reproduce Sexually? p. 822*

6. Most wine grape vines are grafted onto rootstock of another species. How does this help grape yields?
   a. A hardy rootstock can replace a weak rootstock.
   b. A high-producing vine stock can replace a low-producing vine stock.
   c. It allows vintners to select for pest resistance without losing grape quality.
   d. All of the above
   Textbook Reference: *38.3 How Do Angiosperms Reproduce Asexually? p. 833*

7. The induction of flowering by means of exposure to low temperature is called
   a. vernalization.
   b. frigidation.
   c. apomixis.
   d. None of the above
   Textbook Reference: *38. 2 What Determines the Transition from the Vegetative to the Flowering State? p. 831*

8. The cyclic variation in an organism that corresponds to a set light–dark cycle is referred to as an organism's
   a. biorhythm.
   b. circadian rhythm.
   c. period.
   d. photoperiod.
   Textbook Reference: *38. 2 What Determines the Transition from the Vegetative to the Flowering State? p. 828*

9. The production of seeds without fertilization is called
   a. apomixis.
   b. parthenogenesis.
   c. conception.
   d. circadian rhythm.
   Textbook Reference: *38.3 How Do Angiosperms Reproduce Asexually? p. 832*

10. Identify the following structures as occurring in the sporophyte or gametophyte generation.

    a. Embryo sac   _____
    b. Antipodal cells   _____
    c. Polar nuclei   _____
    d. Integument   _____
    e. Receptacle   _____
    f. Ovary   _____
    g. Anther   _____
    h. Pollen grain   _____
    Textbook Reference: *38.1 How Do Angiosperms Reproduce Sexually? p. 821*

11. In the transition from vegetative growth to floral growth the _____ must be transformed to the _____. This involves a shift from _____ growth to _____ growth.
    a. apical meristem; floral meristem; indeterminate; determinate
    b. lateral meristem; floral meristem; indeterminate; determinate
    c. apical meristem; floral meristem; determinate; indeterminate
    d. apical cambium; floral cambium; determinate; indeterminate
    Textbook Reference: *38. 2 What Determines the Transition from the Vegetative to the Flowering State? p. 825*

12. Which of the following best describes the fate of the generative cell of the pollen grain?
    a. It coordinates growth of the pollen tube.
    b. It divides by meiosis to produce two sperm nuclei.
    c. It divides by mitosis to produce two sperm nuclei.
    d. None of the above
    Textbook Reference: *38.1 How Do Angiosperms Reproduce Sexually? p. 822*

13. Which of the following is *not* part of a megagametophyte?
    a. pollen grain
    b. synergids
    c. antipodal cells
    d. All of the above
    Textbook Reference: *38.1 How Do Angiosperms Reproduce Sexually? p. 821*

14. Double fertilization involves all of the following *except* a sperm cell fusing with
    a. an egg cell.
    b. the two polar nuclei.
    c. a pollen grain.
    d. All of the above
    Textbook Reference: *38.1 How Do Angiosperms Reproduce Sexually? p. 822*

15. The shoot apex and the root apex are found during what stage of embryo development?
    a. The heart stage embryo
    b. The egg stage embryo
    c. The inflorescence meristem stage embryo
    d. The torpedo stage embryo
    Textbook Reference: *38.1 How Do Angiosperms Reproduce Sexually? p. 821*

Application Questions

1. Compare and contrast asexual and sexual reproduction in plants. In which category does self-fertilization belong?
   Textbook Reference: 38.1 How Do Angiosperms Reproduce Sexually? p. 819

2. Describe egg formation in angiosperms. Begin with the sporophyte.
   Textbook Reference: 38.1 How Do Angiosperms Reproduce Sexually? p. 820

3. Much effort is spent detasseling corn (removing male flowers). Explain why corn plants cannot be sprayed with a meiosis inhibitor to halt pollen production. (Hint: Male and female flowers occur on the same corn plant.)
   Textbook Reference: 38.1 How Do Angiosperms Reproduce Sexually? p. 821

4. Flowering is stimulated when light sets off a gene cascade. Explain how this may be hormonally controlled.
   Textbook Reference: 38. 2 What Determines the Transition from the Vegetative to the Flowering State? p. 829–830

5. Describe double fertilization. What is the ploidy level of the products of double fertilization?
   Textbook Reference: 38.1 How Do Angiosperms Reproduce Sexually? p. 822

6. When is asexual reproduction beneficial to plants? Under what conditions is sexual reproduction beneficial?
   Textbook Reference: 38.3 How Do Angiosperms Reproduce Asexually? p. 832

7. Define a fruit. Which of the following are fruits: tomato, pear, potato, banana, cucumber, snow pea, peanut, sunflower seed?
   Textbook Reference: 38. 2 What Determines the Transition from the Vegetative to the Flowering State? p. 825

Answers

Knowledge and Synthesis Answers

1. **b.** Flowering is regulated by darkness. Interruptions in darkness can prevent flowering.

2. **c.** The interruptions in the dark cycle by the cleaning crew have affected the signals that tell the plant to begin flowering.

3. **b.** The plants are most likely biennials that spend one season growing vegetatively and one season producing flowers before dying.

4. **a.** The runner (stolon) is asexually propagating this plant.

5. **c.** Some plants have self-incompatibility genes that prevent the growth of the pollen tube through the style. Others have mechanical barriers to their own pollen.

6. **d.** Grafting allows selection of hardy rootstock, well-producing vine stock, and disease resistance.

7. **a.** Vernalization is noted in winter wheat and is used agriculturally to grow high-yielding wheat.

8. **b.** Circadian rhythms roughly match a 24-hour cycle and can be reset by altering light and dark cycles.

9. **a.** Dandelions and other plants produce seeds without meiosis and fertilization (called apomixis). These seeds are genetically identical to the parent plant.

10. **a.** Embryo sac   gametophyte
    b. Antipodal cells   gametophyte
    c. Polar nuclei   gametophyte
    d. Integument   sporophyte
    e. Receptacle   sporophyte
    f. Ovary   sporophyte
    g. Anther   sporophyte
    h. Pollen grain   gametophyte

11. **a.** Apical meristems and floral meristems differ because growth from the apical meristem is indeterminate and growth from the floral meristem is determinate, leading to four whorls of floral structures.

12. **b.** The two sperm nuclei used in double fertilization are derived from the generative cell of the pollen grain.

13. **a.** The megagametophyte is the female gametophyte and is initially called the embryo sac and contains three antipodal cells and two synergid cells at the seven-cell stage.

14. **b.** Double fertilization involves the fertilization of the egg cell and the two polar nuclei by two sperm cells.

15. **d.** The torpedo stage embryo has the shoot apex and the root apex on either end of the hypocotyl.

Application Answers

1. Asexual reproduction does not involve meiosis, fertilization, or genetic recombination. The offspring from asexual reproduction are genetically identical to the parent plant. Sexual reproduction requires a meiotic event and fertilization. Self-fertilization is sexual reproduction, even though genetic recombination is limited. This is because a meiotic event occurs, and fertilization is necessary for reproduction to take place.

2. See Figure 38.1 in the textbook.

3. Meiosis occurs in the megasporocyte as well as the microsporocyte. Inhibition of pollen formation via a meiosis inhibitor will also inhibit egg formation.

4. Light induction begins with the leaves. A single isolated leaf may stimulate floral production throughout the plant. In experiments in which leaves from one plant were grafted on other plants, an "induced" leaf caused a plant to flower even though the plant had not been exposed to the required amount of darkness. Therefore, the leaf must send a signal (most likely a hormone) that begins the gene cascade leading to flower formation.

5. Double fertilization occurs when the two sperm nuclei produced from the generative cell of the pollen grain unite with the egg nucleus and two polar nuclei respectively. The resulting zygote is diploid and the endosperm is triploid.

6. Asexual reproduction is beneficial in stable environments in which many genetically identical plants are sustainable. This process can help to colonize a habitat or a population of plants to spread. The disadvantage is lack of genetic diversity. Lack of diversity can be detrimental if the environment changes rapidly, requiring adaptability of the population.

7. A fruit is the seed, the ovary wall, and may include other structures of a flowering plant. With the exception of the potato, all of the listed structures are fruits.

CHAPTER 39 Plant Responses to Environmental Challenges

Important Concepts

Plants and their pathogens interact, each giving clues to the other.

- The action of a pathogen signals a plant to mount a defense. In turn, the plant's defense signals the pathogen. These signals go back and forth until one or the other "wins."

- The first plant defense system is nonspecific; an attempt to prevent entry by pathogens. Cutin, suberin, and waxes cover the cork and epidermis of the plant to exclude pathogens. Plants cannot repair damaged and invaded tissues, so they seal them off to prevent systemic infection.

- Upon recognition of a pathogen, polysaccharides are deposited in the cell wall and seal off the plasmodesmata to form a barrier between cells. Subsequently, lignin is deposited, which reinforces the barrier and acts as a pathogen toxin.

- Once invaded, plants produce chemical defenses against pathogens. Phytoalexins are nonspecific antifungal and antibacterial compounds produced by the plant. Pathogenesis-related proteins (PR proteins) are enzymes that function to break down the walls of pathogens or serve as signaling molecules.

- In the hypersensitive response, pathogen-containing tissue and surrounding tissues die on infection and become necrotic tissue. As these cells are dying, they release phytoalexins and salicylic acid, which is a relative of aspirin. These compounds "arm" the plant against subsequent invaders.

- Systemic acquired resistance protects a plant against further infection. This is due to the stimulation of PR protein production after an infection event. Salicylic acid produced at the site of an infection is transported to other parts of the plant to signal the presence of an invader. Plants also release methyl salicylate, which can become airborne and serve as a signal to other plant parts or to neighboring plants to mount a defense via PR proteins.

- Gene-for-gene resistance is a highly specific type of resistance and depends on a plant's having an allele for a gene that matches an allele in the pathogen. Dominant *R* genes code for resistance in plants and dominant *Avr* genes in the pathogen result in less-virulent strains of the pathogen by coding for a substance that elicits a defensive response in the plant. Though this mechanism is not completely understood, the gene product interactions typically lead to the hypersensitive response.

- Plants can use interference RNA (RNAi, also known as posttranscriptional gene silencing) to respond to attack by RNA viruses. A plant produces small pieces of short interfering RNA (siRNA) from interactions with the viral RNA. These siRNA degrade the viral mRNA.

Herbivory confers advantages and disadvantages to plants.

- Some plants depend on the grazing of herbivores to increase productivity due to coevolution of the plant and the herbivore. Removal of some leaves triggers increased photosynthesis in the remaining leaves and a better partitioning of resources. Grazing also allows light to reach young leaves in plants such as grasses. Grazing can stimulate some plants to produce leaves longer and flower more abundantly.

- For the majority of plants, grazing is a problem; therefore, defenses against grazing exist. Plants produce specific secondary metabolites to use as a defense from grazers. The effects of these chemicals are diverse, ranging from neurotoxins to hormone mimics. Table 39.1 provides examples of the different types of secondary metabolites in plants.

- Canavanine is a defensive secondary metabolite that plays multiple roles in a plant. It assists with nitrogen storage in the seed, and because it is similar to arginine, it is incorporated into herbivore proteins after consumption. The failure to distinguish between arginine and canavanine during protein synthesis alters the tertiary structure of proteins and can be lethal to the organism consuming the plant. Some organisms have the ability to distinguish between the two and are not harmed by canavanine.

- Defense is the result of a series of chemical signaling events. Figure 39.7 details the events leading from wounding to production of an insecticide. Wounding causes the release of systemin, which travels to other parts of a plant and induces membrane breakdown. By-products of this breakdown are jasmonates, which enter the nucleus and activate production of a protease inhibitor that inhibits digestion in the predator.

- Some plants produce products that make them resistant to attack. These chemicals can be isolated and the genes producing them identified. Resistance can be bred into plants or transferred using recombinant DNA technology.

Plants use a variety of strategies to prevent harming themselves with their own defensive mechanisms.

- Plants protect themselves from their own defenses by isolating toxic secondary metabolites into compartments. Water-soluble toxins are stored in vacuoles; hydrophobic poisons are stored in laticifers along with latex, whereas others are dissolved in the waxes covering the epidermis.

- Some toxic substances are produced only in already-damaged or dying tissue. The precursors of toxic material are stored in different compartments than the enzymes that convert them into active poison, and these combine only once the cell is damaged.

- In plants that use the canavanine amino acid in defense, the plant's enzymes are able to distinguish between canavanine and arginine.

- Insects sometimes find ways to get around the mechanisms used by plants to deter them. One example is milkweed and a beetle that feeds on its leaves.

Some plants must be able to cope with limited or excess water availability.

- Some plants, called xerophytes, are specifically adapted to dry areas. They survive in these areas by completing their entire life cycle in the short period that water is available. Others may have special modifications such as thickened cuticles, epidermal hairs, and stomatal crypts to prevent water loss. Succulents have water-storing leaves, others drop leaves during drought, and yet others have water-storing stems and highly reduced leaves.

- Roots can also be modified for drought conditions. Long taproots can reach water supplies far underground. Shallow, fibrous root systems that grow only during rainy seasons are also adapted to desert growth. Xerophytic plants are able to extract more water when it is available by changing the osmotic and water potential of their root cells. They do this by storing proline in their vacuoles.

- Too much water can be as dangerous to plant survival as too little. Roots submerged in water do not get adequate oxygen to sustain respiration. Some plants that grow in standing water overcome this by producing pneumatophores, which are root extensions that grow above water and provide oxygen to the entire root system. Others rely on fermentation and thus grow very slowly. Yet others have leaf modifications called aerenchyma for buoyancy and oxygen storage.

Temperature extremes pose yet another stress to plant survival.

- High temperatures denature proteins and destabilize membranes. Cold temperatures decrease membrane fluidity, and freezing causes rupturing of membranes if ice crystals form.

- Transpiration provides evaporative cooling to plants in high temperatures, but at the expense of water loss. Adaptations to high temperatures are similar to adaptations to xerophytic conditions. Plants produce heat shock proteins that help stabilize protein structure against denaturation.

- Plants can adjust to cold through a process of cold-hardening, which involves repeated exposure to cool nondamaging temperatures. This alters the saturated and unsaturated fatty acid composition of membranes allowing them to remain fluid and makes them more resistant to rupture. Heat shock proteins are also produced in response to cold temperatures. Some plants have antifreeze compounds that prevent ice crystal formation.

Some plants are adapted to survive in saline environments.

- Halophytes (salt-loving plants) are adapted to saline environments. They are the only plants that accumulate sodium and chloride ions, which they store in vacuoles in leaves. The increased salt concentration of the halophyte means it has a more negative water potential.

- Some plants are able to excrete salt so that it does not reach toxic proportions. Salt glands move salt to the leaf surface, where it can be lost to wind or rain. Salt glands on the leaf assist with water procurement from the roots and with reduction of water loss to evaporation.

- Many modifications, such as succulence, accumulation of praline, and crassulacean acid metabolism, allow for survival in both drought and saline conditions.

Heavy metal accumulation in soils is detrimental to most plants.

- Chromium, mercury, lead, and cadmium are poisonous to most plants. Some geographic areas are rich in these metals as a result of normal geological processes, and others have been contaminated by human activity.

- Plants living in these areas have adapted to allow them to accumulate these heavy metals. These plants are important in bioremediation and for cleanup of these areas. The exact mechanism of survival of plants in these areas is not known.

The Big Picture

- Plants must respond to invasion by pathogens, physical damage by natural events and herbivory, variable water supplies, temperature fluctuations, and variable soil conditions. Natural selection has made plants uniquely suited to the areas they occupy. Some thrive in very harsh environments, whereas others cannot.

- Interactions between plants and pathogens stimulate a series of chemical changes that ward off further infection. Plant strategies to isolate pathogens include sealing plasmodesmata, releasing proteins that interact with pathogens, and signaling other parts of the plant.

- Herbivory can be advantageous for plants adapted to grazing or detrimental to those that are not. In the latter case, a plant produces chemicals to prevent herbivory and mounts defenses around the damaged tissues.

- Specific adaptations allow plants to withstand adverse conditions. Saline environments, water loss, and high temperatures can be survived if water uptake is maximized and water loss is minimized. Sunken stomata, reduction of surface area, and increased water potential of cells are adaptations to these conditions. Extremes in temperature are survived with heat shock proteins that stabilize metabolic proteins during temperature extremes.

Common Problem Areas

- Gene-for-gene resistance can be difficult to understand, but remember that in any pathogen situation, the pathogen and the host are continually interacting; therefore, over time, interplay among the genomes has been selected for.

- There are many examples of the ways plants deal with pathogens and herbivory. This chapter provides examples, but these are not exhaustive.

Study Strategies

- The best strategy is to break this chapter up into components and study them one at a time so that all of the defenses/modifications do not run together. Focus first on pathogen interaction, then herbivory resistance, and so on.

- Think about how form follows function as you look at how plants are adapted to specific harsh environments.

- Review the following animated tutorial and activity on the Companion Website/CD:
  Tutorial 39.1 Signaling Between Plants and Pathogens
  Activity 39.1 Concept Matching

Test Yourself

Knowledge and Synthesis Questions

1. Primary nonspecific defensive strategies in plants include which of the following?
   a. PR proteins
   b. Suberin
   c. Cutin
   d. Both b and c
   Textbook Reference: 39.1 How Do Plants Deal with Pathogens? p. 836

2. Plants exhibit systemic acquired resistance much the same way people acquire resistance to pathogens; however, the mechanism is quite different. Which of the following does *not* have a role in acquired resistance in plants?
   a. Salicylic acid
   b. *R* genes
   c. Methyl salicylate
   d. All of the above
   Textbook Reference: 39.1 How Do Plants Deal with Pathogens? p. 839

3. Gene-for-gene resistances depend on which of the following?
   a. Compatible alleles in plant and pathogen
   b. Incompatible alleles in plant and pathogen
   c. Recessive *Avr* genes
   d. Recessive *R* genes
   Textbook Reference: 39.1 How Do Plants Deal with Pathogens? p. 839

4. Upon infection by a pathogen, plant cells increase synthesis of polysaccharides. The function of these is to
   a. destabilizes the cell walls.
   b. synthesize antibodies to the pathogen.
   c. isolate the pathogen in the invaded tissue.
   d. break down the wax barrier.
   Textbook Reference: 39.1 How Do Plants Deal with Pathogens? p. 836

5. Canavanine is toxic to many herbivores but not to plants. Which of the following can best explain this differential toxicity?
   a. Canavanine is confused with arginine in plants.
   b. Canavanine is incorporated into proteins and causes them to fold properly.
   c. Plants are able to differentiate between arginine and canavanine.
   d. All of the above
   Textbook Reference: 39.2 How Do Plants Deal with Herbivores? p. 841

6. Which of the following best describes how plants produce their own insecticide?
   a. Wounded cells release systemin, systemin directly induces synthesis of protease inhibitors, protease inhibitors act as insecticides.
   b. Wounded cells release jasmonates, jasmonates stimulate systemin synthesis, systemin causes the production of protease inhibitors, protease inhibitors act as insecticides.
   c. Wounded cells release systemin, systemin causes membrane breakdown, membrane breakdown releases jasmonates, jasmonates induce synthesis of protease inhibitors, protease inhibitors act as insecticides.

d. Wounded cells release systemin, systemin causes membrane breakdown, membrane breakdown releases jasmonates, jasmonates act as insecticides.
Textbook Reference: 39.2 How Do Plants Deal with Herbivores? p. 842

7. Which of the following is (are) strategies for coping with drought conditions?
   a. Water-storing tissues
   b. Leaf loss during drought
   c. Sunken stomata
   d. All of the above
   Textbook Reference: 39.3 How do Plants Deal with Climate Extremes? pp. 845–846

8. Halophytes are different from all other types of plants in that they
   a. can accumulate sodium and chloride ions.
   b. have a positive water potential.
   c. contain no sodium and chloride ions.
   d. All of the above
   Textbook Reference: 39.4 How Do Plants Deal with Salt and Heavy Metals? p. 848

9. Which of the following conditions stimulates the production of heat shock proteins?
   a. Abnormally high temperatures only
   b. Abnormally low temperatures only
   c. Both abnormally high or abnormally low temperatures
   d. Heat shock proteins are continually available in a plant.
   Textbook Reference: 39.4 How Do Plants Deal with Salt and Heavy Metals? p. 847

10. The main function of heat shock protein is to
    a. stabilize proteins necessary to a cell's survival.
    b. reinforce membranes that lose fluidity.
    c. cause the plant to enter dormancy.
    d. act as an antifreeze compound.
    Textbook Reference: 39.4 How Do Plants Deal with Salt and Heavy Metals? p. 847

11. Cold-hardening involves all of the following *except*
    a. production of antifreeze proteins.
    b. increased production of saturated fatty acids.
    c. production of phytoalexins.
    d. production of heat shock proteins.
    Textbook Reference: 39.4 How Do Plants Deal with Salt and Heavy Metals? p. 847

12. What keeps plants that produce toxins for defense from poisoning themselves?
    a. The toxic substances are kept throughout all plant tissues.
    b. The toxic substances are stored in laticifers.
    c. The toxic substances are produced by every cell in the plant all the time.
    d. The toxic substances are stored in the same place as enzymes that convert them to the active form.
    Textbook Reference: 39.2 How Do Plants Deal with Herbivores? p. 844

13. Which of the following is *not* a secondary plant metabolite?
    a. Phenolics
    b. Alkaloids
    c. Terpenes
    d. PR proteins
    Textbook Reference: 39.1 How Do Plants Deal with Pathogens? p. 839, Table 39.1

14. The hypersensitive response involves the release of
    a. salicylic acid.
    b. PR proteins.
    c. *R* genes.
    d. phytoalexins.
    Textbook Reference: 39.1 How Do Plants Deal with Pathogens? p. 838

15. A plants immune response to RNA viruses involves
    a. heat shock proteins.
    b. RNA interference.
    c. small interfering RNA.
    d. All of the above
    Textbook Reference: 39.1 How Do Plants Deal with Pathogens? p. 840

Application Questions

1. Describe strategies used by plants to prevent herbivory.
   Textbook Reference: 39.2 How Do Plants Deal with Herbivores? pp. 840–843

2. Explain what is meant by coevolution of plants and herbivores. How can grazing increase the productivity of some plants?
   Textbook Reference: 39.2 How Do Plants Deal with Herbivores? pp. 840–841

3. It has long been a practice in rural areas to nail fencing to trees. Describe how the act of nailing a fence to a tree trunk can introduce pathogens into the tree. What measures does the tree take to ward off disease?
   Textbook Reference: 39.1 How Do Plants Deal with Pathogens? pp. 836, 838

4. At a T-intersection in a small town in the snowbelt region of Pennsylvania, a local business plants several ornamental shrubs. During the winter, snowplows push snow from the intersection around the shrubs. In an attempt to save the shrubs, the business covers the shrubs with burlap. Will this save the shrubs? What conditions in the intersection are likely to cause the greatest damage to the shrubs?
   Textbook Reference: 39.4 How Do Plants Deal with Salt and Heavy Metals? p. 848

5. Bioremediation of mine and ore refinery sites is an area of intense study. How can plants help with the cleanup of sites contaminated with heavy metals?
   Textbook Reference: 39.4 How Do Plants Deal with Salt and Heavy Metals? p. 849

6. Many plants produce toxins that damage eukaryotic predators. Considering that the cell structure of both the predator and the plant is similar, why is the plant not affected by its own toxins?

Textbook Reference: 39.2 How Do Plants Deal with Herbivores? p. 844

7. Pansies are planted in the winter in the southeastern United States. To ensure that the plants are ready to enter cool ground, nurseries cold-harden the plants for several days. Describe this process and explain why it is done.
Textbook Reference: 39.3 How do Plants Deal with Climate Extremes? p. 847

Answers

Knowledge and Synthesis Answers

1. **d.** Cutin and suberin, along with waxes, aid in preventing entry of pathogens into a plant by forming a barrier.

2. **b.** Plants that have been exposed to pathogens wall off invaded tissue. This tissue releases salicylic acid to the rest of the plant, which stimulates PR proteins. The tissue also releases airborne methyl salicylate, which causes release of PR proteins in distant regions of the plant and even in neighboring plants.

3. **a.** The plant must have an *R* allele, and the pathogen must have an *Avr* allele. Compatibility between these alleles leads to resistance.

4. **c.** Polysaccharides seal plasmodesmata and reinforce cell walls. The purpose of this is to contain the pathogen in a minimal number of cells.

5. **c.** Canavanine's mechanism of action is that it is taken up by the predator and incorporated into its protein structures. Because the structure of canavanine is different from that of the arginine it replaces, the predator's proteins take up a structure that is lethal to the predator. The plant itself and some predators have enzyme mechanisms to distinguish canavanine from arginine and prevent its incorporation into proteins.

6. **c.** See Figure 39.5 in the textbook.

7. **d.** Water storage tissues allow the plant to take advantage of any water that is available and hold on to it for lean times. Leaf loss, when photosynthesis cannot be sustained, prevents loss of resources. Sunken stomata prevent excessive water loss to evaporation.

8. **a.** They can accumulate sodium and chloride ions. This makes their water potential more negative, and they are able to extract more water from their surroundings than plants that do not store the ions.

9. **c.** Heat shock proteins are produced in response to extreme high and low temperatures.

10. **a.** Heat shock proteins are chaperonin proteins. They function to stabilize and prevent denaturation of essential proteins.

11. **c.** Cold-hardening involves production of antifreeze proteins, heat shock proteins, and increase in saturated fatty acids in the membrane. The production of phytoalexins is involved in chemical defenses against pathogens.

12. **b.** Plants that produce toxins as a defense typically store them in special locations such as laticifers where they won't damage tissue.

13. **d.** Alkaloids, phenolics, and terpenes are all examples of secondary plant metabolites. PR proteins are proteins that function in defense.

14. **a.** The hypersensitive response involves the release of salicylic acid.

15. **c.** Plants that are invaded by RNA viruses use the RNA of the invading virus to interfere with and block viral replication.

Application Answers

1. Plants prevent herbivory by producing secondary compounds that affect potential herbivores. These compounds may act as neurotoxins, inhibit digestion, or have a whole host of other consequences (see Table 39.1).

2. Coevolution of herbivores and plants occurs when the herbivore has resistance to toxins that the plant may produce, and the plant develops strategies to cope with herbivory. In some cases, regular grazing increases photosynthetic productivity of the plant by reducing shading and allowing better partitioning of resources.

3. Nailed fencing on trees penetrates the nonspecific primary defenses of a tree. In response to pathogens introduced by the nail, the tree begins to isolate the affected tissues. Polysaccharides are synthesized to plug plasmodesmata, and lignin reinforces the cell walls and acts as a toxin to the invading pathogens. Phytoalexins and PR proteins are released, and salicylic acid also may be released. The tissue around the nail begins to die, and pathogens are isolated.

4. The burlap will most likely do little for the shrubs. The roots are in most danger. Because streets are heavily salted in the winter, the snow around them creates a saline environment. If the shrubs are not adapted to a saline environment, little will help them survive the salt accumulation around their roots.

5. Only plants that have evolved a strategy for taking up and coping with heavy metals can grow in these sites. By isolating and specifically growing these types of plants, the slow process of removing the metals from the soil can begin.

6. Plants may isolate the toxins into vacuoles within their cells so that toxins do not come into contact with the active machinery of the cells. Plants may restrict toxin production to already-damaged tissue. Plants may also develop enzymes that differentiate toxins and render them harmless.

7. Cold-hardening is gradual exposure to adverse temperatures. This allows for structural changes in cell membranes to help them cope with extreme temperatures. It also allows for the production of heat shock proteins and other compounds necessary for survival at temperature extremes.

CHAPTER 40 Physiology, Homeostasis, and Temperature Regulation

Important Concepts

Control of the internal environment is necessary for proper cell function.

- Due to their small size, single-celled organisms can receive all their required nutrients from the surrounding media. Small multicellular organisms, in which all the cells are only a few layers from the environment, can receive nutrients from the environment without the help of specialized cells. Larger multicellular organisms can maintain an internal environment that differs from the external environment. Specialized cells and groups of cells help to control the makeup of the internal environment and maintain the differences between the internal and external environments (see Figure 40.1).

- Homeostasis is the state of having a constant, regulated internal environment, and homeostatic mechanisms help maintain this state.

Control systems help to regulate physiology and maintain homeostasis.

- Physiological regulatory systems require both a set point at which the parameter is to be held and feedback that provides information about the state of the system (see Figure 40.2). Differences between the set point and the feedback information indicate to the regulatory mechanism which corrective measures are required to restore a parameter to its set point.

- Physiological control systems must take information from the regulator systems and effect changes. Negative feedback results in a slowdown or reversal of a process, returning the variable controlled by that process back toward its set point or regulated level.

- In contrast to negative feedback controls, positive feedback results in amplification of a response. Regulation by positive feedback is less common than regulation by negative feedback.

- The set point maintained by an organism for a given physiological process can also be changed by feedforward information.

Groups of cells are organized as tissues, organs, and organ systems.

- A group of similar cells with the same form and function constitutes a tissue. There are four types of tissues: epithelial, connective, muscle, and nervous.

- Epithelial tissues make up the lining of the inner and outer body surfaces, such as the lining of the various organs and the skin. Some epithelial cells have secretory functions, whereas others have cilia to help move substances over surfaces. Still other epithelial cells function in protection, absorption, or the provision of information to the nervous system.

- Connective tissue consists of an extracellular matrix that holds together a dispersed group of cells. The protein collagen makes up the bulk of the extracellular matrix. Cartilage, bone, adipose tissue, and blood are all connective tissues.

- Muscle tissue is either skeletal, smooth, or cardiac. Cells of muscle tissue contract to generate force and movement.

- Neurons and glial cells are the two types of cells found in nervous tissue. Neurons generate and conduct electrochemical signals. Glial cells support and protect neurons.

- Organs are groups of tissues that carry out a specific function in the body. Most organs are composed of all four of the major tissue types. The integration of a number of organs to carry out a specific function is known as an organ system, such as the digestive system.

Temperature plays an important role in life.

- Most living cells function only between a temperature of 0°C and 40°C, with most organisms having much narrower limits in their range of survival temperatures. For most cells, the upper limit for survival is 45°C.

- All physiological processes in organisms are sensitive to temperature. Increases in temperature usually increase the rate of a given process. The Q_{10} is a measure of change in a physiological process as the temperature increases or decreases by 10°C (see Figure 40.8). Most biological Q_{10} values range from 2 to 3. A Q_{10} of 2 means that the reaction rate doubles when temperature increases by 10°C.

- Acclimatization is the change in a physiological process that occurs over time in response to a change in the environment. For example, a fish experimentally exposed in summer to decreasing water temperatures will show slower physiological functions. However, the physiological functions of the same fish in winter will not be as slow as they were when temperatures were experimentally reduced in summer. The physiology of the fish has acclimatized to seasonal changes in water temperature. Fish may catalyze reactions with one set of enzymes in the summer and a different set of enzymes in the winter. Acclimatization typically results in an enhanced probability of thriving in the new environmental conditions.

Animals can be classified as ectotherms or endotherms.

- Animals can be classified into two main groups based on their source of heat. Endotherms regulate their body temperature with internal heat production by means of a high metabolic rate and effective insulation. Ectotherms have a low metabolism and little insulation and must rely on the environment to provide body heat.

- Ectotherms have body temperatures that track environmental temperatures. In contrast, endotherms maintain a constant body temperature even as the environmental temperature changes (see Figure 40.9A). A hibernating mammal is an example of a heterotherm, an animal that behaves sometimes as an endotherm (in this case, in summer) and sometimes as an ectotherm (in this case, in winter).

- Both endotherms and ectotherms use behavioral mechanisms to regulate body temperature. Ectotherms shuttle between warm and cool environments during the day to maintain body temperature within a given range (see Figure 40.10A). Endotherms also use behavioral thermoregulation as a first line of thermoregulation (see Figure 40.10B).

- There are a number of routes through which heat can be gained or lost to the environment (see Figure 40.11). Evaporation from the skin or respiratory tract cools the body. Solar radiation from the sun heats the body. Thermal radiation, convection, and conduction can all either warm or cool the body, depending on the thermal environment.

- The total balance of heat production and heat exchange for an animal is its energy budget. To maintain a constant body temperature, the heat entering an animal must equal the heat leaving.

- Both ectotherms and endotherms use changes in blood flow to control body temperature. Increases in blood flow result in increased heat exchange with the environment, whereas decreases in blood flow result in decreased heat exchange (see Figure 40.12).

- Some large fishes have the ability to maintain a body temperature higher than the surrounding water. They use special vascular countercurrent heat exchangers that retain metabolically produced heat in blood that remains in the core (see Figure 40.13).

- A number of ectothermic insects are able to generate heat to raise their body temperature. Some produce heat by contracting their flight muscles. Others, such as honeybees, engage in group thermoregulation whereby individuals form clusters in winter and regulate temperature by adjusting their own metabolic heat production and density of the cluster.

Endotherms have special adaptations to help in controlling body temperature.

- The metabolic rate of an organism is the total energy used by the animal and is usually measured as oxygen consumption.

- For endotherms, the basal metabolic rate per gram of tissue decreases as animals get larger (see Figure 40.15). The reason for this pattern is unclear.

- The range of environmental temperatures at which the metabolic rate of endotherms does not change is known as the thermoneutral zone (see Figure 40.16). The upper and lower critical temperatures are the environmental temperatures at which metabolic rate begins to increase. Increases in heat production occur by nonshivering and shivering thermogenesis when the environmental temperature is below the lower critical temperature. Nonshivering thermogenesis involves a specialized adipose tissue, brown fat (see Figure 40.17). Metabolism increases when environmental temperature is above the upper critical temperature because the animal must expend energy to cool down through panting or sweating.

- Animals that live in hot areas display increases in surface area to promote heat exchange with the environment (see Figure 40.18A). Decreases in surface area are seen in animals that live in cold climates (see Figure 40.18B). This decrease in surface area decreases the area over which heat can be lost.

- Fur and feathers provide insulation in mammals and birds experiencing cold environments. These structures help to retain heat produced in the body and help maintain a constant body temperature.

- Evaporative cooling provides an avenue for heat loss to the environment by sweating or panting.

The hypothalamus is the vertebrate thermostat.

- The hypothalamus, a structure at the bottom of the brain, is involved in the control of homeostasis for many regulatory systems, including thermoregulation. The hypothalamus has a set point temperature, and changes in temperature of the hypothalamus result in thermoregulatory responses to offset any temperature change (see Figure 40.19). The hypothalamus also integrates thermal information received from the skin.

- In mammals, hypothalamic set points are different at different environmental temperatures. Other factors, including the sleep–wake cycle and level of activity, also influence set points.

- The body has the ability to increase the set point temperature in response to infections, resulting in fever. A fever is a rise in body temperature in response to compounds called pyrogens. Some pyrogens come from bacteria or viruses that invade the body, whereas others are produced by cells of the immune system when challenged.

- Hypothermia is the general state of below-normal body temperature. Some animals are able to decrease body temperature either on a daily scale (daily torpor) or over longer time periods (hibernation) (see Figure 40.20). Daily torpor and hibernation are examples of regulated hypothermia that allow animals to survive during periods of low temperature and scarce food.

The Big Picture

- One of the most important roles of tissues, organs, and organ systems is to help maintain homeostasis within the body's cells. Only by maintaining relatively constant intracellular conditions can basic metabolic reactions continue.

- Animals can either generate their own body heat (endotherms) or rely on the environment to determine body temperature (ectotherms). Endotherms typically have much greater insulation than ectotherms and higher metabolic rates. The metabolic rates of endotherms and ectotherms respond differently to changes in environmental temperature. Endotherms increase metabolism as environmental temperature decreases; they also increase metabolic rate at high temperatures when energy is needed for sweating and panting. The metabolic rate of ectotherms is temperature-dependent and falls with decreases in environmental temperature.

- Control of body temperature can occur by behavioral and physiological mechanisms. Both endothermic and ectothermic vertebrates use the brain's hypothalamus as the thermostat for temperature regulation.

Common Problem Areas

- Homeostasis does not necessarily mean holding every body variable absolutely constant, but rather making sure that body variables are held at the correct level for the circumstances. Even body temperature in humans, which we think of as being a constant 37°C, drops slightly in the early morning hours and rises considerably during extended vigorous exercise.

- Understanding the difference between an ectotherm and an endotherm can be difficult. These terms refer to the *source* of heat that determines an animal's temperature. Whereas body temperatures of ectotherms are determined primarily by external sources of heat, those of endotherms are determined by heat generated metabolically.

Study Strategies

- Refer to Figure 40.2 to understand how homeostasis is maintained through regulatory systems.

- Review the organization of physiological systems from cells to tissues to organs to organ systems. This will help you when you study each of the organ systems in later chapters.

- Review the following animated tutorial and activity on the Companion Website/CD:
Tutorial 40.1 The Hypothalamus: The Body's Thermostat
Activity 40.1 Thermoregulation in an Endotherm

Test Yourself

Knowledge and Synthesis Questions

1. Which of the following contributes to the proper maintenance of homeostasis in the bodies of animals?
   a. pH buffering of the blood
   b. Control of blood flow to the skin
   c. Production of hormones
   d. All of the above
   Textbook Reference: *40.1 Why Must Animals Regulate Their Internal Environments? pp. 855–856*

2. In an environment with an ambient temperature lower than an animal's core body temperature, which of the following would be an *inappropriate* physiological or behavioral response if the animal needed to eliminate excess body heat?
   a. Wallowing in a pool of water
   b. Preventing blood flow to peripheral vessels
   c. Sweating
   d. All of the above
   Textbook Reference: *40.3 How Do Animals Alter Their Heat Exchange with the Environment? pp. 862–864*

3. A lizard lives in a desert environment where the temperature at night is low and during the day is high. What might this animal do to maintain the highest stable body temperature?
   a. Stay in a burrow during the night and shuttle between the sun and shade on the surface during the day
   b. Stay on the surface during the night and move to a burrow during the day
   c. Increase metabolism and heat production during the night and decrease it during the day
   d. None of the above
   Textbook Reference: *40.3 How Do Animals Alter Their Heat Exchange with the Environment? pp. 862–863*

4. On a warm day your dog has been running around fetching his Frisbee and returning it to you. You notice that he is panting heavily. Which of the following does *not* describe the condition of your dog?
   a. Your dog is using evaporative cooling to lower core body temperature.
   b. Your dog is operating at a basal metabolic rate.
   c. Your dog is at the upper end of its thermoneutral zone.
   d. All of the above
   Textbook Reference: *40.4 How Do Mammals Regulate Their Body Temperatures? pp. 866–868*

5. Evaporative cooling is an effective way to increase heat loss, but carries with it the physiological drawback of
   a. increased use of ATP and substantial water loss.
   b. lowering the hypothalamic thermal set point.
   c. decreased use of ATP and substantial water gain.
   d. exhaustion of the supply of brown fat.
   Textbook Reference: *40.4 How Do Mammals Regulate Their Body Temperatures? p. 868*

6. Fever
   a. is a higher than normal body temperature that is always dangerous.
   b. decreases the metabolic rate of the body to conserve energy.
   c. is regulated by chemicals that reset the thermostat of the body to a higher set point.
   d. causes the liver to release large amounts of calcium, which seems to inhibit bacterial growth.
   Textbook Reference: *40.4 How Do Mammals Regulate Their Body Temperatures? pp. 869–870*

7. Which of the following is a characteristic unique to connective tissue?
   a. Highly modified cells that show the special property of contractility
   b. Diverse anatomy with distinct specializations for information transfer
   c. Lining of the inner surfaces of the intestines and lungs in mammals
   d. Loose array of cells embedded in an extracellular matrix
   Textbook Reference: *40.1 Why Must Animals Regulate Their Internal Environments? pp. 858–859*

8. Elephants use their ears to release heat to the environment. What mechanisms might they employ to increase heat loss from the ears?
   a. Increased convection due to flapping of the ears
   b. Moving into the sun
   c. Increased blood flow to the ears
   d. Both a and c
   Textbook Reference: *40.3 How Do Animals Alter Their Heat Exchange with the Environment? pp. 862–863*

9. Which organ system is involved in the maintenance of homeostasis in humans?
   a. Endocrine
   b. Digestive
   c. Muscle
   d. All of the above
   Textbook Reference: *40.1 Why Must Animals Regulate Their Internal Environments? pp. 855–856*

10. What is the difference between a negative feedback mechanism and a positive feedback mechanism?
    a. Negative feedback mechanisms only exist in the circulatory system.
    b. Negative feedback mechanisms return a system to a set point, and positive feedback mechanisms amplify a response.
    c. Negative feedback mechanisms move away from a set point, and positive feedback mechanisms stabilize toward a set point.
    d. Negative feedback mechanisms stabilize toward a set point, and positive feedback mechanisms reset the set point.
    Textbook Reference: *40.1 Why Must Animals Regulate Their Internal Environments? pp. 856–857*

11. A number of physiological processes can undergo acclimatization. Which of the following statements about acclimatization is true?
    a. It occurs in response to seasonal temperature changes.
    b. Acclimatization of metabolic rate occurs because enzyme expression changes.
    c. It involves the changing of a set point.
    d. All of the above
    Textbook Reference: *40.2 How Does Temperature Affect Living Systems? p. 861*

12. In fast-swimming cool-water fishes such as sharks and tunas, which of the following contribute(s) to the generation and maintenance of core body temperatures in excess of ambient water temperatures?
    a. Specializations in the size and arrangement of blood vessels
    b. Low rates of activity in the swimming muscles
    c. An ability to acclimatize rapidly to the surrounding water
    d. Metabolic rates that are insensitive to temperature change
    Textbook Reference: *40.3 How Do Animals Alter Their Heat Exchange with the Environment? pp. 864–865*

13. Which of the following statements about hypothalamic function is *false*?
    a. Circulating blood temperature is monitored by the hypothalamus.
    b. Hypothalamic thermal set points never change.
    c. The hypothalamus is a part of the central nervous system.
    d. Different thermoregulatory responses have different hypothalamic set points.
    Textbook Reference: *40.4 How Do Mammals Regulate Their Body Temperatures? pp. 868–869*

14. Some animals use brown fat as a source of heat generation. Which of the following statements about brown fat is true?
    a. Brown fat is involved in nonshivering thermogenesis.
    b. The protein thermogenin is involved in uncoupling proton movement from ATP production.
    c. Brown fat is found in endotherms.
    d. All of the above
    Textbook Reference: *40.4 How Do Mammals Regulate Their Body Temperatures? p. 867*

15. During childbirth, the pressure exerted on the mother's cervix by the emerging infant leads to increased contraction of the uterus. This interaction can be explained by
    a. negative feedback.
    b. feedforward feedback.
    c. positive feedback.
    d. acclimatization.
    Textbook Reference: 40.1 Why Must Animals Regulate Their Internal Environments? pp. 856–857

16. Which of the following statements about nervous tissue is true?
    a. Neurons outnumber glial cells in our nervous system.
    b. Glial cells generate and conduct electrochemical signals.
    c. Cell bodies are long extensions of neurons over which impulses travel to reach other neurons.
    d. Glial cells support and protect neurons.
    Textbook Reference: 40.1 Why Must Animals Regulate Their Internal Environments? p. 859

Application Questions

1. Having measured the metabolism of two animals of similar size at 15°C and 25°C, you find that the metabolic rate of the first animal is 115 and 55 ml of oxygen per hour at 15°C and 25°C, respectively. The metabolic rate of the second animal is 5.5 and 11.5 ml of oxygen per hour at 15°C and 25°C, respectively. Calculate the Q_{10} for the metabolic rate in both animals and determine if they are endotherms or ectotherms.
   Textbook Reference: 40.2 How Does Temperature Affect Living Systems? p. 860

2. You have measured the metabolism of the first species in Question 1 at both low and high environmental temperatures. You find that the metabolic rate at 40°C is higher than at 30°C. What physiological phenomenon explains why the metabolic rate of this animal rises when environmental temperature is above the upper critical limit?
   Textbook Reference: 40.4 How Do Mammals Regulate Their Body Temperatures? p. 868

3. What do mammals use as insulation, and what physiological controls exist to increase insulation?
   Textbook Reference: 40.3 How Do Animals Alter Their Heat Exchange with the Environment? p. 864

4. Animals have the ability to undergo acclimatization. Describe this process in relation to temperature regulation.
   Textbook Reference: 40.2 How Does Temperature Affect Living Systems? p. 860

5. Why is it important for an organism to maintain homeostasis?
   Textbook Reference: 40.1 Why Must Animals Regulate Their Internal Environments? pp. 855–856

Answers

Knowledge and Synthesis Answers

1. **d.** All of the choices are mechanisms to help to maintain the internal environment of an animal.

2. **b.** If the environment were cooler than the animal, the animal would want to lose heat to the environment across the skin. Increasing peripheral blood flow would move heat from the core of the animal to the surface, where it could be lost to the cooler environment.

3. **a.** In the desert, if an ectothermic lizard wants to maintain the most stable body temperature, it should be in a burrow during the night, where temperatures do not drop very much, and then shuttle between the sun and shade during the day.

4. **b.** After exercise, your dog will have a high metabolic rate and will pant in an attempt to cool down. Its body temperature will most likely be above the upper thermoneutral zone.

5. **a.** Evaporative cooling is a costly means to lower body temperature. ATP must be used either during sweating or panting. Water loss can also be high during evaporative cooling.

6. **c.** Fevers are brought about by a resetting of the thermoregulatory set point in the hypothalamus in response to pyrogens.

7. **d.** Connective tissues are blood, cartilage, and bone, all of which contain a loose population of cells embedded in an extracellular matrix.

8. **d.** Elephants use both physiological mechanisms (increased blood flow to the ears) and behavioral mechanisms (flapping of ears) to release heat into the environment.

9. **d.** All of these organ systems are involved in maintaining homeostasis.

10. **b.** Negative feedback loops use information about the state of a system to bring the internal environment back toward the set point values. Positive feedback loops amplify the response of a system, moving the system away from the set point.

11. **d.** Physiological processes acclimate in response to a seasonal change in the environment. Acclimatization may involve changes in enzyme expression and resetting of a physiological set point.

12. **a.** Large fishes such as the tuna are able to keep core body temperature above ambient temperature by using a number of mechanisms. They have a special countercurrent heat exchanger built into their blood vessels. They also have high metabolic rates and thus produce heat in their swimming muscles.

13. **b.** The hypothalamic thermal set points change daily, monthly, and seasonally.

14. **d.** Brown fat is a mechanism by which thermogenin protein uncouples proton movement from ATP production. This occurs in endotherms, and the process of heat production is termed nonshivering thermogenesis.

15. **c.** Increased uterine contraction in response to pressure on the cervix is an example of a positive feedback.

16. **d.** Glial cells outnumber neurons in the nervous system and provide support and protection to neurons.

Application Answers

1. The metabolic rate of the first animal increases with a decrease in environmental temperature, making this animal an endotherm. The metabolic rate of the second animal decreases with a decrease in environmental temperature, making this animal an ectotherm. Using the equation $Q_{10} = (R_T/R_{T-10})$, the Q_{10} for metabolic rate can be calculated. The Q_{10} of the endotherm's metabolic rate is 0.48, whereas the Q_{10} of the ectotherm metabolic rate is 2.1 using the equation provided.

2. An endotherm in a hot environment will attempt to keep core body temperature from increasing. The heat-loss mechanisms available to an endotherm, sweating and panting, require the use of ATP and result in an increased metabolic rate. Endotherms will expend energy to protect body temperature and lose heat to the environment.

3. Mammals use fur and fat as effective insulators. Fur is a good insulator because it traps still, warm air from the body. Increases in the length of the fur will increase effectiveness as an insulator. Fat works in a similar manner because it has a low thermal conductance. Humans do not have sufficient hair on their bodies for it to provide good insulation; therefore, they wear clothes as an insulation layer. Physiologically, mammals can change their blood flow patterns to increase or decrease the amount of warm blood that is received by the skin. When active and needing to get rid of excess heat, mammals transport heat to hairless skin surfaces.

4. Animals undergo acclimatization in response to changes in seasonal conditions such as temperature. Often these changes are brought about by the production of enzymes that function better at the new temperature. Acclimatization results in metabolic functions being less sensitive to long-term changes in temperature than to short-term changes.

5. Homeostasis is the maintenance of a constant internal environment. The body functions with the help of proteins and enzymes. Changes in characteristics of the internal environment such as pH, temperature, glucose level, and oxygen and carbon dioxide levels can affect cellular function. Loss of homeostasis can lead to improper function of proteins and cell membranes and ultimately lead to cell death.

CHAPTER 41 Animal Hormones

Important Concepts

Hormones are chemical signals used in physiological regulation.

- Hormones are secreted by endocrine cells, present as either individual cells or as groups of cells known as glands.

- Hormones are secreted into the extracellular fluid and then most diffuse into the blood. Once in the blood, they are carried to target cells, where they bind with a receptor to trigger the function of the hormone.

- Not all hormones act on a global scale; some are secreted into the extracellular fluid and then taken up by local cells, without diffusing into the bloodstream. These local hormones are known as paracrine hormones. Histamine is an example of a paracrine hormone. Histamine acts locally in response to tissue damage by causing dilation of blood vessels. Hormones that act on the cells that secrete them are known as autocrine hormones.

- Unlike endocrine glands, exocrine glands secrete their products into ducts that empty onto the surface of the skin or into a body cavity. Examples of exocrine glands are sweat glands and salivary glands.

Hormones are involved in the control of development in insects.

- Hormones control molting (shedding of the exo-skeleton) and metamorphosis (transformation to the adult stage) in insects.

- In an elegant set of experiments, Wigglesworth used the blood-sucking bug *Rhodnius* to examine the control of molting. Decapitation of the bugs at various times after feeding showed that there was a substance secreted in the heads of these bugs that stimulates molting.

- The brains of *Rhodnius* and other insects produce prothoracicotropic hormone (PTTH; previously called brain hormone) that is stored in the corpora cardiaca attached to the brain. After a meal, brain hormone diffuses in extracellular fluid to the prothoracic gland, where it stimulates the release of ecdysone, a hormone that stimulates molting.

- In insects with incomplete metamorphosis, juvenile hormone results in further molts as a juvenile. In the absence of juvenile hormone, an insect molts into the adult form. In insects with complete metamorphosis, juvenile hormone results in molting into a larva, declining levels result in molting into a pupa, and in the absence of juvenile hormone a pupa molts into an adult.

Hormones act on the receptors of the target cells.

- Hormones can be divided into three chemical groups. Most hormones are peptides or proteins; these hormones are water-soluble. Steroid hormones, the second chemical group, are derived from cholesterol and are lipid-soluble. Amine hormones, most of which are derived from the amino acid tyrosine, constitute the third chemical group; some of these hormones are water-soluble and some are lipid-soluble.

- Most hormones are released in very small quantities. A hormone works only on cells that have specific receptors for it.

- Lipid-soluble hormones are able to cross the plasma membrane of cells and act on receptors inside the cell, ultimately altering gene expression. Water-soluble hormones cannot pass through the plasma membrane and thus act on receptors at the surface of the cell. These surface receptors are large glycoprotein complexes with binding, transmembrane, and cytoplasmic domains. Binding to surface receptors usually triggers the target cell's response by activating enzymes within the cell.

- A single hormone can act on many different types of cells to produce different effects. Epinephrine, secreted by the adrenal glands, activates the fight-or-flight response and causes changes in heart rate, blood flow, and the immune system. Epinephrine also stimulates the breakdown of glycogen in the liver and fats in adipose tissue.

The pituitary gland illustrates the close connection between nervous and endocrine systems.

- The pituitary gland is attached to the hypothalamus and provides an example of how nervous and endocrine systems interact. The hypothalamus produces two hormones that are secreted by the pituitary gland, and other hormones released by the hypothalamus influence release of pituitary hormones.

- The pituitary gland produces hormones that control many of the other endocrine glands in the body. The pituitary is made up of the anterior pituitary and posterior pituitary, each utilizing different control mechanisms and releasing different hormones.

- The posterior pituitary releases antidiuretic hormone and oxytocin. Both of these hormones are produced by neurons in the hypothalamus and move down the neurons into the posterior pituitary. Antidiuretic hormone (ADH) acts on the kidneys to stimulate reabsorption of water to help maintain blood pressure and solute concentration of the blood. During childbirth, oxytocin stimulates uterine contractions in the mother. After birth, oxytocin stimulates the ejection of milk from mammary glands.

- The anterior pituitary gland releases four tropic hormones (thyrotropin, corticotropin, luteinizing hormone, and follicle-stimulating hormone) that control the function of the thyroid, adrenal cortex, testes, and ovaries. The anterior pituitary also produces growth hormone, prolactin, endorphins, and enkephalins. Growth hormone (GH) stimulates the liver to release chemicals (somatomedins) that stimulate bone and cartilage growth. In female mammals, prolactin stimulates milk production. Endorphins and enkephalins influence pain pathways in the brain.

- The anterior pituitary gland is controlled by hormones produced in the hypothalamus and transported to the anterior pituitary gland by portal blood vessels. Two of the controlling hormones are thyrotropin-releasing hormone (TRH), causing the release of thyrotropin that stimulates the thyroid gland, and gonadotropin-releasing hormone (GnRH), stimulating the release of hormones that control the gonads.

- The anterior pituitary gland is also controlled by negative feedback. Hormones released from target glands may inhibit further release of the particular tropic hormone from the anterior pituitary.

The thyroid gland regulates metabolic rate and lowers blood calcium levels.

- The thyroid gland helps regulate metabolic rate through the actions of thyroxine (T_4) and triiodothyronine (T_3), both of which increase metabolic rate in mammals. Thyrotropin from the anterior pituitary gland stimulates the thyroid gland to produce thyroxine. Poor regulation of thyroxine production (either hyperthyroidism, thyroxine excess, or hypothyroidism, thyroxine deficiency) can result in an enlarged thyroid gland, a condition known as goiter.

- Another thyroid gland hormone, calcitonin, lowers calcium levels in the blood by stimulating osteoblasts, which deposit new bone, and inhibiting osteoclasts, which break down bone.

The parathyroid glands raise blood calcium levels.

- The parathyroid glands produce parathyroid hormone, which increases blood concentrations of calcium by stimulating osteoclasts to dissolve bone. Calcitonin (from the thyroid gland) and parathyroid hormone act antagonistically, countering the effects of one another.

- Vitamin D is a hormone produced by skin cells and is activated when it passes through the liver and kidneys. Vitamin D helps to raise blood calcium levels. For example, in the kidneys, vitamin D works with parathyroid hormone to decrease loss of calcium in urine. In bone, Vitamin D stimulates bone turnover and thus the movement of calcium from bone to blood.

The pancreas regulates blood sugar levels.

- Endocrine cells in the pancreas, called the islets of Langerhans, produce the hormones insulin and glucagon, which regulate glucose levels in the blood. Insulin binds to receptors on cells, allowing glucose to enter the cell from the bloodstream. When insulin is absent or the receptors become insensitive to it, glucose accumulates in the blood, resulting in the disease diabetes mellitus.

- When blood glucose levels fall well below normal, glucagon stimulates the liver to convert glycogen back into glucose and levels of glucose rise in the blood.

- A rapid rise in levels of glucose and amino acids in the blood stimulates the pancreas to release somatostatin, which inhibits the release of insulin and glucagon. Cells of the hypothalamus also produce somatostatin. Somatostatin from this source inhibits the release of growth hormone and thyrotropin from the anterior pituitary gland.

The adrenal glands produce hormones with numerous regulatory functions.

- The adrenal glands have two regions, the adrenal medulla, making up the core of the glands, and the adrenal cortex, which surrounds the medulla.

- Epinephrine and norepinephrine are produced in the adrenal medulla and are involved in the fight-or-flight response.

- The adrenal cortex produces glucocorticoids, mineralocorticoids, and sex steroids from cholesterol. The main glucocorticoid is cortisol, which mediates our response to stress by stimulating cells that are not necessary for the fight-or-flight response to decrease their use of glucose and to use fats and proteins as energy sources. The main mineralocorticoid is aldosterone, which is involved in the reabsorption of sodium and excretion of potassium at the kidney. Compared to the amounts of sex steroids secreted by the gonads, the amounts produced by the adrenal cortex are negligible.

The gonads produce sex steroid hormones.

- In males, testes produce androgens (testosterone), whereas in females, ovaries produce estrogens and progesterone. Early in development, the sex organs of human embryos are similar. At about week 7, presence of the Y chromosome in male embryos causes the undifferentiated gonads to begin producing androgens. In response to androgens, the reproductive system develops into a male system. In the absence of androgens, the reproductive system develops into a female system.

- Sex steroids increase in concentration at the time of puberty. Both luteinizing hormone (LH) and follicle-stimulating hormone (FSH) from the anterior pituitary gland control the production of sex steroids. Increased levels of LH and FSH in females and LH in males stimulate the gonads to produce more sex hormones. These hormones in turn have profound physiological effects on the development of secondary sexual characteristics.

The pineal gland produces melatonin.

- Melatonin is involved in biological rhythms, including photoperiodicity, the phenomenon whereby changing day length across the seasons prompts changes in physiology.

- The release of melatonin by the pineal gland occurs in the dark; light inhibits melatonin release.

Scientists have ways to detect and measure hormones and to characterize their receptors and signal transduction pathways.

- Hormone concentrations are measured by immunoassays.

- A single hormone can have different receptors, and efforts to characterize receptors often rely on biochemical separation techniques.

- The abundance of hormone receptors is sometimes regulated through negative feedback mechanisms. Downregulation involves a decrease in the number of receptors in response to high levels of a hormone. Upregulation can occur when levels of a hormone are suppressed; in response, a target cell may increase the number of receptors for that hormone.

- Binding of a hormone to its receptor activates signal transduction pathways within cells. Amplification of the signal occurs in many of these pathways, thus explaining how tiny quantities of a hormone can produce enormous physiological effects.

The Big Picture

- Hormones regulate functions ranging from growth to sexual maturity. In numerous ways, the endocrine system complements the nervous system as major regulators of body function and homeostasis.

- Circulating hormones travel via the bloodstream to distant sites within the body to affect target tissues.

Other hormones, called paracrine hormones, act locally. Autocrine hormones act on the secreting cells themselves.

- Hormone action is controlled by several different mechanisms. In many instances there is a "hormone cascade" in which a hormone controls the release of a hormone, which controls the release of a hormone, and so on. Negative feedback occurs when a released hormone inhibits (negatively feeds back on) the tissue that may have stimulated its release in the first place. Tropic hormones influence other endocrine glands.

Common Problem Areas

- Appropriate for its title "master gland," the pituitary produces numerous regulatory hormones. The functions of the anterior and posterior pituitary are quite distinct, and keeping them separate in your mind is important to your understanding of the endocrine system.

- Remember that most hormones are broadly distributed throughout the body and so all cells are exposed to the same level of hormone concentration. Hormones have specific actions because different cells have different levels of sensitivity to them, not because the hormone concentration differs throughout the body.

Study Strategies

- The many hormones may seem to constitute a long laundry list. Try to learn the function along with the name of each hormone, as this will help you remember the target tissues.

- Some hormones have extremely specific actions (e.g., follicle-stimulating hormone); whereas others have broad effects on many target tissues (e.g., epinephrine, thyroid hormones). Learning which hormones are "generalists" and which are "specialists" will help you better understand the endocrine system.

- Biologists studying the endocrine system frequently use abbreviations for the names of hormones. Learn these abbreviations along with the full name of a hormone.

- Refer to the following animated tutorial and activities on the Companion Website/CD:
Tutorial 41.1 Complete Metamorphosis
Activity 41.2 Endocrine Glands in Humans
Activity 41.2 Concept Matching: Vertebrate Hormones

Test Yourself

Knowledge and Synthesis Questions

1. A paracrine hormone
   a. circulates in the bloodstream and affects distant cells.
   b. always acts on a wide variety of target tissues.
   c. acts on nearby cells.
   d. None of the above
   Textbook Reference: 41.1 *What Are Hormones and How Do They Work? p. 875*

2. A target cell's response to a hormone depends on
   a. the amount of hormone released.
   b. the number of receptors on that target cell.
   c. how well the hormone binds with the receptor.
   d. All of the above
   Textbook Reference: *41.1 What Are Hormones and How Do They Work? pp. 878–879; 41.4 How Do We Study Mechanisms of Hormone Action? pp. 891–893*

3. "Upregulation" of hormone receptors refers to
   a. increase in abundance of hormone receptors in response to low hormone levels.
   b. increase in abundance of hormone receptors with high neurotransmitter levels.
   c. increase in hormone levels produced by increase in abundance of hormone receptors.
   d. decrease in hormone levels produced by decrease in abundance of hormone receptors.
   Textbook Reference: *41.4 How Do We Study Mechanisms of Hormone Action? p. 892*

4. Hormones have various regulatory functions. Which of the following statements does *not* describe how hormones function?
   a. They act in very low concentration.
   b. Many act at sites distant from where they are produced.
   c. They are transported in blood.
   d. All of the above
   Textbook Reference: *41.1 What Are Hormones and How Do They Work? p. 875*

5. In insects, juvenile hormone
   a. is produced by the corpora cardiaca.
   b. is also known as brain hormone.
   c. prevents maturation.
   d. All of the above
   Textbook Reference: *41.1 What Are Hormones and How Do They Work? p. 877–878*

6. Vitamin D
   a. is a hormone secreted by skin cells.
   b. is activated on passing through the liver and kidneys.
   c. decreases loss of calcium in the urine.
   d. All of the above
   Textbook Reference: *41.3 What Are the Major Mammalian Endocrine Glands and Hormones? p. 886*

7. Which of the following glands has one part that develops from the embryonic mouth cavity and a second part that develops from the floor of the developing brain?
   a. Adrenal gland
   b. Thyroid gland
   c. Hypothalamus
   d. Pituitary gland
   Textbook Reference: *41.2 How Do the Nervous and Endocrine Systems Interact? p. 881*

8. Which of the following glands is involved in biological rhythms, such as photoperiodicity?

   a. Pineal gland
   b. Thyroid gland
   c. Adrenal glands
   d. Ovaries
   Textbook Reference: *41.3 What Are the Major Mammalian Endocrine Glands and Hormones? p. 890*

9. The half-life of hormones in the plasma ranges from
   a. minutes to hours.
   b. minutes to days or weeks.
   c. days to weeks.
   d. hours to months.
   Textbook Reference: *41.4 How Do We Study Mechanisms of Hormone Action? p. 891*

10. The target tissues of hormones are those tissues that
    a. the particular hormone can actually penetrate.
    b. have specific enzymes with which hormones directly interact.
    c. have high concentrations of the "second messenger."
    d. have receptors for the particular hormone.
    Textbook Reference: *41.1 What Are Hormones and How Do They Work? p. 875*

11. Which of the following sets of vertebrate hormones are produced in the anterior pituitary gland?
    a. Somatostatin, antidiuretic hormone, insulin
    b. Prolactin, growth hormone, enkephalins
    c. Oxytocin, prolactin, adrenocorticotropin
    d. Estrogen, progesterone, testosterone
    Textbook Reference: *41.2 How Do the Nervous and Endocrine Systems Interact? p. 882*

12. Hormones that are secreted by one endocrine gland and control the activities of another endocrine gland are called _____ hormones.
    a. growth
    b. obstructive
    c. tropic
    d. selective
    Textbook Reference: *41.2 How Do the Nervous and Endocrine Systems Interact? p. 881*

13. Which of the following has both endocrine and exocrine functions?
    a. Pancreas
    b. Heart
    c. Testes
    d. Adrenal glands
    Textbook Reference: *41.3 What Are the Major Mammalian Endocrine Glands and Hormones? p. 887*

14. Which of the following hormones have antagonistic effects?
    a. Insulin; glucagon
    b. Growth hormone; oxytocin
    c. Oxytocin; prolactin
    d. Cortisol; testosterone
    Textbook Reference: *41.3 What Are the Major Mammalian Endocrine Glands and Hormones? p. 887*

15. Steroid hormones are
    a. water-soluble.
    b. produced by the thyroid gland.
    c. lipid-soluble.
    d. derived from the amino acid tyrosine.
    Textbook Reference: 41.1 What Are Hormones and How Do They Work? p. 878

Application Questions

1. Describe the modes of action of lipid-soluble and water-soluble hormones.
   Textbook Reference: 41.1 What Are Hormones and How Do They Work? p. 878

2. The pituitary gland is made up of the anterior and posterior pituitary. What are the relationships between these two parts of the pituitary gland and the hypothalamus?
   Textbook Reference: 41.2 How Do the Nervous and Endocrine Systems Interact? pp. 880–883

3. Why would breast-feeding soon after birth help a mother's uterus return to near its prepregnant size?
   Textbook Reference: 41.2 How Do the Nervous and Endocrine Systems Interact? p. 881

4. Wigglesworth conducted a number of studies on the insect *Rhodnius* to examine the control of molting. Describe what would happen to a fourth-instar *Rhodnius* if Wigglesworth were to either partially or fully decapitate the insect one week after a blood meal.
   Textbook Reference: 41.1 What Are Hormones and How Do They Work? pp. 876–877

5. Would castration (removal of the testes) lead to an immediate and complete absence of testosterone in the bloodstream? Why or why not?
   Textbook Reference: 41.4 How Do We Study Mechanisms of Hormone Action? p. 891

Answers

Knowledge and Synthesis Answers

1. **c.** Paracrine hormones act on nearby cells. Autocrine hormones act on the very same cells that secrete them. Circulating hormones enter the bloodstream and affect distant cells.

2. **d.** The response of target cells to a hormone depends on how much of the hormone is present and acting on the target, how many receptors are present on the surface of the cell that the hormone can act on, and how well the hormone binds with the receptor.

3. **a.** "Upregulation" of hormone receptors on a cell is the production of more receptors when a hormone is at low levels over time in the blood or other fluids surrounding the cell.

4. **d.** All of the statements are correct. Hormones are found and act at low concentrations in the body. They are transported by the blood or diffuse through interstitial space to reach target cells that may be some distance from the secreting gland.

5. **c.** Juvenile hormone is secreted by the corpora allata of insects and prevents maturation. Prothoracicotropic hormone (brain hormone) is secreted by the corpora cardiaca and is involved with ecdysone in molting.

6. **d.** Vitamin D is a hormone secreted by skin cells that becomes more active once it passes through the liver and kidneys. Vitamin D raises levels of calcium in the blood by reducing loss of calcium in the urine.

7. **d.** The pituitary gland has an anterior lobe that develops from an outpocketing of the embryonic mouth cavity and a posterior lobe that develops from an outpocketing of the floor of the developing brain.

8. **a.** The pineal gland secretes melatonin, a hormone that controls biological rhythms. One such rhythm is photoperiodicity, the phenomenon whereby seasonal changes in day length influence physiological processes in animals.

9. **b.** Hormones tend to act over time periods of minutes to days or weeks. Epinephrine has a half-life of up to three minutes, whereas cortisol has a half-life of many days.

10. **d.** For a hormone to act on a target cell, the target cell must have the receptors for the specific hormone to bind and trigger the hormonal action.

11. **b.** Prolactin, growth hormone, and enkephalins are produced in the anterior pituitary.

12. **c.** Hormones that control endocrine gland function are known as tropic hormones.

13. **a.** The pancreas has both endocrine and exocrine functions. It releases the hormones glucagon, insulin, and somatostatin to extracellular fluid and produces digestive enzymes that travel through ducts to the small intestine.

14. **a.** Insulin lowers blood glucose and glucagon raises blood glucose; thus these hormones have antagonistic effects.

15. **c.** Steroid hormones are lipid-soluble and are produced by the gonads and adrenal glands.

Application Answers

1. The plasma membrane of cells is hydrophobic. Lipid-soluble hormones, which as their name suggests dissolve in lipids (fats), cross the plasma membrane and act on receptors inside the target cells. Water-soluble hormones cannot cross the lipid-based plasma membrane and act on receptors on the outside of target cells. Such binding by water-soluble hormones initiates changes in the cell by activating enzymes. Because lipid-soluble hormones cross the plasma membrane and enter the cell, they tend to take longer to act, but they also remain active for a longer time period than water-soluble hormones.

2. The hormones secreted by the posterior pituitary gland are produced by neurons in the hypothalamus. The neurohormones move down neurons into the posterior pituitary gland, where they are stored and secreted. The

anterior pituitary produces the hormones it secretes. However, hormones from the hypothalamus control the release of the anterior pituitary's hormones.

3. Suckling by a baby stimulates the release of oxytocin from the posterior pituitary, and this leads to milk ejection from the mammary glands. Oxytocin also causes uterine contractions, and after birth this helps the mother's uterus return toward its prepregnancy size.

4. The insect *Rhodnius* can live for long periods of time after it has been decapitated. If the insect is partially decapitated, leaving the corpora allata, the insect will molt into a fifth-instar juvenile. This is because the corpora allata produces juvenile hormone, which prevents maturation into an adult. The animal that was fully decapitated would molt into an adult rather than another juvenile instar. The fully decapitated animal had enough prothoracicotropic hormone (brain hormone) circulating one week after the meal that it would molt; however, there would be no juvenile hormone present, and thus the animal would molt into an adult.

5. Castration would not lead to an immediate absence of testosterone in the bloodstream. First, many hormones, including testosterone, have half-lives of days to weeks. Second, small amounts of sex steroids, such as testosterone, are produced by the adrenal glands.

CHAPTER 42 Animal Reproduction

Important Concepts

Some animals reproduce asexually.

- A variety of animals, mostly invertebrates, employ asexual reproduction, producing genetically identical offspring from the parent. Individuals that reproduce asexually are often sessile or members of a sparse population, situations that make searching for and finding a mate difficult.

- New offspring can be produced either by budding off outgrowths to produce a new individual or by regenerating a new individual from pieces of an animal (see Figure 42.1).

- Animals can produce new individuals asexually through parthenogenesis—the development of unfertilized eggs into new offspring.

- Sometimes sexual behavior is required for asexual reproduction; this is the case in a species parthenogenetic whiptail lizard (see Figure 42.2).

Sexual reproduction produces genetic diversity.

- Sexual reproduction in animals produces genetic diversity when two haploid gametes join to form a diploid cell.

- Sexual reproduction has three stages: gametogenesis, mating, and fertilization.

- Gametogenesis produces haploid gametes using meiotic cell division.

- Gametogenesis occurs in an animal's gonads. Male gametes—sperm—are produced in the testes, and female gametes—eggs or ova—are produced in the ovaries. Gametes are derived from germ cells. These germ cells migrate to the gonads of the embryo.

Males produce sperm by spermatogenesis.

- Spermatogenesis begins with a male germ cell (2n) that proliferates through mitosis to produce spermatogonia (see Figure 42.3A). Spermatogonia mature into primary spermatocytes. The first meiotic division produces secondary spermatocytes (2n), and a second meiotic division produces spermatids (n).

- Mammalian spermatids remain in cytoplasmic contact and share gene products of the X chromosome at this stage.

- The spermatids differentiate and mature into sperm cells. Mature mammalian sperm are not in cytoplasmic contact with one another.

Females produce eggs by oogenesis.

- Oogenesis begins with differentiation of female germ cells (2n) into oogonia (2n) that mature into primary oocytes (see Figure 42.3B).

- The primary oocytes enter prophase of the first meiotic division and in human females remain arrested in prophase for at least 10 years (until puberty is reached); some remain arrested for up to 50 years. No new primary oocytes will be produced during the life of the female.

- Upon exiting prophase, the primary oocyte completes the first meiotic division to produce a secondary oocyte (2n) and a polar body. The second meiotic division produces an ootid (n) and another polar body. This second meiotic division may not be completed until fertilization in some species.

Fertilization is the joining of a haploid sperm with a haploid egg to form a diploid zygote.

- For successful fertilization to occur, the sperm and egg must recognize one another. Such recognition is mediated by specific recognition molecules on the gametes. In mammals, a glycoprotein in the zona pellucida (a layer surrounding the mammalian egg) binds to recognition molecules on the head of the sperm and triggers the acrosomal reaction (see Figure 42.6).

- During the acrosomal reaction, the acrosomal membrane of the sperm breaks down, releasing enzymes that digest a path for the sperm through the protective layers surrounding the egg (see Figure 42.5).

- Fusion of the plasma membranes of the sperm and egg triggers blocks to polyspermy that prevent more than one sperm from entering the egg.

- Soon after entry by a sperm, the egg is metabolically activated and stimulated to begin development. The diploid nucleus of the zygote is created when the haploid nucleus of the egg fuses with the haploid nucleus of the sperm.

Fertilization occurs differently in the water and on land.

- Many aquatic animals have external fertilization of eggs. The eggs and sperm are released into the water, where fertilization may occur.

- Many fish and amphibians that display external fertilization employ special behaviors to bring gametes close to each other to increase the probability that fertilization will occur.

- Terrestrial animals employ internal fertilization in which sperm and egg fuse within the female reproductive tract rather than in the external environment.

- Internal fertilization requires accessory sex organs that pass the sperm on to the female during copulation. The penis is a male accessory sex organ, and the vagina, in some species, is a female accessory sex organ. Glands, ducts, and tubules associated with reproduction are also accessory sex organs. The term *genitalia* refers to external sex organs.

- Some male mites and scorpions produce a sperm-filled spermatophore that they deposit in the environment. When a female comes across the spermatophore, she straddles it and takes the packet of sperm into her reproductive tract. Thus, these animals have internal fertilization without copulation.

Some animals have both male and female reproductive systems.

- Dioecious species have separate male and female individuals.

- In monoecious species, a single individual may produce both eggs and sperm; such individuals are called hermaphrodites. Hermaphrodites can be either simultaneous (individuals are male and female at the same time) or sequential (individuals change sex at some point during their life span).

The move to land required a reproductive system that functions in a dry environment.

- Amphibians were the first vertebrates to live on land, but they still require water to reproduce.

- Many reptiles and all birds have solved the problem of a dry environment by being oviparous; they lay protective shelled eggs termed amniotic eggs. Internal fertilization must occur prior to shell formation. Most nonmammalian viviparous animals are ovoviviparous, with embryos developing in shelled eggs that are retained in females until hatching.

- All mammals except the monotremes are viviparous, retaining the embryo within the uterus.

Human males produce sperm and deliver it to the female.

- Males produce semen containing sperm (*n*) and fluids that facilitate fertilization. In the testis, sperm cells are produced by spermatogenesis in the seminiferous tubules. Sertoli cells surround developing sperm cells and provide protection and nourishment (see Figure 42.10C).

- Sperm develop a flagellum at the back and at the front, an acrosome cap that contains digestive enzymes. Mitochondria provide energy for the flagellum. Sperm are stored in the epididymis, which connects to the vas deferens and then to the urethra, which opens to the outside of the body at the tip of the penis.

- In addition to sperm, semen contains fluids from accessory glands. The bulbourethral glands produce an alkaline and mucoid secretion that helps control pH and lubricates the tip of the penis. The seminal vesicles produce seminal fluid that contains fructose, an energy source for sperm, and mucus and proteins. The prostate gland produces the milky prostate fluid and a clotting enzyme that causes the protein in seminal fluid to turn semen into a gelatinous mass. Prostate secretions are alkaline, so they also help to neutralize the acidity of the male and female reproductive tracts.

- During sexual arousal, the penis becomes engorged with blood, and contraction of smooth muscle in the ducts and at the base of the penis results in emission and ejaculation of semen. After sexual stimulation, the vessels at the base of the penis constrict, leading to loss of the erection. The dilation of blood vessels in the penis during an erection is initiated by the neurotransmitter nitric oxide (NO) and its effects on the second messenger cGMP. After ejaculation, decreases in NO and cGMP lead to the constriction of blood vessels associated with loss of an erection.

- Testosterone, produced in the Leydig cells of the testes, is the male sex hormone that controls sexual function and sperm production (see Figure 42.11).

Human females produce eggs, receive sperm, and gestate the embryo.

- The ovaries release eggs into the body cavity, where they enter the oviduct (also called the fallopian tube) and move toward the uterus (see Figure 42.12A). Sperm swim up the vagina, through the opening to the uterus (cervix), through the uterus, and into the oviduct. Fertilization occurs in the oviduct; embryos develop in the uterus.

- The initial division of the zygote produces a blastocyst that moves into the uterus, where it burrows into the uterine wall (the endometrium). The blastocyst interacts with the endometrium to form the placenta, the organ that provides nutrients and removes wastes from the developing embryo.

- Mature eggs are produced during a 28-day ovarian cycle (see Figure 42.13). Initially, 6 to 12 primary oocytes begin to mature and are surrounded by follicle cells to produce individual follicles. After one week, one primary oocyte continues to grow while the others shrink. Prior to ovulation, the primary oocyte undergoes a meiotic division to become a secondary oocyte and is expelled from the ovary. The remaining follicle cells become the corpus luteum, a glandular structure that produces progesterone and estrogen.

- The uterine cycle, in concert with the ovarian cycle, prepares the endometrium to receive the blastocyst. If a blastocyst does not implant by around day 14 of the uterine cycle, the endometrium is broken down and subsequently expelled in the process of menstruation. The uterine cycles of most mammals other than humans do not include menstruation. Other mammals resorb the endometrium at the end of the uterine cycle; in these species, females display a period of sexual receptivity around ovulation termed estrus.

- Luteinizing hormone (LH) and follicle-stimulating hormone (FSH) control the timing of the ovarian and uterine cycles (see Figure 42.14). The first step of the uterine and ovarian cycle is menstruation. Prior to day 10 of the uterine cycle, FSH and LH levels are kept low by negative feedback from estrogen. Around days 12 to 14, this changes, and estrogen exerts positive feedback, which causes a rise in FSH levels. Rising levels of FSH stimulate the follicles to begin to mature. LH levels also increase, triggering ovulation, the release of the egg from the mature follicle. Following ovulation, the follicle cells form the corpus luteum. Estrogen and progesterone produced by the corpus luteum stimulate the growth and maintenance of the endometrium. In the absence of fertilization, the corpus luteum degenerates, estrogen and progesterone levels decrease, and menstruation occurs to start the cycle again.

- When the egg is fertilized and implants in the uterus, human chorionic gonadotropin is produced, which keeps the corpus luteum functional. Estrogen and progesterone stimulate development and maintenance of the endometrium, thereby preparing the uterus for pregnancy.

- Late in pregnancy estrogen increases contractility of uterine muscle. The uterine contractions of childbirth are triggered by pressure from the fetal head on the maternal cervix. This mechanical stimulus triggers release of oxytocin by the mother and fetus. Oxytocin stimulates increased strength and frequency of uterine contractions in a positive feedback cycle (see Figure 42.16A). Stages of childbirth include dilation of the cervix, delivery of the baby, and delivery of the placenta.

Human sexual responses have four phases.

- There are four phases of response to sexual stimulation in men and women: excitement, plateau, orgasm, and resolution.

- The excitement phase is characterized by increases in heart rate, blood pressure, and muscle tension, and erection of the penis in males and clitoris in females. Breathing rate increases during the plateau phase, and further increases occur in heart rate and blood pressure. Ejaculation occurs during orgasm in males;

females may have several orgasms in quick succession. During the resolution phase, blood drains from the genitals, and body physiology returns to resting conditions.

- In males, a refractory period occurs after orgasm. During this period, which may last for minutes or hours, a full erection and orgasm are not possible.

Humans use several methods to prevent pregnancy.

- Methods of contraception differ in their mode of action and failure rate (see Table 42.1).

- The rhythm method involves avoiding sexual intercourse during the time period when the female is most likely to be fertile and has a 15 to 35 percent failure rate.

- Condoms, diaphragms, and cervical caps work by creating a barrier to sperm. They have a failure rate of from 3 to about 25 percent.

- Several methods employ hormones to prevent ovulation by the female. Oral contraceptives contain synthetic estrogen and progesterone that interfere with the development of the ova and suspend the ovarian cycle. The failure rate of oral contraceptives is less than 3 percent.

- Intrauterine devices work by blocking the implantation of a fertilized egg in the uterus and have a failure rate of less than 6 percent.

- Both males and females can be sterilized. Males can undergo a vasectomy in which the vasa deferentia are cut, whereas females can have their oviducts cut in a procedure called tubal ligation (see Figure 42.17).

- The termination of pregnancy is known as abortion. A spontaneous abortion occurring early in pregnancy is commonly called a miscarriage. RU-486 blocks progesterone receptors in the uterine lining and when used early in pregnancy causes shedding of the endometrium and embryo.

- Controlling female fertility is much easier than controlling male fertility. Males continuously produce millions of sperm. Continuous chemical intervention would be needed to completely block spermatogenesis.

Medical science has found ways to help overcome infertility.

- A number of problems may prevent couples from conceiving. A male may not produce adequate numbers of sperm or the sperm may lack motility. In the female, the environment of the uterus may be poor for hosting sperm or the fertilized egg, or the oviducts may be blocked, preventing passage of gametes.

- Artificial insemination places sperm in the female reproductive tract where fertilization can occur. This technique is commonly used when males have a low sperm count (sperm is collected and concentrated before the procedure or donor sperm is used).

- Assisted reproductive technologies (ARTs) take eggs from a female and then either fertilize them outside the body (in vitro fertilization, IVF) or place a mixture of sperm and eggs in the female oviduct. Sperm that cannot gain access to the plasma membrane of an egg may be injected into an egg in a procedure called intracytoplasmic sperm injection (ICSI).

- Genetic testing may be performed on cells taken from early blastocysts in a procedure known as preimplantation genetic diagnosis (PGD).

Some diseases are sexually transmitted.

- Sexually transmitted diseases have been present since ancient times.

- Over 10 million new cases of STDs occur in the United States each year.

- STDs can be caused by viruses, bacteria, yeasts, or protozoans. The symptoms and incidence of STDs vary widely (see Table 42.2).

- At present, latex condoms are the only contraceptive device effective in preventing the transmission of STDs.

The Big Picture

- Animals reproduce both asexually and sexually. Asexual reproduction includes budding, regeneration, and parthenogenesis. Sexual reproduction has an advantage over asexual reproduction in that it helps to increase genetic diversity. Sexual reproduction has three main stages: gametogenesis, mating, and fertilization.

- Gametogenesis is the process by which the sex cells are produced. In males, haploid sperm develop from diploid spermatogonia by the process of spermatogenesis. In females, the haploid egg develops from diploid oogonia by the process of oogenesis.

- The female has two interrelated cycles, the ovarian and the uterine cycles, during which an egg is produced and released from the ovary and the uterus is prepared for implantation and pregnancy. These two cycles occur in parallel and are coordinated by changes in hormone levels. Luteinizing hormone, follicle-stimulating hormone, estrogen, and progesterone all play a role in the control of these cycles.

- External fertilization is typical of many aquatic animals. Internal fertilization, often with the help of secondary sex organs to ensure fertilization, occurs in terrestrial animals. The eggs and sperm of a given species recognize one another.

- A number of methods are available both for preventing pregnancy and for overcoming infertility.

Common Problem Areas

- Understanding the ploidy level of the developing egg and sperm can be very confusing, but it is essential in determining how genetic inheritance works.

- Following the hormonal changes and cues associated with human ovarian and menstrual cycles can be confusing. Create a chronological sequence of hormones and the events they trigger during both cycles.

Study Strategies

- Recognize that although sperm and egg formation are completely separate processes, each with its own characteristics, the same general events are happening in both. This will help you to recognize and remember the stages of spermatogenesis and oogenesis. Be sure, though, to also understand the differences between spermatogenesis and oogenesis with respect to timing and number of gametes produced.

- To better understand the changes in ploidy in gametes and the products of their combination, create a flowchart that shows how the haploid state changes over one or two cycles.

- Review the following animated tutorials and activities on the Companion Website/CD:
  Tutorial 42.1 Fertilization in a Sea Urchin
  Tutorial 42.2 The Ovarian and Uterine Cycles
  Activity 42.1 Human Male Reproductive Tract
  Activity 42.2 Spermatogenesis
  Activity 42.3 Human Female Reproductive Tract

Test Yourself

Knowledge and Synthesis Questions

1. Asexual reproduction is an effective strategy in stable environments because
   a. gametogenesis is most efficient under these conditions.
   b. the resulting offspring, genetically identical to their parents, are preadapted to their environment.
   c. asexual parthenogenesis produces a large amount of genetic diversity.
   d. animal cells tend to be more totipotent under stable conditions.
   Textbook Reference: 42.1 How Do Animals Reproduce Without Sex? p. 897

2. An important difference between a sperm and an egg is
   a. their size.
   b. the amount of cytoplasm they contain.
   c. whether or not they are motile.
   d. All of the above
   Textbook Reference: 42.2 How Do Animals Reproduce Sexually? p. 899

3. Which of the following would *not* be present in the body or lifestyle of a terrestrial male animal that reproduces sexually?
   a. Internal fertilization
   b. A secondary sexual organ as part of the genitalia
   c. Synchrony of reproductive physiology with a female
   d. Oogenesis
   Textbook Reference: 42.2 How Do Animals Reproduce Sexually? pp. 899–901

4. Which of the following best represents the normal path of a sperm cell as it makes its way from the point of entry into a female's reproductive tract to the location where fertilization typically occurs?
   a. Cervix, vagina, ovary, oviduct
   b. Vagina, cervix, uterus, oviduct
   c. Uterus, cervix, vagina, oviduct
   d. Vagina, uterus, cervix, oviduct
   Textbook Reference: 42.3 How Do the Human Male and Female Reproductive Systems Work? p. 910

5. The function of the seminal vesicle is to
   a. produce a solution of fructose to provide energy for the sperm.
   b. secrete alkaline fluids that neutralize the acidity of a female's reproductive tract.
   c. initiate the muscular contractions that lead to emission.
   d. produce lubrication for the tip of the penis.
   Textbook Reference: 42.3 How Do the Human Male and Female Reproductive Systems Work? p. 908

6. Which of the following is an example of positive feedback control in the reproductive cycle of males or females?
   a. The increased response of the hypothalamus and anterior pituitary gland in response to estrogen.
   b. The decreased response of the hypothalamus and anterior pituitary gland in response to estrogen.
   c. The inhibition of luteinizing hormone by high levels of testosterone.
   d. The stimulation of luteinizing hormone by low levels of testosterone.
   Textbook Reference: 42.3 How Do the Human Male and Female Reproductive Systems Work? pp. 911–912

7. Which of the following statements about oogenesis is *false*?
   a. The polar bodies degenerate after the second meiotic division.
   b. The ovum produced is haploid.
   c. The major growth phase of the primary oocyte occurs in prophase I.
   d. The primary oocyte is haploid.
   Textbook Reference: 42.2 How Do Animals Reproduce Sexually? pp. 899–901

8. Which of the following is *not* an accessory sex organ?
   a. Penis
   b. Prostate gland
   c. Gonad
   d. Breast
   Textbook Reference: 42.2 How Do Animals Reproduce Sexually? p. 904

9. For approximately how long during the human female's menstrual cycle are progesterone concentrations high enough to maintain the uterus in a proper condition for pregnancy?
   a. The entire duration of the cycle
   b. No portion of the cycle
   c. During the first half of the cycle
   d. During the second half of the cycle
   Textbook Reference: 42.3 How Do the Human Male and Female Reproductive Systems Work? pp. 911–912

10. Which of the following statements about the birth control pill is *false*?
    a. It works by preventing ovulation.
    b. It works by preventing implantation.
    c. The ovarian cycle is suspended by the birth control pill.
    d. It contains low doses of estrogen and progesterone.
    Textbook Reference: 42.4 How Can Fertility Be Controlled and Sexual Health Maintained? pp. 914–915

11. Which of the following is *false*?
    a. Sex hormones control negative feedback of the hypothalamus.
    b. Contraception through complete chemical blockage of sperm production is possible.
    c. Surgical transfer of gametes from one area of the reproductive tract to another is possible.
    d. Reproductive organs can be susceptible to cancer.
    Textbook Reference: 42.4 How Can Fertility Be Controlled and Sexual Health Maintained? p. 916

12. If you compared the genetic makeup of a female animal produced by parthenogenesis with that of its mother, which of the following would you expect?
    a. About 100 percent genetic similarity
    b. About 50 percent genetic similarity
    c. No genetic similarity
    d. Parthenogenetic animals do not have mothers
    Textbook Reference: 42.1 How Do Animals Reproduce Without Sex? p. 898

13. Which of the following statements about oviparity is *false*?
    a. Only birds and reptiles are oviparous.
    b. The large amount of yolk in the egg provides the nutrients for the developing embryo.
    c. The shell protects the egg from dehydrating.
    d. Both oxygen and carbon dioxide can diffuse through the shell.
    Textbook Reference: 42.2 How Do Animals Reproduce Sexually? p. 905

14. During spermatogenesis a single male germ cell produces _____ sperm cell(s).
    a. 1
    b. 2
    c. 3
    d. 4
    Textbook Reference: 42.2 How Do Animals Reproduce Sexually? pp. 899–900

15. Which of the following statements about fertilization is *false*?
    a. The egg permits several sperm to enter it.
    b. The plasma membranes of sperm and egg fuse.
    c. The egg is activated and stimulated to begin development.
    d. Species-specific recognition occurs between egg and sperm.
    Textbook Reference: 42.2 How Do Animals Reproduce Sexually? pp. 901–903

16. Which of the following statements is *false*?
    a. Oxytocin triggers the uterine contractions of childbirth in a positive feedback cycle.
    b. Condoms are the only contraceptive devices effective against the transmission of STDs.
    c. RU-486 blocks estrogen receptors.
    d. Men, but not women, have a refractory period following orgasm.
    Textbook Reference: *42.4 How Can Fertility Be Controlled and Sexual Health Maintained? p. 915*

Application Questions

1. *Daphnia* (water fleas) can reproduce sexually or asexually. When and why might these animals switch between sexual and asexual reproduction?
   Textbook Reference: *42.1 How Do Animals Reproduce Without Sex? p. 897*

2. While visiting the shore of a small pond, you notice two frogs mating in the water. Based on what you know about external fertilization, what will ensure that the female's eggs will be fertilized?
   Textbook Reference: *42.2 How Do Animals Reproduce Sexually? p. 903*

3. How is it possible for a sea star to reproduce both sexually and by regeneration? How will the genetic makeup of offspring from each method differ?
   Textbook Reference: *42.1 How Do Animals Reproduce without Sex? pp. 897–898*

4. Given what you know about the hormones that regulate the ovarian and uterine cycles in human females, explain the mechanism by which birth control pills containing synthetic estrogen and progesterone prevent pregnancy.
   Textbook Reference: *42.4 How Can Fertility Be Controlled and Sexual Health Maintained? pp. 914–915*

5. Why are multiple births (for example, septuplets) associated with assisted reproductive technologies (ARTs)?
   Textbook Reference: *42.4 How Can Fertility Be Controlled and Sexual Health Maintained? p. 916*

Answers

Knowledge and Synthesis Answers

1. **b.** The parents that have survived to reproduce asexually are able to survive in the current stable environment. Therefore, the offspring should be preadapted for this stable environment.

2. **d.** There are many differences between an egg and sperm. They are produced by different sexes, sperm is motile and an egg is not, and an egg is much larger than sperm.

3. **d.** In males, sperm is produced during spermatogenesis. An egg is produced during oogenesis. Males have a secondary sexual organ such as the penis that aids in internal fertilization. External fertilization is not practical on land.

4. **b.** A sperm is ejected by the male into the vagina. From the vagina the sperm move through the cervix into the uterus and finally the oviduct, where fertilization occurs.

5. **a.** The seminal vesicles are involved in producing the seminal fluid. One of the components of the seminal fluid is an energy source for the sperm in the form of fructose.

6. **a.** During days 12 through 14, a positive feedback occurs in response to estrogen. The hypothalamus and anterior pituitary gland are stimulated to release LH and FSH.

7. **d.** During oogenesis, the primary oocyte is diploid; after the first meiotic division into the secondary oocyte the cell becomes haploid.

8. **c.** The gonads are not secondary sex organs; they are the primary sex organs producing the sperm and the egg.

9. **d.** High levels of progesterone are needed to maintain the uterus in the proper condition for pregnancy. The levels of progesterone are high only during the second half of the uterine cycle.

10. **b.** The birth control pill interferes with the maturation of the follicles and the ova, inhibiting the release of an egg.

11. **b.** Males constantly make sperm, so any chemical block of sperm production would have to be constant. Chemical blocks of sperm would also have serious medical side effects.

12. **a.** Parthenogenesis is a form of asexual reproduction in which offspring develop from unfertilized eggs; there is no sexual recombination of genes, so a female animal born parthenogenetically will be nearly identical to its mother genetically. (This kind of reproduction occurs in aphids, for example.) In many hymenopterans (bees, wasps, and ants), only males are born from unfertilized eggs. Such males are haploid, so they are not genetically identical to their diploid mothers.

13. **a.** Birds and reptiles are not the only oviparous species—a group of mammals, the monotremes, are also egg-layers.

14. **d.** During spermatogenesis a single male germ cell will undergo two meiotic divisions, resulting in the production of four haploid sperm cells.

15. **a.** During fertilization an egg uses blocks to polyspermy to prevent more than one sperm from entering it.

16. **c.** RU-486 blocks progesterone receptors and causes sloughing off of the endometrium and embryo.

Application Answers

1. Female *Daphnia* tend to produce new females by parthenogenesis when conditions are favorable. When conditions turn unfavorable, they produce males instead of females. After males are produced, the

females reproduce sexually with them, producing a resting egg. Mating with the males allows for genetic mixing of the population and the production of a resting egg that will produce a new parthenogenetic female when conditions turn favorable again.

2. For successful external fertilization in water, the two frogs must engage in a behavior that will bring the male's sperm in close proximity to the female's eggs. The second important thing that must occur is that the male and female must both produce large numbers of gametes to ensure that eggs are fertilized. Finally, the male and female must be at the same reproductive stage for successful mating to occur.

3. If a sea star loses one of its arms, it can regenerate a new one to replace it. If the detached arm remains alive, it can regenerate an entire new sea star. The new sea star will be genetically identical to the original one. Sea stars also produce eggs and sperm that join during sexual reproduction to produce new individuals. The new sea stars will share a genetic makeup with both parents, increasing genetic diversity.

4. During normal cycling, estrogen and progesterone serve as negative feedback signals (except at very high levels) to the hypothalamus and pituitary gland and thus keep follicle-stimulating hormone (FSH) and luteinizing hormone (LH) at low levels. The low doses of estrogen and progesterone in birth control pills function in the same manner; these hormones prevent ovulation (and thus pregnancy) by keeping FSH and LH low.

5. Normally, a woman ovulates a single secondary oocyte each month; if this oocyte is fertilized, then a single birth may result. Many ARTs use hormones ("fertility drugs") to trigger superovulation, the release of several secondary oocytes from the ovary. If more than one of these oocytes is fertilized, then multiple births may result.

CHAPTER 43 Animal Development: From Genes to Organisms

Important Concepts

Fertilization is the joining of the sperm and the egg.

- Early development of animals has been studied extensively in a few model organisms: sea urchins, frogs, chickens, and humans.

- Development begins when a haploid sperm joins with a haploid egg. However, fertilization does more than produce a diploid zygote; fertilization activates development.

At fertilization, the egg and sperm contribute differently to the zygote.

- The egg contributes most of the cytoplasm and organelles to the zygote.

- The sperm contributes its haploid nucleus and a centriole that becomes the centrosome and eventually forms the mitotic spindles.

- In an unfertilized frog egg, nutrients are concentrated in the vegetal hemisphere of the egg in the lower half, whereas the haploid nucleus is found in the upper half, of the animal hemisphere.

- Upon fertilization in the frog egg, the cytoplasm of the egg undergoes rearrangement (see Figure 43.1). A sperm enters in the animal hemisphere at what will become the ventral side of the frog and causes rotation of the cortical cytoplasm to create a gray crescent, which will then become the dorsal region of the frog. This helps to establish the left-to-right axis and the anterior-to-posterior axis of the developing animal.

- During the movement of the cytoplasm, the transcription factor β-catenin is degraded in parts of the egg so that it becomes concentrated in the dorsal side of the embryo (see Figure 43.2). This happens because the protein GSK-3 moves to the ventral side of the zygote where it targets β-catenin for degradation. On the dorsal side, proteins inhibit the action of GSK-3 and, thus, the concentration of β-catenin is higher.

Cleavage is a rapid series of cell divisions early in development.

- The first divisions of the cells occur rapidly with little growth or differentiation in a process known as cleavage. During this time, the embryo divides into smaller and smaller cells, producing first a morula and then a blastula. The blastula has a central cavity known as the blastocoel. Individual cells in the blastula are known as blastomeres.

- The amount of yolk in the egg determines the plane along which cell division occurs (see Figure 43.3). Cleavage furrows are impeded by yolk. Animals such as sea urchins and frogs have small amounts of yolk and have complete cleavage. The eggs of fishes, reptiles, and birds have large amounts of yolk and have incomplete cleavage in which the embryo develops from a disc of cells (known as the blastodisc) on top of the yolk. Some insects undergo superficial cleavage, a variation of incomplete cleavage.

- The orientation of mitotic spindles determines the cleavage planes and thus the arrangement of daughter cells. In the sea urchin and frog, the mitotic spindles are parallel or perpendicular to the animal–vegetal axis, and the resulting cleavage pattern is radial. In mollusks, the mitotic spindles are at oblique angles to the animal–vegetal axis and the resulting cleavage pattern is spiral.

In mammals, cells separate into two groups between the 16-cell and 32-cell stage.

- Mammals have a unique rotational cleavage pattern. Cleavage in mammals is relatively slow and asynchronous. The zygote produces both the embryo and extraembryonic structures that interact with the mother. Unlike in other species where cleavage is directed by molecules already present in the egg, in mammals genes are expressed during cleavage and play a large role.

- During cleavage from the 16-cell to the 32-cell stage, the cells separate into an inner cell mass that will become the embryo and a trophoblast that will become part of the placenta and attach to the uterus during implantation. At the 32-cell stage the mass of mammalian cells is called a blastocyst (see Figure 43.4).

- Mammalian fertilization occurs in the oviduct, and the zygote moves down into the uterus where it undergoes implantation. The zona pellucida inhibits early implantation and is lost just prior to implantation in the uterus. Implantation anywhere other than the uterus is a dangerous condition called ectopic pregnancy.

Blastomeres become determined during late cleavage.

- During cleavage, cytoplasm is distributed to the cells of the blastula in such a way that cells in different regions have different levels of nutrients and informational molecules. Specific blastomeres will become specific tissues and organs, and this can be mapped out in the developing cells as a fate map of the blastula (see Figure 43.6). The blastomeres become committed to specific fates at different times for different species.

- Animals with mosaic development have blastomeres that are set to contribute to specific parts of the embryo. In animals with regulative development, the blastomeres can be removed, and other cells will compensate for the loss (see Figure 43.7).

Gastrulation involves movement and differentiation of cells.

- During gastrulation, the germ layers of the tissues form and position themselves in the embryo. Gastrulation makes possible the inductive interactions between cells, which trigger differentiation and organ formation.

- Three germ layers develop during gastrulation. The inner layer or endoderm will become tissues lining the gut, respiratory, and circulatory systems; the outer layer or ectoderm will become the skin, hair, nails, glands, and the nervous system; and the middle layer or mesoderm will become the tissues of many organs, including blood vessels, muscle, bone, liver, and heart.

Gastrulation in sea urchins involves invagination of the vegetal pole.

- The vegetal pole in sea urchins first flattens. The flat portion of the vegetal pole then moves inward as an invagination to form the endoderm and primitive gut, known as the archenteron. Some cells from the vegetal pole break free to become the primary mesenchyme cells that make up the mesoderm (see Figure 43.8).

- Secondary mesenchyme cells are produced from cells that break off at the tip of the archenteron and enter the blastocoel.

- The opening of the archenteron or blastopore will become the anus, and the place where the tip of the archenteron meets the ectoderm will become the mouth.

The dorsal lip controls embryonic organization during gastrulation in the frog.

- The frog embryo has more yolk than the sea urchin, and gastrulation is more complex (see Figure 43.9). Initially, cells near the gray crescent begin to bulge into the blastocoel. These initial cells (bottle cells) move along the interior of the blastula and pull the outer surface of cells along with them, creating a dorsal lip; this process is called involution. The first cells moving

in are prospective endoderm, and they form the archenteron.

- As gastrulation continues, cells from the animal hemisphere move toward the site of involution; this process is known as epiboly. As epiboly continues, the dorsal lip forms a complete circle around a plug of yolk-rich cells. At the end of gastrulation, the cells have become fate-determined and are layered as the ectoderm, endoderm, and mesoderm.

- Hans Spemann examined the timing and fate of cells during gastrulation in salamanders to determine if they were totipotent or able to direct development (see Figure 43.10). Spemann and Hilde Mangold, his student, found that when the dorsal lip was transplanted to another gastrula, it resulted in a second site of gastrulation and eventually two embryos attached at the belly. They concluded that the dorsal lip acts as the primary embryonic organizer (see Figure 43.11).

- β-catenin is a candidate for initiating the signal cascade involved in organizer activity. When β-catenin activity is knocked out by antisense RNA, the embryo does not go through gastrulation; overexpression of β-catenin can induce a second axis of embryo formation.

- Presence of β-catenin and a complex series of interactions between growth and transcription factors create the organizer and lead to induction of the body plan. Primary embryonic organizer activity is started from the vegetal cells below the gray crescent. The transcription factors *goosecoid* and *siamois* are critical in the signal cascade. Tcf-3 proteins act as repressors for *siamois*. When the Tcf-3 protein is blocked by β-catenin, expression of *siamois* occurs, and the goosecoid protein is produced (see Figure 43.12).

- As the organizer migrates from the dorsal lip, it inhibits various growth factors along the way to achieve different patterns of differentiation along the anterior-posterior axis.

In reptile and bird gastrulation, the primitive groove is the blastopore, and Hensen's node corresponds to the amphibian dorsal lip.

- Cleavage produces a blastodisc on top of the large amount of yolk in the egg.

- The blastula is a circular layer of cells composed of an outer epiblast layer and an inner hypoblast layer. The epiblast will form the embryo proper and the hypoblast will form extraembryonic membranes. The fluid-filled space between the two layers is the blastocoel.

- The primitive streak is formed by the movement of cells in the epiblast toward the midline (see Figure 43.13). Along the primitive streak, a primitive groove forms and cells move through this blastopore, becoming endoderm and mesoderm in the blastocoel.

- In the chicken there is no archenteron; endoderm and mesoderm move forward and form gut structures. A group of cells at the anterior end of the primitive groove, known as Hensen's node, acts in the same manner as the dorsal lip of the frog blastopore. Cells that move over Hensen's node differentiate into the notochord and structures of the head.

Gastrulation in placental mammals is similar to that in reptiles and birds.

- In placental mammals, the cells of the embryo segregate into an epiblast and a hypoblast with a blastocoel in between, just as in birds and reptiles. Extraembryonic membranes, including that which contributes to the placenta, develop from the hypoblast. The embryo and some extraembryonic membranes develop from the epiblast.

- Gastrulation in mammals occurs just as in birds, with the formation of a primitive groove through which epiblast cells migrate to form endoderm and mesoderm.

The nervous system begins to develop during neurulation.

- After gastrulation, the organs begin to develop through the process of organogenesis. Initial development of the nervous system, neurulation, begins early in organogenesis.

- During neurulation, a rod of connective tissue, called the notochord, develops in the blastocoel from chorda-mesoderm cells derived from the dorsal mesoderm. The notochord provides structural support to the developing embryo and induces overlying cells to form the nervous system. Eventually the notochord will be replaced by the vertebral column.

- A neural plate forms above the notochord from the ectoderm, which will begin to fold into a cylinder forming the neural tube (see Figure 43.14). The anterior end of the neural tube will become the major parts of the brain, with the rest forming the spinal cord. Failure of this process leads to serious birth defects such as spina bifida and anencephaly.

- The repeating pattern of the vertebrate body plan forms from blocks of somite tissue derived from mesoderm located on both sides of the notochord (see Figure 43.15). The somites will become muscles, ribs, and vertebrae.

- Neural crest cells break from the neural tube and migrate inward to become peripheral nerves.

Homeotic genes control differentiation in the anterior–posterior directions during development.

- Body segments differentiate as the embryo develops. Hox genes control this differentiation of body segments.

- Hox genes are found on different chromosomes and are expressed along the anterior–posterior axis in their order on the chromosomes (see Figure 43.16).

- Dorsal–ventral differentiation is controlled by a separate set of genes. For example, the *sonic hedgehog* gene, expressed in the notochord, prompts cells in the overlying neural tube to become part of the ventral spinal cord.

- *Pax* genes (a family of homeobox genes), such as *Pax3*, are also important in nervous system and somite development.

- After segmentation occurs, organs and organ systems develop rapidly.

Birds and reptiles develop extraembryonic membranes.

- Extraembryonic membranes surround the embryos of reptiles, birds, and mammals.

- In birds, the yolk sac is derived from the hypoblast and surrounds the entire yolk to help retrieve nutrients from the yolk and deliver them to the embryo (see Figure 43.17).

- Cells from the ectoderm and mesoderm form the amnion, which helps provide an aqueous environment, and the chorion, which develops just under the shell. The chorion regulates water, oxygen, and carbon dioxide exchanges across the shell.

- The allantois is derived from endoderm and is a membrane that forms a sac to store metabolic waste products.

The placenta of mammals provides exchanges between the mother and the embryo.

- In mammals, the first extraembryonic membrane to form, the trophoblast, interacts with the endometrium to attach to the uterine wall and begin implantation.

- The hypoblast cells interact with the trophoblast to form the chorion, which, along with tissues from the uterine wall, forms the placenta (see Figure 43.18).

- The amnion of the developing mammal surrounds the embryo to produce a closed, fluid-filled environment.

- The allantois of mammals has the function of removing nitrogenous wastes, but its function is relatively minor in many mammals, including humans. The tissues of the allantois help to form the umbilical cord, which carries major blood vessels that provide a route for exchanges of nutrients, wastes, carbon dioxide, and oxygen between the mother and fetus.

- Amniocentesis and chorionic villus sampling are two methods used to detect genetic markers for diseases (see Figure 43.19).

Human pregnancy is divided into trimesters.

- In humans, pregnancy lasts about 266 days, compared with 21 days in mice and 600 days in elephants.

- During the first trimester of human development, the embryo develops into a miniature version of the adult, called the fetus. Cell division and tissue differentiation are rapid during the first trimester. At this stage, the developing fetus is most sensitive to environmental disruptors. The corpus luteum continues to produce estrogen and progesterone during this trimester. Most organs have started to form by the end of the first trimester.

- During the second trimester of development, the first fetal movements are felt, the fetus reaches a size of 600 g, and the mother begins to show physical evidence of her pregnancy.
- The third trimester is marked by rapid growth of the fetus and the maturation of the internal organs.
- Development continues after birth, with changes in the brain being particularly evident up until adolescence.

The Big Picture

- Animal development involves a number of steps: fertilization, cleavage, gastrulation, and organogenesis (for example, neurulation).
- Fertilization is the joining of the egg and sperm to form a zygote. Following fertilization, the cells of the embryo begin to divide in a process known as cleavage. Eggs undergo either complete cleavage or incomplete cleavage, depending on the amount of yolk present. Those that undergo complete cleavage form a blastula, and those that undergo incomplete cleavage form a flat blastodisc.
- Cleavage in mammals is different from cleavage in other groups. It is much slower, and the products of embryonic genes play a role in directing its course; in other animals, cleavage is directed entirely by molecules present in the egg prior to fertilization, not the products of embryonic genes.
- During gastrulation, the germ layers—endoderm, ectoderm, and mesoderm—form and move into specific positions. In the sea urchin, an archenteron that will become the gut forms, and specific cells form the three germ layers. In frogs, a dorsal lip is formed, and the germ cells move into place. The dorsal lip is considered the primary embryonic organizer in amphibians. The blastodisc of reptiles, birds, and mammals goes through a much different pattern of gastrulation.
- Neurulation occurs early in organogenesis. The notochord induces overlying ectoderm to form a neural plate. The neural plate forms into a neural tube that will become the brain and spinal cord. Somites produce the repeating segments of the vertebrate body plan. A number of genes play a large role in the differentiation of tissues along the body axes.
- Reptiles and birds have four major extraembryonic membranes: yolk sac, amnion, chorion, and allantois. Mammals have a special placenta for the exchange of nutrients, gases, and wastes between the mother and the developing fetus.

Common Problem Areas

- All fertilized cells must divide to grow into adult animals. Sorting out the features of division that are common to all animals from those specific to certain taxonomic groups can be difficult and confusing until you begin to recognize the basic patterns.

- Many students find gastrulation difficult to picture and understand, especially given the different ways in which gastrulation occurs in different organisms.
- Students are often confused about the structure and role of the placenta. Is it embryonic? Is it maternal? And how do nutrients, gases, and wastes travel between the mother and embryo or fetus?

Study Strategies

- There is a great deal of terminology in this chapter. Yet there are also many terms that are common to worms and wolves alike. Create a list of new terms from this chapter. Then determine which are the more general terms applicable to numerous animals (e.g., "somites," "gastrulation") and learn these first. Then tackle those referring to development in specific animals (e.g., "dorsal lip of the blastopore," "primitive streak").
- A key point in understanding animal development is that all specialized tissues arise from the three germ layers: endoderm, ectoderm, and mesoderm. More important than just committing these three layers to memory is learning the specialized tissues that derive from them. This will enable you to make predictions about tissue functions and to compare body plans among animals.
- Review the following animated tutorials and activity on the Companion Website/CD:
  Tutorial 43.1 Gastrulation
  Tutorial 43.2 Tissue Transplants Reveal the Process of Determination
  Activity 43.1 Extraembryonic Membranes

Test Yourself

Knowledge and Synthesis Questions

1. The sperm and the egg make different contributions to the zygote. Which statement about their contributions is *false*?
   a. The sperm contributes most of the organelles.
   b. The egg contributes most of the cytoplasm.
   c. Both the sperm and the egg contribute a haploid nucleus.
   d. All of the above
   Textbook Reference: 43.1 How Does Fertilization Activate Development? p. 921

2. In which animal is the cleavage pattern rotational?
   a. Frog
   b. Mammal
   c. Sea urchin
   d. All of the above
   Textbook Reference: 43.1 How Does Fertilization Activate Development? p. 923

3. Why do some species have complete cleavage and others incomplete cleavage?
   a. Incomplete cleavage occurs in species with small volumes of cytoplasm.

b. Complete cleavage, found in mammals, is a more evolved characteristic.

c. Incomplete cleavage occurs in species with large amounts of yolk.

d. Complete cleavage occurs only in eggs that have been fertilized by two sperm.

Textbook Reference: 43.1 How Does Fertilization Activate Development? p. 923

4. A fate map can be used to map out the tissues and organs that will eventually develop from specific germ layers. What will ectoderm eventually become?
a. The lining of the gut
b. The epidermal layer of skin
c. Muscle
d. The heart

Textbook Reference: 43.2 How Does Gastrulation Generate Multiple Tissue Layers? p. 926

5. The location of the _____ determines the anterior–posterior axis of the embryo.
a. primitive streak
b. blastopore
c. vegetal hemisphere
d. hypoblast

Textbook Reference: 43.2 How Does Gastrulation Generate Multiple Tissue Layers? p. 927

6. What is the order of the germ layers from the inside to the outside?
a. Mesoderm, ectoderm, endoderm
b. Endoderm, ectoderm, mesoderm
c. Ectoderm, mesoderm, endoderm
d. Endoderm, mesoderm, ectoderm

Textbook Reference: 43.2 How Does Gastrulation Generate Multiple Tissue Layers? p. 927

7. The dorsal lip of the blastopore organizes embryo formation in frogs. What is the equivalent in chickens?
a. Epiblast
b. Hypoblast
c. Hensen's node
d. Bottle cells

Textbook Reference: 43.2 How Does Gastrulation Generate Multiple Tissue Layers? p. 932

8. Birds develop extraembryonic membranes during development. Which statement about avian extraembryonic membranes is *false*?
a. The yolk sac surrounds the yolk and provides nutrients.
b. The amnion and chorion are derived from ectoderm and mesoderm.
c. The allantois stores nutrients.
d. The chorion exchanges gases and water between the embryo and the environment.

Textbook Reference: 43.4 What Is the Origin of the Placenta? pp. 934–935

9. Which of the following statements about the mammalian blastocyst is *false*?
a. The trophoblast gives rise to the embryo proper.

b. Maternal genes are expressed during cleavage.
c. The blastocyst implants in the mother's uterus.
d. Early mammalian cleavage is relatively slow.

Textbook Reference: 43.4 What Is the Origin of the Placenta? p. 935

10. The primary embryonic organizer is most likely initiated by _____.
a. the yolk
b. TCF-3 protein
c. β-catenin
d. None of the above

Textbook Reference: 43.2 How Does Gastrulation Generate Multiple Tissue Layers? pp. 929–930

11. The _____ eventually develop into vertebrae, ribs, and muscles and are found along the sides of the

_____.
a. somites; notochord
b. neural tube cells; notochord
c. blastopore cells; dorsal lip
d. neural crest cells; dorsal lip

Textbook Reference: 43.3 How Do Organs and Organ Systems Develop? p. 933

12. Why did Hans Spemann call the dorsal lip of the blastopore the embryonic organizer?
a. It is the point where gastrulation begins.
b. It becomes part of the nervous system.
c. It becomes part of the notochord.
d. It leads to the establishment of the embryonic axes.

Textbook Reference: 43.2 How Does Gastrulation Generate Multiple Tissue Layers? pp. 928–929

13. During its development, the human embryo is contained within a fluid-filled chamber bounded by the membranous
a. yolk sac.
b. amnion.
c. chorion.
d. allantois.

Textbook Reference: 43.4 What Is the Origin of the Placenta? pp. 935–936

14. The third trimester of human prenatal development is characterized by
a. rapid growth.
b. formation of major organs.
c. greatest sensitivity to damage from drugs and radiation.
d. gastrulation.

Textbook Reference: 43.5 What Are the Stages of Human Development? p. 937

15. Which of the following statements about human development is *false*?
a. The hormone human chorionic gonadotropin is secreted by the blastocyst soon after implantation.
b. Fetal movements are first felt by the mother during the second trimester.
c. Substantial developmental changes occur in the brain between birth and adolescence.

d. Given the time needed to develop their relatively large brains, humans have the longest gestation period among mammals.
Textbook Reference: 43.5 What Are the Stages of Human Development? pp. 936–938

Application Questions

1. In the following diagram of a sea urchin gastrula, label the ectoderm, endoderm, primary mesenchyme, secondary mesenchyme, blastopore, blastocoel, and archenteron.

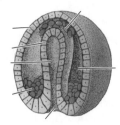

Textbook Reference: 43.2 How Does Gastrulation Generate Multiple Tissue Layers? p. 927

2. Label the listed structures in the following longitudinal cross section of a late frog gastrula: blastopore, area where brain will form, yolk plug, notochord, neural plate, archenteron, ectoderm, mesoderm, and endoderm. Also indicate the (A) anterior, (P) posterior, (D) dorsal, and (V) ventral areas of the embryo.

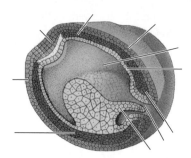

Textbook Reference: 43.2 How Does Gastrulation Generate Multiple Tissue Layers? p. 928

3. Compare cleavage in birds and in mammals. How do differences in cleavage between these two groups influence gastrulation?
Textbook Reference: 43.1 How Does Fertilization Activate Development? pp. 923–925

4. How does the development of the extraembryonic membranes of birds and mammals differ and how are they similar? What is the main reason for the differences?
Textbook Reference: 43.4 What Is the Origin of the Placenta? pp. 934–936

5. Compare the trophoblast and inner cell mass of the human blastocyst with respect to function. From which of the two do embryonic stem cells come from?
Textbook Reference: 43.4 What Is the Origin of the Placenta? pp. 935–936

Answers
Knowledge and Synthesis Answers

1. **a.** The egg donates most of the organelles to the zygote. In some species, the sperm may contribute a centriole.

2. **b.** Cleavage in mammals occurs in a rotational pattern.

3. **c.** Incomplete cleavage occurs because the cleavage furrows cannot completely penetrate the yolk. Incomplete cleavage occurs in bird and reptile species with large amounts of yolk.

4. **b.** Ectoderm will eventually become the epidermal layer of skin.

5. **b.** The blastopore, which will eventually become the anus, marks the posterior of the embryo.

6. **d.** The germ layers that are formed during gastrulation are the inner layer of endoderm, the middle layer of mesoderm, and the outer layer of ectoderm.

7. **c.** In chickens, Hensen's node is the equivalent of the dorsal lip of the frog blastopore.

8. **c.** During development, an avian embryo produces wastes, but because they develop in a shell, the wastes must be stored in the egg. The allantois forms a sac that is used for the storage of metabolic wastes.

9. **a.** In the mammalian blastocyst, the trophoblast forms the fetal part of the placenta. A disc-shaped portion of the inner cell mass becomes the embryo.

10. **c.** β-catenin is thought to be the initiator of organizer activity during early development.

11. **a.** Somites are located along the notochord in the developing vertebrate. These cells will develop into the vertebrae, ribs, and muscles.

12. **d.** The dorsal lip leads to the establishment of the embryonic axes.

13. **b.** The amnion makes up the membrane that surrounds the developing mammalian fetus.

14. **a.** Rapid growth characterizes the third trimester. Gastrulation, formation of major organs, and greatest sensitivity to drugs and radiation occur during the first trimester.

15. **d.** The human gestation period is 9 months, and it is not the longest among mammals. Length of gestation is related to overall body size and not relative brain size.

Application Answers

1.

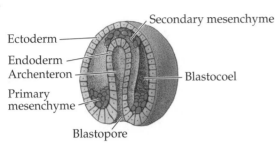

Secondary mesenchyme

Ectoderm

Endoderm

Archenteron

Primary mesenchyme

Blastocoel

Blastopore

2.

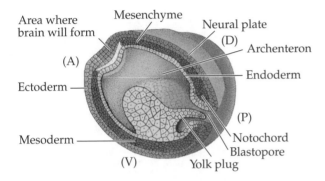

Area where brain will form

Mesenchyme

Neural plate

(D)

Archenteron

(A)

Ectoderm

Endoderm

Mesoderm

(P)

Notochord

Blastopore

(V)

Yolk plug

3. Due to the large amount of yolk in the avian egg, cleavage in birds is incomplete, and the result is a blastodisc that sits on top of the yolk (see Figure 43.3). Unlike bird embryos, mammalian embryos undergo complete cleavage in a rotational pattern, resulting in a structure called a blastocyst (see Figure 43.4). The pace of cleavage is also slower in mammals than in birds, and embryonic genes play a role in directing cleavage in mammals, whereas in birds, cleavage is directed by molecules already present in the egg. Despite these differences, gastrulation is very similar in birds and mammals. The avian blastodisc develops into a flattened blastula with epiblast and hypoblast surrounding a fluid-filled blastocoel; the mammalian inner cell mass also splits into an epiblast and a hypoblast. Although the details vary (see the next question), the developmental fates of these structures are fundamentally the same in both groups: The hypoblast will give rise to extraembryonic membranes, and the embryo proper will develop from the epiblast.

4. Both the birds and mammals develop extraembryonic membranes. The extraembryonic membranes of the birds are used to obtain food from the large reserve of yolk (yolk sac), to exchange water and gases with the environment across the shell (amnion and chorion), and to store nitrogenous wastes (allantois). Mammals have extraembryonic membranes that form a placenta involved in exchanging nutrients, oxygen, and carbon dioxide with the mother. The hypoblast cells that form the yolk sac in birds along with the trophoblast cells and the chorion all contribute to the placenta. An allantois also develops, but has a limited function in humans.

5. The trophoblast secretes enzymes that allow the blastocyst to burrow into the uterine lining (endometrium) of the mother and to implant there. The trophoblast then forms chorionic villi, the embryo's major contribution to the placenta. In contrast, the inner cell mass forms the embryo proper and the amnion, yolk sac, and allantois. Embryonic stem cells come from the inner cell mass.

CHAPTER 44 Neurons and Nervous Systems

Important Concepts

The nervous system consists of neurons and glial cells.

- The nervous system is made up of two major types of cells: neurons and glial cells.

- Neurons transmit information as electric signals (action potentials). Afferent neurons carry sensory information into the nervous system. Efferent neurons carry information from the nervous system to effectors, such as muscles or glands. Interneurons facilitate communication between sensors and effectors.

- Simple animals, such as the sea anemone, possess a nerve net (see Figure 44.1A).

- The nervous system of invertebrates is made up of clusters of neurons distributed throughout the body. These clusters are known as ganglia.

- Vertebrates have a central nervous system—the brain and spinal cord—and a peripheral nervous system—the neurons in the rest of the body.

- Information is passed from one neuron to another neuron at a synapse. The first neuron is known as the presynaptic neuron, and the second as the postsynaptic neuron.

- A neuron is composed of four parts: the cell body, dendrites, axons, and axon terminals (see Figure 44.3A). The cell body contains the nucleus, with dendrites coming off it and receiving information from other neurons. The axon conducts action potentials away from the cell body. Axon terminals interact with the neuron's target cells to form a synapse.

- Chemical synapses are more common than electrical synapses. At a chemical synapse, an action potential arriving at the axon terminals of a presynaptic neuron causes the release of neurotransmitter, which travels across the synapse to bind with receptors on the postsynaptic neuron.

- Glial cells serve many functions in the nervous system, such as supplying nutrients to neurons, maintaining proper ionic balance, and removing waste particles. Glial cells that insulate axons in the peripheral nervous system are Schwann cells, and those that insulate axons in the central nervous system are oligodendrocytes.

Astrocytes found around blood vessels in the brain form a barrier to some toxins as well as to certain drugs.

Neurons have an electrically charged cell membrane at rest.

- Resting neurons have a negative charge inside and a positive charge outside, resulting in a difference in electrical charge across the membrane known as the membrane potential. In an unstimulated neuron, this voltage difference is called a resting potential. Electrodes can be used to measure resting potentials (see Figure 44.5).

- The electrical charge across the membrane at rest is due to differences in concentrations of the charged ions sodium (Na^+), chloride (Cl^-), potassium (K^+), and calcium (Ca^{2+}).

- The lipid bilayer of the plasma membrane is impermeable to ions. Ions move across the plasma membrane through channels or by ion pumps.

- The sodium–potassium pump transports Na^+ out of the cell and K^+ into it, and thereby maintains higher concentrations of Na^+ ions outside the cell and higher concentrations of K^+ ions inside it (see Figure 44.6A). Because the pump is an enzyme complex that needs ATP to perform its work, it is sometimes called sodium–potassium ATPase.

- At rest, neurons have a specific charge due to K^+ movement to the outside of the cell, resulting in a resting potential. K^+ channels are the most common open channels, allowing K^+ to diffuse out of the cell down the concentration gradient that has been set up by the Na^+–K^+ pump (see Figure 44.6B).

- Ion channels are selective pores in the plasma membrane that allow specific ions to diffuse across the membrane. Ion channels can be voltage-gated (responding to changes in the voltage across the membrane), chemically gated (responding to the presence of a specific chemical), or mechanically gated (responding to mechanical force applied to the membrane).

- Patch clamping is a technique that allows the recording of voltage differences due to the movements of ions through channels in an isolated patch of plasma membrane (see Figure 44.8). Sakmann and Neher developed patch clamping in the 1980s and received a Nobel Prize in 1991.

Neurons generate action potentials to conduct nerve impulses.

- Changes in resting membrane potential ("depolarization") produce action potentials or nerve impulses.

- Membranes can be depolarized or hyperpolarized (see Figure 44.9). Depolarization occurs when the inside of a neuron becomes less negative compared to the resting potential. Hyperpolarization occurs when the inside of a neuron becomes more negative compared to the resting potential.

- Changes in polarity of the plasma membrane due to opening and closing of ion channels are passed down along an axon to transmit a signal as an action potential.

- Action potentials are very short-lived changes in membrane potential (see Figure 44.10). Action potentials are generated when the membrane reaches a threshold potential, Na^+ voltage-gated channels open, and Na^+ enters the cell to make the inside of the axon positive. Voltage-gated K^+ channels then open, allowing positive-charged K^+ to leave the axons to help return the membrane potential back to the resting level. As the K^+ channels open, the Na^+ channels close and cannot be opened for about 1 to 2 milliseconds, which is the refractory period. The Na^+–K^+ pump helps return the concentration of ions back to the resting levels.

- Action potentials travel down an axon by a positive feedback mechanism that stimulates adjacent regions of an axon to generate the action potential. The Na^+ ions that enter during an action potential flow to adjoining regions of the axon, stimulating depolarization and the movement of the action potential along the axon (see Figure 44.11).

- The refractory period during which the Na^+ channels cannot act can be explained by the presence of two gates in the channel, an activation gate and an inactivation gate. The refractory period keeps an action potential moving in one direction, away from the cell body.

- An action potential is an all-or-none response; the depolarization must reach a threshold level for an action potential to occur. An action potential is also a self-regenerating response; once an action potential occurs at one location on an axon, it stimulates the adjacent area to generate an action potential.

Action potentials can jump down an axon.

- In the nervous systems of invertebrates, the conduction velocity of axons is increased by increasing the diameters of axons. In the nervous systems of vertebrates, conduction velocity of axons is increased by a different means, myelination.

- Myelination occurs by glial cells wrapping themselves around some axons; Nodes of Ranvier are gaps in the myelin wrapping at which depolarization can occur. Depolarization jumps from node to node, increasing the speed of transmission of an action potential along the axon in a process known as saltatory conduction (see Figure 44.12).

Neurons communicate across synapses.

- Transfer of information from one cell to another by either a chemical or electrical message occurs in a small junction called a synapse. Presynaptic nerve cells send a message across a synapse to postsynaptic cells.

- A chemical synapse uses a chemical messenger, or neurotransmitter, to communicate between the presynaptic and postsynaptic cells. The neurotransmitter is produced in the axon terminal and packaged into vesicles. The neurotransmitter is released into the synaptic cleft when the vesicle binds with the presynaptic membrane. The neurotransmitter crosses the synaptic cleft and binds with receptors on the surface of the postsynaptic cell. In electrical synapses, the action potential spreads directly from presynaptic to postsynaptic cell.

Acetylcholine is a common synaptic neurotransmitter.

- The neurotransmitter acetylcholine is the chemical messenger carrying information between motor neurons and muscle cells at neuromuscular junctions (see Figure 44.13). Acetylcholine is enclosed in a vesicle at the presynaptic synapse, which fuses with the membrane to release the acetylcholine into the 20- to 40-nm-wide gap between the two cells, the synaptic cleft.

- Ca^{2+} channels open when the action potential reaches the axon terminal, causing Ca^{2+} to rush in and regulate the fusing of the acetylcholine-containing vesicles to the presynaptic membrane.

- Acetylcholine released into the synaptic cleft binds with receptors in the postsynaptic membrane, the motor end plate, which opens Na^+ channels, resulting in depolarization of the motor end plate (see Figure 44.14).

- Acetylcholinesterase breaks down acetylcholine in the synapse junction to halt the action of the released acetylcholine.

Synapses can either depolarize or hyperpolarize postsynaptic membranes.

- Synapses in vertebrates can be excitatory and depolarize the postsynaptic membrane or inhibitory and hyperpolarize the postsynaptic membrane.

- Neurons may receive synaptic inputs from many neurons. Excitatory and inhibitory postsynaptic potentials are summed by adding simultaneous potentials (spatial summation) or by summing rapid firing of one postsynaptic potential (temporal summation) (see Figure 44.15).

- The region of the cell body at the base of the axon, called the axon hillock, is the "decision-making" area of a neuron. If the axon hillock is depolarized, the axon will fire an action potential.

Two main types of neurotransmitter receptors exist.

- The two general categories of neurotransmitter receptors are ionotropic and metabotropic. Ionotropic receptors are ion channels on the postsynaptic membrane that are activated by the binding of the neurotransmitter. They allow fast, short-lived responses.

- Metabotropic receptors are transmembrane proteins that do not act like ion channels. Instead, metabotropic receptors act by initiating second messengers, such as G proteins, to activate ion channels (see Figure 44.16). When mediated by metabotropic receptors, postsynaptic cell responses are usually slower and longer-lived than those generated by ionotropic receptors.

Electrical synapses connect some neurons.

- Electrical synapses are formed by direct contact between adjacent neurons; these synapses contain numerous gap junctions.

- Two neurons forming an electrical synapse are joined by connexons, which are tunnels (pores) between the two neurons that allow ions to pass between the two cells.

- Electrical synapses are very fast connections and can transmit an action potential in either direction, making them good for rapid communication.

- Electrical synapses are less common than chemical synapses in vertebrate nervous systems. With electrical synapses, there can be no temporal summation of synaptic inputs. Electrical synapses also require large areas of contact and cannot be inhibitory.

The action of a neurotransmitter is determined by the receptor to which it binds.

- There are over 50 neurotransmitters, including amino acids, peptides, gases (for example, nitric oxide), purines, and monoamines (see Table 44.1). One neurotransmitter can act on a number of different receptors, and its particular action depends on the receptor to which it binds.

- Acetylcholine has nicotinic receptors and muscarinic receptors. Both types are found in the central nervous system, where the nicotinic receptors are ionotropic and tend to be excitatory, whereas the muscarinic receptors are metabotropic and tend to be inhibitory. Receptors for acetylcholine also occur outside the central nervous system.

The neurotransmitter glutamate may have a role in learning and memory.

- Two types of glutamate ionotropic receptors—NMDA and AMPA—may play a role in learning. NMDA receptors allow for a slow, long-lasting influx of Na^+, whereas the AMPA receptors allow for rapid influx (see Figure 44.17). The NMDA receptors also let Ca^{2+} into the postsynaptic cell, resulting in long-term cellular changes.

- Both of these receptors are excited by glutamate.

- Long-term potentiation is an enhanced response in the postsynaptic neuron with repeated stimulation (see Figure 44.18). Long-term potentiation may be involved in memory.

- Mice with NMDA receptors that stayed open longer than usual learned and remembered tasks better than normal mice.

Neurotransmitters must be removed from the synapse in order to turn off their action.

- The actions of neurotransmitters can be stopped in several ways. First, enzymes may destroy the neurotransmitter. Second, the neurotransmitter may simply diffuse away from the synaptic cleft. Third, nearby cell membranes may take up the neurotransmitter using active transport.

- Some gases used in chemical warfare cause death by paralysis by inhibiting acetylcholinesterase, the enzyme that destroys acetylcholine at the neuromuscular junction. The drug Prozac, used to treat depression, acts by slowing the reuptake of the neurotransmitter serotonin, thus prolonging serotonin's action at the synapse.

The Big Picture

- The neuron, with support from surrounding glial cells, is the functional unit of the nervous system. The neuron is composed of a cell body, dendrites, and an axon. The membrane of a neuron has a difference in voltage across it. Nerve impulses are passed down the axon of a neuron as action potentials. An action potential is a temporary disruption of the "battery-like" state of the resting neuron membrane due to the opening and closing of sodium and potassium voltage-gated channels. The action potential is an all-or-none response that occurs when the depolarization of an axon reaches a threshold level.

- At the end of the axon synapse, information flow is controlled through excitation or inhibition of synapses. There are many different neurotransmitters that transmit a nerve impulse from the presynaptic cell to the postsynaptic cell. The action of a specific neurotransmitter will depend on the receptor to which it binds.

Common Problem Areas

- One of the most difficult challenges when learning about the nervous system is understanding how the resting membrane potential and action potential are produced. Remember that the neuron at rest is like a battery and that the charge of the membrane is dependent on the ions that are on either side and moving across the membrane.

- Students often have difficulty with the function of a synapse and how the action potential is transmitted across the synapse. Remember that in many cases the signal goes from an electrical signal, to a chemical signal, and back to an electrical signal.

Study Strategies

- Students should understand the concept of "pre-"and "postsynaptic"neurons. Try to remember the sequence of events and relate them to the function of the neuron transmitting a stimulus from one cell to another.

- One of the major problems that students have is in understanding how the membrane potential is set up. Students should understand what each important anion and cation does at rest and during the action potential. Then, piece by piece, put together a sense of how the neuron membrane is charged at rest, and what happens during an action potential.

- Review the following animated tutorials and activity on the Companion Website/CD:
  Tutorial 44.1 The Resting Membrane Potential
  Tutorial 44.2 The Action Potential
  Tutorial 44.3 Synaptic Transmission
  Activity 44.1 Concept Matching: Neurotransmitters

Test Yourself

Knowledge and Synthesis Questions

1. The extensions of postsynaptic neurons that provide the main receptive surface for presynaptic neurons are
   a. nuclei.
   b. somas.
   c. axons.
   d. dendrites.
   Textbook Reference: *44.1 What Cells Are Unique to the Nervous System? p. 945*

2. The substance that wraps the axon of many neurons and provides for increased conduction speed is
   a. dendrase.
   b. histamine.
   c. acetylcholine.
   d. myelin.
   Textbook Reference: *44.1 What Cells Are Unique to the Nervous System? p. 946*

3. The long extensions from the neurons that provide the pathway for action potentials to synapse are
   a. dendrites.
   b. cell bodies.
   c. axons.
   d. presynaptic membranes.
   Textbook Reference: *44.1 What Cells Are Unique to the Nervous System? p. 945*

4. The threshold of a neuron is the
   a. amount of inhibitory neurotransmitter required to inhibit an action potential.

b. membrane voltage at which an axon potential will be suppressed.
   c. amount of excitatory neurotransmitter required to elicit an action potential.
   d. membrane voltage at which the membrane potential develops into an action potential.
   Textbook Reference: *44.2 How Do Neurons Generate and Conduct Signals? p. 952*

5. When a membrane is at the resting potential, the concentration of
   a. sodium and potassium ions is higher on the inside of its membrane.
   b. sodium and potassium ions is higher on the outside of its membrane.
   c. sodium ions is higher on the inside of its membrane and of potassium ions is higher on the outside.
   d. sodium ions is higher on the outside of its membrane and of potassium ions is higher on the inside.
   Textbook Reference: *44.2 How Do Neurons Generate and Conduct Signals? pp. 946–949*

6. Glial cells are specialized to do all of the following *except*
   a. receive neural impulses.
   b. insulate axons.
   c. supply neurons with nutrients.
   d. help maintain a proper ionic environment for the neuron.
   Textbook Reference: *44.1 What Cells Are Unique to the Nervous System? p. 946*

7. The cells that create the blood–brain barrier, keeping toxic substances from entering the brain are _____ and belong to a type of neural tissue called _____.
   a. endothelial cells; Schwann cells
   b. astrocytes; glial cells
   c. glial fibers; axons
   d. dendrites; synapses
   Textbook Reference: *44.1 What Cells Are Unique to the Nervous System? p. 946*

8. A particular disease of the nervous system specifically involves the Ca^{2+} ion channels at the chemical synapses of motor neurons where neurotransmitter is stored and released. In other words, this disease affects the
   a. axon terminals of the presynaptic cell and the release of acetylcholine.
   b. axon terminals of the postsynaptic cell and the release of K^+ ions.
   c. electrical synapses.
   d. axon terminals of the presynaptic cell and the release of K^+ ions.
   Textbook Reference: *44.3 How Do Neurons Communicate with Other Cells? pp. 955–956*

9. Which of the following statements about electrical synapses is *false*?

a. Connexons form molecular tunnels between two cells.

b. Electrical synapses cannot be inhibitory.

c. Electrical synapses do not allow for temporal summation.

d. Electrical transmission is very slow.

Textbook Reference: *44.3 How Do Neurons Communicate with Other Cells? p. 958*

10. One group of neurotransmitter receptors is the metabotropic receptors. Which of the following statements about metabotropic receptors is *false*?

a. Metabotropic receptors are transmembrane proteins.

b. Metabotropic receptors are coupled with A proteins.

c. Metabotropic receptors are not ion channels.

d. Responses in the postsynaptic cell mediated by metabotropic receptors are usually slower than those mediated by ionotropic receptors.

Textbook Reference: *44.3 How Do Neurons Communicate with Other Cells? p. 958*

11. The rapid depolarization of a neuron during the first half of an action potential is due to the

a. exit of K^+ ions from the cell through gated potassium channels.

b. rapid reversal of ion concentration caused by the action of the sodium–potassium pump.

c. entry of Na^+ ions into the cell through gated sodium channels.

d. movement of both Na^+ and K^+ ions through appropriate open channels.

Textbook Reference: *44.2 How Do Neurons Generate and Conduct Signals? pp. 951–952*

12. The refractory period of a neuron results from

a. the period when the sodium–potassium pump is nonfunctional.

b. activation of voltage-gated chloride channels.

c. closing of inactivated voltage-gated sodium channels.

d. the action potential reaching the synapse.

Textbook Reference: *44.2 How Do Neurons Generate and Conduct Signals? p. 952*

13. Which of the following statements about the process of summation in a neuron is *false*?

a. Slight perturbations of the membrane potential spread across the postsynaptic cell body.

b. Axons that terminate closer to the axon hillock have more influence on the summation process than those that don't.

c. Summation essentially consists of comparing the total of excitatory and inhibitory postsynaptic inputs.

d. The concentration of voltage-gated sodium channels is highest in the dendrites of the postsynaptic cell.

Textbook Reference: *44.3 How Do Neurons Communicate with Other Cells? p. 957*

14. Which of the following statements about neuro-transmitters is *false*?

a. Gases, such as carbon monoxide and nitric oxide, may act as neurotransmitters.

b. Each neurotransmitter has a single type of receptor.

c. Amino acids and their derivatives, monoamines, function as neurotransmitters.

d. Neurotransmitters have different effects in different tissues.

Textbook Reference: *44.3 How Do Neurons Communicate with Other Cells? pp. 958–959*

15. The electrical events labeled as EPSPs are the result of _____ of the _____ membrane.

a. hyperpolarization; postsynaptic

b. depolarization; postsynaptic

c. hyperpolarization; presynaptic

d. depolarization; presynaptic

Textbook Reference: *44.3 How Do Neurons Communicate with Other Cells? p. 957*

Application Questions

1. You are trying to determine the role of the potassium channels in a neuron. To do this you have knocked out all the functional potassium channels, and you depolarize the membrane potential. What will happen to the membrane potential after depolarization?

Textbook Reference: *44.2 How Do Neurons Generate and Conduct Signals? pp. 946–953*

2. The active ingredients in many nerve gases belong to a class of chemicals called anticholinesterases. Suggest a possible synaptic mechanism to explain how these chemicals can damage an animal's nervous system.

Textbook Reference: *44.3 How Do Neurons Communicate with Other Cells? pp. 960–961*

3. The following diagram represents a simple reflex loop. Label the sensory neuron, motor neuron, and effector in the neural circuit. Also indicate which part of the circuit is the sensory side and which is the motor side.

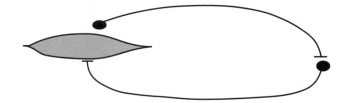

Textbook Reference: *44.1 What Cells Are Unique to the Nervous System? pp. 943–944*

4. Explain how an action potential travels faster down an axon wrapped in myelin than one without myelin.

Textbook Reference: *44.2 How Do Neurons Generate and Conduct Signals? pp. 954–955*

5. Why do certain substances, such as anesthetics and alcohol, have rapid effects on the brain, whereas others cannot reach the brain?

Textbook Reference: *44.1 What Cells Are Unique to the Nervous System? p. 946*

Answers

Knowledge and Synthesis Answers

1. **d.** The neuron is composed of a cell body, an axon, and dendrites. The dendrites form synapses with pre-synaptic cells to create the junction where information from one neuron is transferred to another neuron.

2. **d.** The glial cells that coat the axon of some neurons form myelin.

3. **c.** The neuron is composed of the cell body, the dendrite, and the axon. The axon carries action potentials away from the cell body to the synapses.

4. **d.** For an action potential to occur in an axon, the membrane must be depolarized above a certain level. This level is known as the threshold.

5. **d.** The resting potential of a neuron membrane occurs when the sodium ion concentration is higher on the outside and the potassium ion concentration is higher on the inside.

6. **a.** Glial cells perform many functions in the nervous system, but they do not receive neural impulses.

7. **b.** The blood–brain barrier is formed by astrocytes that wrap around the blood vessels traveling through the brain. Astrocytes are a special kind of glial cell.

8. **a.** If the disease acts on a chemical synapse where the neurotransmitter is stored and released, it is affecting the axon terminals of the presynaptic cell. Ca^{2+} channels are involved in regulating the release of acetylcholine by allowing Ca^{2+} to enter the presynaptic cell and promoting the fusing of acetylcholine-containing vesicles to the membrane.

9. **d.** Electrical synapses join two cells with protein tunnels known as connexons. These synapses provide for very fast transmission between cells.

10. **b.** Metabotropic receptors are coupled with G proteins that bind GDP. When the receptor binds a neuro-transmitter, the GDP is turned into GTP, and a subunit of the G protein separates and activates effector proteins. Responses mediated by metabotropic receptors are relatively slow when compared to those mediated by ionotropic receptors.

11. **c.** The first step in an action potential is the influx of Na^+ ions leading to a depolarization of the axon membrane. Na^+ ions rush into the cell due to the higher concentration outside of the cell and the negative membrane potential.

12. **c.** After the spike of the depolarization, the sodium voltage-gated channels close. One of the properties of these channels is that they will open again only after a short delay. This short delay is known as the refractory period when the sodium voltage-gated channels are inactive.

13. **d.** Dendrites, and most of the cell body, have few gated sodium channels. These channels mediate the action potentials that travel down the axon, where their levels are high.

14. **b.** Each neurotransmitter has multiple types of receptors.

15. **b.** Excitatory postsynaptic potentials (EPSPs) make it easier for an action potential to occur, so they depolarize the postsynaptic membrane.

Application Answers

1. The potassium voltage-gated channels are responsible for setting up the resting potential of a membrane. Potassium ions have a tendency to diffuse out of the cell, leaving a negative charge inside. Knocking out the function of the potassium voltage-gated channels would result in the cell's being unable to maintain resting potential. If the cell was depolarized by the opening of sodium voltage-gated channels, it might not repolarize because the potassium channels that help to repolarize the membrane would not be functioning.

2. The neurotransmitter at many synapses is acetyl-choline. Acetylcholine transmits the action potential from a presynaptic cell to a postsynaptic cell. Acetyl-cholinesterase is found in the synaptic cleft, and it cleaves acetylcholine to help remove it from the synaptic cleft after an action potential. A nerve gas composed of anticholinesterases would block the action of acetylcholinesterase, causing acetylcholine to build up in the synaptic cleft. This buildup would mean that the receptors on the postsynaptic cell would remain bound with acetylcholine.

3.

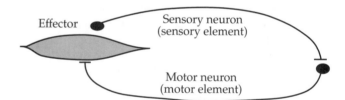

4. The conduction of an action potential down a myelinated axon is known as saltatory conduction. The myelin acts to insulate areas of the axon, preventing depolarization. The areas of the axon between the myelin are known as nodes of Ranvier. Depolarization can only occur at these nodes. As the action potential moves down a myelinated axon, the influx of sodium ions at one node diffuses down the axon. This results in the depolarization of the next node of Ranvier. Depolarization can occur only in the downstream nodes because the upstream nodes are in a refractory period. As a result, the action potential moves quickly down the axon to the synapse.

5. Astrocytes are glial cells that help form the blood–brain barrier by surrounding tiny, very permeable blood vessels in the brain. However, because the barrier is made of plasma membranes, lipid-soluble substances, such as anesthetics and alcohol, can pass through it.

CHAPTER 45 Sensory Systems

Important Concepts

Sensory cells detect stimuli and transmit the information to the CNS.

- Sensory cells are modified neurons specialized to detect internal and external stimuli and to transmit information concerning these stimuli as action potentials to different sites in the central nervous system (CNS).

- Sensory cells are involved in maintaining homeostasis by transmitting information about the status of the internal environment to the CNS.

- Sensory transduction is the process by which a mechanical, thermal, or chemical stimulus to a sensory cell is transformed ("transduced") into an action potential. Sensory transduction can occur by direct activation of receptor proteins that open or close ion channels to change membrane potential. In this situation, the sensory receptor protein is ionotropic. Alternatively, the stimulus can activate a second messenger within the cell that couples with a G protein to eventually open ion channels. In this situation, the sensory receptor protein is metabotropic.

- A receptor potential is a change in the resting membrane potential of a sensory receptor cell in response to a stimulus. Receptor potentials produce action potentials by either causing the release of a neurotransmitter that induces an associated neuron to generate action potentials or by generating action potentials within the sensory cell itself.

- Sensory organs are groups of sensory cells that, along with other cells, collect, filter, and amplify stimuli. Eyes, ears, and noses are examples of sensory organs. Sensory systems include the sensory cells, the associated structures, and the networks of neurons that process the information.

- Many sensory cells have diminished responses to a stimulus over time in a process known as adaptation. Adaptation allows organisms to ignore background conditions and focus on new information.

Chemoreceptors produce action potentials in response to a chemical stimulus.

- Chemical stimuli in the external and internal environments stimulate chemoreceptors. Chemoreceptors are responsible for smell and taste, and for monitoring levels of particular chemicals (for example, carbon dioxide) inside the body.

- Insects use chemical signals, called pheromones, to attract mates by remotely stimulating their target's chemoreceptors (see Figure 45.3). The concentration of the pheromone released by a female creates a gradient that provides information about her specific location.

- Olfaction is the sense of smell. The olfactory sensors of vertebrates are neurons with axons extending to the olfactory bulb of the brain; the dendrites of these neurons are exposed as hairs to the environment within the epithelium of the nasal cavity (see Figure 45.4). Olfactory receptor proteins are found on the hairs, and each receptor binds with specific odorants. The ability to discriminate many different odors is due to the high number of specific receptors. Binding of an odorant results in depolarization of the cell through a G protein that activates a second messenger that then opens sodium channels. The strength of a smell is related to the number of odorant molecules that bind to receptors.

- Amphibians, reptiles, and some mammals have a vomeronasal organ, a paired structure located in the nasal epithelium. In mammals, the vomeronasal organ senses pheromones and conveys information to the accessory olfactory bulb in the brain. In snakes, the forked tongue presents odorant molecules from the environment to the chemoreceptors of the vomeronasal organ on the roof of the mouth; thus, in snakes, the tongue is used in smell and not in taste.

- Gustation, the sense of taste, relies on clusters of chemoreceptor cells called taste buds (see Figure 45.5). Binding of the stimulus to receptor proteins on the microvilli of sensory cells causes a change in membrane potential and the release of neurotransmitters that stimulate sensory neurons at the base of the taste bud.

- Humans can perceive five general tastes: sweet, sour, salty, bitter, and umami (meaty).

Mechanoreceptors detect mechanical force.

- Mechanical force causes distortion of the membranes of mechanoreceptors, causing ion channels to open and an action potential to be generated.

- The skin has several different types of mechanoreceptors (see Figure 45.6). Meissner's corpuscles are nerve endings that are very sensitive but adapt rapidly; they detect light touches to the skin. Merkel's discs adapt slowly and provide information about objects touching the skin. Pacinian corpuscles and Ruffini endings are deeper in the skin and respond to vibrations. The density of tactile mechanoreceptor cells varies with region of the body.

- Stretch receptors are mechanoreceptors that sense position and stress on limbs. Muscle spindles are stretch receptors in skeletal muscles that perceive muscle stretch. Golgi tendon organs are found in the tendons and ligaments and provide information about forces in contracting muscles. Collectively, these mechanoreceptors provide information on body posture and limb position, as well as muscle and tendon tension.

The auditory system contains mechanoreceptors specialized for sound reception.

- The auditory system takes in sound as pressure waves and transforms them into action potentials.

- The pinna is the outer portion of the mammalian ear that collects and directs sound waves into the auditory canal. The end of the auditory canal is covered with a tympanic membrane, which vibrates and transmits sound waves to bones (ossicles) in the middle ear. The ossicles are the malleus (hammer), incus (anvil), and stapes (stirrup), and together they transmit sound waves to the membrane called the oval window.

- Sound travels through the oval window into the fluid-filled cochlea, where pressure waves are turned into action potentials. The cochlea, in the inner ear, is a three-canal chamber with two membranes: Reissner's membrane and the basilar membrane divide the chamber. The organ of Corti contains hair cells with stereocilia that are in contact with a tectorial membrane that transduces movement of the basilar membrane into action potentials to the auditory nerve (see Figure 45.8).

- The upper and lower chambers of the cochlea are separated at the distal end by the basilar membrane. Sound waves enter through the oval window as pressure waves. The waves move down the cochlea, reach the end, and move back down, where they are absorbed by the flexible membrane of the round window.

- If the oval window vibrates at faster frequencies, the wave cannot make it all the way down the cochlea, and it crosses the basilar membrane, causing it to flex. Flexing of the membrane stimulates hairs and the organ of Corti. Different pitches of sound cause the basilar membrane to flex at different locations, stimulating different hairs (see Figure 45.9).

Displacement is detected by hair cells.

- Hair cells are mechanoreceptors that release neurotransmitters in response to the bending of stereocilia on the hairs. In mammals, the inner ear uses hair cells in the semicircular canals and the vestibular apparatus to maintain equilibrium (see Figure 45.11).

- The inner ear of vertebrates has hair cells that detect a body's position relative to gravity. In the semicircular canal of the ear, flow of semicircular fluid shifts the cupulae containing the hair cells when the head changes position. In the vestibular apparatus, layers of calcium carbonate resting on hair cells are moved by gravity when the head changes position. Both of these stimuli to the hair cells inform the organism of its position in relation to gravity.

- The lateral line system of fish is composed of hair cells that line canals located just under the skin. This system detects displacement of water around the fish (see Figure 45.12).

Light-sensitive pigments are used in photoreceptors.

- Rhodopsin is a family of pigments made up of the protein opsin and the light-absorbing prosthetic group 11-*cis*-retinal (see Figure 45.13). The 11-*cis*-retinal absorbs photons of light and changes conformation to all-*trans*-retinal, causing opsin to change conformation and become photoexcited rhodopsin.

- Photoexcited rhodopsin triggers a cascade that ultimately leads to changes in membrane potential and the photoreceptor's response to light. Excited rhodopsin activates the G protein transducin. Transducin converts cyclic GMP (which opens Na^+ channels) to 5'-GMP, allowing Na^+ channels to close and the cell to hyperpolarize.

- In vertebrate eyes, rod cells are photoreceptor cells that contain an inner segment, a synaptic terminal, and an outer segment with many rhodopsin molecules (see Figure 45.14). Rods are found in the retina, along with a layer that transduces visual information into action potentials.

- Rod cells become hyperpolarized in response to light and transmit this by decreasing the levels of neurotransmitter released (see Figure 45.15).

Invertebrates display a variety of visual systems.

- Flatworms have photoreceptor cells organized into eye cups, which are used to orient away from light sources.

- Arthropods have compound eyes with 800 to 10,000 ommatidia (optical units) per eye. The ommatidia contain light-sensitive photoreceptors called retinula cells (see Figure 45.16). The inner border of the retinula cells are covered by microvilli that contain rhodopsin. The compound eye communicates a low-resolution image to the CNS.

- Cephalopod mollusks (and vertebrates) have eyes that form detailed images.

The image-forming eyes of vertebrates have several components.

- Cephalopod mollusks and vertebrates independently evolved image-forming eyes with very similar structures (see Figure 45.17).

- The vertebrate eye is surrounded by the sclera, formed from connective tissue. The cornea is the transparent sclera through which light passes.

- The iris controls the amount of light entering the eye through the pupil.

- The lens focuses images onto the retina at the back of the eye by changing from a spherical to a flattened shape. We focus on near and far objects by the contraction and relaxation of ciliary muscles that round up or flatten the lens to bring objects into focus (see Figure 45.18). Fishes, amphibians, and reptiles move the lenses of their eyes closer to or farther from their retinas.

The vertebrate eye focuses light onto photoreceptors.

- The center of the retina is the fovea, an area with the highest density of photoreceptor cells. A blind spot with no photoreceptors in the retina is located where the blood vessels and optic nerve exit the back of the eye.

- The human retina contains both rods (which are more light-sensitive) and cones (which absorb light of various wavelengths, allowing for color vision). Cones have different opsin molecules that absorb blue, green, yellow, or red (see Figure 45.20).

- There are five layers of neurons in the retina, with the photoreceptive rods and cones in the last layer (see Figure 45.21).

- The first layer of cells consists of ganglion cells (which create the action potential) and the axons of the optic nerve. Bipolar cells are stimulated by neurotransmitters from the photoreceptors to transmit the signal from the photoreceptor to the ganglion cells by the release of a neurotransmitter.

- The two other layers are the horizontal cells (which connect adjoining groups of photoreceptors and bipolar cells) and amacrine cells (which connect adjoining groups of bipolar cells and ganglion cells).

The Big Picture

- Sensory structures work by converting some form of stimulus—mechanical, chemical, light—into action potentials in the nervous system, which are then interpreted by the central nervous system as a perceived sense.

- Receptors are named on the basis of their sensitivity. Chemoreceptors respond to chemical stimulation, mechanoreceptors respond to mechanical stimulation, and photoreceptors respond to light.

- Different animals have different sensitivities of senses, as well as different types of senses.

Common Problem Areas

- The action potentials produced in the neurons of the ear, eye, knee, or stomach are identical. The action potentials coming from the eye, for example, are interpreted as light because of the region of the brain that receives and analyzes them.

- Many of the more complexly structured receptors have both neural and nonneural components. The nonneural components (e.g., the ear pinnae) help channel or otherwise alter or filter the stimulus that will arrive at the neural component of the receptor.

Study Strategies

- The route by which sound travels in the ear can be very confusing. View the cochlea in the uncoiled form as in Figure 45.9. This will help you to visualize how pressure waves of different wavelengths produce different sounds.

- All of the senses have what appear to be very different mechanisms for the transmission of information to the brain. To help sort all of this out, remember that there are only a few types of receptors that respond to stimuli and that they all generate action potentials. The steps in sensory transduction are also similar between all the different sensory systems.

- Review the following animated tutorial and activities on the Companion Website/CD:
  Tutorial 45.1 Sound Transduction in the Human Ear
  Activity 45.1 Structures of the Human Ear
  Activity 45.2 Structure of the Eye
  Activity 45.3 Structure of the Retina

Test Yourself

Knowledge and Synthesis Questions

1. An electrode is inserted into a chemosensory nerve leading away from a taste bud in the mouth of a dog. A mild acid solution is then flushed continuously over the sensors associated with this nerve. Initially, the nerve responds to this stimulation but over time ceases to carry action potentials. This observation would best be explained by
   a. translocation.
   b. adaptation of the sensory cells.
   c. depletion of neurotransmitter in the sensory nerve.
   d. second messenger influences that increase cell membrane potentials.
   Textbook Reference: 45.1 *How Do Sensory Cells Convert Stimuli into Action Potentials? p. 967*

2. Which of the following statements about sensory cells is *false*?
   a. Mechanoreceptors detect stimuli that distort membranes.

b. Chemoreceptors monitor aspects of the internal environment.

c. Chemoreceptors are involved in smell, taste, and hearing.

d. Photoreceptors exhibit a conformational change when stimulated by light.

Textbook Reference: *45.2 How Do Sensory Systems Detect Chemical Stimuli? p. 967*

3. Silkworm moths use chemosensory signals for mate attraction. These signals are known as _____.
   a. rhodopsin
   b. hormones
   c. pheromones
   d. G proteins

Textbook Reference: *45.2 How Do Sensory Systems Detect Chemical Stimuli? p. 968*

4. Stretch receptors in the aorta and carotid artery sense changes in arterial pressure. These receptors are

   _____.
   a. chemoreceptors
   b. thermoreceptors
   c. electroreceptors
   d. mechanoreceptors

Textbook Reference: *45.3 How Do Sensory Systems Detect Mechanical Forces? p. 970*

5. All but which of the following systems employs hair cells as its transducer?
   a. Meissner's corpuscle
   b. Lateral line
   c. Organ of Corti
   d. Vestibular apparatus

Textbook Reference: *45.3 How Do Sensory Systems Detect Mechanical Forces? pp. 974–975*

6. Which of the following statements about human gustation is *false*?
   a. Taste bud cells are relatively short lived because of the high degree of abrasion they encounter.
   b. Taste buds are confined to the oral cavity.
   c. Changes in the membrane potential of the taste bud sensory cells cause them to release neurotransmitter onto the dendrites of sensory neurons.
   d. Humans perceive only three categories of tastes: sweet, sour, and bitter.

Textbook Reference: *45.2 How Do Sensory Systems Detect Chemical Stimuli? pp. 969–970*

7. Which of the following statements about the photosensitive molecule rhodopsin is *false*?
   a. Opsin is converted from the 11-*cis* to the all-*trans* form upon absorbing a photon of light.
   b. The retinal is the light-absorbing group.
   c. Photoexcited rhodopsin triggers a cascade of reactions that ultimately alters the membrane potential of a photoreceptor cell.
   d. None of the above

Textbook Reference: *45.4 How Do Sensory Systems Detect Light? p. 976*

8. Which of the following cells in the human visual system send information directly to the brain?
   a. Amacrine cells
   b. Bipolar cells
   c. Ganglion cells
   d. Rods and cones

Textbook Reference: *45.4 How Do Sensory Systems Detect Light? pp. 980–981*

9. Through which of the following cell layers must a photon of light pass before striking a cone cell in the eye of a hawk?
   a. Amacrine
   b. Bipolar
   c. Ganglion
   d. All of the above

Textbook Reference: *45.4 How Do Sensory Systems Detect Light? pp. 980–981*

10. Which of the following statements about animal vision is *false*?
    a. Photoreceptors in simple invertebrates produce image vision.
    b. There are several types of cones in vertebrate retinas, each of which detects different wavelengths.
    c. Ommatidia make up the compound eye of an arthropod.
    d. Although retinal actually absorbs the photon, the wavelength it absorbs best is determined by opsin.

Textbook Reference: *45.4 How Do Sensory Systems Detect Light? pp. 977–981*

11. Which of the following about sensory receptor proteins is *false*?
    a. Ionotropic receptor proteins are either ion channels themselves or they directly influence the opening of ion channels.
    b. Photoreceptors are ionotropic.
    c. Mechanoreceptors are ionotropic.
    d. Chemoreceptors are metabotropic.

Textbook Reference: *45.1 How Do Sensory Cells Convert Stimuli into Action Potentials? pp. 965–966*

12. Which of the following is *not* a membrane found in the cochlea?
    a. Reissner's membrane
    b. Tectorial membrane
    c. Tympanic membrane
    d. Basilar membrane

Textbook Reference: *45.3 How Do Sensory Systems Detect Mechanical Forces? pp. 972–973*

13. Which of the following about receptor potentials is *false*?
    a. They are changes in the resting membrane potential of a sensory cell in response to a stimulus.
    b. The receptor potential spreads from the cell body of a sensory cell to the axon hillock, where action potentials are generated.
    c. They must be converted to action potentials to travel long distances.

d. A receptor potential always releases a neurotransmitter to induce an action potential.

Textbook Reference: 45.1 How Do Sensory Cells Convert Stimuli into Action Potentials? pp. 966–967

14. Which of the following statements about the detection of chemical stimuli is *false*?
    a. Snakes use their tongue to smell.
    b. Many mammals have a vomeronasal organ to detect pheromones.
    c. A greater frequency of action potentials is associated with perception of a more intense smell.
    d. Taste buds are confined to the oral cavity in terrestrial and aquatic animals.

    Textbook Reference: 45.2 How Do Sensory Systems Detect Chemical Stimuli? p. 969

15. Which of the following statements about sensory systems is *false*?
    a. Rattlesnakes have pit organs to detect infrared wavelengths.
    b. Cephalopod mollusks and vertebrates independently evolved image-forming eyes.
    c. Nocturnal animals have a high percentage of cones in their retinas while diurnal animals have a high percentage of rods.
    d. Fish detect water movements with their lateral lines.

    Textbook Reference: 45.4 How Do Sensory Systems Detect Light? p. 981

Application Questions

1. This past winter was an awful year for colds. You had a cold for a number of weeks, during which time your sense of smell was diminished. What was the cause of your loss of smell?

   Textbook Reference: 45.2 How Do Sensory Systems Detect Chemical Stimuli? pp. 968–969

2. After a loud rock concert your hearing appears to be dampened. What portion of the ear has been altered to dampen your hearing?

   Textbook Reference: 45.3 How Do Sensory Systems Detect Mechanical Forces? p. 975

3. You have just given a presentation in your biology class that had many elaborate red- and green-colored slides. Afterward, a male friend tells you that he could not see any of the differences you were reporting. Why could your friend not see the differences?

   Textbook Reference: 45.4 How Do Sensory Systems Detect Light? pp. 980–981

4. Label each of the following structures on the diagram of the human eye shown below: sclera, cornea, iris, pupil, lens, retina, fovea, and blind spot.

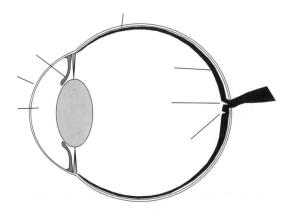

Textbook Reference: 45.4 How Do Sensory Systems Detect Light? p. 979

5. A friend of yours has an infection that is causing vertigo (dizziness). In which sensory organ is the infection? In which particular part of the organ is the infection?

   Textbook Reference: 45.3 How Do Sensory Systems Detect Mechanical Forces? pp. 974–975

Answers

Knowledge and Synthesis Answers

1. **b.** When a sensor cell is stimulated by an unchanging, steady-state stimulus, it will become adapted. This allows the sensory system to ignore the unchanging stimulus while still being able to respond to new information.

2. **c.** Chemoreceptors are involved in smell and taste but not hearing.

3. **c.** Pheromones are chemical signals that a female releases into the environment. The male uses chemoreceptors to follow the pheromone to the source.

4. **d.** The stretch receptors of the aorta, which detect changes in pressure, are examples of mechanoreceptors.

5. **a.** The Meissner's corpuscle of the skin does not use hair cells to sense a stimulus. The cell membranes of the Meissner's corpuscle deform in response to light touching of the skin.

6. **d.** Humans can perceive five tastes: sweet, salty, sour, bitter, and umami (a meaty taste). The combination of taste and smell provides the complex subtle flavors of the food we eat.

7. **a.** Rhodopsin contains two groups: the protein opsin and the light-sensitive group retinal. Retinal, not opsin, is converted from the 11-*cis* to the all-*trans* form upon absorbing a photon of light. Opsin does change conformation in response to a change in the rhodopsin to signal the detection of light.

8. **c.** The ganglion cells transmit information from the bipolar cells to the brain. The axons of the ganglion cells connect with the optic nerve.

9. **d.** The photoreceptive cells are located at the back of the retina. Light must pass through a layer of ganglion cells, a layer of amacrine and bipolar cells, and a horizontal cell layer.

10. **a.** The photoreceptors in simple invertebrates such as the flatworm only detect light–dark contrast.

11. **b.** Photoreceptors are metabotropic because they influence ion channels indirectly, through G proteins and second messengers.

12. **c.** Although the tympanic membrane is found in the human ear, it is not found in the cochlea. The tympanic membrane is the membrane that transmits sounds from the auditory canal and the middle ear.

13. **d.** The receptor potential does not always release a neurotransmitter to induce an associated neuron to generate an action potential. Sometimes the receptor potential generates action potentials within the sensory cell itself.

14. **d.** Taste buds are confined to the oral cavity in terrestrial animals. However, some aquatic animals, such as fish, have taste buds in their skin.

15. **c.** Nocturnal animals have a high percentage of rods in their retinas, and diurnal animals have a high percentage of cones.

Application Answers

1. Your sense of smell depends on olfactory cilia receptors binding with odorant molecules. This triggers an action potential that is sent to the olfactory bulb. The olfactory cilia line the surface of the nasal epithelium. Usually there is a thin layer of protective mucus covering this epithelium. However, when you have a cold the production of mucus increases. This increased mucus covers the epithelium and the olfactory cilia, making it more difficult for odorant molecules to penetrate the mucus and reach the cilia. Thus you are not receiving as much stimulation, and your sense of smell is decreased.

2. Sound is transmitted into the ear and into the fluid-filled cochlea. Sounds that are too loud will eventually damage the hair cells of the organ of Corti. Because these hair cells transmit the sound to the organ of Corti, any damage will decrease the response of the organ of Corti to sounds. This damage to the hair cells is cumulative and permanent.

3. Your friend is red-green color blind. The cones in our eyes allow us to see color. We have cones for red, green, and blue. Your friend either has red and green cones that do not function properly, or lacks them altogether.

4.

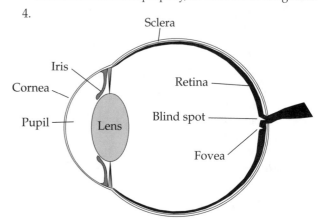

5. Your friend's infection is in the ear, specifically the inner ear that contains the organs of equilibrium (in addition to the cochlea).

CHAPTER 46 The Mammalian Nervous System: Structure and Higher Function

Important Concepts

The mammalian nervous system consists of the central nervous system and the peripheral nervous system.

- The brain and the spinal cord together constitute the central nervous system (CNS).
- The nerves of the peripheral nervous system transport information between every tissue in the body and the CNS.
- A nerve is a bundle of axons that carries information in both directions between the organs of the body and the CNS.
- The peripheral nervous system is composed of afferent nerves that carry information such as hearing and vision to the CNS, and efferent nerves that carry information from the CNS to muscles and glands (see Figure 46.1). Efferent pathways can be further classified as voluntary or involuntary (autonomic).
- The CNS also receives chemical information from circulating hormones.

The CNS develops from the embryonic neural tube.

- The neural tube is a tube of neural tissue running down the length of a vertebrate embryo in its early developmental stages (see Figure 46.2). The anterior end of the neural tube will develop into the main parts of the brain—the hindbrain, the midbrain, and the forebrain—with the spinal cord developing from the rest of the neural tube.
- The hindbrain is made up of the medulla and pons, which are involved in control of physiological functions, and the cerebellum, which is involved in coordinating commands leaving the brain with the actual actions of the muscles.
- The midbrain receives and processes auditory and visual information.
- The forebrain consists of a central diencephalon region and a surrounding telencephalon region. The diencephalon contains the thalamus, a relay station for sensory information, and the hypothalamus, which is involved in maintenance of homeostasis. The cerebral hemispheres (cerebrum) make up the telencephalon.

The spinal cord is the main neural highway between the body and the brain.

- The spinal cord is a bidirectional neural pathway for information flow between the peripheral nervous system and the brain.
- The spinal cord also converts some afferent information from the peripheral nervous system into efferent information sent back to the peripheral nervous system in a process known as a spinal reflex (see Figure 46.3).
- A monosynaptic reflex, such as the knee-jerk reflex, involves only an afferent neuron, an efferent neuron, and one synapse. Leg muscle stretch receptors trigger the sensory neuron to conduct action potentials to the spinal cord upon stretching. The action potential is passed through the synapse of the sensory and motor neuron located in the central gray matter of the spinal cord. More complicated reflexes involve interneurons and more synapses.
- Flexor muscles, which flex limbs, and extensor muscles, which straighten limbs, control limb movement. The motor neurons that stimulate these muscles are antagonistic and polysynaptic, with each sensory neuron stimulating one motor neuron while inhibiting the other motor neuron at two distinct synapses.

The reticular system and limbic system interact with the forebrain.

- The reticular system contains groups of neurons forming a nucleus. Such nuclei are located through the medulla, pons, and midbrain.
- Many of the connections involved in functional control of the body occur in the reticular system, along with the control of sleep and waking.
- The limbic system is involved in instincts and emotions (see Figure 46.4). A portion of the limbic system known as the hippocampus helps to transfer short-term memory to long-term memory. The amygdala, another part of the limbic system, is involved in fear and fear memory.

The cerebrum is the largest portion of the mammalian brain.

- The cerebral hemispheres are covered by a convoluted cerebral cortex, a sheet of gray matter that processes sensory information and higher-order information in the association areas (see Figure 46.5). The convolutions are ridges known as gyri and valleys known as sulci. Underneath the gray matter is white matter containing axons that connect cell bodies in the cortex and other areas of the brain. Each cerebral hemisphere is broken down into four main regions: the temporal lobe, the frontal lobe, the occipital lobe, and the parietal lobe. Each region has an association area that is involved in various functions.

- The temporal lobe processes auditory information, and the association areas are involved with recognition, identification, and naming of objects. Damage to the temporal lobe results in a condition in which individuals are unable to identify a stimulus, even though they can perceive it. Integration of spoken language can also be diminished with damage to the temporal lobe.

- The frontal lobe contains the primary motor cortex made up of axons that project to muscles in the body. The primary motor cortex can be mapped according to the locations that control movements of various body parts (see Figure 46.7). The association areas of the frontal lobe are involved in planning and personality.

- The central sulcus separates the parietal lobe from the frontal lobe. The primary somatosensory cortex is located in the parietal lobe. As with the motor cortex, the location of the body that is sensed by the somatosensory cortex can be mapped on the brain (see Figure 46.7). Contralateral neglect syndrome results from damage to the right parietal lobe and is characterized by the inability to detect stimuli from the left side of the body (see Figure 46.9).

- Visual information is processed by the occipital lobe of the cerebrum. The association areas for the occipital lobe are involved in translation of visual stimuli into language.

The sympathetic and parasympathetic systems make up the autonomic nervous system.

- The sympathetic and parasympathetic divisions of the autonomic nervous system have antagonistic effects on the organ systems they innervate and play a large role in the maintenance of homeostasis.

- The sympathetic division is involved in the fight-or-flight response of increased heart rate, blood pressure, and cardiac output. The parasympathetic division slows down heart rate and decreases blood pressure and cardiac output; however, it accelerates digestive activities.

- Preganglionic neurons of the autonomic efferent pathway have cell bodies in the brain stem or spinal cord. They use the neurotransmitter acetylcholine at synapses where the ganglion lies outside the CNS. The axon that runs from the ganglion to the target cells is known as the postganglionic neuron.

- Norepinephrine is the neurotransmitter in postganglionic neurons of the sympathetic system, whereas acetylcholine is the neurotransmitter in postganglionic neurons of the parasympathetic system.

- Both acetylcholine and norepinephrine stimulate the pacemaker cells in the heart to control heart rate.

- The parasympathetic system has preganglionic neurons that come from the brain stem and the last segment of the spinal cord (the sacral region). In the sympathetic system, preganglionic neurons come from the middle (lumbar) and upper (thoracic) regions of the spinal cord (see Figure 46.10).

Visual images are constructed in the occipital cortex.

- In the retina, one ganglion cell receives and integrates the information from many groups of photoreceptors that make up a circular receptive field with a center and a surround (see Figure 46.11). On-center receptive fields excite ganglion cells and are connected by the bipolar cells. The off-center receptive fields inhibit ganglion cells and connect though horizontal cells. Receptive fields of ganglion cells can overlap. Glutamate is the neurotransmitter of photoreceptors cells, and although usually excitatory, it can be inhibitory under certain conditions.

- Information from the retina is transferred by the optical nerve to the thalamus and then to the visual cortex located in the occipital cortex.

- The visual cortex is composed of cells with specific receptive fields associated with areas of the retina that respond to specific light patterns. Simple cells are stimulated by static bars of light with specific orientations (see Figure 46.12). Complex cells receive input from several simple cells that have receptive fields in different locations on the retina. Mental images of the world are determined by analyzing the edges of the patterns of light falling on the retina.

- The optic nerve from the two eyes joins at the optic chiasm (see Figure 46.13). At the optic chiasm, half of the axons from one eye cross over to the opposite side of the brain. Binocular vision occurs because the overlap in the field of view for each eye is transmitted to the same location in the visual cortex, the binocular cells. Binocular cells receive overlapping visual information from both eyes and interpret the disparity between the overlapping information from the two eyes to produce a three-dimensional image.

Sleep involves complex cerebral function.

- An electroencephalogram (EEG) measures the electrical activity of neurons in the cerebral cortex; at any one time, large numbers of neurons are monitored in the regions below the electrodes. An electromyogram (EMG) records electrical activity of muscles, and an electrooculogram (EOG) records eye movements.

- There are two main states of sleep in humans: rapid-eye movement (REM) sleep and non-REM sleep (see Figure 46.14).

- Non-REM sleep has four states progressing from stage 1 to the restorative stages 3 and 4. During non-REM sleep the neurons in the thalamus and cerebral cortex become hyperpolarized due to the opening of K^+ channels. During hyperpolarization, the cells fire off in synchronized bursts due to Ca^{2+} channels deactivating, which produces a slow-wave pattern.

- During REM sleep, dreams and nightmares occur along with near complete paralysis of skeletal muscles. Neurons that were hyperpolarized during non-REM sleep return to waking levels, allowing information to be processed. Afferent and efferent pathways are inhibited during REM sleep.

- During a night of sleep, the brain cycles between REM sleep and non-REM sleep, with 80 percent of sleep being non-REM sleep.

Learning and memory involve long-lasting synaptic changes.

- Learning occurs when behavior is modified as a result of experience. Long-lasting synaptic changes must occur for learning to take place. Long-term potentiation (LTP) occurs when high-frequency electrical stimulation makes circuits of the hippocampus more sensitive to subsequent stimulation.

- Long-term depression (LTD) occurs when hippocampal circuits become less sensitive as a result of repetitive, low-level stimulation.

- Associative learning in animals involves linking two unrelated stimuli, as in the conditioned reflex of Pavlov's dogs. The dogs were conditioned to associate eating with the ringing of a bell, so that even in the absence of food the mere ringing of a bell stimulated salivation. Another example of a conditioned reflex involves the eye-blink reflex of rabbits. By repeatedly pairing a tone and a puff of air, rabbits are conditioned to blink in response to the tone alone. Further experiments showed that a specific part of the cerebellum was necessary for rabbits to learn the conditioned reflex.

- Memory is the phenomenon whereby the nervous system retains what has been experienced. Immediate memories are very short-term vivid memories of what has just happened. Short-term memories last between 10 and 15 minutes and contain less information than immediate memories. Repetition or reinforcement enhances the transfer of short-term memory to long-term memory.

- Declarative memory is memory of people, places, events, and things. Procedural memory is memory of how to perform motor tasks, such as riding a bicycle.

The language center is located in the left cerebral hemisphere.

- In most people, the ability to produce and interpret language occurs in the left hemisphere of the cerebrum.

- The two cerebral hemispheres are connected by the corpus callosum. This connection allows the two hemispheres to communicate. Severing the corpus callosum results in an individual being unable to express in language the knowledge that is in the right hemisphere.

- A number of areas have been located that are important for language (see Figure 46.15). The frontal lobe contains Broca's area, which is involved in speech. Wernicke's area is located in the temporal lobe and involves sensory aspects of language. The angular gyrus is thought to be involved in integrating spoken and written language.

- Language involves the flow of information between the various regions mentioned earlier. Damage to any of the areas involved in language will break the chain of information and result in aphasia, the condition of not being able to use or understand written or spoken words.

The Big Picture

- The mammalian nervous system can be divided both anatomically and functionally. Anatomically, it consists grossly of the central and peripheral nervous systems. Functionally, it consists of the sympathetic and parasympathetic nervous systems. Afferent nerves carry information toward the central nervous system from the peripheral nervous system, and efferent nerves carry information from the central to the peripheral nervous system.

- The brain can be divided anatomically and functionally into many different regions, each responsible for its many vital actions. The "higher" brain centers of the cerebrum are responsible for conscious thought and deliberate (voluntary) movements. The "lower" brain centers such as the cerebellum, pons, and medulla regulate involuntary movements and are involved primarily in maintaining homeostasis throughout the body.

Common Problem Areas

- It is easy to confuse "afferent" and "efferent," and it is important in your understanding of the nervous system to differentiate them. Think of *a*fferent as *a*rriving, and *e*fferent as *e*xiting the reference point—usually the CNS.

- It is important to recognize that although the neurotransmitters acetylcholine and norepinephrine always have antagonistic effects on each other, there is no universal pattern as to which stimulates tissue and which inhibits tissue. For example, acetylcholine causes the smooth muscle in blood vessels in many regions of the body to relax, but causes the smooth muscle in the stomach and intestines to contract.

- It is often difficult for students to appreciate that, when dealing with sensory input to the brain, specific regions of the body "map onto" specific regions of the cerebrum. Similarly, with regard to control of movements, specific regions of the brain "map onto" specific regions of the body. Moreover, the amount of brain matter devoted to a particular body region depends on the amount of muscle control and sensors contained in that body region. Thus, a relatively small area of the cerebral hemispheres is devoted to the upper leg (which has a relatively limited range of movement and sensation), whereas the tongue, with its many sensory receptors and high degree of mobility, commands more of the tissue of the cerebrum.

Study Strategies

- Several key terms in this chapter occur repeatedly: parasympathetic, sympathetic, afferent, efferent, agonist, antagonist, preganglionic, postganglionic. Until you master this vocabulary, it will be difficult to put together a comprehensive picture of the nervous system. Create your own list of the crucial terms you see repeatedly, and make sure that you understand their meanings.

- Brain anatomy is complex. Start by dividing up the structures into forebrain, midbrain, and hindbrain regions. Then identify each of the subcomponents. Your understanding will be more complete if you learn the general functions of each brain section as you go along. That is, rather than learning the anatomy of the brain and then starting over to learn the functions, learn the two at the same time.

- Review the following animated tutorials and activities on the Companion Website/CD:
  Tutorial 46.1 Information Processing in the Spinal Cord
  Tutorial 46.2 Information Processing in the Retina
  Activity 46.1 The Human Cerebrum
  Activity 46.2 Brain Language Areas
  Activity 46.3 Brain Structures

Test Yourself

Knowledge and Synthesis Questions

1. A man has damage to his brain that affects his ability to recognize the faces of people he knows. Where did the damage occur?
   a. Hypothalamus
   b. Temporal lobe
   c. Parietal lobe
   d. Frontal lobe
   Textbook Reference: 46.1 How Is the Mammalian Nervous System Organized? p. 990

2. The secretion of hormones from an endocrine gland is most likely under the control of which of the following components of the nervous system?

   a. Autonomic
   b. Voluntary
   c. Dendritic
   d. Limbic
   Textbook Reference: 46.2 How Is Information Processed by Neuronal Networks? p. 992

3. The motor cortex of the cerebrum is
   a. mapped from the head region on the lower side of the cortex to the lower part of the body on the upper side of the cortex.
   b. areas of the brain where the most sensitive body areas are represented by the largest areas of the cortex.
   c. located in the parietal lobe.
   d. None of the above
   Textbook Reference: 46.1 How Is the Mammalian Nervous System Organized? p. 990

4. Which of the following statements about the sympathetic division of the autonomic nervous system is *false*?
   a. It increases heart rate.
   b. It relaxes the urinary bladder.
   c. It stimulates digestion.
   d. It increases blood pressure.
   Textbook Reference: 46.2 How Is Information Processed by Neuronal Networks? p. 992

5. Which of the following is *not* part of the central nervous system?
   a. Brain stem
   b. Spinal gray matter
   c. Cerebellum
   d. Neuronal cell body of a sensory afferent
   Textbook Reference: 46.1 How Is the Mammalian Nervous System Organized? p. 985

6. Observations of patients with aphasia indicate that
   a. only Broca's area is essential for normal language skills.
   b. language skills depend on proper neural flow between the temporal lobes and motor cortex.
   c. the right hemisphere dominates the production and use of language in humans.
   d. most people can control language functions from either cerebral hemisphere.
   Textbook Reference: 46.3 Can Higher Functions Be Understood in Cellular Terms? p. 1000

7. A friend wakes you from sleep, and you have the sensation of having just experienced a vivid dream. Which of the following would be *false*?
   a. Your hands and feet have been twitching slightly.
   b. Your eyes have been twitching.
   c. Most of your voluntary body muscles have been inactive.
   d. Your cerebral cortex was not as active as when you were awake.
   Textbook Reference: 46.3 Can Higher Functions Be Understood in Cellular Terms? p. 998

8. If you want to find a complex cell in a cat's visual system, you would look in
   a. Broca's area.
   b. the spinal cord.
   c. the occipital cortex.
   d. the reticular system.
   Textbook Reference: *46.1 How Is the Mammalian Nervous System Organized? p. 991*

9. Which of the following statements about the peripheral nervous system is *false*?
   a. Afferent portions of the peripheral nervous system carry information to the CNS.
   b. Efferent portions of the peripheral nervous system carry information from the CNS.
   c. The peripheral nervous system communicates only with the circulatory and digestive systems.
   d. The efferent portion is broken down into voluntary and involuntary divisions.
   Textbook Reference: *46.1 How Is the Mammalian Nervous System Organized? pp. 985–986*

10. Which of the following statements about the developing CNS is *false*?
    a. The CNS develops from a solid neural cylinder.
    b. The midbrain becomes part of the brain stem.
    c. The forebrain develops into both the diencephalon and the telencephalon.
    d. The hindbrain develops into the medulla, the pons, and the cerebellum.
    Textbook Reference: *46.1 How Is the Mammalian Nervous System Organized? pp. 986–987*

11. In which of the following cortical lobes can association areas and long-term memory be found?
    a. Temporal
    b. Parietal
    c. Occipital
    d. All of the above
    Textbook Reference: *46.1 How Is the Mammalian Nervous System Organized? p. 990*

12. Which of the following animals would likely have the smallest ratio of telencephalon size to body size?
    a. Fishes
    b. Amphibians
    c. Mammals
    d. Reptiles
    Textbook Reference: *46.1 How Is the Mammalian Nervous System Organized? p. 987*

13. Which of the following statements about the knee-jerk reflex is *false*?
    a. It is a monosynaptic reflex.
    b. The leg extends in response to a hammer tap.
    c. Chemoreceptors sense the hammer tap.
    d. The afferent nerve travels from the receptor to the spinal cord.
    Textbook Reference: *46.1 How Is the Mammalian Nervous System Organized? p. 987*

14. The ridges of the cerebral cortex are called _____ and the valleys of the cerebral cortex are the _____.
    a. gyri; interneuron
    b. gyri; sulci
    c. sulci; gyri
    d. interneuron; gyri
    Textbook Reference: *46.1 How Is the Mammalian Nervous System Organized? p. 989*

15. Which of the following describes the processing of visual information by the retina?
    a. Convergence of information
    b. Telencephalization
    c. Long-term depression
    d. Long-term potentiation
    Textbook Reference: *46.2 How Is Information Processed by Neuronal Networks? p. 995*

Application Questions

1. What are the differences between the brain functions involved in reading a written sentence out loud and repeating a sentence one has just heard?
   Textbook Reference: *46.3 Can Higher Functions Be Understood in Cellular Terms? pp. 1000–1001*

2. You have been sitting up all night watching movies in the living room of your apartment, and your roommate enters from a bedroom, apparently sleepwalking. What type of sleep is your roommate in at the moment, and how can you tell?
   Textbook Reference: *46.3 Can Higher Functions Be Understood in Cellular Terms? p. 998*

3. You have been eyeing hard candy in a dish and finally decide to unwrap a piece and eat it. As you begin to suck on the candy, your salivary glands begin to secrete saliva. What parts of the nervous system are involved in the sequence of events?
   Textbook Reference: *46.2 How Is Information Processed by Neuronal Networks? pp. 992–993*

4. Humans have the ability to see things in three dimensions. What allows us to have binocular vision?
   Textbook Reference: *46.2 How Is Information Processed by Neuronal Networks? pp. 996–997*

5. During a boxing match, a sharp punch to the jaw may cause a knockout; a loss of consciousness. Which area of the brain do you think was affected by the punch?
   Textbook Reference: *46.1 How Is the Mammalian Nervous System Organized? p. 988*

Answers

Knowledge and Synthesis Answers

1. **b.** The temporal lobe is involved in recognition of people and objects. A person who has had damage to the temporal lobe will not be able to identify someone by face and must use other cues.

2. **a.** Secretion of hormones by endocrine glands is involuntary. The autonomic nervous system consists of

efferent pathways that link the CNS with many physiological functions.

3. **a.** The motor neurons in the head region control specific parts of the body. Parts of the body can be mapped on the motor cortex, from the head region on the lower side to the lower part of the body at the top.

4. **c.** The sympathetic division of the autonomic nervous system inhibits digestion rather than stimulating it.

5. **d.** The cell bodies of the sensory neurons are located in the periphery and send their axons to the CNS.

6. **b.** Speaking a written or heard word requires neural flow from Wernicke's area in the temporal lobe to Broca's area in the frontal lobe. Therefore, language skills depend on proper neural flow between the temporal lobes and the motor cortex.

7. **d.** During REM sleep the body is paralyzed, except for twitching of muscles. During this stage of sleep the brain is as active as when you are awake.

8. **c.** The occipital cortex is the higher brain region controlling vision.

9. **c.** The peripheral nervous system is in contact with every tissue in the body sending and receiving information to the CNS.

10. **a.** The CNS develops from a hollow tube composed of neural tissues.

11. **d.** Association areas and long-term memory are present in all these regions of the cerebral cortex.

12. **a.** Fishes have the most undeveloped telencephalon of the vertebrates listed, resulting in the smallest ratio.

13. **c.** The knee-jerk reflex is an example of a mono-synaptic reflex. Stretch receptors sense the hammer tap on the tendon.

14. **b.** The ridges on the surface of the cerebral cortex are the gyri and the valleys are the sulci.

15. **a.** In the retina, the information from over 100 million photoreceptors is integrated by about 1 million ganglion cells; this type of processing is called convergence of information.

Application Answers

1. Both speaking written language and repeating heard language follow similar paths in the brain. The main difference has to do with the initial region of the brain perceiving the word. In reading a word, the area at the back of the cerebrum is used to visualize it. The spoken word uses an area of the cerebrum just behind the area used for speech. Once the word has been processed by the initial centers, the path used for speaking the word is the same. Wernicke's area is stimulated, followed by Broca's area, and then the motor area.

2. Your roommate is sleepwalking during non-REM sleep. During REM sleep the skeletal muscles of the body become paralyzed, and the sleeper is not able to move. You can tell that your roommate is not in REM sleep because you will not see the eye movement typical of this stage of sleep.

3. The peripheral system contributed to both seeing the candy and the movements of the arms and legs that you used to pick it up. The parasympathetic branch of the autonomic nervous system stimulated salivation. The central nervous system was involved in the recognition and decision to unwrap and eat the candy.

4. The right side of your brain receives visual information from the left visual field, and the left side receives information from the right visual field. In the visual cortex, the cells are organized in columns that alternate between receiving information from the right and left eyes. At the borders of the columns, the inputs from the right and left eyes overlap. The cells that receive the overlap are called binocular cells, and they interpret the disparity between what the two eyes sense. This provides a three-dimensional image.

5. The sharp punch caused the boxer's head to turn sharply and this likely twisted the medulla and reticular activating system. The reticular system is a network of neurons in the brain stem that, unless inhibited by other regions of the brain, activates the cerebral cortex and causes consciousness. A sharp blow might affect the reticular system and cause temporary loss of consciousness.

CHAPTER 47 Effectors: How Animals Get Things Done

Important Concepts

Muscle cells are responsible for tissue contraction.

- Vertebrates have three types of muscle: skeletal, cardiac, and smooth. Skeletal muscle is responsible for voluntary movements and the unconscious movements associated with breathing. Cardiac muscle is responsible for the beating of the heart. Smooth muscle is controlled by the autonomic nervous system, and it is responsible for the contraction that occurs in many hollow organs, such as the gut and bladder.

- In all three types of muscle tissue, contraction is due to the interaction between the contractile proteins actin and myosin.

Muscle contraction is described by the sliding filament theory.

- Skeletal muscles are striated voluntary muscles made of large muscle fibers (see Figure 47.1). Muscle fibers have many nuclei and are composed of actin and myosin. Molecules of actin are organized into thin filaments and those of myosin into thick filaments. Bundles of actin and myosin filaments are arranged into myofibrils.

- The contracting unit of myofibrils is the sarcomere, which contains actin and myosin filaments and has very distinct repeating patterns.

- In a sarcomere, the actin filaments are anchored by the Z lines, and myosin filaments are found at the center in the A band (see Figure 47.2).

- In relaxed muscle, the H zone and I band are the regions in which there is no overlap of actin and myosin. During muscle contraction, the Z lines move toward each other, and the H zone and I band shrink in size due to the sliding of actin filaments along the myosin filaments. This is the sliding filament theory of muscle contraction.

The proteins myosin and actin are the key to muscle contraction.

- Myosin molecules are made of two polypeptide chains wrapped around each other, each with a globular head at one end, much like two twisted golf clubs. A myosin filament is composed of many myosin molecules (see Figure 47.3).

- Actin filaments are composed of two monomer chains in a helical arrangement, appearing as two linear strings of pearls wrapped around each other. The proteins tropomyosin and troponin are associated with actin (see Figure 47.3).

- Myosin heads change conformation when they bind to actin filaments at myosin binding sites, forming a cross-bridge connection. The conformational change in the myosin pulls the actin in toward the middle of the sarcomere. ATP then binds to an ATP binding site on myosin, resulting in the bound actin's release from the myosin and the myosin's return to the original conformation. Many myosin molecules cycle through binding with actin to shorten a sarcomere.

Calcium ions regulate myosin and actin filament movements.

- Action potentials form at motor neuron synapses with hundreds of muscle fibers all contracting at the same time. All the fibers activated by a single motor neuron constitute a motor unit. Action potentials spread deep into the cytoplasm or sarcoplasm of the muscle through transverse tubules or T tubules that are in contact with sarcoplasmic reticulum throughout the sarcoplasm (see Figure 47.5). The sarcoplasmic reticulum takes up and releases Ca^{2+} ions into the sarcoplasm, thereby controlling relaxation and contraction of the myofibrils.

- In relaxed muscle, tropomyosin and troponin cover the myosin binding sites on actin filaments, inhibiting muscle contraction. Calcium regulates contraction by binding with troponin, causing the tropomyosin to change conformation and expose the actin–myosin binding sites on the actin filaments (see Figure 47.6).

Cardiac muscle causes beating of the heart.

- Cardiac muscle is found in the heart and is composed of branched muscle cells that form a strong meshwork (see Figure 47.7B). Cardiac muscle cells are smaller than skeletal muscle cells and each cell has only a single nucleus. Intercalated discs add additional strength by holding the cells together. Gap junctions within the intercalated discs allow cardiac muscle cells to be electrically coupled.

- Heartbeats originate at the pacemaker cardiac muscle cells and spread rapidly through gap junctions in the muscle. The heartbeat is said to be myogenic because it is generated by the heart muscle itself.

- The mechanism of excitation–contraction coupling in cardiac muscle cells is called Ca^{2+}-induced Ca^{2+} release.

Smooth muscle causes contraction in many internal organs.

- Smooth muscles are involuntary muscles comprised of long, simple cells, each with a single nucleus (see Figure 47.7C). Smooth muscles are controlled by acetylcholine and norepinephrine of the autonomic nervous system (see Figure 47.8). When smooth muscle is stretched it contracts with strength proportional to the stretch of the muscle.

- Smooth muscle cells are arranged in sheets. Gap junctions allow electrical contact between the cells and promote coordinated contraction.

- In smooth muscle, contraction is controlled by a calmodulin–Ca^{2+} complex. This complex activates myosin kinase, an enzyme that phosphorylates the myosin head to cause contraction. Myosin phosphatase works in the opposite direction by dephosphorylating myosin.

Skeletal muscles show graded contractions, ranging from twitches to tetanus.

- An action potential in a skeletal muscle fiber causes twitch or contraction of the muscle. Twitches can occur as discrete contractions, or if they occur frequently enough, the twitches are summed together (see Figure 47.9).

- Maximum muscle tension, or tetanus, occurs when there is a high rate of stimulation by action potentials. Temporal summation occurs when there is fast twitching of individual fibers, and spatial summation occurs when more than one motor unit stimulates contraction.

- During tetanic contraction, actin and myosin bonds cycle to help keep a muscle fiber from stretching. ATP levels control the length of a tetanic contraction because it is the molecule that provides the energy for myosin to break the bond with actin.

- Muscle tone reflects the small but changing number of motor units active in a muscle at any given time.

The strength and endurance of muscles depend on fiber type.

- Slow-twitch muscle fibers are highly resistant to fatigue because they have a lot of myoglobin (an oxygen-binding protein similar to hemoglobin), mitochondria, and blood vessels. They are also called "red" or "oxidative" muscle (see Figure 47.10).

- Fast-twitch muscle fibers rapidly develop maximum tension but fatigue quickly. They have lower levels of

mitochondria, myoglobin, and blood vessels and are called "white" or "glycolytic" muscle (see Figure 47.10).

The strength of a muscle fiber is related to its length.

- The amount of force a sarcomere can generate depends on its resting length (see Figure 47.11).

- Stretching a muscle causes the sarcomeres to shorten, resulting in less overlap between actin and myosin filaments and less force.

Exercise enhances the strength and endurance of muscle.

- Anaerobic exercise, such as weight lifting, increases strength. Such exercise induces the formation of new actin and myosin filaments in existing muscle fibers and thus produces bigger fibers and bigger muscles.

- Aerobic exercise, such as jogging, increases endurance. Such exercise increases myoglobin, number of mitochondria, density of capillaries, and enzymes involved in energy utilization.

Muscles have three systems for obtaining ATP.

- Muscles use the immediate, glycolytic, and oxidative systems to obtain ATP needed for contraction (see Figure 47.12).

- The immediate system utilizes preformed ATP and creatine phosphate.

- The glycolytic system follows the immediate system within seconds and metabolizes carbohydrates to lactate and pyruvate.

- The oxidative system is fully activated within about one minute and completely metabolizes carbohydrates or fats to water and carbon dioxide.

Invertebrates use hydrostatic skeletons and exoskeletons for support and movement.

- Many soft-bodied invertebrates have a fluid-filled body cavity that acts as a hydrostatic skeleton. Earthworms have circular muscles and longitudinal muscles that oppose each other to act on the hydrostatic skeleton and control elongation and shortening of body segments (see Figure 47.13).

- Arthropods have an exoskeleton, or cuticle, composed of chitin that offers protection and provides sites for muscle attachment.

- As an animal that has an exoskeleton grows, it undergoes the process of molting. With each molt the old exoskeleton is shed, revealing a new one that has developed underneath.

The vertebrate endoskeleton provides a frame for support and movement.

- Endoskeletons are growing, living tissue that provide sites for muscle attachment and support for the body.

- The human skeleton is composed of a central axial skeleton and an appendicular skeleton (see Figure 47.15).

- The pliable portions of the endoskeleton, such as the framework of the nose, and the surface of the joints of the endoskeleton are composed of cartilage containing the protein collagen.
- The bone in the endoskeleton is strong and solid and is composed mainly of collagen and calcium phosphate. Bone is constantly being remodeled by two types of cells, osteoblasts, which lay down new bone, and osteoclasts, which break down bone (see Figure 47.16). When an osteoblast becomes enclosed by the matrix it is laying down, it stops forming matrix and exists within a lacuna; at this stage, the cell is called an osteocyte. Osteocytes communicate with one another and influence the activities of osteoblasts and osteoclasts.
- Developing bone is created as membranous bone growing on a scaffolding of connective tissue or as cartilage bone hardening from an initial cartilage state. Compact bone is solid, whereas cancellous bone is lightweight, with many cavities. Compact bone is known as Haversian bone; it is composed of concentric rings with blood vessels running through the middle. These structural units are known as Haversian systems (see Figure 47.18).

Movable bones come together at joints.

- There are six types of joints where bones meet, allowing for a variety of bone movement (see Figure 47.19).
- The muscles attached to bones at joints work antagonistically (see Figure 47.20). The flexor muscles bend the joints, whereas the extensor muscles straighten them.
- Two types of connective tissues hold joints and bones together. Ligaments hold bone to bone, and tendons hold muscle to bone.

Effectors can be cells other than muscle cells.

- Nematocysts are specialized cells found in cnidarians that are fired at prey animals either to entangle or poison them.
- Chromatophores are specialized cells that contain pigment and allow an animal to change color. Color change through chromatophores allows animals to blend with their backgrounds and communicate with conspecifics. Such color change may be under neural or hormonal control or both.
- Glands, such as endocrine and exocrine glands, also act as effector organs.
- Some animals have electric organs, believed to have evolved from muscle tissue, that are able to produce large electrical currents.
- Some animals have light-emitting organs. The emission of light by organisms is called bioluminescence and it functions in communication or predator–prey interactions.

The Big Picture

- Actin and myosin are the "universal" proteins for motion. Whether they are located in a unicellular animal or the leg muscle of a human, the molecular interactions of actin and myosin produce movement in the structures in which they reside.
- The sliding filament theory provides a model for how actin and myosin filaments slide past each other, resulting in the shortening (contraction) or lengthening (relaxation) of muscle cells. The whole process of actin and myosin interaction is tightly regulated by the movement of calcium ions into and out of the intracellular spaces of muscle cells, which in turn is activated by the arrival of action potentials in motor neurons. Muscle contraction requires the use of energy in the form of ATP.
- Muscles act in concert with skeletons that are either external (invertebrates) or internal (vertebrates). In vertebrates, bones are articulated, forming joints that provide for specialized directional movements of the tissues supported by the bones (e.g., limbs). Muscles controlling joint movement are often located in pairs acting antagonistically, with one set of muscles causing bending, or flexion, of the joint and the other causing straightening, or extension, of the joint.
- Animals have a diversity of effectors, including muscles, glands, specialized cells (chromatophores and nematocysts), and organs (electric and light-emitting) used in communication and predator–prey interactions.

Common Problem Areas

- The sarcomere is the functional unit of the muscle cell, and until you understand its fine structure—Z lines, H lines, etc.—it will be difficult for you to appreciate how the sarcomere shortens through the actions of actin and myosin.
- The way that movements of multiple muscle groups cause both flexion and extension of a joint can be confusing. Thinking of joints in terms of levers and pulleys may help you understand their actions.

Study Strategies

- The interactions of myosin, actin, troponin, tropomyosin, and calcium and the role of action potentials in stimulating muscle contraction make up a complex, multistep process.
  - First, break the process down into its constituents and learn their location and general structures.
  - Second, determine how actin and myosin move relative to each other through a series of power strokes.
  - Finally, ascertain how calcium ions released from the sarcomeres by action potentials initiate and maintain the whole process of muscle contraction.

- Review the following animated tutorials and activities on the Companion Website/CD:
  Tutorial 47.1 Molecular Mechanism of Muscle Contraction
  Tutorial 47.2 Smooth Muscle Action
  Activity 47.1 The Structure of a Sarcomere
  Activity 47.2 The Neuromuscular Junction
  Activity 47.3 Joints

Test Yourself

Knowledge and Synthesis Questions

1. Which of the following statements is *false*?
   a. Strength training induces the formation of new actin and myosin filaments in existing muscle fibers.
   b. Satellite cells are muscle stem cells.
   c. Aerobic exercise increases the number of mitochondria in muscle cells and the density of capillaries in muscle.
   d. A common effect of strength training is the production of more muscle fibers.
   Textbook Reference: 47.2 What Determines Muscle Strength and Endurance? p. 1014

2. A motor unit is best described as
   a. all the nerve fibers and muscle fibers in a single muscle bundle.
   b. one muscle fiber and its single nerve fiber.
   c. a single motor neuron and all the muscle fibers that it innervates.
   d. the neuron that provides the central nervous system with information on the state of contraction of the muscle.
   Textbook Reference: 47.1 How Do Muscles Contract? p. 1008

3. Which of the following statements is *false*?
   a. Cardiac muscle is striated.
   b. Smooth muscle does not contain actin.
   c. Skeletal muscle is considered voluntary.
   d. Smooth muscle is found in the digestive tract and the walls of the bladder.
   Textbook Reference: 47.1 How Do Muscles Contract? pp. 1005, 1010–1012

4. The oxygen-binding molecule in skeletal muscle is
   a. myoglobin.
   b. hemoglobin.
   c. ATP.
   d. myokinase.
   Textbook Reference: 47.2 What Determines Muscle Strength and Endurance? p. 1014

5. The action potential that triggers a muscle contraction travels deep within the muscle cell by means of
   a. sarcoplasmic reticulum.
   b. transverse (or T) tubules.
   c. synapses.
   d. motor end plates.
   Textbook Reference: 47.1 How Do Muscles Contract? p. 1009

6. A sarcomere is best described as a
   a. moveable structural unit within a myofibril bounded by H zones.
   b. fixed structural unit within a myofibril bounded by Z lines.
   c. fixed structural unit within a myofibril bounded by A bands.
   d. moveable structural unit within a myofibril bounded by Z lines.
   Textbook Reference: 47.1 How Do Muscles Contract? pp. 1006–1007

7. ATP provides the energy for muscle contraction by allowing for
   a. an action potential formation in the muscle cell.
   b. the breaking of actin-myosin bonds.
   c. the formation of actin-myosin bonds.
   d. release of calcium by sarcoplasmic reticulum.
   Textbook Reference: 47.1 How Do Muscles Contract? p. 1007

8. Ca^{2+} binds to _____ in skeletal muscle and leads to exposure of the binding site for _____ on the _____ filament.
   a. troponin; myosin; actin
   b. troponin; actin; myosin
   c. actin; myosin; troponin
   d. tropomyosin; myosin; actin
   Textbook Reference: 47.1 How Do Muscles Contract? p. 1009

9. Tropomyosin is moved by which of the following proteins?
   a. Calmodulin
   b. Acetylcholine
   c. Actin
   d. Troponin
   Textbook Reference: 47.1 How Do Muscles Contract? p. 1010

10. Summation of frequent muscle twitches to give maximum contraction is called
    a. motor unit summation.
    b. twitch.
    c. facilitation.
    d. tetanus.
    Textbook Reference: 47.2 What Determines Muscle Strength and Endurance? p. 1012

11. _____ are responsible for the dynamic remodeling of bone that is always underway.
    a. Osteoblasts
    b. Osteoblasts and osteoclasts
    c. Osteoclasts and osteocytes
    d. Osteoblasts, osteoclasts, and osteocytes
    Textbook Reference: 47.3 What Roles Do Skeletal Systems Play in Movement? p. 1017

12. The primary difference between an endoskeleton and an exoskeleton is
    a. the presence of both circular and longitudinal muscles.

b. whether or not the skeleton is on the inside of the body.

c. the presence or absence of joints.

d. the amount of fluid in the body.

Textbook Reference: 47.3 What Roles Do Skeletal Systems Play in Movement? pp. 1016–1017

13. A soccer player suffers a knee injury that damages the tissue holding his upper and lower leg bones together. The damaged tissue is probably a

a. muscle.

b. tendon.

c. ligament.

d. cartilage.

Textbook Reference: 47.3 What Roles Do Skeletal Systems Play in Movement? p. 1019

14. Sites of elongation between the ossified regions of long bones are called

a. glue lines.

b. fulcrums.

c. hinge joints.

d. epiphyseal plates.

Textbook Reference: 47.3 What Roles Do Skeletal Systems Play in Movement? p. 1018

15. Which of the following is considered an effector?

a. Chromatophore

b. Endocrine gland

c. Electric organ

d. All of the above

Textbook Reference: 47.4 What Are Some Other Kinds of Effectors? pp. 1020–1021

Application Questions

1. Smooth muscle contracts involuntarily, whereas skeletal muscle contraction is under voluntary control. What is the difference between the mechanisms that control smooth muscle and skeletal muscle contraction?

Textbook Reference: 47.1 How Do Muscles Contract? pp. 1008–1010, 1012

2. White muscle and red muscle are found in different parts of the body and are used for different types of movement. What are the physiological and morphological characteristics that distinguish the two types of muscle?

Textbook Reference: 47.2 What Determines Muscle Strength and Endurance? p. 1013

3. Smooth muscle contracts involuntarily in the digestive tract, blood vessels, and urinary bladder. Describe the two main ways that smooth muscle contraction and the membrane potential of smooth muscle are controlled.

Textbook Reference: 47.1 How Do Muscles Contract? pp. 1010–1012

4. Earthworms have a hydrostatic skeleton. Describe how a hydrostatic skeleton is used to move an earthworm through the soil.

Textbook Reference: 47.3 What Roles Do Skeletal Systems Play in Movement? p. 1016

5. Why does weight-bearing exercise help to prevent osteoporosis?

Textbook Reference: 47.3 What Roles Do Skeletal Systems Play in Movement? pp. 1017–1018

Answers

Knowledge and Synthesis Answers

1. **d.** Strength training typically produces bigger, rather than more, muscle fibers.

2. **c.** A motor unit is a single motor neuron and all the muscle fibers that it innervates.

3. **b.** Although contraction of smooth muscle is controlled differently from that of skeletal muscle, smooth muscle does contain actin and myosin.

4. **a.** Myoglobin is the main oxygen-carrying molecule in skeletal muscle.

5. **b.** The action potential arriving to the muscle travels into the muscle through the transverse (or T) tubules.

6. **d.** A sarcomere is a structural unit within a myofibril bounded by Z lines that contain actin and myosin.

7. **b.** ATP provides energy that is used to break actin–myosin bonds.

8. **a.** Calcium is released from the sarcoplasmic reticulum and binds with troponin, resulting in the myosin-binding site on actin being exposed.

9. **d.** The binding of calcium with troponin causes a conformational change in tropomyosin.

10. **d.** Tetanus is the maximum level of muscle contraction.

11. **d.** Osteoblasts, osteoclasts, and osteocytes are all involved in remodeling bone.

12. **b.** Endoskeletons (such as those of mammals) are found inside the body, and exoskeletons (such as those of insects) are found outside the body.

13. **c.** Ligaments hold bones together.

14. **d.** Epiphyseal plates are the sites of elongation in long bones.

15. **d.** Chromatophores, endocrine glands, and electric organs are all effectors.

Application Answers

1. Smooth muscle contraction and skeletal muscle contraction are regulated by the presence of Ca^{2+}. In smooth muscle, Ca^{2+} joins with the protein calmodulin to activate myosin kinase in the sarcoplasm. Myosin kinase then phosphorylates the myosin head, allowing the myosin head to bind with actin. In skeletal muscle, Ca^{2+} binds with troponin on the actin filaments. This binding of Ca^{2+} causes a conformational change in tropomyosin and uncovers the myosin binding sites on the actin, allowing myosin to bind with actin.

2. Fast-twitch fibers are known as white muscle and have few mitochondria, small amounts of myoglobin, and few blood vessels. Slow-twitch fibers are known as red muscle. Red muscle has many mitochondria, large amounts of myoglobin, and many blood vessels.

3. Smooth muscle is sensitive to stretching. It will depolarize in response to any stretching and contract, with the strength of the contraction proportional to the amount of stretch in the muscle. Parasympathetic inputs of acetylcholine also cause depolarization of the muscle membrane resulting in contraction, whereas norepinephrine hyperpolarizes the muscle membranes.

4. The hydrostatic skeleton of the earthworm is an incompressible fluid-filled cavity surrounded by longitudinal and circular muscles. Contraction of the longitudinal muscles causes the segments to contract and the body to shorten. Contraction of the circular muscles causes the body segments to elongate and the body to lengthen. Alternating contractions between the longitudinal and the circular muscles move the animal in a push and pull manner. Bristles on the body help hold the animal in place after elongation, and the body is pulled forward during shortening.

5. Osteoporosis is a decrease in bone density that results when the destruction of bone by osteoclasts outpaces the formation of new bone by osteoblasts, leading to thin, brittle bones. Weight-bearing exercises place stress on bone, ultimately altering the interplay of osteoblast and osteoclast activity to induce thickening of bone.

CHAPTER 48 Gas Exchange in Animals

Important Concepts

Respiratory gas exchange occurs by diffusion.

- The respiratory gas oxygen (O_2) is required by cells to produce energy in the form of ATP.
- The respiratory gas carbon dioxide (CO_2) is one of the waste by-products of ATP production and must be eliminated from an animal's body.
- The transfer of these gases occurs by simple diffusion in the respiratory systems of animals. Partial pressures are used to express the concentrations of gases in a mixture.

Fick's law of diffusion describes the exchange of gas in the respiratory system.

- The partial pressure gradient is the difference between the partial pressure of a gas at two locations.
- The partial pressure gradients of O_2 and CO_2 drive the movement of O_2 into the body and CO_2 out of the body.
- Rate of diffusion, described by Fick's law of diffusion, depends in part on a diffusion coefficient that varies according to temperature, the medium, and the diffusing molecules. Rate of diffusion also depends on the cross-sectional area over which the gas is diffusing and the partial pressure gradient.

Several physical parameters affect respiratory gas exchange.

- There are major differences in the O_2 capacity of air and water.
- Water, because of its density, is more expensive to move than air. Water also contains far less O_2 than does air, and O_2 also diffuses far more slowly in water than in air.
- Temperature affects respiration of aquatic animals because there is less O_2 in warm water than in cold, and an animal's need for oxygen increases with temperature (see Figure 48.2).
- The tendency for a gas to move by diffusion across gills, skin, or lungs depends on its partial pressure. The sum of all the gases' partial pressures in a gas mixture is the total partial pressure, which in the atmosphere equals the atmospheric pressure.

- Oxygen makes up 20.9 percent of the atmospheric pressure. As elevation increases and atmospheric pressure decreases, the total amount of oxygen in air decreases.
- CO_2 diffuses across the respiratory organs in both air and water. In air, diffusion is rapid because of the large partial pressure gradient between the blood and the atmosphere. In contrast to the situation for oxygen of a decreasing partial pressure gradient with altitude, the partial pressure gradient for CO_2 does not change with altitude. In aerated water, CO_2 diffuses easily, but if the water is stagnant, a large partial pressure gradient may not exist, and CO_2 diffusion will be slower. Nevertheless, in stagnant water, lack of oxygen becomes a problem for water-breathing animals long before problems with CO_2 exchange occur.

Animals have adaptations to maximize gas exchange.

- Animals maximize respiration by decreasing the distance that gases must diffuse between the blood and the external environment. They achieve this with very thin respiratory surface membranes.
- Animals also actively move the environmental respiratory fluid over the gills and through the lungs and perfuse the internal side of the respiratory organ to carry the respiratory gases.
- Some aquatic amphibians and insects have external gills with large surface areas for exchange of respiratory gases with water (see Figure 48.3A).
- Internal gills have a large surface area and are protected from the environment by an animal's body cavity. They must be actively ventilated with water to achieve gas exchange.

Diverse types of respiratory organs have evolved in animals.

- Insects have tracheae—air tubes that end at the tissue cells as air capillaries and open to the environment through spiracles. Gases diffuse through the tracheae into the air capillaries, but are also moved by movement of the animal's body parts (see Figure 48.4).
- Some aquatic insects carry a bubble of air with them underwater. As they use the O_2 in the air bubble, the

P_{O_2} in the bubble decreases, resulting in a P_{O_2} gradient between the water and the air bubble. Oxygen diffuses into the air bubble from the water down the P_{O_2} gradient, restoring oxygen in the bubble.

- Fish have internal gills with a large surface area that they ventilate with unidirectional flowing water to maximize external P_{O_2} levels. Each gill has hundreds of gill filaments with folds, or lamellae, that act as the respiratory gas exchange surfaces. Countercurrent flow of blood in the lamellae and water over the lamellae maximize the P_{O_2} gradient between the water and the blood (see Figures 48.5 and 48.6). As blood flows through the lamellae it is always in contact with water that has a higher oxygen level, resulting in a continuous oxygen gradient to maximize the uptake of oxygen into the blood.

- Birds create a unidirectional flow of air through air sacs and lungs, which make up the avian respiratory system. Gas exchange occurs in the parabronchi and air capillaries of the lungs; gas exchange does not occur in the air sacs (see Figure 48.7). Air requires two cycles of inhalation and exhalation to move from the posterior air sac, to the lung, to the anterior air sac, and out of the respiratory tract (see Figure 48.8).

The lungs of most vertebrates are ventilated by tidal ventilation.

- Lungs evolved in fishes as outpocketings of the digestive tract. In all vertebrates except birds, lungs are dead-end sacs in which ventilation is tidal. In tidal ventilation, air flows in and exhaled gases flow out by the same route.

- The tidal breathing of mammals can be described by a series of lung volumes (see Figure 48.9). The normal volume of air moved during one cycle is known as the tidal volume. The vital capacity is made up of the resting tidal volume, plus the additional volume of air that can be taken in with a large breath (inspiratory reserve volume) and the additional volume of air that can be forced out with a large exhale (expiratory reserve volume). Not all of the air can be forced out of the lungs. Some air remains in the bronchi and trachea, making up the dead space; this air is called the residual volume. The total lung capacity is the sum of the vital capacity and the residual volume.

- Tidal breathing limits the partial pressure gradient needed to drive the diffusion of oxygen from air into blood. Fresh air is not moving into the lungs during some of the breathing cycle, and when it does enter the lungs, it mixes with stale air.

The lung is the respiratory organ in mammals.

- Mammalian lungs have a very large surface area and a very short gas diffusion pathway.

- Air enters at the oral cavity or nasal passage, travels down the pharynx through the larynx, and then through the trachea, which branches into two bronchi,

one leading to each lung. In the lung, more branching occurs to produce bronchioles and finally the site of gas exchange—small air sacs, the alveoli (see Figure 48.10).

- Oxygen diffuses across the thin-walled alveoli (less than 2 μm) into many surrounding capillaries.

- The alveoli have surface tension due to an aqueous layer that makes it difficult to inflate them. This is overcome by a surfactant produced by the cells forming the alveolar walls. The aqueous layer is needed to ensure the diffusion of gases across the membrane.

- Mucus, produced along the lung's larger airways, catches and removes inhaled dust and microorganisms. Cilia along the airways move the mucus up the airways and into the throat, acting as a mucus escalator.

- Lungs are inflated by negative pressure created by contraction of the diaphragm muscle in the thoracic cavity. Closed pleural membranes line the thoracic cavity and allow for the development of negative pressure. During exhalation, the diaphragm relaxes and the thoracic cavity contracts (see Figure 48.11). At times of strenuous exercise, intercostal muscles also help change the volume of the thoracic cavity.

Oxygen is transported in the blood by hemoglobin in vertebrates.

- Gases diffuse between the respiratory organ and the circulatory system, where they are transported in blood. The liquid blood plasma transports only a small amount of dissolved O_2, with the vast majority of O_2 transported by the oxygen-binding pigment hemoglobin found in the red blood cells of vertebrates.

- Hemoglobin contains four polypeptide subunits, each with a heme group that reversibly binds O_2. The amount of O_2 bound to hemoglobin depends on the partial pressure of O_2. At the high blood P_{O_2} that is characteristic of blood leaving the lungs, hemoglobin is almost 100 percent saturated with O_2. At the tissues, the P_{O_2} is lower and hemoglobin releases approximately 25 percent of its bound O_2. The relationship between the P_{O_2} of the environment and the percentage of hemoglobin bound with oxygen is known as the S-shaped oxygen-binding curve (see Figure 48.12). At low levels of P_{O_2}, only one of the hemoglobin subunits can bind O_2. Once it binds a single O_2 molecule, the conformation of the hemoglobin changes, making it easier for the other three sites on the hemoglobin to bind oxygen in a process known as positive cooperativity.

- In muscle, myoglobin is the O_2-carrying molecule and functions primarily to store O_2 intracellularly. The O_2 affinity is higher in myoglobin than in hemoglobin, but myoglobin has only one unit to bind one O_2 molecule.

- Fetal hemoglobin has a different structure from the adult form, with a higher affinity for O_2.

- The Bohr effect is the shifting of the hemoglobin binding curve to the right, in response to a decrease in pH of the blood, resulting in more oxygen being released at a given environmental P_{O_2} at the lower pH.

- Hemoglobin's affinity for O_2 is lowered by 2,3 bisphosphoglyceric acid (BPG), which binds with hemoglobin to cause a conformational change in the hemoglobin, resulting in a right shift in the oxygen-binding curve. BPG is a metabolite of glycolysis.

Carbon dioxide is transported as dissolved CO_2 and as bicarbonate ions.

- Tissues produce CO_2 as a by-product of metabolism. CO_2 must be moved from the tissues and excreted into the environment.

- Significant amounts of CO_2 dissolve in the plasma and are carried in this form to the lungs.

- Most of the CO_2 is transported as bicarbonate ions. CO_2 reacts with water to make bicarbonate ions and a proton. Carbonic anhydrase speeds up the conversion of CO_2 to carbonic acid, which then dissociates into bicarbonate ions.

Breathing is controlled by the brain and blood pH levels.

- Breathing is controlled by the autonomic nervous system from neurons in the medulla. Brain areas above the medulla modify breathing in response to speaking, eating, and emotional states.

- In air-breathing animals, chemoreceptors on the medulla are sensitive to CO_2 levels and the pH of cerebral spinal fluid. Breathing increases when CO_2 levels increase and pH decreases. Control of breathing is relatively insensitive to blood O_2 levels, but chemoreceptors in the arteries leaving the heart (carotid and aorta) can signal the medulla in response to low levels of O_2.

- In water-breathing animals, O_2 is the primary feedback stimulus for breathing.

The Big Picture

- Cellular metabolism requires O_2 and produces CO_2 as a waste product that must be eliminated. Animals have evolved a variety of structures and mechanisms for exchanging these gases with the environment. Tracheal systems in insects, gills in fishes, and lungs in terrestrial vertebrates are three examples of gas-exchange organs. Very short diffusion distances and very large surface areas—adaptations designed to maximize gas exchange, characterize all.

- Oxygen is transported by the protein hemoglobin in vertebrates. The P_{O_2} levels surrounding hemoglobin determine oxygen binding. If P_{O_2} is high, hemoglobin accepts O_2, but it readily gives up its O_2 when the P_{O_2} falls. These properties allow hemoglobin to bind O_2 in the gas-exchange organ, transport it to the metabolizing tissues, and then release its oxygen for consumption in cellular metabolism.

Common Problem Areas

- It is easy—though incorrect—to think that the exchange of O_2 and CO_2 is a "two-way street," with O_2 taking the immediate place of CO_2 in the lungs and CO_2 then taking the place of O_2 in the tissues. In fact, the mechanisms for exchange of these two gases are completely different.

- Understanding how O_2 is picked up, transported by, and released from hemoglobin can be quite confusing. Think of the oxygen dissociation curve as a "tool." Think first of the situation at the top of the curve where P_{O_2} is high (the lungs or gills, for example). If the P_{O_2} is high, hemoglobin will maximally bind oxygen. Think, then, of transporting that blood to a region with a given P_{O_2}, and read from the curve what must happen to O_2 saturation under the new P_{O_2} level. The O_2 given from hemoglobin is now available for tissue respiration.

Study Strategies

- Use Fick's law of diffusion as a guideline for understanding gas exchange. The various components of the law—distance for diffusion, partial pressure gradient, surface area over which gas exchange occurs—provide a good framework for understanding why gas-exchange organs have evolved with certain characteristics in common. For example, gas-exchange organs tend to have large surface areas, have very short diffusion distances, and experience large partial gradients for O_2 across their surfaces. All of this makes sense in the context of Fick's law of diffusion. Fick's law of diffusion also governs the movement of O_2 and CO_2 in the body, and it can help you remember where and why these gases are picked up and released in the body.

- Review the following animated tutorials and activities on the Companion Website/CD:
  Tutorial 48.1 Airflow in Birds
  Tutorial 48.2 Airflow in Mammals
  Activity 48.1 Human Respiratory System
  Activity 48.2 Oxygen-Binding Curves
  Activity 48.3 Concept Matching

Test Yourself
Knowledge and Synthesis Questions

1. Which of the following statements is *false*?
   a. A given volume of air is easier to move across the respiratory organs than the same volume of water.
   b. Air holds more oxygen per unit volume than does water.
   c. Water breathers have a difficult time ridding themselves of CO_2 because CO_2 does not dissolve well in water.
   d. Temperature increases affect the O_2 content of water more than they do that of air.

   Textbook Reference: *48.1 What Physical Factors Govern Respiratory Gas Exchange? p. 1026*

2. External gills, tracheae, and lungs share which of the following sets of characteristics?
   a. Part of gas-exchange system; exchange both CO_2 and O_2; increase surface area for diffusion
   b. Used by water breathers; based on countercurrent exchange; use negative pressure breathing
   c. Exchange only O_2; are associated with a circulatory system; found in vertebrates
   d. Found in insects; employ positive-pressure pumping; based on crosscurrent flow
   Textbook Reference: 48.2 What Adaptations Maximize Respiratory Gas Exchange? p. 1028

3. Which of the following represents a larger volume of air than is normally found in the resting tidal volume of a human lung?
   a. Residual volume
   b. Inspiratory reserve volume
   c. Expiratory reserve volume
   d. All of the above
   Textbook Reference: 48.2 What Adaptations Maximize Respiratory Gas Exchange? p. 1032

4. Which of the following is shared by bird and mammal lungs?
   a. Both need two cycles of inhalation and exhalation for air to move through the lungs.
   b. Both contain alveoli as the terminal end of the lung.
   c. Both have an anatomical dead space.
   d. Both exchange O_2 and CO_2 with blood in capillaries.
   Textbook Reference: 48.2 What Adaptations Maximize Respiratory Gas Exchange? pp. 1030–1032

5. According to Fick's law of diffusion, which of the following does *not* play a role in the diffusion of oxygen across a membrane?
   a. Surface area
   b. Volume
   c. Difference in concentration or partial pressure
   d. Diffusion distance
   Textbook Reference: 48.1 What Physical Factors Govern Respiratory Gas Exchange? p. 1026

6. Because of the relatively high altitude of Antonito, Colorado, the town has a normal barometric pressure of about 600 mm Hg rather than 760 mm Hg as at sea level. The partial pressure of oxygen in Antonito's air is approximately _____ mm Hg.
   a. 75
   b. 126
   c. 160
   d. 76
   Textbook Reference: 48.1 What Physical Factors Govern Respiratory Gas Exchange? pp. 1025–1026

7. Air flows into the lungs of mammals during inhalation because the
   a. pressure in the lungs falls below atmospheric pressure.
   b. volume of the lungs decreases.
   c. pressure in the lungs rises above atmospheric pressure.

d. diaphragm moves upward toward the lungs.
   Textbook Reference: 48.3 How Do Human Lungs Work? pp. 1034–1035

8. The movement of O_2 and CO_2 between the blood in the tissue capillaries and the cells in tissues depends most directly upon
   a. active transport of O_2 and CO_2.
   b. total atmospheric (barometric) pressure differences across the cell membranes.
   c. diffusion of O_2 and CO_2 down a concentration gradient.
   d. diffusion of O_2 and CO_2 down a partial pressure gradient.
   Textbook Reference: 48.1 What Physical Factors Govern Respiratory Gas Exchange? pp. 1025–1026

9. The alveoli of the lungs do not contain "air" because
   a. we normally do not ventilate our lungs at a high enough rate.
   b. the lungs have too many alveoli to ventilate.
   c. there is dead space in the trachea and bronchi.
   d. the trachea and bronchi are too small in volume.
   Textbook Reference: 48.3 How Do Human Lungs Work? p. 1034

10. Which of the statements about hemoglobin is *false*?
    a. Hemoglobin allows the blood to carry a large amount of O_2.
    b. Hemoglobin contains a single polypeptide chain with very high affinity for O_2.
    c. Hemoglobin is packaged inside red blood cells.
    d. Fetal hemoglobin is structurally different from adult hemoglobin.
    Textbook Reference: 48.4 How Does Blood Transport Respiratory Gases? p. 1036

11. As blood becomes fully O_2 saturated, hemoglobin is combining with _____ molecule(s) of oxygen.
    a. 1
    b. 2
    c. 4
    d. 8
    Textbook Reference: 48.4 How Does Blood Transport Respiratory Gases? p. 1036

12. The Bohr shift describes the
    a. outward movement of Cl^- from the blood cell in exchange for HCO^-_3 moving into the cell.
    b. leftward shift of the entire oxygen equilibrium curve when temperature rises.
    c. rightward shift of the entire oxygen equilibrium curve when pH rises.
    d. rightward shift of the entire oxygen equilibrium curve when pH falls.
    Textbook Reference: 48.4 How Does Blood Transport Respiratory Gases? p. 1037

13. The presence of CO_2 in blood will lower pH because CO_2 combines with _____, with the rate of reaction increased by _____.
    a. H_2O to form H^+ and HCO^-_3; carbonic anhydrase
    b. H_2O to form only HCO^-_3; carbonic anhydrase

c. H_2O to form only H^+; carbonic ions

d. H^+ to form HCO^-_3; oxyhemoglobin

Textbook Reference: 48.4 How Does Blood Transport Respiratory Gases? p. 1038

14. The largest proportion of CO_2 carried by the blood is in the form of

a. bicarbonate ions (HCO^-_3) carried in the plasma.

b. molecular CO_2 dissolved in the plasma.

c. bicarbonate ions (HCO^-_3) carried within the red blood cells.

d. molecular CO_2 chemically bound to hemoglobin.

Textbook Reference: 48.4 How Does Blood Transport Respiratory Gases? p. 1038

15. Which of the following is involved in the control of breathing?

a. Neurons in the medulla

b. Chemoreceptors on the surface of the medulla

c. Chemoreceptors in large blood vessels leaving the heart

d. All of the above

Textbook Reference: 48.5 How Is Breathing Regulated? pp. 1039–1040

16. Which of the following statement is *false*?

a. Air-breathing animals are remarkably insensitive to falling levels of O_2.

b. In water-breathing animals, O_2 is the primary feedback stimulus for breathing.

c. In humans, chemoreceptors for CO_2 are located on the medulla.

d. Areas of the brain above the medulla have no impact on breathing rate.

Textbook Reference: 48.5 How Is Breathing Regulated? pp. 1039–1040

Application Questions

1. In a dredge, you have just captured an aquatic animal that has no specialized respiratory surfaces. Describe how the organism can get enough oxygen to all of its cells to survive.

Textbook Reference: 48.2 What Adaptations Maximize Respiratory Gas Exchange? p. 1028

2. The gills of fish employ a countercurrent exchange mechanism to ensure that a large amount of oxygen is extracted from the water. Describe how the countercurrent exchange works.

Textbook Reference: 48.2 What Adaptations Maximize Respiratory Gas Exchange? pp. 1029–1030

3. What is the difference between the myoglobin oxygen-binding curve and the hemoglobin oxygen-binding curve?

Textbook Reference: 48.4 How Does Blood Transport Respiratory Gases?, p. 1037

4. Why would you be more likely to find birds than mammals at high altitudes?

Textbook Reference: 48.2 What Adaptations Maximize Respiratory Gas Exchange? pp. 1030–1032

5. Why would your tropical fish face severe respiratory problems (and possibly death) if the heater in their tank malfunctioned and caused extremely high water temperatures?

Textbook Reference: 48.1 What Physical Factors Govern Respiratory Gas Exchange? pp. 1026–1027

Answers
Knowledge and Synthesis Answers

1. **c.** Water breathing is more difficult than air breathing because the higher density of water makes it more expensive to move across the respiratory surfaces than air, it has less oxygen than air, and the oxygen content is very dependent on the temperature of the water. CO_2, however, dissolves easily into water and is easy to get rid of.

2. **a.** The external gill, tracheae, and lungs are all examples of gas-exchange systems used by various animals to exchange both CO_2 and O_2. One of the special properties of all respiratory systems is that they increase surface area for diffusion.

3. **d.** All have a larger volume than the resting tidal volume.

4. **d.** The lungs of birds and mammals are both sites for the exchange of O_2 and CO_2 with blood in capillaries. The lungs of birds take two cycles for air to move through them, whereas the mammalian lung only takes one cycle. The mammalian lung is the only one that ends in alveoli and has an anatomical dead space.

5. **b.** The volume of the gas is not included in Fick's law of diffusion. All the other parameters are important in determining the rate of diffusion of gases in the respiratory system of animals.

6. **b.** Air at sea level with an atmospheric pressure of 700 mm Hg has a partial pressure of oxygen of 160 mm Hg ($700 \times 21\%$). Therefore, the partial pressure of O_2 at Antonito, Colorado, is $600 \times 21\% = 126$ mm Hg.

7. **a.** Inhalation in the mammalian lung occurs by means of negative pressure produced by contraction of the diaphragm. Thus the pressure in the lungs falls below atmospheric pressure.

8. **d.** Movement of O_2 and CO_2 from the blood to the tissues always occurs by diffusion of O_2 and CO_2 down their partial pressure gradients.

9. **c.** The alveoli do not contain "air" with 20.9 percent O_2 because incoming air is mixed with air left in the dead space of the trachea and bronchi that has had some of the O_2 removed by the lungs.

10. **b.** Hemoglobin is composed of four subunits, which act with positive cooperativity to bind O_2. Oxygen-binding myoglobin found in muscle has only one unit.

11. **c.** Hemoglobin has 4 subunits, each of which binds to 1 molecule of O_2 for a total of 4 molecules of O_2 bound to 1 hemoglobin molecule.

12. **d.** The Bohr effect describes the action of pH on the oxygen-binding curve. Decreases in pH result in a net right shift of the entire oxygen equilibrium curve.

13. **a.** Carbon dioxide combines with H_2O in the plasma to form H^+ and HCO^-_3. The enzyme carbonic anhydrase catalyzes the reaction.

14. **a.** The majority of CO_2 is carried as bicarbonate ions (HCO^-_3) in the plasma.

15. **d.** All are involved in the control of breathing.

16. **d.** Areas of the brain above the medulla modify the breathing rate with respect to speech, eating, coughing, and emotional state.

Application Answers

1. Having no respiratory surfaces, the animal must rely on simple diffusion from the environment into its cells. For simple diffusion to be an effective means of oxygen transport, the animal must be very small, with all of its cells only a few cell layers from the environment. The metabolic rate of this animal is most likely very low to accommodate the lack of any special respiratory surfaces.

2. In the countercurrent exchange in the gills of fish, water and blood move in opposite directions. This results in blood with low oxygen content entering the gills and flowing past water that has a higher oxygen content. Oxygen diffuses down the partial pressure gradient into the blood. As blood moves along the lamella, it picks up oxygen, but it always has a lower oxygen content than the water moving in the opposite direction. In this way the blood is able to optimize oxygen uptake during the entire trip through the lamella.

3. The oxygen-binding curve of myoglobin is shifted to the left in relationship to the oxygen-binding curve of hemoglobin. This allows myoglobin to pick up O_2 and hold it at lower partial pressures of O_2.

4. Birds have an extremely efficient respiratory system characterized by the continuous unidirectional flow of air through the lungs. In contrast, mammals use tidal ventilation, a less-efficient method in which air flows in and exhaled gases flow out the same route. Tidal ventilation results in the mixing of fresh air with residual air in the lungs. O_2 availability decreases with altitude, and thus we would expect to see birds rather than mammals at very high elevations.

5. Your tropical fish would be facing the double bind of needing more oxygen as the water temperature increases (due to their increasing metabolic rate with temperature) yet facing lower levels of oxygen in the warmer water (warm water holds less dissolved oxygen than does cold water).

CHAPTER 49 Circulatory Systems

Important Concepts

Small organisms do not need circulatory systems.

- An animal's circulatory system transports nutrients and wastes to and from tissues and cells.

- In the smallest multicellular animals, every cell within the body is no more than one to two cells from the environment. Such small animals do not require a circulatory system because materials can move efficiently into and out of cells by diffusion alone.

- Some aquatic invertebrates have a gastrovascular cavity in which exchange of gases and nutrients occurs. Other organisms are flattened to increase the surface-to-volume ratio, ensuring that cells are close to the outside environment.

- Large animals need a circulatory system to carry wastes, gases, and nutrients to and from all the cells in the body.

Circulatory systems may be classified structurally as open or closed.

- In an open circulatory system, blood and other tissue fluids are not separated (see Figure 49.1A and B). A pump may be present to help move body fluids through the various large compartments, assisted by movements of the animal.

- In a closed circulatory system, blood is pumped through a series of interconnected closed vessels (see Figure 49.1C). There are numerous advantages to closed systems compared to open ones: rapid transport of fluid in the closed vessels, the ability to divert the flow of blood between different tissues, and direct hormone and nutrient transport to the tissues.

- The closed system typical of vertebrates is composed of arteries and arterioles, which carry blood from the heart to the body; veins and venules, which carry blood from the body to the heart; and an intervening capillary bed. In fishes, blood is pumped from the heart to the gills to the tissues and back to the heart. In birds and mammals, the systemic circuit flows from the heart to the body tissues, and the pulmonary circuit flows from the heart to the gas exchange organ.

The gills and systemic tissues play an important role in the fish circulatory system.

- Fishes pump blood with a heart that has one atrium and one ventricle. Blood leaves the heart and enters the gills, where gas exchange occurs. Some arterial blood pressure is lost in passing through the gills.

- Blood leaving the gills collects in the aorta and travels to the systemic organs and tissues, eventually returning to the heart.

- African lungfish have a well-vascularized lung for gas exchange that is formed embryonically from an outpocketing of the gut. A heart with a partially divided atrium helps to separate oxygenated blood returning from the lung and deoxygenated blood returning from the tissues as they enter the single ventricle. A large proportion of oxygenated blood bypasses the gills and goes directly to the tissues. The lungfish heart and vascular system exhibit adaptations that partially separate systemic and pulmonary circuits.

Amphibians have partial separation of pulmonary and systemic blood.

- Amphibians have two atria, one that receives deoxygenated blood from the tissues and one that receives oxygenated blood from the lungs.

- Anatomical features of the ventricle help direct deoxygenated blood preferentially to the lungs and skin (which also acts as a gas exchange organ in amphibians) and oxygenated blood to the systemic body tissues. Even though the two circuits are not fully divided, blood tends to flow in parallel.

Reptiles control their intracardiac shunt to regulate the distribution of oxygenated and deoxygenated blood leaving the heart.

- Turtles, snakes, and lizards have two atria and a ventricle that is internally divided into subchambers that allow for control of the distribution of blood to the lungs and body.

- During air breathing, blood from the right side of the ventricle is preferentially pumped to the lungs because there is a higher resistance in the systemic circuit than in the pulmonary circuit. Also, there is an asynchronous ejection from the heart because pumping occurs first from the right and then the left side of the ventricle. At the stage in the cardiac cycle in which the blood from the left side leaves the heart, the resistance is higher in the pulmonary circuit. Thus the last blood leaving the heart tends to be deoxygenated blood and is preferentially directed to the pulmonary circulation.

- When a reptile stops breathing, vessel constriction in the lungs causes pulmonary circuit resistance to increase, causing blood from the right side to bypass the pulmonary circulation and flow to the systemic circuit.

- Alligators and crocodiles are unique in that they have a four-chambered heart, resembling that of a bird or mammal. However, alligators and crocodiles retain a left aorta that arises from the right side of the heart. There is a small passageway between the left and right aortas through which blood can bypass the pulmonary circulation. Thus, crocodiles and alligators can operate their heart like that of a mammal, distributing blood equally to the pulmonary and systemic circulations, or they can partially bypass the lungs by directing blood from the left aorta into the right aorta and on to the systemic circulation.

Birds and mammals have a completely parallel circulatory system.

- Birds and mammals have a four-chambered heart that allows for complete separation and no mixing of systemic and pulmonary blood.

- This cardiovascular arrangement maximizes O_2 transport by eliminating mixing between pulmonary and systemic circulations.

- The pulmonary circulation operates at low pressure, protecting the delicate lung membranes, whereas the systemic circuit operates at high pressures, allowing the perfusion of tissues and organs distant from the heart.

The human heart contains separate right and left sides pumping to the lungs and tissues, respectively.

- The human heart has two atria and two ventricles (see Figure 49.2). The atrioventricular valve between the atrium and ventricle stops backflow of blood from the ventricle into the atrium. The pulmonary and aortic valves reside between the ventricles and arteries and help maintain one-way flow through the heart.

- Deoxygenated blood flows from the body through the superior and inferior vena cava into the right atrium and into the right ventricle by passive filling during heart relaxation. During heart contraction, blood flows from the right ventricle to the pulmonary artery and the lungs, where blood becomes oxygenated.

- Oxygenated blood from the lungs flows back to the heart via pulmonary veins and into the left atrium, then into the left ventricle, and finally out through the aorta to the systemic tissues.

- Resistance in the blood vessels of the body's systemic tissues is higher than in the lungs, so the left ventricle must be larger and more muscular than the right ventricle to generate more force while pumping the same volume of blood.

- Systole is the contraction of the ventricle, and diastole is the relaxation of the ventricle. At the very end of diastole, the atria contract. Together these two phases make up the cardiac cycle.

- During the cardiac cycle the pressure is highest during ventricle contraction (systole) and lowest during relaxation (diastole; see Figure 49.3).

- Blood pressure can be measured using a sphygmomanometer (see Figure 49.4).

The human heartbeat is generated in the heart's pacemaker.

- The pacemaker, or sinoatrial node, is a small section of highly modified muscle tissue located at the junction of the superior vena cava and right atrium.

- The pacemaker rhythmically produces action potentials that pass on to the rest of the heart and stimulate a heartbeat (see Figure 49.6).

- Gap junctions connect cardiac muscle cells and thus allow the action potentials to spread rapidly from cell to cell.

- The action potential moves from the atrium to the ventricles through the atrioventricular node (AV node). The AV node slows down the transmission of the action potential from the atrium to the ventricle. The action potential passes rapidly down the bundle of His and to the Purkinje fibers fanning out in the ventricular tissue. From the Purkinje fibers the action potential spreads through both ventricles, causing contraction to occur slightly after the contraction of the atrium.

- The autonomic nervous system also plays a role in the control of the heartbeat. Sympathetic activity increases heart rate, and parasympathetic activity decreases heart rate (see Figure 49.5).

- The electrocardiogram (ECG or EKG) measures the electrical activity of the heart as a complex wave pattern (see figure 49.8). The wave has five parts: P (depolarization and contraction of the atrium), Q, R, and S (depolarization of the ventricles), and T (relaxation and repolarization of the ventricles).

Blood is a tissue composed of fluid plasma and many different cell types.

- The fluid matrix of blood is known as plasma and contains water, salts, and proteins. The hematocrit (also called packed-cell volume) is the percentage of blood volume composed of cells. A normal hematocrit is about 42 percent.

- A number of different cell types in blood carry out a wide range of functions (see Figure 49.9). The cell types include the erythrocytes, basophils, eosinophils, neutrophils, lymphocytes, monocytes, and platelets.

- Erythrocytes, or red blood cells, are shaped like biconcave discs when mature. Their function is to transport respiratory gases. In mammals, but not other vertebrates, mature red blood cells lack a nucleus and other organelles.

- Erythrocytes are produced in stem cells located in the marrow of bone. Once they are released into the bloodstream they survive for about 120 days. The hormone erythropoietin is produced in the kidneys and regulates the production of red blood cells.

- The spleen serves as a reservoir for red blood cells.

- Platelets, which help to start blood clotting, are also found in blood and are produced by the breaking off of cell fragments from megakaryocytes in the bone marrow. Platelets bind to the edges of a break in a vessel wall, not only providing a mechanical plug but also starting a chemical chain reaction that leads to the conversion of the circulating protein prothrombin into thrombin. Thrombin stimulates the formation of sticky fibrin threads that form a meshwork and cover the hole in the vessel (see Figure 49.10).

- Plasma contains many gases, ions, nutrients, proteins, and other molecules. The plasma also contains the hormones that are circulated by the blood to target cells. The composition of plasma is very similar to that of interstitial fluid, except that proteins are present in higher concentrations in plasma.

The vascular system is composed of arteries, capillaries, and veins.

- Arteries carry blood away from the heart to the body under relatively high pressure. Veins carry blood toward the heart from the body under low pressure (see Figure 49.11). The very fine capillaries interconnect the arteries with the veins in the tissues, dissipate pressure, and are sites for all exchanges of gases, nutrients, and wastes.

- Arteries and the smaller arterioles are elastic, allowing them to stretch during systole and rebound during diastole. In vertebrates, the walls of vessels contain smooth muscle that controls the vessel's resistance by causing changes in its diameter. Arteries and arterioles are known as resistance vessels.

Capillaries are sites for exchange of gases, nutrients, and wastes, and the locations of considerable fluid loss and recovery.

- Capillaries connect the arterioles with the venules in the tissues. Capillaries have a large cross-sectional area compared with arteries and veins and extremely thin walls to allow the diffusion of gases, nutrients, and wastes (see Figure 49.11).

- Blood moves slowly as it enters the capillaries. Blood pressure from the arterial side forces fluids, ions, and small molecules out of the capillaries through small holes called fenestrations in a process known as filtration. Near the venule end of the capillaries, the difference in osmotic potential between the plasma

inside the capillaries and the surrounding fluids creates an osmotic pressure resulting in fluid recovery back into the capillaries. Thus the net flow of water between the plasma and tissue fluid is determined by the difference in blood pressure and osmotic pressure (see Figure 49.13). The two opposing forces, blood pressure and colloidal osmotic pressure, are known as Starling's forces.

- Edema is the tissue swelling that results from the accumulation of fluid in extracellular spaces. This condition is associated with protein starvation (which leads to low concentrations of proteins in the blood) and inflammation (which causes increased permeability of capillaries).

- Some capillaries are selective about which ions and molecules can pass through them, whereas others are less selective. The blood–brain barrier consists of highly selective brain capillaries.

Blood returns to the heart through veins.

- Blood returns to the heart at low pressure in expandable veins. Because of the low pressure, veins that act against gravity have valves, and blood flowing toward the heart is helped by the movement of the surrounding skeletal muscle (see Figure 49.14). Gravity can cause pooling of blood in the veins if a person is inactive for a long period of time.

- Veins are called capacitance vessels because of their ability to stretch and store blood.

- When veins return a greater volume of blood to the heart, the heart contracts more forcefully. This is explained by the Frank–Starling law, which states that if cardiac muscle cells are stretched (as happens when the volume of returning blood increases), they contract more forcefully.

The lymphatic system returns tissue fluid not reclaimed by the capillaries to the blood.

- Once interstitial fluid enters the lymphatic system, it is called lymph. In humans, lymph is moved through the lymphatic vessels up the body by surrounding skeletal muscle contractions. Lymph is returned to the blood through the thoracic ducts that empty into large veins at the base of the neck.

- Lymph nodes exist in mammals and birds and contribute to the defenses of the body. Foreign materials and microorganisms are removed by the phagocytic action of white blood cells in the lymph nodes.

Cardiovascular disease affects the arteries and veins.

- Hardening of the arteries is known as atherosclerosis.

- Plaque deposits begin to form at sites of endothelial damage, and lipids tend to accumulate on the deposits. Connective tissue invades the deposits and makes the wall of the artery less elastic. Arteries narrow as plaque deposits grow. Blood platelets stick to the plaque, and this can lead to the formation of a blood clot, or thrombus.

- The coronary arteries that supply the heart with blood are highly susceptible to blockage, resulting in a heart attack (myocardial infarction).

- An embolus is a piece of a thrombus that breaks off, travels in the vessels, and then lodges in a vessel. When this happens in the brain, it is known as a stroke. A stroke results in denial of blood flow to a particular region of the brain, causing paralysis, loss of cognitive abilities, or death.

- Genetics and age are the main factors that determine your risk of developing atherosclerosis. Environmental factors also contribute to risk and include high-fat and high-cholesterol diets, smoking, and lack of exercise.

The circulatory system is under hormonal, neural, and local control.

- Blood flow in the capillary beds of many tissues is controlled locally by autoregulatory mechanisms. Precapillary sphincters at the arterial end of the capillaries control the flow of blood, with contraction of these sphincters limiting or stopping blood flow through the capillary bed (see Figure 49.16). Low levels of O_2 and high levels of CO_2 can open the precapillary sphincters. The increased supply of blood brings in more O_2 and removes more CO_2; this response is called hyperemia.

- Norepinephrine from sympathetic neurons and epinephrine released from the adrenal medulla cause arteries and arterioles to contract, reducing blood flow and increasing central blood pressure.

- Autonomic control of heart rate and constriction of blood vessels occurs from the medulla. Stretch receptors (known as baroreceptors) in the aorta and carotid arteries help regulate blood pressure. With rising blood pressure they inhibit sympathetic output and cause the heart to slow and arterioles to dilate. When pressure is low, sympathetic output is stimulated, increasing heart rate and constriction of arterioles. Chemoreceptors in the aorta and carotid arteries sense changes in blood composition.

- In response to a fall in arterial pressure (signaled by decreased activity of baroreceptors), the posterior pituitary releases antidiuretic hormone (ADH, also known as vasopressin). ADH stimulates the kidneys to reabsorb more water, and this results in increased blood volume and pressure.

Diving mammals conserve blood oxygen by slowing their heart rate during dives.

- Marine mammals exhibit a diving reflex during which the heart rate slows and blood flow to all tissues except the brain drops dramatically. During dives, the reduced blood to tissues switches the tissues to anaerobic metabolism and suppresses the metabolism of the tissue. During a dive, such mammals are hypometabolic, having a metabolic rate below their basal rate.

- Marine mammals also have greater stores of oxygen in their bodies than other mammals due to greater blood volumes, concentrations of hemoglobin in blood, and concentrations of myoglobin in muscle.

- Humans also have a diving reflex that results in a mild slowing of the heart rate when the face is under water. The reflex is controlled by the vagus nerve and parasympathetic nervous system.

The Big Picture

- Cardiovascular systems are required when animals grow too big to acquire nutrients and eliminate waste by diffusion alone. Animals have evolved a wide variety of cardiovascular systems, which can be classified generally as open or closed. Open systems have relatively low pressures and have large sinuses that bathe tissues and organs in circulating fluid. Closed systems have an interconnected series of vessels that keep blood separated from tissue but allow gases, nutrients, and wastes to diffuse through their walls.

- During the evolution of vertebrates, the heart has changed from the simple two-chambered heart of fishes, to the complex four-chambered heart of birds and mammals. However, the hearts of amphibians and reptiles do provide for complex intracardiac shunting and bypass of blood flow to the lungs, which is an adaptation that allows them to stop breathing or to breathe intermittently.

- The human heart is really two pumps in one. The left pump provides oxygenated blood under high pressure to the systemic tissues, and the right pump sends deoxygenated blood under low pressure to the lungs. A cardiac pacemaker sets the cardiac rhythm.

- The flow of blood through capillaries is highly regulated at the local level by changes in their local surroundings; blood flow through arteries and veins is regulated by the combined action of neural activity and circulating hormones.

Common Problem Areas

- It is frequently thought that the hearts of amphibians and reptiles are primitive structures, awaiting "repair through evolution" to the more highly evolved bird and mammal heart. In fact, the hearts of amphibians and reptiles are highly adapted to suit their particular lifestyle, specifically their intermittent breathing and relatively low metabolic rate. By being able to shunt blood within the heart, these animals can save the energy that would otherwise be wasted in sending blood to the temporarily nonventilated lungs.

- Students often fail to appreciate that the human heart is actually two entirely distinct pumps—the right heart pumping blood to the lungs and the left heart pumping blood to the body. The two pumps just happen to be packaged in the same structure—the heart. In fact, cardiologists refer to the "left heart" and "right heart," emphasizing the separate nature of these two pumps. Figuring out the flow of blood through the human heart becomes easier when it is considered from this perspective.

Study Strategies

- Determining the pattern of blood flow through the human heart will be easier if you mentally "unfold" the circulation into a great circle, in which there are two pumps at opposite sides of the circle with a vascular bed intervening between each pump. Once you have mastered this concept, you will see that the only way to place the two pumps in proximity in the same structure (the heart) is to twist the circle into a folded figure eight—the typical diagram of the circulation in the textbook.

- Following from the preceding, learn the various chambers and valves in the context of a sequence of structures in each of the right and left pumps of the heart, rather than as a list of terms printed on a complex, anatomically accurate diagram of the heart.

- Refer to the following animated tutorial and activities on the Companion Website/CD:
  Tutorial 49.1 The Cardiac Cycle
  Activity 49.1 Vertebrate Circulatory System
  Activity 49.2 The Human Heart
  Activity 49.3 Structure of a Blood Vessel

Test Yourself

Knowledge and Synthesis Questions

1. Which of the following is *not* one of the reasons that closed circulatory systems are more efficient than open circulatory systems?
   a. Closed systems rely exclusively on simple diffusion for transport, whereas open systems rely on pumping mechanisms.
   b. Transport within closed systems is more rapid than in open systems.
   c. Blood can easily be directed to specific areas in closed systems, but not in open systems.
   d. Closed systems operate better under higher pressure than do open systems.
   Textbook Reference: 49.1 Why Do Animals Need a Circulatory System? p. 1047

2. How does the pattern of blood flow change when reptiles undergo periods of intermittent breathing?
   a. Blood is shunted from the systemic circuit to the pulmonary circuit.
   b. Blood is shunted specifically to the brain.
   c. Blood is shunted from the pulmonary circuit to the systemic circuit.
   d. The pattern does not change during intermittent breathing.
   Textbook Reference: 49.2 How Have Vertebrate Circulatory Systems Evolved? pp. 1048–1049

3. In which of the following would you record the highest blood pressure?
   a. The ventricle supplying blood to the gills of a fish
   b. The anterior dorsal artery of an ant
   c. The pulmonary vein of a frog
   d. The ventricle supplying blood to the systemic circuit of a bird
   Textbook Reference: 49.2 How Have Vertebrate Circulatory Systems Evolved? pp. 1047–1049

4. Blood consists of a fluid fraction consisting of _____ and a solid fraction consisting of _____.
   a. plasma; water, erythrocytes, and platelets
   b. erythrocytes; leukocytes, macrophages, and platelets
   c. plasma; erythrocytes, platelets, and leukocytes
   d. leukocytes; erythrocytes and platelets
   Textbook Reference: 49.4 What Are the Properties of Blood and Blood Vessels? pp. 1055–1057

5. Red blood cells are
   a. biconcave cells containing hemoglobin.
   b. spherical cells containing hemoglobin.
   c. spherical cells capable of ameboid motion and containing hemoglobin.
   d. biconcave cells that contain platelets.
   Textbook Reference: 49.4 What Are the Properties of Blood and Blood Vessels? pp. 1055–1056

6. Blood clotting pathways cause the conversion of _____ to _____.
   a. vitamin K; prothrombin
   b. fibrin; fibrinogen
   c. thrombin; prothrombin
   d. None of the above
   Textbook Reference: 49.4 What Are the Properties of Blood and Blood Vessels? pp. 1056–1057

7. Which of the following regions of the vascular bed is the actual site of gas exchange with surrounding tissue?
   a. Arteries
   b. Capillaries
   c. Lymphatic vessels
   d. Veins
   Textbook Reference: 49.4 What Are the Properties of Blood and Blood Vessels? pp. 1057–1059

8. Which is the correct sequence of parts through which cardiac action potentials pass?
   a. Purkinje fibers, AV node, SA node, bundle of His, atrial fibers
   b. AV node, atrial fibers, SA node, bundle of His, Purkinje fibers
   c. SA node, bundle of His, atrial fibers, AV node, Purkinje fibers
   d. SA node, atrial fibers, AV node, bundle of His, Purkinje fibers
   Textbook Reference: 49.3 How Does the Mammalian Heart Function? pp. 1053–1054

9. The purpose of the AV node is to _____, and the purpose of the Purkinje fibers is to _____.
   a. create simultaneous atrial and ventricular depolarization; speed up transmission of the cardiac impulse into the ventricle
   b. delay ventricular depolarization relative to atrial depolarization; insulate the cardiac impulse from the general ventricular fibers
   c. delay ventricular depolarization relative to atrial depolarization; ensure the cardiac impulse spreads rapidly and evenly throughout the ventricles
   d. delay atrial depolarization relative to ventricular depolarization; transmit the cardiac impulse to very small localized groups of ventricular fibers

   Textbook Reference: *49.3 How Does the Mammalian Heart Function? pp. 1053–1054*

10. The atrial walls are _____ than the ventricular wall, and pressure generated in the atrial chambers is _____ than in the ventricles.
    a. thinner; higher
    b. thinner; lower
    c. thicker; higher
    d. thicker; lower

    Textbook Reference: *49.3 How Does the Mammalian Heart Function? pp. 1050–1051*

11. The left ventricle exceeds the right ventricle in the
    a. amount of blood that enters during heart contraction.
    b. volume expelled during contraction.
    c. pressure developed during contraction.
    d. All of the above

    Textbook Reference: *49.3 How Does the Mammalian Heart Function? p. 1051*

12. Which of the following structures of the lymphatic system acts primarily as a filter for detecting and destroying microorganisms in lymph traveling through major lymph vessels?
    a. Lymph nodes
    b. Thymus
    c. Lymph capillaries
    d. Tonsils, but not the appendix

    Textbook Reference: *49.4 What Are the Properties of Blood and Blood Vessels? p. 1060*

13. The net loss of fluid from blood capillaries will increase if
    a. plasma filtration decreases.
    b. the osmotic pressure of plasma increases.
    c. blood pressure increases in the capillaries.
    d. the colloid osmotic pressure of interstitial fluid decreases.

    Textbook Reference: *49.4 What Are the Properties of Blood and Blood Vessels? pp. 1058–1059*

14. Which of the following statements about the control of circulation in humans is *false*?
    a. Sympathetic nerve input to skeletal muscle causes the blood vessels in the muscle to dilate.
    b. Blood flow can be regulated by autonomic nerve signals emanating from cardiovascular control centers in the medulla of the brain.
    c. Carotid artery chemoreceptors detect low O_2 levels in the blood and promote increased blood pressure.
    d. Hormones such as angiotensin and vasopressin cause venules to constrict.

    Textbook Reference: *49.5 How Is the Circulatory System Controlled and Regulated? pp. 1061–1063*

15. Marine mammals that dive exhibit
    a. lower blood volumes than other mammals.
    b. hypometabolism, induced by the diving reflex.
    c. lower concentrations of hemoglobin than other mammals.
    d. a reflex that speeds the heart and increases blood flow to digestive tissues.

    Textbook Reference: *49.5 How Is the Circulatory System Controlled and Regulated? pp. 1063–1064*

Application Questions

1. What are the important differences between the circulatory system of a fish and that of a reptile?

   Textbook Reference: *49.2 How Have Vertebrate Circulatory Systems Evolved? pp. 1047–1049*

2. The human heart pumps blood to the pulmonary and the systemic circuit. Describe the path that blood takes as it moves through the circulatory system, starting at the right ventricle.

   Textbook Reference: *49.3 How Does the Mammalian Heart Function? pp. 1050–1051*

3. You are about to take a final exam in your freshman biology course and are very nervous. What responses will your circulatory system have to your nervousness?

   Textbook Reference: *49.5 How Is the Circulatory System Controlled and Regulated? pp. 1061–1063*

4. Blood moves through the arteries and veins at a faster velocity than the blood moving through the capillaries. What is the cause and effect of blood slowing down as it moves through the capillaries?

   Textbook Reference: *49.4 What Are the Properties of Blood and Blood Vessels? pp. 1057–1059*

5. Many animals with open circulatory systems are relatively inactive. Explain how insects, with their open circulatory systems, maintain their high levels of activity.

   Textbook Reference: *49.1 Why Do Animals Need a Circulatory System? p. 1047*

Answers

Knowledge and Synthesis Answers

1. **a.** Both closed systems and open systems rely on pumping mechanisms to distribute fluid throughout the body.

2. **c.** During intermittent breathing, the blood is shunted from the pulmonary circuit to the systemic circuit. The

resistance in the pulmonary circuit increases, resulting in the shunt.

3. **d.** Of the vertebrate groups, mammals and birds tend to have the highest blood pressures. Therefore, the ventricle supplying blood to the systemic circuit of a bird has the highest pressure.

4. **c.** The fluid portion of blood is the plasma; three components of the solid fraction are erythrocytes, platelets, and leukocytes.

5. **a.** Red blood cells are biconcave cells that contain the oxygen-binding hemoglobin.

6. **d.** Blood clotting involves none of the answer options listed. Prothrombin is converted into thrombin and fibrinogen is converted into fibrin.

7. **b.** Gas exchange at the tissues occurs across the capillaries.

8. **d.** The cardiac action potential passes through the SA node, atrial fibers, AV node, bundle of His, and finally the Purkinje fibers.

9. **c.** The AV node delays the ventricular depolarization relative to atrial depolarization, so atrial contraction occurs before ventricular contraction. The Purkinje fibers ensure that the action potential spreads rapidly and evenly throughout the ventricles.

10. **b.** The atrium has thinner walls and generates lower pressures than the ventricles.

11. **c.** The left ventricle generates a greater pressure in the blood flowing to the systemic circuit than the right ventricle with blood flowing to the pulmonary circuit.

12. **a.** The lymph nodes filter and destroy microorganisms that are traveling through the lymphatic system.

13. **c.** Changes in blood pressure will change the amount of filtration that occurs at the capillaries. Thus, increases in blood pressure will result in greater rates of filtration.

14. **d.** Angiotensin and vasopressin act to control constriction of the arterioles, not the venules.

15. **b.** The diving reflex of marine mammals induces hypometabolism, having a metabolic rate lower than the basal rate.

Application Answers

1. Fish have a two-chambered heart, with blood flowing from the heart of the fish to the gills and then directly to the body. Reptiles have a more complex three-chambered heart, with a subdivided ventricle that directs outflow to the pulmonary circuit and to the systemic circuit. Thus the circulatory system of fish delivers blood to the respiratory organ and the body tissues in succession, whereas the circulatory system of reptiles delivers blood to the respiratory organ and the body tissues in parallel.

2. Blood from the right ventricle exits through the pulmonary valve into the pulmonary artery and then goes to the lungs. From the lungs, the blood returns to the left atrium and passes through the atrioventricular valve into the left ventricle. Blood from the left ventricle is pumped to the systemic circuit through the aortic valve and into the aorta. From the aorta, blood moves to the tissues in arteries and arterioles, passes through tissues in the capillaries, and returns via the venules, veins, and ultimately the superior vena cava and inferior vena cava to the right atrium of the heart. From the right atrium, blood is pumped back where we started—into the right ventricle through the atrioventricular valve.

3. Stress and nervousness elicit the fight-or-flight response. Signals from the medullary cardiovascular control center stimulate sympathetic inputs from the autonomic nervous system. In response to sympathetic inputs, the adrenal gland releases epinephrine into the bloodstream. Epinephrine stimulates an increased heart rate and arterial pressure. The blood flow to the smooth muscles decreases due to constriction of blood vessels. By constriction of smooth muscles, blood is diverted away from areas such as the digestive tract to the skeletal muscle needed for fight or flight.

4. Blood is pumped from the heart and into the arteries, where it has a relatively fast velocity. Upon entering the capillaries much of the velocity is lost. The capillaries have a much higher total cross-sectional area than the arteries. Blood flow velocity is inversely proportional to the cross-sectional area; thus, the more area through which the blood flows, the slower it will go. This is similar to a stream that widens to become a river. Water in the upstream areas moves faster than the water in wider downstream areas. In the same way, the blood moves slowly through the capillaries and faster through the arteries and veins. The slow velocity of blood through the capillaries allows for greater exchange of materials with the tissues.

5. Despite having an open circulatory system, insects are able to maintain their high levels of activity because they do not rely on their circulatory systems for the exchange of respiratory gases. Instead, insects rely on their system of air-filled tubes, called tracheae, to serve as sites for gas exchange.

CHAPTER 50 Nutrition, Digestion, and Absorption

Important Concepts

All animals must eat to acquire energy for survival.

- Animals are heterotrophs, meaning that they must get their energy and nutrients from the fats, carbohydrates, and proteins of other animals and plants. The energy that animals acquire from food is used to make ATP, a high-energy phosphate compound that provides energy for cellular work. The conversion of ATP during cellular work creates heat as a by-product.

- The calorie and kilocalorie (Calorie) are the measures of heat energy.

- Metabolic rate is a measure of the total energy used by an animal. Basal metabolic rate is the resting metabolic rate or the energy consumption needed for all essential physiological functions.

- Energy needed for metabolism can be obtained from food taken in or from stored food. Fats, carbohydrates, and proteins are the components of food that provide energy. Fat has the highest energy content and is the most important form of energy stored in an animal's body.

- Energy budgets compare calories consumed with calories expended and allow scientists to apply a cost–benefit approach when analyzing behaviors.

Fuel molecules are stored in the body.

- Carbohydrates are stored as glycogen in the liver and muscles.

- Fat has more energy per gram than carbohydrates and can be stored more compactly, making it the most important form of stored energy in animals.

- An animal is undernourished when it takes in too little food to meet its energy requirements. An animal is overnourished when it consistently takes in more food than it needs to meet its energy requirements.

Essential molecules must be obtained from a food source.

- The acetyl group supplies the carbon skeleton for larger organic molecules. Animals cannot produce acetyl groups through their own metabolism and so must obtain them from another source (see Figure 50.4).

- Amino acids needed to make proteins are obtained by breaking down proteins in food. Some amino acids can be synthesized, but the essential amino acids can only be obtained from outside sources.

- Essential amino acids are needed for proper protein production. In humans the essential amino acids are isoleucine, leucine, lysine, methionine, phenylalanine, threonine, tryptophan, and valine. All eight essential amino acids can be found in meat, eggs, soybean products, and milk. Vegetarian diets should include complementary dietary mixtures, such as grains and legumes, to avoid protein malnutrition (see Figure 50.5).

- Ingested proteins are not used in their ingested form because, as macromolecules, they are not absorbed by cells in the gut. Also, their structures and functions are species-specific. Additionally, foreign proteins in the body would be recognized as invaders and would be attacked. Instead, proteins are broken down into peptides and amino acids.

- Essential fatty acids, such as linoleic acid, must also be acquired in food and are necessary for producing membrane phospholipids.

- The nutrients needed for survival are categorized as either macronutrients or micronutrients, depending on the quantity needed (see Table 50.1).

- Vitamins are carbon compounds that are required in small amounts and often function as coenzymes. Fat-soluble vitamins accumulate in the liver, whereas water-soluble vitamins are eliminated in urine. Vitamin D is needed to help in the uptake of calcium into the body.

Diseases can be caused by nutrient deficiency.

- Malnutrition is caused by the lack of any essential nutrient in the diet. Chronic malnutrition leads to a particular deficiency disease. For example, protein deficiency leads to the disease kwashiorkor, which is characterized by swelling, distension of the abdomen, and degeneration of the liver and other organs. When left untreated, protein deficiency will eventually cause death.

- Humans require 13 vitamins. Shortages in vitamins can lead to a large number of diseases (see Table 50.2).

- Nutrient deficiencies may be due to the body's inability to absorb essential nutrients even when they are present in the diet. For example, pernicious anemia is caused by the inability to absorb vitamin B_{12} due to inadequate production of intrinsic factor, a peptide needed for absorption of the vitamin.

There are several different ways for animals to feed.

- Two types of organisms feed on dead matter: saprobes (also called decomposers or saprotrophs) absorb nutrients from decaying organic matter, whereas detritivores actively feed on dead matter. Predators feed on other organisms, and there are three main types: herbivores feed on plants, carnivores feed on animals, and omnivores feed on plants and animals.

- Teeth are composed of a hard outer surface (enamel), a bony layer (dentine), and a pulp cavity that contains blood vessels and nerves (see Figure 50.7A).

- The shapes and organization of mammalian teeth reflect different diets (see Figure 50.7B). Because plant tissue is difficult to break down, herbivores have teeth that are good for shearing, crushing, tearing, and grinding, to break down vegetation. Because carnivorous animals must capture, hold, and eat other animals, their teeth are adapted for stabbing, ripping, and shredding their prey. Omnivorous animals have teeth adapted for multiple purposes.

Digestion occurs in either a cavity or a tube in most animals.

- Food is generally digested extracellularly by enzymes that break it down. Most enzymes are produced in an inactive form called a zymogen and are converted to the active form at the site of use. Enzymes do not digest the gut because of a protective layer of mucus that lines it. There are several classes of digestive enzymes:

  - Carbohydrases hydrolyze carbohydrates.

  - Proteases hydrolyze proteins.

  - Peptidases hydrolyze peptides.

  - Lipases hydrolyze fats.

  - Nucleases hydrolyze nucleic acids.

- In simple animals, a single opening that functions as both a mouth and an anus connects a gastrovascular cavity to the environment.

- More complex animals have tubular guts with a mouth that takes in food and an anus through which digestive wastes are eliminated. Food enters the mouth and, in some animals, is broken up by grinding mechanisms that may include teeth or mandibles; many birds grind their food using small stones or gravel in an upper region of the gut called the gizzard. After grinding, food moves to a storage chamber, such as the stomach or crop, where digestion may or may not occur. It then enters the midgut or intestine, where most of the nutrients are absorbed. The hindgut absorbs water and ions and stores waste until it is expelled from the anus during defecation.

- The parts of the gut that absorb nutrients have increased surface area in the form of fingerlike projections known as villi, which in turn have microscopic projections called microvilli (see Figure 50.9).

The gut wall is composed of the same layers of tissue types throughout its length.

- Four layers of different cell types form the wall of the gut (see Figure 50.11).

- The innermost layer is a layer of mucosa that surrounds the cavity of the gut, or lumen. Cells of the mucosal epithelium have secretory functions; some secrete mucus, while others secrete enzymes, hormones, or hydrochloric acid. In some regions of the gut, cells of this layer have absorptive functions.

- The submucosa, lying under the mucosa, contains blood and lymph vessels that transport the nutrients. This layer also contains a network of nerves that regulates gut activities.

- Smooth muscle surrounds the submucosa in both circular and longitudinal layers. These layers help to move the contents of the digestive tract by peristalsis.

- The peritoneum is a membrane that lines the abdominal cavity and covers the gut and other abdominal organs.

Food passes through the gut by smooth muscle contraction.

- Food enters the mouth and is swallowed, passing through the pharynx into the esophagus (see Figure 50.12A).

- Food is then moved along by peristalsis, which is a wave of muscle contraction that moves food through the gut (see Figure 50.12B).

- The esophageal sphincter helps keep food from reentering the esophagus from the stomach, and the pyloric sphincter controls the movement of stomach contents into the intestines.

Several enzymes help break down food.

- The enzyme amylase, secreted by the salivary glands, begins the breakdown of starch in the mouth.

- In the stomach, the major enzyme is pepsin, which breaks down proteins into shorter peptides. Pepsin is initially secreted by gastric glands of the stomach in the zymogen form, pepsinogen (see Figure 50.13). Low pH activates the conversion of pepsinogen to pepsin through autocatalysis, a positive feedback process.

- The secretion of hydrochloric acid by the gastric glands in the wall of the stomach maintains a low pH.

- Ulcers of the stomach are locations where the mucosal lining has been damaged by the bacterium *Helicobacter pylori*.

- Small amounts of a few substances, such as alcohol, aspirin, and caffeine, are absorbed across the walls of the stomach.
- The acidic mixture of digesting food and gastric juices is known as chyme, which slowly passes into the intestines through the pyloric sphincter.

Most chemical digestion occurs in the small intestine.

- Carbohydrate and protein digestion continue in the small intestine.
- In humans, the small intestine can be more than six meters in length, with villi and microvilli contributing to a large surface area necessary for the absorption of nutrients. The small intestine is divided into the duodenum, jejunum, and ileum.
- Fat digestion begins in the small intestine with the help of secretions and enzymes from the liver and pancreas.
- The liver secretes bile through a side branch of the hepatic duct into the gallbladder, where the bile is stored. When lipids are present, bile is secreted by the gallbladder into the duodenum through the common bile duct (see Figure 50.15). Bile molecules have hydrophobic ends, which attach to the tiny fat droplets to stabilize particles of fat in the form of micelles. This increases the exposed surface area of the lipids for digestion by fat-digesting enzymes, the lipases (see Figure 50.16A).
- The zymogen trypsinogen is secreted by the pancreas and is converted into active trypsin by enterokinase. The pancreas produces several zymogens that are released through the pancreatic duct, which joins the common bile duct and empties into the duodenum. The zymogens are converted to the active form by trypsin. The pancreas also helps maintain a slightly alkaline pH in the duodenum by secreting bicarbonate ions.

The small intestine is the site of most nutrient absorption.

- Final digestion of proteins and carbohydrates occurs among the microvilli of the small intestine. Proteins are broken down into amino acids by peptidases. The enzymes maltase, lactase, and sucrase break down disaccharides into monosaccharides. The mucosal cells secrete the peptidases and maltase, lactase, and sucrase.
- Absorption occurs via active transport for ions (sodium, calcium, and iron), amino acids, and sugars.
- Fats move through the plasma membranes as fatty acids and monoglycerides (see Figure 50.16B). In the mucosal cells they are converted into triglycerides and combine with cholesterol and phospholipids to form chylomicrons that pass into the lymphatic system. Bile is not absorbed across the membrane of the duodenum, but is actively resorbed in the ileum and returned to the liver.

- The hepatic portal vein carries blood from the digestive organs to the liver where nutrients are either stored or converted to needed molecules.

The large intestine is the site of water and ion absorption.

- Material entering the large intestine has had most of the nutrients removed, but still contains important ions and water. The water and ions are absorbed across the large intestine, leaving behind semisolid feces.
- In the colon, absorption of too much water can lead to constipation, whereas absorption of too little water can produce diarrhea.

Herbivores employ fermenting microorganisms to help break down plant matter.

- Cellulose is the main component of the food of herbivores. Nevertheless, most herbivores cannot produce cellulase enzymes, which break down plant cellulose. Microorganisms in the digestive tracts of herbivores produce the cellulase enzymes that break down cellulose into fatty acids.
- Ruminants have a large four-chambered stomach in which fermentation occurs (see Figure 50.17). The rumen and reticulum, the first two chambers, contain many microorganisms that break down cellulose. The contents of the rumen are periodically regurgitated into the mouth for rechewing; this process is often described as *chewing the cud*.
- From the rumen, food passes through the omasum, where water absorption occurs, into the abomasum, or true stomach. In the abomasum, hydrochloric acid and proteases kill the microorganisms and digest them.
- Some herbivores have a microbial fermentation chamber called the cecum extending from the large intestine. Many of these species produce two types of feces, one that is pure waste and the other that still contains some nutrients. Coprophagy is the reingesting of this second type of feces to digest and absorb additional nutrients. In humans, the cecum has become the appendix, an organ with no digestive function.

Digestion is regulated by unconscious reflexes and hormones.

- Salivation and swallowing are unconscious reflexes. The digestive tract has an intrinsic nervous system that coordinates motility and digestive activities.
- Many digestive functions are controlled by hormones (see Figure 50.18). The duodenum secretes secretin, a hormone that stimulates the pancreas to release bicarbonate ions into the small intestine to control pH.
- Fats and proteins stimulate the mucosa of the small intestine to secrete cholecystokinin, which stimulates the release of bile and pancreatic digestive enzymes.
- Gastrin is secreted by the stomach into the blood, where it is transported to the upper regions of the stomach to stimulate stomach secretions and movement.

The liver interconverts fuel molecules.

- When fuel molecules are abundant in the blood, the liver stores them as glycogen or fat. When fuel molecules are low in the blood, the liver delivers them back into the blood.

- The liver can convert monosaccharides into glycogen or fat, and vice versa. The liver can also convert amino acids and certain other molecules into glucose; this process is called gluconeogenesis.

Lipoproteins allow fats to be transported in the circulatory system.

- Lipoproteins are particles of fat and cholesterol covered with proteins. Chylomicrons are large lipoproteins produced by the mucosal cells of the intestine. Other lipoproteins are produced by the liver.

- Very-low-density lipoproteins (VLDLs) contain mostly triglyceride fats, which they transport to cells of adipose tissue; these lipoproteins are only 3 percent cholesterol. Low-density lipoproteins (LDLs) contain 50 percent cholesterol, which they carry around the body for storage or use. High-density lipoproteins (HDLs) remove cholesterol from tissues and carry it to the liver; they are 15 percent cholesterol.

Insulin and glucagon help regulate blood glucose levels.

- The absorptive period is the period after a meal when digestion occurs. The postabsorptive period is the period when the gut is empty and the body uses energy reserves.

- During the absorptive period, the liver takes up glucose from the blood and converts it to glycogen and fat, fat cells take up glucose, and the body preferentially uses glucose as the energy source.

- During the postabsorptive period, the liver breaks down glycogen into glucose, fatty acids are supplied to the blood by adipose tissues, and the body preferentially uses fatty acids as an energy source.

- The pancreatic hormones insulin and glucagon control metabolic fuel use (see Figure 50.19).

- Insulin is released in response to high blood glucose levels. Insulin stimulates cells to burn glucose, fat cells to use glucose to make fat, and liver cells to convert glucose to glycogen and fat.

- At postabsorptive blood glucose levels, insulin secretion is diminished, and the liver and fat cells break down glycogen.

- At low levels of blood glucose, the pancreas releases glucagon, which stimulates the liver to break down glycogen and produce glucose.

Regulation of food intake involves the hypothalamus and hormones.

- Hunger and satiety are influenced by the region of the brain known as the hypothalamus. Both the ventro-medial hypothalamus and lateral hypothalamus play a role in food intake.

- Leptin, a hormone produced by fat cells, acts on the hypothalamus and is involved in control of hunger and satiety. Ghrelin, a hormone produced by the stomach, acts on the hypothalamus to stimulate appetite.

Some toxins become concentrated in animals.

- Plants and animals may contain toxic compounds. Human activities also introduce toxic compounds into the environment. Environmental toxicology is a new field that investigates the impacts of poisons on the environment.

- Water-soluble toxins can be metabolized quickly and filtered out of the body. Lipid-soluble toxins are metabolized slowly and tend to accumulate in adipose tissues.

- Bioaccumulation is the concentration of toxins in animals occupying higher and higher positions in the food chain. Long-lived species at the top of the food chain run the risk of heavy pesticide burdens in their bodies.

- The cytochrome P450 enzymes are involved in detoxification of natural chemicals. Some synthetic chemicals are not broken down by P450s and therefore, they bioaccumulate.

- Some synthetic chemicals may also act as hormone mimics.

The Big Picture

- Animals have evolved a wide variety of mechanisms for acquiring needed energy and raw materials for metabolism. All animals are heterotrophs; they can achieve energy and nutrient input by eating plants, animals, or both.

- The digestive system is specialized for efficient digestion of the type of food it must break down. The relatively short guts of carnivores have acid and protein-reducing enzymes. The longer guts of herbivores have numerous holding chambers for the laborious process of breaking down plant material. The teeth of carnivores and herbivores also reflect dietary differences.

- Food passing down the length of the gut is first broken down mechanically into small pieces and then broken down chemically into small molecules that can be easily absorbed across the gut. As nutrients and water are absorbed from ingested plant and animal material, the remaining material forms the feces, which are eventually eliminated from the body.

- The process of digestion is under both neural and hormonal control. The brain plays a role in regulating food intake, and the digestive tract has an intrinsic nervous system to regulate digestive functions. Numerous hormones control the peristaltic movements of the gut and the secretion of enzymes and other chemicals involved in digestion. Hormones are also involved in regulating food intake.

Common Problem Areas

- The study of digestive function may seem uninteresting or, worse, unpleasant to students. Yet, some of the best and clearest examples of negative feedback and homeostasis are to be found in the study of digestion.

- Sometimes students think that there is a single digestive mechanism for food and are surprised to learn of the complexity of the process. Be sure to recognize that different food components—proteins, carbohydrates, fats—are digested, absorbed, and transported by very different mechanisms.

Study Strategies

- Although the guts of higher animals are structurally complex, they can be divided functionally into a foregut, midgut, and hindgut. If you break down the structure of the gut into these sections and determine which physiological process occurs in each section, you will have divided the task into smaller pieces.

- Because different types of food are digested and absorbed by different mechanisms, learn how a particular type of food is digested and match the appropriate enzymes to that process. For example, it will be much easier to remember how proteins are digested, rather than to recount a long list of digestive enzymes and then try to pick the one that might break down protein. In short, learn the process, not the list!

- Review the following animated tutorials and activities on the Companion Website/CD:
  Tutorial 50.1 The Digestion and Absorption of Fats
  Tutorial 50.2 Insulin and Glucose Regulation
  Activity 50.1 Mineral Elements Required by Animals
  Activity 50.2 Vitamins in the Human Diet
  Activity 50.3 Mammalian Teeth
  Activity 50.4 Human Digestive System

Test Yourself

Knowledge and Synthesis Questions

1. Certain amino acids are essential to the diet of animals because
   a. they prevent overnourishment.
   b. they are cofactors and coenzymes that are required for normal physiological function.
   c. an animal cannot directly synthesize them through the transfer of an amino group to an appropriate carbon skeleton.
   d. animals need these substances in order to make stored fats that are used during hibernation and migration.
   Textbook Reference: 50.1 What Do Animals Require from Food? p. 1072

2. Which of the following descriptions describes the digestive characteristics of a sheep?
   a. It is a saprobe; it engulfs food and performs intracellular digestion.

   b. It is an autotroph; it synthesizes organic nutrients and performs extracellular digestion.
   c. It is an herbivore; it ingests food and performs extracellular digestion.
   d. It is a detritivore; it ingests food and performs intracellular digestion.
   Textbook Reference: 50.2 How Do Animals Ingest and Digest Food? pp. 1075–1076

3. Protection of the walls of the stomach against the action of its own digestive juices
   a. results from the presence of an antienzyme chemical formed by the gastric glands.
   b. results from the nervous reactions of the lining of the stomach.
   c. is controlled by a center in the medulla of the brain.
   d. results from the mucous coating that covers its inner surface.
   Textbook Reference: 50.3 How Does the Vertebrate Gastrointestinal System Function? p. 1080

4. Chylomicrons are produced in the
   a. mouth.
   b. stomach.
   c. lumen of the small intestines.
   d. epithelial cells of the small intestines.
   Textbook Reference: 50.3 How Does the Vertebrate Gastrointestinal System Function? pp. 1083–1084

5. The gallbladder
   a. produces bile.
   b. is part of the liver.
   c. stores bile produced by the liver.
   d. produces cholecystokinin.
   Textbook Reference: 50.3 How Does the Vertebrate Gastrointestinal System Function? p. 1082

6. The pancreas
   a. produces exocrine products involved in chyme digestion.
   b. is exclusively an endocrine gland that produces salivary amylase.
   c. contains villi to increase surface area.
   d. produces urobiligen (a bile pigment).
   Textbook Reference: 50.3 How Does the Vertebrate Gastrointestinal System Function? p. 1083

7. Hydrochloric acid
   a. is secreted by the gastric glands of the liver.
   b. is secreted by the gastric glands of the stomach.
   c. produces a low pH in the small intestine.
   d. Both a and c
   Textbook Reference: 50.3 How Does the Vertebrate Gastrointestinal System Function? pp. 1080–1081

8. Bile produced in the liver is associated with which of the following?
   a. Emulsification of fats into tiny globules in the small intestine
   b. Digestive action of pancreatic amylase
   c. Emulsification of fats into tiny globules in the stomach

d. Digestion of proteins into amino acids
Textbook Reference: 50.3 How Does the Vertebrate Gastrointestinal System Function? p. 1082

9. Most of the chemical digestion of food in humans is completed in the
   a. small intestine.
   b. appendix.
   c. ascending colon.
   d. stomach.
   Textbook Reference: 50.3 How Does the Vertebrate Gastrointestinal System Function? p. 1082

10. Which of the following does *not* contribute to the large surface area available for nutrient absorption in the small intestines?
    a. Villi
    b. Intestinal length
    c. Microvilli
    d. Bile duct
    Textbook Reference: 50.3 How Does the Vertebrate Gastrointestinal System Function? p. 1084

11. Waves of muscle contractions that move the intestinal contents are
    a. caused by contraction of skeletal muscle.
    b. regulated by liver secretions.
    c. called peristalsis.
    d. voluntary.
    Textbook Reference: 50.3 How Does the Vertebrate Gastrointestinal System Function? pp. 1079–1080

12. What is the function of enterokinase?
    a. It converts pepsinogen to pepsin.
    b. It converts trypsinogen to trypsin.
    c. It digests proteins.
    d. It activates HCl.
    Textbook Reference: 50.3 How Does the Vertebrate Gastrointestinal System Function? p. 1083

13. Digestive enzymes responsible for breaking down disaccharides include
    a. pepsin, trypsin, and trypsinogen.
    b. amylase, pepsin, and lipase.
    c. sucrase, lactase, and maltase.
    d. pepsin, trypsin, and chymotrypsin.
    Textbook Reference: 50.3 How Does the Vertebrate Gastrointestinal System Function? p. 1083

14. Which of the following is characteristic of the large intestine?
    a. It has almost no bacterial populations.
    b. It contains chyme.
    c. It absorbs much of the water remaining in waste materials.
    d. It is the site of most of digestion.
    Textbook Reference: 50.3 How Does the Vertebrate Gastrointestinal System Function? p. 1084

15. The innermost layer of the digestive tract is the
    a. peritoneum.
    b. mucosa membrane.

c. submucosa membrane.
d. lumen.
Textbook Reference: 50.3 How Does the Vertebrate Gastrointestinal System Function? pp. 1078–1079

16. Which of the following hormones stimulates gluconeogenesis?
    a. Glucagon
    b. Insulin
    c. Estrogen
    d. All of the above
    Textbook Reference: 50.4 How Is the Flow of Nutrients Controlled and Regulated? pp. 1086–1087

17. Which of the following statements about mammalian teeth is true?
    a. Enamel covers the crown and the root of a tooth.
    b. Teeth lack nerves and blood vessels.
    c. The teeth of omnivores are more specialized than the teeth of carnivores and herbivores.
    d. Carnivores have large canines, whereas herbivores have large premolars and molars.
    Textbook Reference: 50.2 How Do Animals Ingest and Digest Food? p. 1076

18. Toxins ingested by animals
    a. come exclusively from human activities.
    b. always bioaccumulate.
    c. are broken down by enzymes of the stomach.
    d. may cause cancer or infertility if they bioaccumulate.
    Textbook Reference: 50.5 How Do Animals Deal with Ingested Toxins? p. 1089

Application Questions

1. You have just eaten a small pizza with a lot of cheese. Because this food is high in lipids, your digestive system will have to perform specific tasks to digest and absorb the fats. What are the steps that your digestive system will take to digest and absorb the fats in this meal?
   Textbook Reference: 50.3 How Does the Vertebrate Gastrointestinal System Function? pp. 1082–1084

2. During the absorptive period, your body is digesting and absorbing nutrients, whereas during the postabsorptive period, your body is using up stored fuel. What are the main differences in fuel use and control mechanisms during these two periods?
   Textbook Reference: 50.4 How Is the Flow of Nutrients Controlled and Regulated? pp. 1086–1087

3. Ingested food typically contains many useful proteins. However, these proteins are not used in the form in which they are ingested, but are broken down into smaller peptides and amino acids. Why does the digestive system break down these ingested proteins?
   Textbook Reference: 50.1 What Do Animals Require from Food? p. 1072

4. Humans have been polluting the environment for many years, dumping toxins into many of Earth's

ecosystems. What effect does this have on toxin levels in the bodies of animals at the top of the food chain?

Textbook Reference: 50.5 How Do Animals Deal with Ingested Toxins? p. 1089

5. In ruminants, such as cows and bison, microorganisms that break down cellulose are found in two chambers of the greatly enlarged stomach. In other herbivores, including rabbits, the microorganisms that break down cellulose are found in the cecum, a chamber off the large intestine. In which type of herbivore would the absorption of nutrients from cellulose be more efficient?

Textbook Reference: 50.3 How Does the Vertebrate Gastrointestinal System Function? pp. 1084–1085

Answers

Knowledge and Synthesis Answers

1. **c.** Essential amino acids must be acquired through diet because an animal cannot directly synthesize all the amino acids needed for protein production.

2. **c.** A sheep is an herbivore that ingests plant material for nutrition. Digestion occurs extracellularly in the digestive tract of the animal.

3. **d.** The stomach is protected from digestive enzymes and low pH by the mucus secreted over its inner surface.

4. **d.** The epithelial cells of the small intestine produce chylomicrons by combining triglycerides with cholesterol and phospholipids, allowing lipids to pass into the lymphatic system.

5. **c.** Bile is produced by the liver, stored in the gallbladder, and released into the small intestine to aid in lipid digestion.

6. **a.** The pancreas produces exocrine products such as lipases, nucleases, amylases, and trypsin, all of which are involved in chyme digestion.

7. **b.** Hydrochloric acid is a strong acid secreted by gastric glands in the lining of the stomach. It lowers the pH of the stomach fluid.

8. **a.** Bile aids in the digestion of lipids in the small intestine by changing fat droplets into small fat particles.

9. **a.** The small intestine is the main site for chemical digestion in humans.

10. **d.** The length of the intestines and the presence of microvilli and villi increase the surface area of the small intestine.

11. **c.** Food is moved through the digestive tract by wavelike contractions of smooth muscle called peristalsis.

12. **b.** Enterokinase converts the zymogen trypsinogen to trypsin.

13. **c.** The digestive enzymes sucrase, lactase, and maltase are involved in the breakdown of disaccharides into the monosaccharides glucose, galactose, and fructose.

14. **c.** The large intestine is the site of water and ion absorption. The large populations of bacteria found in the large intestine contribute useful vitamins to their hosts.

15. **b.** The membranes of the digestive tract are, from the inside to the outside: mucosa, submucosa, and circular and longitudinal muscles. The peritoneum surrounds the digestive tract.

16. **a.** Glucagon increases the rate of gluconeogenesis, and insulin reduces the rate of gluconeogenesis.

17. **d.** Carnivores have large canines for holding and killing prey animals. Herbivores have large premolars and molars for grinding plant material.

18. **d.** Toxins ingested by animals may come from natural sources, such as defensive chemicals in the plants or animals they consume, or from human activities. Whereas enzymes in the liver can detoxify many natural toxins, they cannot detoxify some synthetic toxins. Lipid-soluble compounds tend to bioaccumulate and can cause cancer or infertility in top predators.

Application Answers

1. The digestion of fats begins in the small intestine. Lipases are water soluble, but lipids tend to aggregate into large droplets in an aqueous environment. Bile helps to increase the exposed surface area of the lipids by forming small micelles with the lipids. This allows lipases to break down the fats into free fatty acids and monoglycerides. The fatty acids dissolve into the plasma membrane of the intestinal epithelial cells and will be absorbed across this surface. In mucosal cells, the free fatty acids and monoglycerides are turned into triglycerides, which are incorporated into chylomicrons. The chylomicrons can then pass out of the mucosal cells and into the lymphatic system (see Figure 50.16).

2. During the postabsorptive period, cells tend to metabolize fats, keeping the blood glucose reserves for the nervous system. Insulin levels are low during the postabsorptive period. During the absorptive period the body attempts to store the nutrients and fuel that are being absorbed. The preferred fuel source at this time is glucose, and insulin is released to help facilitate glucose uptake.

3. Proteins in digested food are not absorbed and used in their original form for a number of reasons. First, the proteins are too large to be easily absorbed by the cells of the digestive tract, whereas amino acids are readily absorbed. Second, proteins are species-specific, and thus a protein that is optimal for a prey species may not be optimal for the predator species. Lastly, the proteins of another species would be considered foreign and attacked by the immune system.

4. Toxins in the environment tend to bioaccumulate in top predators. Lipid-soluble toxins accumulate in body tissues of predators over their life spans. Therefore, as predators eat more contaminated food, concentrations of toxins in their bodies increase, resulting in increased body burdens.

5. In vertebrates, most nutrients are absorbed in the small intestine. Because ruminants break down cellulose in chambers of their stomach (which comes before the small intestine in position along the gut), the nutrients from cellulose are available for absorption in the small intestine, and absorption is relatively efficient. In non-ruminant herbivores, however, cellulose is broken down in the cecum, after it has passed through the small intestine, so nutrient absorption is less efficient. Some non-ruminant species produce two types of feces and eat the one containing cecal material to gain greater access to nutrients.

CHAPTER 51 Salt and Water Balance and Nitrogen Excretion

Important Concepts

Excretory organs control the solute concentrations of animals.

- Excretory organs control the level of solutes and the volume of tissue fluid and excrete urine. Urine composition depends on the environment surrounding an animal.

- Water movement in the body occurs either by pressure differences or differences in solute potential that produce water movement by osmosis. There is no active transport of water in excretory systems.

- Excretory systems use filtration across membranes, reabsorption, and secretion to control water and solute balance.

- Osmolarity is the number of moles of osmotically active solutes per liter of solvent.

- Osmoconformers do not actively regulate the osmolarity of their tissues; rather, they allow them to come to equilibration with their environment. Marine invertebrates tend to be osmoconformers unless the concentrations in the environment are extreme.

- Osmoregulators actively regulate the osmolarity of their tissues, even as environmental osmolarity changes. The brine shrimp is an osmoregulator that maintains its tissue fluid osmolarity below that of the environment in high osmolarity waters and above that of the environment in low osmolarity waters (see Figure 51.1).

- Ionic conformers allow the ionic composition of their extracellular fluid to match that of the environment. Ionic regulators employ active transport mechanisms to keep ions at optimal concentrations in their extracellular fluid. Birds use a nasal salt gland to excrete salt from consumed seawater (see Figure 51.2).

Nitrogen is a potentially toxic metabolic waste.

- The end products of the metabolism of proteins and nucleic acids contain nitrogen and are called nitrogenous wastes.

- Ammonia (NH_3) is a highly toxic nitrogenous waste that must be either excreted or converted to the less toxic urea and uric acid (see Figure 51.3).

- Ammonotelic animals excrete ammonia to the aquatic environment, usually across gill membranes.

- Ureotelic animals excrete nitrogenous waste as water-soluble urea.

- To conserve water, uricotelic birds and reptiles excrete nitrogenous waste as insoluble uric acid.

- Most species produce more than one nitrogenous waste.

Reflecting their diversity, invertebrates employ a number of excretory mechanisms.

- Flatworms have a protonephridium made up of tubules and flame cells that conserve ions and excrete water to produce dilute urine (see Figure 51.4).

- Annelid worms filter the blood across the capillaries into the coelom, where the fluid enters a metanephridia in each segment through a funnel-like opening called a nephrostome. Tubules of the metanephridia actively secrete and absorb various ions and end in nephridiopores, which open to the environment to excrete dilute urine containing nitrogenous wastes and other solutes (see Figure 51.5).

- Insects use Malpighian tubules as their excretory organ. Malpighian tubules actively transport uric acid and sodium and potassium ions from the tissue into the tubules, with water following passively (see Figure 51.6). The tubules lead into the hindgut, where uric acid precipitates and water is reabsorbed. Insects eliminate semi-dry matter containing uric acid and other wastes.

The vertebrate excretory system is composed of nephrons in a kidney.

- Vertebrate excretory organs are kidneys, and nephrons are the functional units of kidneys.

- Vertebrates exhibit diverse excretory adaptations.

Marine and freshwater fishes employ two opposing mechanisms for controlling water and ion balance.

- Freshwater fishes live in an environment hypotonic to their own body fluids, with water tending to move into the body by osmosis. To counter this, they excrete large amounts of dilute urine.

- Marine bony fishes are osmoregulators, maintaining their extracellular fluids at one-third to one-half the osmolarity of seawater. They live in an environment hypertonic to their body fluids, so water tends to be drawn out of the body by osmosis. To cope with living in this environment, they produce a small amount of concentrated urine and actively secrete salts across the gills and the renal tubules.

- Cartilaginous fishes, almost all of which are marine, act as osmoconformers, allowing urea and trimethylamine oxide concentrations to increase, which increases the osmolarity of their tissues. Cartilaginous fishes are not ionic conformers; they have a rectal gland to excrete salts.

Terrestrial amphibians and reptiles must avoid desiccation.

- Amphibians living in or near fresh water have a large water influx, and in response they produce large amounts of dilute urine.

- In contrast, amphibians living in terrestrial environments have low metabolic rates and skin with reduced water permeability, resulting in low water turnover. Some also estivate (burrow underground and reduce metabolic rate) and use their bladders as canteens to store water for later use.

- Reptiles are amniotes; they lay shelled eggs, which allows them to live in a fully terrestrial environment (i.e., they do not have to return to water to reproduce). They have scaly, dry skin to decrease evaporative water loss, and they excrete nitrogenous wastes as uric acid with little water.

Birds and mammals conserve water in their terrestrial habitat.

- Birds and mammals are also amniotes; however, whereas birds lay shelled eggs, most mammals have embryos that develop in the reproductive tract of the mother. Both groups have skins and surface coverings to reduce water loss.

- Mammals, and to a certain extent birds, have the ability to produce urine that is more concentrated than their blood.

Nephrons are the functional units of vertebrate kidneys.

- The nephron of the vertebrate kidneys is composed of the glomerulus, renal tubules, and peritubular capillaries, which are further broken down into vascular and tubule components (see Figure 51.7).

- Three main processes lead to the formation of urine: filtration, tubular reabsorption, and tubular secretion. Filtration occurs at the glomerulus and tubular reabsorption and secretion occur along the renal tubule.

- There are two sets of capillary beds in series in a nephron. The glomerulus is composed of the first set of capillary beds with blood entering at the afferent arteriole and exiting at the efferent arteriole. From the efferent arteriole the peritubular capillaries of the second capillary bed emerge and surround the tubule component of the nephron (see Figure 51.7).

- The renal tubule begins with Bowman's capsule, which encloses the glomerulus. Within Bowman's capsule, podocyte cells wrap around the capillaries of the glomerulus. Blood is filtered by the glomerulus into Bowman's capsule. The endothelial walls of the capillaries in the glomerulus have pores that allow water and small molecules to leave the blood. Only small molecules and water can enter Bowman's capsule to form filtrate.

- Arterial blood pressure is the driving force for filtration at the glomerulus. The porous capillary beds in the glomerulus also contribute to the high filtration rate.

- The filtrate entering the tubules is similar to blood plasma, but as the fluid moves through the tubules, ions and molecules are actively reabsorbed and secreted. This controls the composition of the urine.

Mammalian kidneys produce concentrated urine.

- Humans have two kidneys that filter blood and produce urine (see Figure 51.9). Urine exits the kidney through the ureter and travels to the bladder, where it is stored until it is excreted through the urethra. The timing of urination is controlled by two sphincters in the urethra, one is made of smooth muscle and is involuntary, and the other is made of skeletal muscle and is voluntary.

- The human kidney is shaped like a kidney bean, with a central medulla and a surrounding cortex.

- The glomerulus and adjoining proximal convoluted tubules are located in the cortex. The descending limb of the renal tubule runs down into the medulla, and the ascending limb returns to the cortex in what is known as the loop of Henle. From the ascending limb, the tubules become the distal convoluted tubule, which connects to the collecting ducts and runs down the medulla.

- Nephrons with long loops of Henle that go deep into the medulla are important to the formation of concentrated urine.

- Blood vessels also run in a similar pattern to the tubules. An afferent arteriole carries blood to the glomerulus where it is drained into the efferent arteriole, which gives rise to the peritubular capillaries. Capillaries run through the medulla, parallel to the loop of Henle, to form a vasa recta.

Filtration occurs at the glomerulus and reabsorption occurs largely at the proximal convoluted tubule.

- The typical glomerulus filters about 180 liters of blood per day in an adult human. However, 98 percent of this fluid is reabsorbed into the blood, so that only 2 percent of glomerular filtrate is excreted as urine.

- The majority of the filtrate is reabsorbed in the proximal convoluted tubules. Sodium ions and other solutes are actively transported out of the proximal convoluted tubule and water follows passively, all to be taken up by the peritubular capillaries.

The loop of Henle sets up a concentration gradient in the medulla.

- The loop of Henle produces a concentration gradient in the medulla by acting as a countercurrent multiplier system (see Figure 51.10). The purpose of this gradient is to move water across fluid compartments by osmosis because there is no mechanism for active transport.

- The concentration of the fluids surrounding the loop of Henle is raised by the active reabsorption of Cl^- and the passive flow of Na^+ into the tissue from the thick ascending limb. The thick ascending limb is impermeable to water, so only solutes leave the tubules.

- The thin descending limb is permeable to water but not sodium and chloride ions. Water moves out of the tubules into the surrounding tissues due to the higher concentration of Na^+ and Cl^- in the surrounding tissues.

- In the thin ascending limb, the fluid is more concentrated than the fluid in the surrounding tissues, so Na^+ and Cl^- move out of the tubule. Water cannot move out because of the low permeability of the thin ascending limb.

- The fluid reaching the distal convoluted tubule is less concentrated than blood plasma.

- Aquaporins, a class of membrane proteins that form water channels, explain why some regions of the renal tubule are highly permeable to water (such as the proximal convoluted tubule) while other regions are not (such as the ascending limb of the loop of Henle).

Concentration of urine occurs in the collecting duct.

- The collecting duct passes through the medulla, with its high solute concentration set up by the loop of Henle.

- Water moves out of the collecting duct and into the tissue, resulting in concentrated urine. The permeability of the collecting ducts can be adjusted hormonally to regulate the amount of water excreted.

- Longer collecting ducts result in a greater ability to concentrate urine.

The kidneys help to regulate the pH of the blood.

- Buffers can either absorb or release hydrogen ions.

- Bicarbonate ions are the major buffer in the blood. The kidneys control the level of H^+ and bicarbonate ions in the blood by varying their rate of secretion into the urine (see Figure 51.12).

Dialysis can be used to treat kidney failure.

- Kidney (renal) failure severely disrupts homeostasis, causing high blood pressure (due to the retention of salts and water), uremic poisoning (due to the retention of urea), and acidosis (due to the retention of metabolic acids).

- A dialysis machine can replace kidney function until a kidney transplant can be arranged (see Figure 51.13).

Enzymes and hormones regulate glomerular filtration rate.

- Glomerular filtration rate is a function of the blood flow and pressure to the kidneys.

- The kidneys have autoregulatory mechanisms that help maintain their high filtration rate.

- In response to decreasing arteriole resistance or decreased arterial blood pressure, the afferent renal arterioles open, or dilate, to increase flow to the glomerulus. If this does not increase glomerular filtration rate sufficiently, the kidneys release the enzyme renin into the blood. Renin converts an angiotensin precursor into active angiotensin.

- Angiotensin constricts the efferent renal arterioles to increase pressure through the glomerulus. Angiotensin also constricts peripheral blood vessels, stimulates the release of aldosterone from the kidney, and stimulates thirst.

- Aldosterone stimulates the reabsorption of sodium by the kidney in an effort to maintain blood volume.

Antidiuretic hormone regulates blood pressure and blood osmolarity.

- Antidiuretic hormone (ADH) is produced in the hypothalamus and released from the posterior pituitary gland (see Figure 51.14). ADH controls the permeability of the collecting ducts to water by stimulating production of aquaporins over the long term, and insertion of aquaporins into the membrane over the short term.

- Stretch receptors in the aorta and carotid arteries inhibit the release of ADH in response to increased blood pressure, causing water to leave the body and blood volume to decrease.

- Osmosensors stimulate the release of ADH in response to increased osmolarity, resulting in increased water reabsorption and dilution of the blood.

- ADH also causes constriction of peripheral blood vessels.

- Atrial natriuretic peptide (ANP) is released by the heart in response to high blood pressure. At the kidneys, this hormone decreases sodium reabsorption, resulting in decreased water reabsorption and a lowered blood volume.

The Big Picture

- Different animals experience different types of stresses related to water and salt balance. Animals living in marine environments have to overcome problems with water loss due to osmosis, whereas animals living in freshwater environments have to overcome problems with water gain due to osmosis. Some invertebrates circumvent these problems by allowing their body fluid ion concentration to reflect that of their surroundings.

Other invertebrates, and all vertebrates, retain their body fluids at more constant levels, expending energy to eliminate excess water and/or salts.

- All animals produce metabolic wastes (notably nitrogenous waste) as a by-product of metabolism. Invertebrates have evolved a variety of structures to eliminate wastes, whereas the vertebrates have evolved variations on the kidney. The nephron is the "functional unit" of the vertebrate kidney. By eliminating wastes without losing valuable nutrients and salts, the kidneys cleanse the blood of metabolic waste products.

- Kidneys work by differentially processing nitrogenous waste, ions, and water. In the kidney, large volumes of plasma are filtered from the blood into the interior of the nephron. Then, as the fluid passes through the various specialized regions of the nephron, the desirable components to be retained (specific ions, nutrients, and water) are pulled back into the tissues, leaving behind in the nephron the waste products in an ever more concentrated form of urine. As urine leaves the nephron, water content is adjusted in the collecting ducts through hormonal alteration of their water permeability. If the collecting ducts are highly water permeable, then water will be drawn by osmosis out of the collecting ducts into the region of high salt concentration created by the action of the loop of Henle.

Common Problem Areas

- Depending on whether they are living in fresh water or saltwater, fishes may excrete copious amounts of dilute urine or small amounts of more concentrated urine, or take up ions at their gills. Given that living in either of these environments creates osmoregulatory problems, think about whether the tissues have higher or lower solute concentration relative to the surrounding water, and then predict water and ion movements on that basis.

- Understanding how the nephron functions, and in particular how the loop of Henle works as a "countercurrent multiplier system," is one of the greater challenges in this textbook.

Study Strategies

- Nephron function can be understood more easily if you place yourself in the position of a single Na^+ ion in the blood entering the nephron. Follow your pathway— and your potential pathways—as you traverse the nephron, eventually ending up in the urine. Note especially that it may take you some time to pass through the loop of Henle as you leave the ascending limb, only to recycle back into the descending limb—a process that could be repeated many times before finally escaping the countercurrent multiplier. Repeat this imaginary journey again as a water molecule.

- Review the following animated tutorial and activities on the Companion Website/CD:
  Tutorial 51.1 The Mammalian Kidney

Activity 51.1 Annelid Metanephridia
Activity 51.2 The Vertebrate Nephron
Activity 51.3 The Human Excretory System
Activity 51.4 The Major Organ Systems

Test Yourself
Knowledge and Synthesis Questions

1. What mechanisms does a marine bony fish use to maintain homeostasis?
   a. It excretes only small amounts of water and pumps sodium out of its body at the gills.
   b. It excretes large amounts of water and pumps sodium into its body.
   c. It converts nitrogenous wastes to urea.
   d. None of the above
   Textbook Reference: 51.4 How Do Vertebrates Maintain Salt and Water Balance? p. 1099

2. Which of the following statements about the excretory system of insects is *false*?
   a. Active transport moves materials from the coelomic fluid into the Malpighian tubules.
   b. The Malpighian tubules can produce a highly concentrated urea solution, allowing insects to inhabit some of Earth's driest habitats.
   c. Reabsorption of salts takes place mostly in the gut.
   d. Water reabsorption is by osmotic movement only.
   Textbook Reference: 51.3 How Do Invertebrate Excretory Systems Work? p. 1098

3. What pathway is taken by water and solutes as they travel through a nephron?
   a. Glomerulus, to Bowman's capsule, to proximal tubule, to loop of Henle, to distal tubule, to collecting ducts
   b. Bowman's capsule, to glomerulus, to distal tubule, to loop of Henle, to proximal tubule, to collecting ducts
   c. Glomerulus, to Bowman's capsule, to distal tubule, to loop of Henle, to proximal tubule, to collecting ducts
   d. Glomerulus, to Bowman's capsule, to proximal tubule, to collecting ducts, to distal tubule, to loop of Henle
   Textbook Reference: 51.4 How Do Vertebrates Maintain Salt and Water Balance? pp. 1099–1101

4. Na^+ and Cl^- are actively transported out of the tubules to help set up the countercurrent multiplier. Which of the following are sites of active Na^+ and Cl^- transport in the nephron?
   a. Proximal tubule, ascending limb of the loop of Henle
   b. Descending limb of the loop of Henle, ascending limb of the loop of Henle
   c. Ascending limb of the loop of Henle, proximal tubule
   d. Collecting duct, descending limb of the loop of Henle
   Textbook Reference: 51.5 How Does the Mammalian Kidney Produce Concentrated Urine? pp. 1103–1104

5.–9. Use the following diagram to complete the statements about the mammalian nephron.

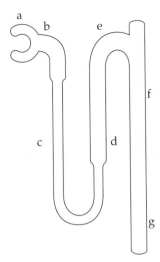

5. The composition of the filtrate would be most like plasma in the tubule next to letter _____.

6. The NaCl concentration in the extracellular fluid would be greatest in the area of letter _____.

7. The osmolarity of the filtrate next to letters _____ is similar to the osmolarity of blood plasma.

8. The urine would be most concentrated in the collecting duct next to letter _____.

9. Most of the glomerular filtrate is reabsorbed into the blood in peritubular capillaries next to letter _____.

   Textbook Reference: *51.5 How Does the Mammalian Kidney Produce Concentrated Urine? pp. 1103–1104*

10. The sole mechanism for water reabsorption by the renal tubules is
    a. active transport.
    b. osmosis.
    c. cotransport with sodium ions.
    d. cotransport with bicarbonate ions.

    Textbook Reference: *51.1 What Roles Do Excretory Organs Play in Maintaining Homeostasis? pp. 1093–1094*

11. Several hormones help to regulate water and solute uptake and release in the nephron. Antidiuretic hormone (ADH) promotes _____ in response to _____.

    a. active transport of Cl^-; increased solute concentration
    b. active transport of Na^+; increased blood pressure
    c. increased permeability of the collecting duct to water; decreased blood pressure
    d. decreased permeability of the collecting duct to water; increased solute concentration

    Textbook Reference: *51.6 What Mechanisms Regulate Kidney Function? pp. 1107–1108*

12. Which of the following is *not* a normal constituent of the glomerular filtrate?
    a. Red blood cells
    b. Urea

c. Sodium ion
d. Glucose

Textbook Reference: *51.4 How Do Vertebrates Maintain Salt and Water Balance? pp. 1100–1101*

13. If the afferent arteriole that supplies blood to the glomerulus becomes dilated,
    a. the protein concentration of the filtrate decreases.
    b. hydrostatic pressure in the glomerulus decreases.
    c. the glomerular filtration rate increases.
    d. All of the above

    Textbook Reference: *51.6 What Mechanisms Regulate Kidney Function? pp. 1107–1108*

14. Osmoreceptors in the brain detect changes in blood ion concentration that can reflexively result in _____ by the kidneys.
    a. water retention
    b. water loss
    c. ion retention
    d. All of the above

    Textbook Reference: *51.6 What Mechanisms Regulate Kidney Function? pp. 1107–1108*

15. If the human kidneys filter 150 liters of plasma in a 24-hour period, what is the typical amount of urine produced and eliminated in that time period?
    a. 0.15 liters
    b. 1.5 liters
    c. 15 liters
    d. 30 liters

    Textbook Reference: *51.5 How Does the Mammalian Kidney Produce Concentrated Urine? p. 1103*

Application Questions

1. Based on your knowledge of how the nephron works, explain why mammals produce concentrated urine whereas reptiles cannot.

   Textbook Reference: *51.5 How Does the Mammalian Kidney Produce Concentrated Urine? pp. 1103–1104*

2. What would happen to the composition of excreted urine if all active transport processes in the nephron came to a stop?

   Textbook Reference: *51.5 How Does the Mammalian Kidney Produce Concentrated Urine? pp. 1103–1104*

3. The drug urizadole inhibits antidiuretic hormone secretion from the pituitary. How will this change glomerular filtration rate, urine flow, and urine concentration in patients taking this medication?

   Textbook Reference: *51.6 What Mechanisms Regulate Kidney Function? pp. 1107–1108*

4. You have just eaten a large number of very salty potato chips, and the osmolarity of your blood has increased. What response will your body have in order to bring your blood osmolarity back to homeostasis?

   Textbook Reference: *51.6 What Mechanisms Regulate Kidney Function? pp. 1107–1108*

5. Earthworms and humans are very different, yet their excretory systems share many characteristics. List a few of the ways in which the excretory systems are similar in form and function.

   Textbook Reference: 51.3 How Do Invertebrate Excretory Systems Work? p. 1097; 51.4 How Do Vertebrates Maintain Salt and Water Balance? pp. 1098–1101

Answers

Knowledge and Synthesis Answers

1. **a.** Marine bony fishes live in an environment in which salts will tend to be drawn into the body and water will tend to be drawn out of the body. To counter these effects, these fishes excrete small amounts of water and actively pump sodium out of the body across the gills.

2. **b.** All are true of insects except that they excrete uric acid, not urea.

3. **a.** The route of water and solutes through the nephron is from the glomerulus, to Bowman's capsule, to proximal tubule, to loop of Henle, to distal tubule, to collecting ducts.

4. **a.** The proximal tubule is the site of active transport of Na^+ out of the tubule. Na^+ also moves out of the tubule at the ascending limb of the loop of Henle, but this is passive transport, with Cl^- being actively transported out.

5. **a.** At Bowman's capsule the filtrate is most similar to plasma.

6. **g.** The sodium concentration is the highest in the extracellular fluid near the middle of the medulla.

7. **a., b.,** and **e.** The filtrate has a similar osmolarity to plasma in the cortex of the kidney, including the proximal convoluted tubule, the glomerulus and Bowman's capsule, and the distal convoluted tubule.

8. **g.** The highest concentration of the filtrate in the collecting ducts will be near their ends, deep in the medulla.

9. **b.** The bulk of the water and solute that are reabsorbed from the filtrate are reabsorbed at the proximal convoluted tubule.

10. **b.** The sole mechanism for water reabsorption in the renal tubules is by osmosis.

11. **c.** Antidiuretic hormone acts on the collecting ducts by increasing permeability to water. Antidiuretic hormone secretion is stimulated by a decrease in blood pressure.

12. **a.** Red blood cells are too large to be filtered out of the blood at the glomerulus and thus will not be found in the filtrate.

13. **c.** Changes in the afferent arteriole pressure affect glomerular filtration rate. Dilation of the afferent arteriole increases pressure, which will increase filtration rate.

14. **d.** Osmoreceptors in the brain detect changes in blood ion concentration, resulting in reflexive increases or decreases in both ion and water retention, depending on the status of blood osmolarity.

15. **b.** About 1 percent of the filtrate is excreted as urine, so about 1.5 liters of the original 150 liters would be urinated.

Application Answers

1. The loop of Henle acts as a countercurrent ion multiplier in the medulla of mammals. This allows mammalian kidneys to produce concentrated urine. Reptiles lack the countercurrent multiplier of the loop of Henle.

2. Active transport of Na^+ and Cl^- in the nephron provides the ions that set up the countercurrent multiplier, allowing for the production of concentrated urine in mammals. If active transport in the nephrons were to stop, the urine produced would eventually be isotonic with blood plasma because the countercurrent multiplier would disappear.

3. Inhibition of antidiuretic hormone by urizadole will affect the permeability of the collecting ducts to water. This will result in increased urine flow because water will not be able to be reabsorbed across the collecting ducts. Blockage of ADH will have no effect on the glomerular filtration rate.

4. In response to increased blood osmolarity, osmo-sensors in the hypothalamus stimulate the release of ADH from the pituitary into the blood. ADH acts to increase the permeability of the collecting ducts to water so that increased amounts of water can be resorbed to bring down the blood osmolarity. The osmosensors will also stimulate thirst, causing you to increase your water intake.

5. The metanephridia of the earthworm and the nephron of humans are both composed of tubules through which filtrate flows. In the tubules of both, certain molecules are actively transported out or secreted in. Both systems produce urine with an osmolarity different from that of the body fluid. The urine of the earthworm is hypotonic, whereas that of humans is usually hypertonic.

CHAPTER 52 Ecology and the Distribution of Life

Important Concepts

Ecology is the scientific study of the interactions among organisms and between organisms and their physical environment.

- Ecology is studied at many levels. Communities are systems embracing all the organisms living together in the same area. Ecosystems embrace all the organisms in an area plus their physical environment. The biosphere embraces all regions of the planet where organisms live.

- The environment comprises both abiotic factors (physical and chemical) and biotic factors (living organisms). Organisms both influence and are influenced by their environment.

- Because of human dependence on both unmanaged and managed ecosystems, an understanding of ecological science is essential to human well-being and perhaps even to human survival.

Climates vary geographically primarily because different places receive different amounts of solar energy.

- The climate of a region is the average of the atmospheric conditions found in a given region over the long run. Weather is the short-term state of those conditions.

- Solar energy varies with latitude. Regions near the poles receive less energy per unit of ground area than regions near the equator because of the lower angle of the sun. At high latitudes there is also more variation over the course of a year in both day length and the angle of arriving solar energy than at latitudes closer to the equator.

- Rising air expands and cools, releasing moisture, whereas descending air is compressed and warmed, taking up more moisture.

- Unequal heating of the atmosphere at low and high latitudes produces vertical and latitudinal movements of air masses. These movements cause very moist climates to occur at the equator and at 60° north and south latitudes (where air rises) and arid climates at about 30° north and south latitudes and near the poles (where air descends).

- In the Tropics, the intertropical convergence zone is caused by the replacement of warm, rising air by air that flows toward the equator from north and south. The position of this zone shifts latitudinally during the course of the year, with the result that the Tropics and Subtropics experience rainy and dry seasons.

- The spinning of Earth on its axis causes air masses moving latitudinally to be deflected to the right in the Northern Hemisphere and to the left in the Southern Hemisphere. Thus, winds blowing toward the equator at low latitudes veer to become the northeast and southeast trade winds, whereas winds blowing away from the equator at mid-latitudes are deflected to become the prevailing westerlies.

- The movement of air over mountains often results in a rain shadow—a dry area on the leeward side of the range.

- Ocean currents are driven primarily by prevailing winds but are deflected by continents. The poleward movement of ocean water warmed in the Tropics is a major mechanism of heat transfer to high latitudes.

- Organisms must adapt to changes in their environment. In addition to short-term changes in behavior, the evolution of many morphological and physiological adaptations has enabled them to function in a variable physical environment.

- Either because of deteriorating local conditions or because of the need to find a better place to reproduce, most organisms move to a new place during their lifetimes, a phenomenon known as dispersal. In response to predictable seasonal changes in the environment, organisms may evolve life cycles that include migration or resting states that occur before adverse conditions materialize.

A biome is a large, terrestrial ecological unit defined by the growth forms of its dominant vegetation.

- The distribution of biomes on Earth is strongly influenced by annual patterns of temperature and precipitation. Often the boundary between two biomes is somewhat arbitrary because one biome gradually merges into another.

- Tundra is found at high latitudes in the Arctic and in high mountains. It is a treeless biome dominated by short perennial plants. In the Arctic, permanently frozen soil called permafrost underlies tundra vegetation.

- Boreal forests are located toward the equator from Arctic tundra and at lower elevations on temperate-zone mountains. The short summers favor evergreen trees, which are the dominant vegetation. Temperate evergreen forests occur on the west coast of continents at middle to high latitudes. They are home to Earth's tallest trees.

- Temperate deciduous forests are dominated by deciduous trees, which produce leaves that photosynthesize rapidly during the warm, moist summers and are lost during the cold winters.

- Temperate grasslands occur in areas that are relatively dry for much of the year. Grasses dominate this biome because they are well adapted to grazing and fire.

- Cold deserts are found in dry regions at middle to high latitudes. Seasonal changes in temperature are great. They are dominated by a few species of low-growing shrubs.

- Hot deserts, characterized by very warm and dry conditions year-round, are found in two belts around the 30°N and 30°S latitudes. Succulent plants that store large quantities of water in their stems are conspicuous in some hot deserts.

- The chaparral biome, found on the west side of continents at mid-latitudes, is dominated by evergreen shrubs and low trees that are adapted to survive periodic fires. Winters are cool and wet; summers are hot and dry.

- Thorn forests and savannas are found in semiarid climates on the equatorial side of hot deserts. Savanna, a grassland punctuated by scattered trees, is maintained by grazing, browsing, and burning. In their absence, it reverts to dense thorn forest dominated by small, spiny shrubs and trees.

- Tropical deciduous forests are dominated by trees that lose their leaves during the long, hot dry season. Because their soils are less leached of nutrients than the soils of wetter areas, most of these forests have been cleared for grazing cattle and growing crops.

- Tropical evergreen forests, found in equatorial regions where annual rainfall exceeds 250 cm annually, have the greatest species richness of all biomes and the highest productivity of all ecological communities. Nevertheless, their soils are poor and usually cannot support long-term agriculture unless massively fertilized. Epiphytes—plants that grow on other plants, deriving their nutrients and water from the atmosphere—thrive in tropical mountain forests.

- The distribution of biomes is determined not only by climate but also by other factors, particularly soil fertility and fire.

Biogeography is the scientific study of the distributional patterns of populations, species, and ecological communities.

- Biogeographers divide Earth into a number of biogeographic regions on the basis of the taxonomic similarities among the organisms living in them. The biotas of the biogeographic regions differ because barriers such as oceans restrict the dispersal of organisms.

- Species found in only one place are said to be endemic to that location. Remote islands contain many endemic species.

- Three scientific advances changed the field of biogeography: the acceptance of the theory of continental drift, the development of phylogenetic taxonomy, and the development of the theory of island biogeography.

- Continental drift has influenced the evolution and mixing of species throughout the history of life on Earth.

- The development of phylogenetic taxonomy has given biogeographers the ability to convert a taxonomic phylogeny into an area phylogeny, a process that involves replacing the names of the taxa with the names of the places where those taxa live or lived. This method is used to help explain how the current distribution of particular groups of species came about.

- According to the theory of island biogeography, equilibrium species richness (the number of species living in an area) on islands is determined by the rates of arrival of new species and extinction of species already present.

- The island biogeography model predicts that the equilibrium number of species should increase with island size and decrease with distance from the mainland. Biogeographers have confirmed the predictions of the model both by observation and experiment.

- A split distribution of a species can be accounted for in two ways.
  - A barrier may appear that splits a species' distribution (a vicariant event).
  - A species may cross an existing barrier and establish a new population (dispersal).

- In the process of biotic interchange, two different biota merge following the fusion of two formerly separated land masses. An example of this process was the formation of the Central American land bridge connecting North and South America. After it formed, many species of mammals that had evolved on one continent colonized the other. This dispersal caused many species (especially South American forms) to become extinct and many new species to evolve.

- Both vicariance and dispersal influence most biogeographic patterns. To determine which is more important in a particular case, scientists apply the parsimony principle. That is, they prefer the explanation that requires the smallest number of unobserved events to account for the pattern.

Aquatic ecosystems occupy three-fourths of Earth's surface.

- Earth's oceans form one interconnected water mass, with only partial barriers to dispersal. But though the oceans are connected, living successfully in a particular region requires that organisms have specific physiological tolerances and morphological characteristics. As a consequence, most marine organisms have restricted ranges.

- Biogeographic regions occur in the oceans as well as on land. Boundaries between biogeographic regions in the oceans typically occur where the temperature and salinity of surface water changes abruptly as a result of ocean currents.

- Primary food production by photosynthesis varies across the ocean. Over most of the oceans, where the dominant photosynthesizers are unicellular organisms, water temperature, salinity, and currents determine the physical environment to which organisms must adjust.

- Few marine species that live in shallow water are able to disperse across wide deep-water barriers.

- Fresh waters are divided into river basins and thousands of relatively isolated lakes. Although freshwater lakes, ponds, and streams contain only about 2.5 percent of Earth's water, they are the habitat for about 10 percent of all aquatic species. An important reason for discontinuities in the ranges of aquatic animals is that most organisms that live in fresh water cannot survive in the oceans, and vice versa.

The Big Picture

- The island biogeographic model, which relates the area of an island to the equilibrium number of species inhabiting the island, has applications in conservation biology (Chapter 57). For example, habitat destruction often leaves remnants called "habitat islands" that, if small, inevitably lose some of the species that existed in the previously more extensive habitat.

- Much evidence supporting the theory of continental drift (described in Chapter 21) is based on the distributions of organisms on Earth. By the same token, the acceptance by geologists and biologists of the reality of continental drift revolutionized the field of biogeography.

- The area phylogeny approach used by biogeographers to understand how the present distribution of groups of organisms came about is a good example of the broad applicability of the modern methods of phylogenetic analysis that were described in Chapters 24 and 25.

Common Problem Areas

- The graphical representations of the concepts of the island biogeographic model are troublesome for some students. Try redrawing the graphs for yourself, paying particular attention to how the axes of the graphs are labeled.

Study Strategies

- In attempting to master the information concerning the many biomes described in this chapter, focus on the graphs that summarize the essential points about each one.

- You also should study Figure 52.5 to get a sense of the geographical location and extent of each biome. A useful exercise to help you learn the relative locations of biomes is to imagine a journey (e.g., from equatorial South America to Alaska, or from Maine to California), naming the biomes that you would pass through.

- Review the following animated tutorials and activity on the Companion Website/CD:
  Tutorial 52.1 Rain Shadow
  Tutorial 52.2 Biomes
  Tutorial 52.3 Biogeography Simulation
  Activity 52.1 Major Biogeographic Regions

Test Yourself

Knowledge and Synthesis Questions

1. Which of the following is *not* usually included within the domain of ecology?
   a. Interactions between conspecifics
   b. Modifications of the environment by organisms
   c. Evolution of different social organizations
   d. Modification of the environment by physical processes
   Textbook Reference: 52.1 What Is Ecology? p. 1113

2. In comparing an acre of land in Colombia to an acre of land in Michigan, which of the following would *not* differ?
   a. The angle of the sun reaching the ground in the month of July
   b. The solar energy flux in the month of July
   c. The annual solar energy flux
   d. The total hours of daylight per year
   Textbook Reference: 52.2 How Are Climates Distributed on Earth? p. 1114

3. If Earth did not spin on its axis, from what direction would the northeast trade winds blow?
   a. Northeast
   b. South
   c. North
   d. East
   Textbook Reference: 52.2 How Are Climates Distributed on Earth? pp. 1114–1115

4. Ocean circulation patterns are influenced by all of the following *except*
   a. circulation of Earth's atmosphere.
   b. deflection by land masses.
   c. upwelling of deeper water.
   d. rotation of Earth on its axis.
   Textbook Reference: 52.2 How Are Climates Distributed on Earth? p. 1115

5. The following diagram shows a mountain with a sea breeze blowing as indicated by the arrow. Circle the letter for the area with air that would be *both* relatively warm and dry.

   Textbook Reference: 52.2 How Are Climates Distributed on Earth? pp. 1114–1115

6. In what area of the diagram in Question 5 is a process occurring that is similar to the process that occurs in the intertropical convergence zone?
   Textbook Reference: 52.2 How Are Climates Distributed on Earth? pp. 1114–1115

7. The biome that is maintained by browsers, grazers, or fire is the
   a. tundra.
   b. savanna.
   c. cold desert.
   d. tropical deciduous forest.
   Textbook Reference: 52.3 What Is a Biome? p. 1125

8. Match the letters of the following biomes with the descriptions that follow.
   a. Tundra
   b. Boreal forest
   c. Hot desert
   d. Chaparral
   e. Tropical deciduous forest
   ____ Mostly coniferous, wind-pollinated and wind-dispersed tree species
   ____ Leaves lost during dry season; agriculturally desirable land
   ____ Cool winters, hot dry summers; maritime climate
   ____ Succulent plants prominent; found at 30°N and 30°S latitudes
   ____ Distribution altitudinally or latitudinally determined; permafrost present
   Textbook Reference: 52.3 What Is a Biome? pp. 1116–1125

9. The following climograph shows yearly variation in rainfall and temperature for four biomes. Select the correct curve for each of the following biomes.

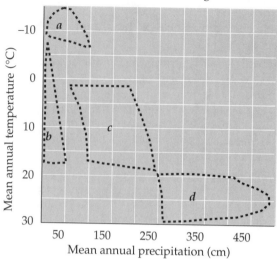

   ____ Tundra
   ____ Tropical evergreen forest
   ____ Temperate deciduous forest
   ____ Cold desert
   Textbook Reference: 52.3 What Is a Biome? pp. 1116–1125

10. The tropical evergreen forest biome has relatively constant _____, but _____ often varies seasonally; the temperate deciduous forest biome has relatively constant _____, but _____ varies seasonally.
    a. temperature; rainfall; rainfall; temperature
    b. rainfall; temperature; temperature; rainfall
    c. rainfall; temperature; rainfall; temperature
    d. temperature; rainfall; temperature; rainfall
    Textbook Reference: 52.3 What Is a Biome? pp. 1118–1125

11. Which of the following biogeographical regions represents the largest area?
    a. Nearctic
    b. Palearctic
    c. Neotropical
    d. Oriental
    Textbook Reference: 52.4 What Is a Biogeographic Region? p. 1128

12. According to the island biogeographic model, which of the statements about the following graph showing the effect of species number on arrival and extinction rates is *not* true?

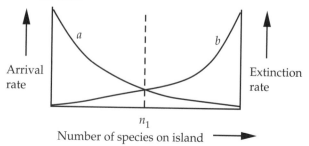

a. Curve *a* is the arrival rate curve.
b. Curve *b* is the extinction rate curve.
c. The arrival rate equals the extinction rate at *n*1.
d. The extinction rate is zero at *n*1.
Textbook Reference: *52.4 What Is a Biogeographic Region? p. 1132*

13. The species numbers of many islands of different sizes are plotted against their distance from the mainland. Data points for islands of similar size were connected to form the four curves shown in this figure.

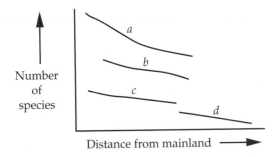

Circle the letter of the curve corresponding to the group of islands with the smallest size.
Textbook Reference: *52.4 What Is a Biogeographic Region? p. 1132*

14. The island biogeographic model makes predictions about the effects of species number on the rate of extinction. In the following figure, the solid curve shows this relationship for a large island. Which of the curves shows the expected relationship for a small island?

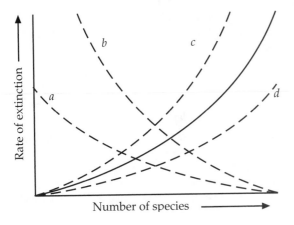

a. Curve *a*
b. Curve *b*
c. Curve *c*
d. Curve *d*
Textbook Reference: *52.4 What Is a Biogeographic Region? p. 1132*

15. The following diagram shows three islands. Islands A and B were connected in the past, C was always separate. A species of land snail is found on all three islands.

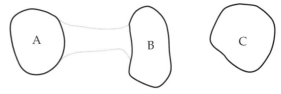

Which of the following does *not* correctly describe the distribution of this snail relative to the three islands?
a. Vicariant distribution relative to A and B
b. Dispersal distribution relative to A and C
c. Dispersal distribution relative to B and C
d. Vicariant distribution relative to A and C
Textbook Reference: *52.4 What Is a Biogeographic Region? p. 1133*

16. If the species of snail mentioned in Question 15 is found *only* on the three islands shown in the preceding diagram, which statement would be correct concerning the endemism of this species?
a. Endemic relative to A and B
b. Endemic relative to C
c. Endemic relative to A, B, and C
d. All of the above
Textbook Reference: *52.4 What Is a Biogeographic Region? p. 1129*

17. Which of the following statements about the biogeography of marine biomes is *false*?
a. Species adapted for life in shallow water are seldom able to disperse across wide deep-water barriers.
b. In interpreting the biogeography of the marine biome, oceanic currents are more important than physical barriers.
c. Organisms adapted for life in nutrient-poor waters usually flourish in nutrient-rich areas.
d. Continental drift does not explain the distribution of many marine organisms.
Textbook Reference: *52.5 How Is Life Distributed in Aquatic Environments? p. 1135*

18. Which of the following comparisons of freshwater and marine habitats is *not* true?
a. About 10 percent of all aquatic species live in freshwater habitats.
b. There are many more insect species in freshwater habitats than in marine habitats.
c. The global volume of marine habitats is much greater than that of freshwater habitats.
d. Freshwater species richness is less than marine species richness in proportion to the relative extent of the two habitats.
Textbook Reference: *52.5 How Is Life Distributed in Aquatic Environments? p. 1136*

19. Which of the following is most likely to be true of a family of freshwater fish with species distributed on several continents?
    a. They are a recently derived group.
    b. They have an ancient lineage that was widely distributed in Laurasia or Gondwana.
    c. Their distribution was not affected by continental drift.
    d. Their distribution can be explained by movements of glaciers.
    Textbook Reference: 52.5 How Is Life Distributed in Aquatic Environments? p. 1136

Application Questions

1. Why is an understanding of ecology essential for the future well-being of humanity?
   Textbook Reference: 52.1 What Is Ecology? pp. 1113–1114

2. Compare dispersal and migration. Do they differ in adaptive function?
   Textbook Reference: 52.2 How Are Climates Distributed on Earth? pp. 1115–1116

3. Tropical evergreen forests have greater overall productivity and species richness than tropical deciduous forests, yet the latter are more easily converted to productive agricultural land than the former. Why?
   Textbook Reference: 52.3 What Is a Biome? pp. 1126–1127

4. More living species of the horse family live in Africa than in any other continent. Yet biogeographers do not believe that the horse family evolved in Africa. What kinds of evidence have biogeographers relied on to explain how the current distributions of horses came about?
   Textbook Reference: 52.4 What Is a Biogeographic Region? p. 1131

5. About four million years ago, a land bridge formed between North and South America. Describe the effects of this event on the mammalian fauna that had evolved in South America prior to the formation of the bridge.
   Textbook Reference: 52.5 How Is Life Distributed in Aquatic Environments? p. 1134

Answers

Knowledge and Synthesis Answers

1. **d.** All except the modification of the environment by physical factors are normally included within the domain of ecology.

2. **d.** All areas on Earth receive equal hours of daylight per year. The seasonal distribution of those hours, the solar energy input, and the sun angle do vary latitudinally.

3. **c.** If Earth did not spin on its axis, air flowing south toward the equator would not be deflected to the right.

Therefore, the northeast trade winds would blow directly from the north instead of from the northeast.

4. **c.** The upwelling of deeper water, especially on the western side of continents, is an effect of the pattern of ocean current circulation and not a cause of the pattern.

5. **e.** The air would have lost most of its moisture while rising on the windward side of the mountain, and in descending to area *e* it would have warmed again.

6. **b.** In the intertropical convergence zone, warm air rises and loses much of its moisture. These same events occur when moist air rises over a mountain.

7. **b.** If savanna vegetation is not grazed, browsed, or burned, it typically reverts to dense thorn forest.

8. **b.** Mostly coniferous, wind-pollinated and wind-dispersed tree species
   e. Leaves lost during dry season; agriculturally desirable land
   d. Cool winters, hot dry summers; maritime climate
   c. Succulent plants prominent; found at 30°N and 30°S latitudes
   a. Distribution altitudinally or latitudinally determined; permafrost present

9. **a.** Tundra
   d. Tropical evergreen forest
   c. Temperate deciduous forest
   b. Cold desert

10. **a.** The tropical evergreen forest biome has relatively constant **temperature,** but **rainfall** often varies seasonally; the temperate deciduous forest biome has relatively constant **rainfall** (or precipitation, because snow may occur in winter) but **temperature** varies seasonally.

11. **b.** See Figure 52.8 in the textbook.

12. **d.** The extinction rate is zero and equals the arrival rate at n^1, which is therefore the equilibrium species number on the island.

13. **d.** Each curve corresponds to a group of similar-sized islands whose species number is plotted against their distance from the mainland. For each curve, species number decreases with distance from the mainland, and the curve that is lowest relative to the vertical axis would be the group of smallest islands.

14. **c.** With less space available, populations of the different species would be smaller. This would subject them to higher extinction rates than would be expected on larger islands.

15. **d.** A vicariant distribution requires that the species once had a continuous distribution in what are now separate areas.

16. **c.** The snail is endemic to the group of islands, but not to any single one or pair of them because it is also found on the remaining islands or island.

17. **c.** Usually organisms adapted for one zone do not do well in a different zone.

18. **d.** Even though freshwater habitats occupy a very small global area, they have about 10 percent of all aquatic species, which, by proportion, gives them a very large species richness.

19. **b.** The most likely explanation for families of freshwater fish with representatives on several continents is that their lineages originated prior to the breakup of Laurasia or Gondwana.

Application Answers

1. A grasp of ecological science is essential for understanding (and hence for preserving) the ecosystems that provide essential services for human society (such as clean air and water). Also, as we alter or destroy natural ecosystems in the interest of providing food and other resources for the growing human population, an understanding of ecological principles may enable us to avoid many of the unanticipated and possibly disastrous consequences of our actions.

2. Both dispersal and migration are movements of individuals of a population (or a whole population) from one location to another. Dispersal is the movement of members of a population away from the site where they were born or where they have been living either to find a better place to reproduce or to avoid deteriorating local conditions. Dispersal may be active but is sometimes passive, as in the case of dispersal of seeds by wind, water, or animals. Migration is the active movement of members of a population in response to cyclical, predictable environmental events, such as seasonal changes. It follows from these definitions that migration is more likely to be a regularly repeated feature of an individual's life cycle than is dispersal, which only occurs once in the life cycle of many organisms.

3. The reason for the greater agricultural productivity of deciduous tropical forests is that their soils are richer in nutrients than the soils of tropical evergreen forests. The higher precipitation characteristic of tropical evergreen forests causes their soils to lose mineral nutrients very quickly if they are not recycled into the living vegetation, which is in fact where most of the nutrients are located in this type of forest.

4. The two main sources of evidence used by biogeographers to explain the present distributions of horses are the fossil record and the area phylogeny method. Fossils clearly indicate that the earliest ancestors of horses evolved in North America. Inasmuch as Przewalski's horse inhabits central Asia (as the area phylogeny in Figure 52.10 shows) and represents the earliest lineage to diverge from the lineages leading to the other living species, it is reasonable to assume that the ancestor of all living species of horses first dispersed from North America to Asia. Similar reasoning leads to the conclusion that further speciation events occurred as horses moved from Asia to Africa and that Africa was the site of the speciation of zebras.

5. The land bridge permitted the mammalian faunas of both North America and South America to disperse to the other continent. Many colonizing North American species were able to compete successfully with South American forms such as marsupials and ground sloths and to drive them extinct. Subsequently, the northern invaders formed new species found today only in South America.

CHAPTER 53 Behavior and Behavioral Ecology

Important Concepts

Biologists ask several kinds of questions about behavior.

- The study of the function and evolution of behavior is known as ethology.

- One type of question asked by ethologists focuses on describing behavior patterns and how and when individuals perform them.

- A second type of question addresses the proximate mechanisms of behaviors: What is the neuronal, hormonal, and anatomical basis of the behavior? What are the relative roles of genes and experience?

- A third type of question concerns the ultimate causes of behavior: How does the behavior contribute to the fitness of the organism that possesses it?

Genes and the environment interact to shape behavior.

- Some behaviors, especially those that are stereotypic and species-specific, may require little or no learning. Even though a behavior is unlearned; however, it may still not be performed if the environmental conditions needed to stimulate it are absent.

- Genes do not encode behaviors. Rather, gene products such as enzymes, in the proper environment, can lead to the development of the proximate mechanisms underlying a behavior.

- Determination of inheritance of behaviors can be made by deprivation experiments, in which all relevant experience is denied. If the animal still exhibits the behavior, it is assumed that the behavior can develop without opportunities to learn it.

- Several kinds of genetics experiments can also help to distinguish between genetic and environmental influences on behavior. One approach, not based on theoretical science, is selective breeding (e.g., of dogs). A second type of experiment is interbreeding (hybridization), in which investigators cross two species with known behaviors and study the behavior of the offspring. A third technique is the gene knockout experiment, in which biologists inactivate a particular gene (e.g., in the mouse) and investigate the behavioral effects of its loss.

- Genetic control of behavior can be highly adaptive for species in which there is no opportunity for learning—as in species with nonoverlapping generations. It is also adaptive when mistakes are costly or fatal. Examples of such behaviors are predator avoidance and mating rituals.

- A releaser is an object, event, or condition necessary for an animal to exhibit a genetically based behavior. The response to the releaser may depend on the motivational state of the animal.

- It is often adaptive for behavior that is predominantly genetically controlled to include a role for learning. For example, wasps use spatial cues from the environment to help them locate their nests (see Figure 53.3).

- In imprinting, an animal learns a set of stimuli during a limited critical period. Imprinting of offspring on parents or of parents on offspring is a learned response that helps in recognition. The critical period for imprinting is often determined by the developmental or hormonal state of the animal.

- Birds use songs in territorial displays and in courtship. In some species, imprinting of the species-specific song is required in the nestling in order for it to sing the song as an adult, even though the song is never sung by the juvenile bird. Imprinting forms a memory of the song that is recalled as the bird approaches adulthood. The adults need both the initial imprinting of the song as a juvenile and the ability to match the song with auditory feedback in order to sing the correct song (see Figure 53.5).

- The influence of hormones on behavior has also been investigated through experiments on song development in birds. In male birds, testosterone levels control singing by inducing certain regions of the brain to grow during the breeding season (see Figure 53.6). During the nonbreeding season, these regions of the brain associated with singing are reduced in size. Female birds that are treated with testosterone during the spring will also develop the species-specific song. In these birds, testosterone has stimulated growth in those areas of the brain associated with singing.

333

Behavioral responses to the environment influence fitness.

- The habitat of an animal is the environment in which it normally lives. In choosing their habitat, animals use cues that are good predictors of general conditions suitable for future survival and reproduction.

- The success of already-settled individuals may indicate habitat quality (see Figure 53.7).

- The territory of an animal is an area from which other individuals of its own species (and sometimes individuals of other species) are excluded. By establishing a territory, an animal may improve its fitness by gaining exclusive use of the resources of part of its habitat.

- Natural selection molds behavior (such as defending a territory) in accordance with costs and benefits (see Figure 53.8). There are three aspects to the cost of behaving:

  1. Energetic cost—the amount of energy the animal expends during the behavior.

  2. Risk cost—the amount of risk the behavior entails for the animal.

  3. Opportunity cost—the benefits the animal forfeits by not engaging in other behaviors instead.

- Territories fall into several categories. Some animals defend all-purpose territories in which nesting, mating, and foraging take place. Other animals, such as seabirds, cannot establish feeding territories but defend nest sites. Still other animals defend territories that are used only for mating.

- Foraging theory is an attempt to understand how animals select what kind of food to eat and where and when to search for it. A plausible hypothesis, supported by experiment, is that the foraging choices of many animals result in their maximizing their rate of energy intake (see Figure 53.10).

- Because minerals or non-nutritive foods are important in the diets of many animals, they may sometimes forage in a way that deviates from the energy-maximization model. The high value humans place on spices (a non-nutritive food) may have evolved because most spices used in cooking have antimicrobial activity (see Figure 53.12).

- One important choice an animal makes is mate selection. Because the quantity of sperm produced by males is typically far greater than the number of eggs produced by females, males and females approach courtship very differently. Males of most species can increase their reproductive success by mating with many females. In contrast, the reproductive success of females depends primarily on the quality of genes she receives from her mate, the resources he controls, or the amount of care he provides for their offspring. As a result, males usually initiate courtship and frequently fight for access to a female. Females seldom fight over males and often reject courting males.

- A male may attract females by courtship behavior that signals his good health and genetic quality. The behavior also may show that he is a good provider of parental care.

- By their ability to recognize the signals at which males cannot cheat, females have favored the evolution of "reliable" signals of male quality.

- Responses to the environment must be timed appropriately. Circadian rhythms are daily cycles in activity, sleep, foraging, and other physiological processes and behaviors that are controlled by an endogenous clock (see Figure 53.14).

- The length of one cycle in a rhythm is defined as one period. Any point in the cycle is known as a phase. Two cycles can be in phase if the rhythms match or be phase-advanced or phase-delayed if they do not match.

- Circadian rhythms can be reset by environmental cues, such as the light–dark cycle, during entrainment. In constant conditions an animal's circadian clock is said to be free-running and will have a natural period that is different from the 24-hour period of the day. Because human biorhythms can be shifted by no more than 30 to 60 minutes per day, we experience "jet lag" when crossing time zones.

- In response to daily cycles of light and dark, animals have evolved different sensory capabilities, depending on when they are typically active. Nocturnal animals (active at night) tend to depend on their abilities to hear, smell, and use tactile information, whereas diurnal animals (active during the day) tend to be highly visual.

- Day length (photoperiod) can trigger seasonal rhythms, which are important in controlling the timing of breeding and hibernating. Animals such as hibernators and equatorial migrants, for whom changes in day length are not a reliable cue, rely instead on circannual rhythms—built-in neural calendars that keep track of the time of the year.

- Animals find their way around their environment by a variety of mechanisms, including piloting, distance-and-direction navigation, and bicoordinate navigation.

- Piloting is orientation by the recognition of landmarks. It is the mechanism used by some species that migrate or that are capable of homing (the ability to return to a specific location).

- Homing and migrating species that are able to take direct routes to their destinations through environments they have never experienced must use mechanisms of navigation other than piloting. Examples of such species are homing pigeons and the many species of migrating birds that are able to fly great distances and return to the same breeding ground each season.

- Distance-and-direction navigation involves knowledge of direction and distance to a destination. The position of the sun and stars can be a source of directional information. Pigeons, for example, have the ability to determine direction by means of a time-compensated solar compass (see Figure 53.17). The stars offer two sources of information about direction: moving constellations and a fixed point (the point directly over the axis on which Earth turns).

- Bicoordinate navigation (also known as true navigation) requires knowing the map coordinates of both the current position and the destination. The behavior of many species (such as albatrosses) suggests that they are capable of this type of navigation.

- In addition to the sun and stars, birds use a magnetic sense (not understood neurophysiologically) as a source of navigational information, and they may also use other environmental cues.

Many animal behaviors involve some form of communication between individuals of a species.

- Communication is used by animals to transmit information. If the transmission of information benefits both the sender and the receiver, the behaviors of individuals may become elaborated through evolution as displays or signals.

- Visual signals provide rapid, directional communication. One drawback to visual signals is the need for light, except in the case of species that have evolved their own light-emitting mechanisms.

- Pheromones are chemical signals used to communicate among individuals of a species. Because of the diversity of their molecular structures, pheromones can communicate very specific, information-rich messages. Because pheromones remain in the environment for some time after they are released, they are useful for such functions as marking territories but unsuitable for the rapid exchange of information.

- Sound has advantages over sight in that it can travel in complex environments, such as a forest, and over long distances, such as the sound produced by whales in the water. Sound also communicates directional information.

- Touch is commonly used in communication. Honeybees dance to communicate the location of a food source in the environment (see Figure 53.19). Round dances are used to communicate that food is less than about 80 meters from the hive. A waggle dance is used to communicate distance and direction to food that is more than 80 meters from the hive. Speed of the waggle indicates distance to the food source; direction of the straight run of the waggle dance indicates the direction of the food source relative to the sun.

- Some fish use electrical fields to sense their surroundings. In the glass knife fish, different individuals emit different frequencies that relate to their status in the population.

Social behavior evolves when individuals that cooperate with others of the same species have higher rates of survival and reproductive success than those achieved by solitary individuals.

- Social systems are dynamic; individuals constantly communicate with one another and adjust their relationships.

- Individuals in a social system experience different costs and benefits, depending on their age, sex, physiological condition, and status.

- Social systems are best understood not by asking how they benefit the species as a whole, but by asking how the individuals that join together benefit by the association.

- Group living may benefit both predator and prey species. For predators, it may improve hunting efficiency or increase the size of the prey that can be captured. For prey, it may provide increased protection against predators.

- Living in a group imposes costs as well as benefits. Individuals in groups may compete for food, interfere with one another's foraging, injure one another's offspring, inhibit one another's reproduction, or transmit diseases to their associates.

- The most widespread social system is the family, an association of one or more parents and their dependent offspring. Most complex social systems in animals have evolved via an extended family.

- Natural selection sometimes favors altruistic acts—behaviors that reduce the reproductive chances of the individual performing the act but increase the fitness of the helped individual. Typically, such behavior is directed toward a relative (nondescendant kin) of the altruist, with whom it shares alleles. By helping its relatives, an individual can increase the representation of some of its own alleles in the population. An altruistic behavior pattern can evolve if it increases the inclusive fitness of the altruist—that is, if the benefit of increasing the reproductive success of nondescendant kin is greater than the cost of decreasing the altruist's own reproductive success.

- Eusocial species are those whose social groups include sterile individuals. The evolution of eusociality in the Hymenoptera (ants, bees, and wasps) has been facilitated by their sex determination system, in which males are haploid and females are diploid, with the result that sisters are genetically more similar to one another than to their own offspring. In eusocial species in which both sexes are diploid (termites, naked mole-rats), inbreeding and the difficulty of establishing independent colonies have favored the evolution of eusociality.

Decisions that organisms make about habitats, food, and associates influence the structure and functioning of ecological systems.

- Habitat and food selection influence the distributions of organisms. For example, animals that live in cold and arid environments may store food outside their bodies and thereby be able to live in environments where food would otherwise be seasonally unavailable. Also, predators can influence the species composition and abundance of their prey by their food choices.

- Interspecific territoriality may influence community structure. For example, individuals of a dominant species may exclude those of a subordinate species from particular habitats that they would occupy if the dominant individuals were removed.

- Social animals may achieve extraordinary abundance. Examples include ants, termites, and humans. The increase in human abundance has occurred because of specialization, leading to advances in agriculture, medicine, and hygiene.

The Big Picture

- The distinction between proximate and ultimate explanations is of fundamental importance in the study of behavior and can be extended to many other fields of biology. Behaviorists concerned with proximate questions wish to unravel the underlying neural and hormonal mechanisms of behavior and to sort out the relative roles of genes and experience in shaping behavior. On the other hand, behaviorists asking ultimate questions attempt to understand how the behavior that they are studying is related to the animal's survival and reproductive success. In this effort, they not only draw on concepts from evolution, ecology, and animal behavior, but also make important contributions to those fields.

- Animals exhibit a wide range of species-specific behaviors that can be either genetically based or environmentally determined. Genetically based behaviors may require triggers or releasers to stimulate an animal. Nevertheless, patterns of behavior typically have genetic and environmental influences.

- Hormones are important in controlling the behavior of animals. Sex steroids are frequently involved in stimulating the different behaviors of males and females of a species. Daily, seasonal, or annual rhythms are often carried out through the actions of hormones.

- Communication is an integral part of animal behavior. Communications can be chemical (involving pheromones), visual (involving displays of fins, feathers, fur, etc.), auditory (sounds created by either general or specialized structures and received by auditory sensors), tactile, or electric. Such communication is an important component of territory-marking and reproductive behavior.

- As you learned from Chapter 22 (The Mechanisms of Evolution), the cost–benefit approach applied in this chapter to the analysis of behavior can be extended to any kind of adaptation of organisms.

- Because, according to standard Darwinian theory, only traits that contribute to individual fitness are favored by natural selection, explaining the evolution of altruistic behavior has presented a major challenge to evolutionary biologists. The concept of inclusive fitness has proved to be very fruitful in understanding altruism in animals and has also been applied to many aspects of human social behavior (though not without controversy).

Common Problem Areas

- It can sometimes be difficult to distinguish between behaviors founded in genetics (instinctive behaviors) and those that are learned during the course of an animal's lifetime. Further muddying the waters, the ability to learn new behavior is, of course, a heritable trait!

- Hormones control many behaviors, and hormone production is often a function of the time of day or the time of year. Thus, a particular hormone that stimulates (or inhibits) behaviors under one condition at one point in time may not have the same effect at another time. Be sure to learn the conditions and caveats that accompany hormone actions.

- The categories of animal orientation—piloting, distance-and-direction navigation, and bicoordinate navigation—can be confusing. Also refer to Figure 53.17 to consolidate your understanding of how the sun is used as a time-compensated compass.

Study Strategies

- As you learn about various animal behaviors, it may be useful to design "mind experiments" in your imagination to help you to distinguish between a genetically determined and an environmentally determined behavior. This also may help you to work out the nature of the behavior.

- Learn to distinguish the various behavior cycles associated with internal controls via "biological clocks" as opposed to behaviors that are directly stimulated by an immediate event in an animal's surrounding environment.

- To study the properties of the various sensory modes of communication (visual, chemical, etc.), make a table that compares them with respect to characteristics such as cost of production, effective signaling distance (and how signaling distance is affected by environmental conditions), durability, and information content.

- Review the following animated tutorials and activities on the Companion Website/CD:
    Tutorial 53.1 The Costs of Defending a Territory
    Tutorial 53.2 Foraging Behavior
    Tutorial 53.3 Circadian Rhythms
    Tutorial 53.4 Time-Compensated Solar Compass

Activity 53.1 Honeybee Dance Communication
Activity 53.2 Concept Matching

Test Yourself

Knowledge and Synthesis Questions

1. On April mornings, you step outside your front door and notice the singing of a male robin. At that time you also observe that the female of the pair is building a nest in a nearby tree. A month later you observe the pair feeding their nestling offspring. Which of the following is a question about the ultimate cause of the behavior of these birds?
   a. Did the male bird begin to sing in April because the photoperiod had increased to a critical length?
   b. What were the relative roles of genes and experience in causing the female robin to build her nest out of particular building materials and to place it in a particular location?
   c. What combination of internal physiological factors and external cues stimulates the parents to feed their nestlings?
   d. Do robins that begin nesting in April raise more offspring than they would raise if they delayed nesting until June, and if so, why?
   Textbook Reference: 53.1 What Questions Do Biologists Ask about Behavior? p. 1141

2. Which of the following statements about genetically determined behaviors is *false*?
   a. Inherited behaviors that involve mating and courtship are highly adaptive.
   b. In interbreeding experiments, closely related species are bred to determine the behavior of the offspring.
   c. Web spinning in spiders is species-specific.
   d. Most behaviors have only a genetic component.
   Textbook Reference: 53.1 What Questions Do Biologists Ask about Behavior? p. 1141; 53.2 How Do Genes and Environment Interact to Shape Behavior? p. 1144

3. Which of the following statements about hormonal influences on bird song is *false*?
   a. Male songbirds castrated as nestlings do not sing proper songs as adults.
   b. Female songbirds can be induced to sing by treatment with testosterone.
   c. Females learn their species song, but do not normally express it under natural conditions.
   d. Testosterone is not necessary for singing in males once they have learned their song.
   Textbook Reference: 53.3 How Do Behavioral Responses to the Environment Influence Fitness? pp. 1146–1147

4. Two species of mice live in the same geographical region, but Species 1 prefers open fields, whereas Species 2 lives in forests. When presented in an experiment with simulated "fields" and "forests," individuals of each species born in a laboratory preferred the environment in which they normally lived. This experiment illustrates the concept of _____.

   a. habitat selection
   b. optimal foraging strategy
   c. territoriality
   d. Both a and b
   Textbook Reference: 53.3 How Do Behavioral Responses to the Environment Influence Fitness? pp. 1147–1148

5. Birds spend some of their time scanning the horizon for predators. While scanning, they cannot be foraging for food. Which of the following does this situation illustrate?
   a. Cooperative hunting
   b. Energetic cost
   c. Risk cost
   d. Opportunity cost
   Textbook Reference: 53.3 How Do Behavioral Responses to the Environment Influence Fitness? p. 1148

6. Suppose a predator has two different prey: Species 1 and Species 2. You perform a series of experiments in which the density of Species 1 is varied while the density of Species 2 is kept constant. At each Species 1 density, you determine how many prey of Species 1 and 2 the predator takes. The two curves (*a* and *b*) plotted in the following graph show the outcome of this study.

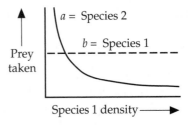

   Select the statement that correctly interprets these curves.
   a. Curve *a* indicates that Species 2 is the preferred prey.
   b. Curve *b* indicates that Species 1 is the preferred prey.
   c. Curve *b* indicates that Species 1 is not the preferred prey.
   d. Neither of these curves provides insight into the prey preference of the predator.
   Textbook Reference: 53.3 How Do Behavioral Responses to the Environment Influence Fitness? pp. 1149–1150

7. Experimental data suggest that human enjoyment of spices in food is probably
   a. of no adaptive significance, because a taste for spices appears to be entirely a learned (or culturally determined) behavior.
   b. adaptive, because most spices help to prevent the growth of microbes in food.
   c. adaptive, because spices mask the bad taste of spoiled food, thereby making it palatable.
   d. maladaptive, because spices often bind to mineral nutrients in the diet and thus prevent their absorption.
   Textbook Reference: 53.3 How Do Behavioral Responses to the Environment Influence Fitness? pp. 1151–1152

8. Animals exhibit daily rhythms in their behavior and physiology. Which of the following statements about circadian rhythms is *false*?
   a. Animals that are active at night do not display circadian rhythms.
   b. A circadian clock can be entrained by environmental cues.
   c. Free-running circadian clocks are seldom exactly 24 hours long.
   d. Genetic mutations can cause changes in the length of the free-running circadian clock.

   Textbook Reference: 53.3 *How Do Behavioral Responses to the Environment Influence Fitness? pp. 1152–1153*

9. Which of the following animals would *not* be expected to have an endogenous circannual rhythm?
   a. A ground squirrel that hibernates
   b. A white-tailed deer living in eastern North America
   c. A cave salamander
   d. A migratory bird that summers in North America and winters in equatorial South America

   Textbook Reference: 53.3 *How Do Behavioral Responses to the Environment Influence Fitness? pp. 1153–1154*

10. Which of the following is an example of piloting?
    a. Silkworms following a trail of pheromones
    b. Marine migration over a featureless ocean
    c. Bees learning the direction of a food source from a dance
    d. Gray whale migration along the pacific coast of America

    Textbook Reference: 53.3 *How Do Behavioral Responses to the Environment Influence Fitness? p.1154*

11. There are a number of different navigational methods used by animals. Which of the following definitions of a navigational method is correct?
    a. Distance-and-direction navigation requires knowledge of latitude and longitude.
    b. Bicoordinate navigation requires knowledge of direction and of distance to a destination.
    c. The stars offer two sources of information about direction; a fixed point and moving constellations.
    d. Piloting involves the magnetic mineral magnetite.

    Textbook Reference: 53.3 *How Do Behavioral Responses to the Environment Influence Fitness? pp. 1154–1155*

12. Which of the following statements about pheromones is true?
    a. Pheromones are only used for communication between individuals of the same species.
    b. Mammalian pheromones can communicate information about the size, sex, and reproductive status of signaler.
    c. Because of their durability in the environment, pheromones are unsuitable for rapid exchange of information.
    d. All of the above

    Textbook Reference: 53.4 *How Do Animals Communicate with One Another? p. 1157*

13. You observe an example of an apparently altruistic act by an animal that seems to reduce its near-term likelihood of reproductive success. What would be the *least* plausible explanation for its behavior?
    a. The act aids the reproductive success of individuals sharing a high proportion of genes with the altruistic individual.
    b. The act is only apparently altruistic; over the long term, the behavior actually contributes to individual fitness.
    c. The act increases the inclusive fitness of the animal performing it.
    d. The act is advantageous because it helps the species to survive and reproduce, even if the altruist itself does not.

    Textbook Reference: 53.5 *Why Do Animal Societies Evolve? pp. 1158–1159*

14. Some birds give a species-specific vocalization called an "alarm call" when they see a predator, although this call may direct the predator toward them. Other members of their species respond to these calls by taking cover. This would be an example of altruistic behavior that is beneficial to the calling bird if
    a. the bird giving the vocalization survives the attack.
    b. the inclusive fitness of the bird giving the vocalization is increased.
    c. all birds survive the attack.
    d. the birds benefiting are offspring of the bird giving the vocalization.

    Textbook Reference: 53.5 *Why Do Animal Societies Evolve? p. 1159*

15. Which of the following characteristics is/are shared by all eusocial species?
    a. A sex determination system in which males are haploid and females are diploid
    b. Queens that mate with a single male
    c. Presence of sterile classes
    d. All of the above

    Textbook Reference: 53.5 *Why Do Animal Societies Evolve? pp. 1159–1160*

16. Which of the following statements would explain why one or more social species have become very abundant?
    a. Social living allows social species to exploit the services of other organisms in harvesting resources.
    b. Social living permits individuals to be specialized in their activities, thereby increasing the efficiency of the group in utilizing resources.
    c. Both a and b
    d. None of the above

    Textbook Reference: 53.6 *How Does Behavior Influence Populations and Communities? pp. 1161–1162*

17. Which of the following statements about territoriality is *false*?
    a. Individuals of a territorial species only defend their territories against other individuals of the same species.

b. Territories do not always include foraging areas.

c. Territorial defense is likely to impose three kinds of costs: energetic, risk, and opportunity.

d. None of the above

Textbook Reference: 53.3 How Do Behavioral Responses to the Environment Influence Fitness? pp. 1148–1149; 53.6 How Does Behavior Influence Populations and Communities? p. 1161

Application Questions

1. Describe the experiments done with bluegill sunfish preying on water fleas and the implications of these studies for foraging theory.

Textbook Reference: 53.3 How Do Behavioral Responses to the Environment Influence Fitness? pp. 1150–1151

2. A pigeon has been trained to feed at food bins at the eastern end of a circular cage from which it can see the sky. There are food bins at the N, NE, E, SE, S, SW, W, and NW ends of the cage. Knowing this, answer the following questions:

a. If the cage is covered and a fixed light source is presented at the east end of the cage at the time of sunrise, where will the pigeon search for food at noon?

b. If a mirror is used to shift the apparent position of the sun at noon from the south to the northwest, where will the pigeon search for food?

c. The bird has been placed in a light-controlled environment for three weeks and phase-delayed by six hours. If the pigeon is returned to the cage under natural lighting conditions at sunset, where will it search for food?

Textbook Reference: 53.3 How Do Behavioral Responses to the Environment Influence Fitness? pp. 1155–1156

3. You go to a nature center that has a honeybee hive that you can see into. As you look at all the busy bees, you notice that one bee is doing a dance. What is the purpose of this dance, and what might it be telling the other bees?

Textbook Reference: 53.4 How Do Animals Communicate with One Another? pp. 1157–1158

4. For each of the two *y*-axes in the following graph, draw a labeled curve that correctly summarizes observations made on goshawks attacking wood pigeons, as described in the textbook.

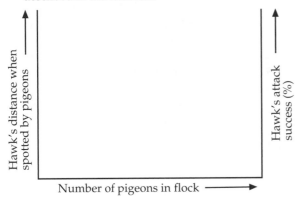

Textbook Reference: 53.5 Why Do Animal Societies Evolve? pp. 1158–1159

5. Based on the unusual sex determination system in Hymenoptera, describe W. D. Hamilton's hypothesis for the evolution of eusociality.

Textbook Reference: 53.5 Why Do Animal Societies Evolve? p. 1160

6. How has the social organization of humans contributed to the great abundance of the human population at present?

Textbook Reference: 53.6 How Does Behavior Influence Populations and Communities? p. 1162

Answers

Knowledge and Synthesis Answers

1. **d.** The first three questions (answer choices a, b, and c) concern the *proximate* mechanisms that underlie a behavior. The fourth question is concerned with the *ultimate* cause of a behavior—the selection pressures that shaped its evolution.

2. **d.** Behaviors tend to involve a complex interaction between learning and genetics.

3. **d.** Young male songbirds must hear the species song and then later be able to hear themselves sing. In the adult, the presence of testosterone is needed to increase the size of regions in the brain associated with singing during the breeding season. Absence of testosterone will result in the inability to perform the correct song.

4. **a.** The environment in which a species normally lives is its habitat.

5. **d.** The forfeited benefits of behaviors that could not be achieved as a result of performing a different behavior, like scanning, comprise the opportunity cost of the performed behavior.

6. **b.** Because the number of Species 2 taken by the predator increases markedly at low Species 1 densities, you can conclude that the predator includes Species 2 in its diet only when insufficient Species 1 individuals are available. The number of Species 1 taken is independent of Species 1 density.

7. **b.** Experiments indicate that the majority of spices commonly used in cooking inhibit the growth of more than one kind of food-borne bacteria.

8. **a.** All animals exposed to a daily cycle display circadian rhythms, regardless of when they are active.

9. **b.** Hibernators, cave-dwellers, and equatorial migratory species lack the necessary daily cues at some stage during the year, so they most likely have endogenous circannual rhythms. Thus the white-tailed deer would be the only animal that could use circadian cues.

10. **d.** Piloting is a means of navigation using visual cues and landmarks. Following the coast during migration is considered piloting.

11. **c.** Many animals use the stars for navigation. The stars present two types of information: a fixed reference point (the point directly above Earth's axis of rotation) and moving constellations, which appear to revolve around the fixed point.

12. **d.** By definition, pheromones are chemicals used for communication among individuals of a single species. Mammalian pheromones can communicate information about several characteristics of the signaler, including all those mentioned in answer b. Though pheromones differ in their durability in the environment depending on their function, they are all much more durable than most visual, auditory, or tactile signals and are therefore relatively poorly suited for rapid exchange of information.

13. **d.** For an altruistic behavior to evolve by natural selection, it must increase the inclusive fitness of the individual performing the behavior. Any explanation of an altruistic act based on its purported benefit to the species as a whole violates this principle.

14. **b.** Altruistic behavior is beneficial to the performer when the improvement in the reproductive success of nondescendant kin exceeds the reduced reproductive success of the individual performing the act. If this condition is met, then the behavior has improved the inclusive fitness of the performer.

15. **c.** By definition, eusocial species live in social groups with sterile castes. The sex determination mechanism in which males are haploid and females are diploid is found in the Hymenoptera (ants, bees, and wasps) but not in termites and naked mole-rats. Within the Hymenoptera, there are species in which queens mate with many males.

16. **c.** The social ants and termites frequently exploit other species in harvesting resources. Human beings, as well as many eusocial species, exhibit specialization of behavior that improves the efficiency of resource utilization (e.g., food procurement).

17. **a.** Although individuals of most territorial species only defend their territories against conspecifics, some animals benefit by defending their territories against individuals of other species as well.

Application Answers

1. Honeybees dance to inform other members of the hive about the distance and location of a food source. If food is less than about 80 meters from the hive, the bees will perform a round dance. If food is farther from the hive, the bees will perform a waggle dance. Only the waggle dance contains information about direction.

2. a. The pigeon will eat from the northern bin at noon. Normally at noon the sun is in the south, and the bird would eat from the eastern bin that is 90° to the left of the sun. With the light in the east, the bird would eat in the north, which is 90° to the left of the light.

b. The bird will go to the bin that is 90° from the sun at noon, and will feed from the southwestern food bin.

c. At sunset the eastern bin with food would normally be 180° from the sun in the west. After the bird has been phase-delayed six hours, it will think that the setting sun is the noon sun. At noon the bird usually eats from the eastern bin, which is located 90° to the left of the sun. Therefore, the bird will eat from the southern bin.

3. In studies in which bluegill sunfish were provided with small, medium, and large water fleas at three different prey densities, the fish took equal proportions of the three prey sizes at low densities, but mostly large water fleas at high prey densities. This agrees with predictions from foraging theory on how a predator should behave to maximize its energy intake.

4. Your curves should show a positive relationship between the hawk's distance when spotted and pigeon flock size and a negative relationship between the hawk's attack success and pigeon flock size (see below).

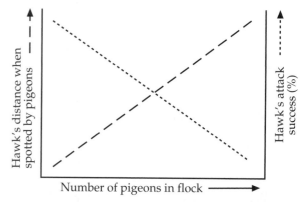

5. All species of the Hymenoptera (ants, bees, and wasps) have a sex determination system in which males are haploid and females are diploid; thus females share 75 percent of their genes with their sisters but only 50 percent of their genes with any offspring they could produce. Hamilton's hypothesis predicts that in species whose sexes are determined in this way, females can increase their inclusive fitness by foregoing reproduction and helping to raise sisters. This hypothesis cannot explain the evolution of eusociality in species without this mode of sex determination or in the many species of hymenopterans in which the queen mates with many males.

6. Social living allowed members of human groups to specialize in different activities. Occupational specialization led to advances in such areas as agriculture, hygiene, and medicine. By increasing food production and decreasing mortality from disease, these advances enabled our species to become enormously abundant.

CHAPTER 54 Population Ecology

Important Concepts

Ecologists use a variety of methods to study populations.

- A population consists of the individuals of a species within a given area.

- The structure of a population refers to the age and size distribution of the individuals in the population and the way those individuals are spread over the environment.

- The density of a population is the number of individuals of the population per unit of area.

- To study the dynamics (changes in structure and density) of a population, demographers measure the rates at which demographic events (births, deaths, immigration, and emigration) are occurring within it.

- Most field studies of animal populations require tagging or marking individuals in some way. No single form of marking can be used on all species. In addition to reporting an animal's location, recently developed tracking devices can record and transmit much important biological information about an individual. Molecular analysis of structures such as bird feathers may reveal the past movements of individuals over long distances (see Figure 54.2).

- Population density can be measured in different ways. Ecologists usually measure the density of organisms in terrestrial environments as the number of individuals per unit of area, but other measures of density may be more useful for aquatic organisms, plants, and some animals.

- In most cases the density and structure of a population must be estimated by statistical methods. For stationary organisms, ecologists estimate these parameters by using sampling plots or transects. For mobile organisms, investigators often use the capture-mark-recapture method.

- Ecologists use estimates of population density to estimate birth, death, and migration rates for a population. The number of individuals present in a population at any given time is equal to the number present at some time in the past, plus the number born between then and now, minus the number that died, plus the number that immigrated into the population, minus the number that emigrated from the population.

- A life table for a population is constructed by determining for a cohort (a group of individuals born at the same time) the number still alive at specific times (survivorship) and, in some cases, the number of offspring they produced during each time interval (fecundity). Life tables are used to predict future trends in populations (see Table 54.1).

- Survivorship curves show the number of individuals in a cohort still alive at different times over the life span. Graphs of survivorship in real populations often resemble one of three types (see Figure 54.3).

  1. Most individuals survive for most of their potential life span and die at about the same age.

  2. Survivorship is about the same throughout most of the life span.

  3. Survivorship of the young is very low, but is high for most of the remainder of the life span.

- The age distribution of a population describes the proportions of individuals in all age categories. It reveals much about the recent history of births and deaths in the population. In populations of long-lived species such as humans, the timing of births and deaths may influence age distributions for many years (see Figure 54.4).

Ecological conditions affect life histories.

- The life history of an organism describes how it allocates its time and energy among growth, reproduction, and other activities. Life histories of different organisms vary dramatically. For example, organisms' life cycles differ in the timing of reproduction and number of offspring produced. Ecologists study life histories because they influence how populations grow and are distributed.

- An understanding of how life history traits influence the growth of populations is important for the successful management of species that humans wish to harvest.

- Ecological conditions, such as the presence or absence of predation, influence the evolution of life histories.

Many factors influence population densities.

- All populations have the potential to grow exponentially. In exponential growth, the per capita rate of increase remains constant per unit of time, whereas the number of individuals added to the population accelerates. This occurs because, as the population grows, more and more individuals are alive to reproduce (see Figure 54.7).

- Exponential growth can be expressed mathematically as

$$\Delta N/\Delta t = (b - d)\, N$$

where $\Delta N/\Delta t$ is the rate of change in the population size (ΔN = change in number of individuals; Δt = change in time), b and d are the per capita birth and death rates, respectively, and N is the population size. (In this equation, b includes both births and immigrations; likewise d includes both deaths and emigrations.)

- The difference between the per capita birth rate (b) and the per capita death rate (d) is the net reproductive rate (r). Where there are no limits on population growth, r has its highest value, designated r_{max}, the intrinsic rate of increase. Therefore, the rate of growth of a population under optimal conditions is

$$\Delta N/\Delta t = r_{max}\, N$$

- The environmental carrying capacity (K) is the maximum population size that the environment can support. It is determined by resource availability, as well as by disease, predators, and, in some cases, social interactions. Because growth tends to slow as the carrying capacity is approached, an S-shaped growth curve is characteristic of many populations growing in environments with limited resources. This type of growth can be expressed mathematically by the logistic growth equation:

$$\Delta N/\Delta t = r\, ((K - N)/K)\, N$$

Population growth stops when $N = K$ because then $(K - N)/K = 0$ (see Figure 54.8).

- Density-dependent environmental factors—such as resource depletion, predation, and disease—cause per capita birth or death rates of a population to change in response to changes in population density. As a result, population growth slows down as density increases.

- Factors (such as adverse weather) that change per capita birth or death rates of a population irrespective of its density are said to be density-independent.

- Fluctuations in the density of a population are determined by the combined effects of all density-dependent and density-independent factors affecting it (see Figure 54.9).

- Populations of species with short-lived individuals that have high reproductive rates typically fluctuate more than those of species with long-lived individuals that have low reproductive rates.

- Population fluctuations may be strongly influenced by years of good reproduction.

- Populations of species that depend on just a few resources are more prone to fluctuations than those of species that use a greater variety of resources.

- Some species are more common than others for several reasons. Factors favoring high population density are utilization of abundant resources, small body size (see Figure 54.11), recent introduction into a new region (see Figure 54.12), and complex social organization. In some cases, species have a small population because they have arisen only recently by polyploidy or a founder event, or they are declining toward extinction (see Figure 54.13).

Spatially variable environments affect population dynamics.

- Most populations are divided into geographically separated subpopulations that live in habitat patches—areas of a particular kind of environment that are surrounded by other kinds. The larger population to which the subpopulations belong is called the metapopulation. Because the individual subpopulations are much smaller than the meta-population, they are prone to fluctuations leading to extinction. In the rescue effect, immigrants from other subpopulations prevent declining subpopulations from becoming extinct. This effect can be reduced by barriers to dispersal (see Figures 54.14 and 54.15).

- Both local conditions and distant events, perhaps even occurring on a different continent, may affect the dynamics of a population (see Figure 54.16).

Humans use the principles of population dynamics to manage and control populations.

- To maximize the number of individuals that can be harvested from a population, the population should be held far enough below carrying capacity to have high birth and growth rates.

- Demographic traits, particularly reproductive capacity, determine how heavily a population can be exploited. For commercial reasons, many fish and whale populations have been overharvested in violation of basic principles of population management (see Figure 54.17).

- The populations of undesirable species are most effectively controlled by lowering the carrying capacity of the environment for those species. For non-native pest species, the introduction of predators and parasites from the pest's original habitat is sometimes an effective control measure but is sometimes disastrous (see Figure 54.18).

- The size of the human population contributes to most of the environmental problems we are facing today, from pollution to extinctions of other species. Earth's carrying capacity for humans depends on our use of resources and the effects of our activities on other species.

The Big Picture

- An understanding of concepts of population ecology such as exponential growth and carrying capacity is fundamental to your ability to think intelligently about issues such as human population growth, the preservation of biodiversity, and the control of undesirable species.

- Darwin's realization that all populations have the inherent capacity for exponential growth was crucial to the development of his theory of natural selection.

Common Problem Areas

- Students frequently have problems interpreting graphs depicting concepts of population ecology. Look carefully at how the axes of a graph are labeled. Ask yourself questions such as: Are the scales arithmetic or logarithmic? Is the fate of a cohort of the population being traced, or is the growth pattern of the entire population being described?

Study Strategies

- Redrawing graphs from the textbook is a good way to reinforce your understanding of the concepts being presented.

- Take advantage of laboratory activities involving computer simulations of population growth or other aspects of population ecology.

- Review the following animated tutorials and activities on the Companion Website/CD:
  Tutorial 54.1 Exponential Growth Simulation
  Tutorial 54.2 Logistic Growth Simulation
  Tutorial 54.3 Habitat Fragmentation
  Activity 54.1 Logistic Growth

Test Yourself

Knowledge and Synthesis Questions

1. The aspects of population structure studied by ecologists include the
   a. distribution of genotypes within a population.
   b. age distribution of a population.
   c. spacing of population members.
   d. Both b and c
   Textbook Reference: 54.1 How Do Ecologists Study Populations? p. 1167

2. In an effort to measure the size of a bluegill population in a pond, you capture 40 bluegills and mark each with a tag on the tail fin and then release them. A week later you return to the pond and catch 40 more, of which 8 have a tag on the tail fin. What is the best estimate of the bluegill population in the pond?
   a. 100
   b. 160
   c. 200
   d. 320
   Textbook Reference: 54.1 How Do Ecologists Study Populations? p. 1169

3. Based on the following life table, during what time interval is survivorship greatest?

Age (years)	Number Alive	
0	800	a. 0–1 years
1	770	b. 1–2 years
2	550	c. 2–3 years
3	125	d. 3–4 years
4	75	e. 4–5 years
5	0	

Textbook Reference: 54.1 How Do Ecologists Study Populations? pp. 1169–1170

4. For a population that is stable in size, the following age distribution indicates that the population's

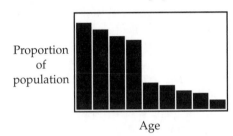

   a. birth and death rates are both high.
   b. birth and death rates are both low.
   c. birth rate is low but its death rate is high.
   d. birth rate is high but its death rate is low.
   Textbook Reference: 54.1 How Do Ecologists Study Populations? pp. 1170–1171

5. In the following graph, which of the expressions (a through d) of the exponential growth equation should be increased for curve 1 to become more like curve 2?

$$\frac{\Delta N}{\Delta t} = (b - d)\, N$$

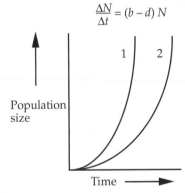

   a. $\Delta N / \Delta t$
   b. d
   c. b
   d. $(b - d)$
   Textbook Reference: 54.3 What Factors Influence Population Densities? p. 1173

6. Species A and B have intrinsic rates of increase (r_{max}) of A = 0.25 and B = 0.50. In reference to the graph shown in Question 5, the population growth curve for A should be more like curve _____.
   Textbook Reference: 54.3 What Factors Influence Population Densities? p. 1173

7. In the logistic population growth curve shown below, the rate of growth is greatest at which point?

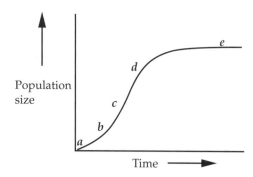

Textbook Reference: *54.3 What Factors Influence Population Densities? pp. 1173–1174*

8. At which point in the graph shown above would there be zero population growth ($\Delta N/\Delta t = 0$)?

    Textbook Reference: *54.3 What Factors Influence Population Densities? pp. 1173–1174*

9. Based on the graph shown below, which of the following events is *least* likely to be true of this population?

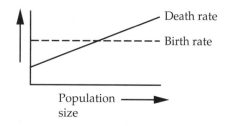

    a. Parasites spread between population members.
    b. Increased competition for food causes some individuals to delay reproduction.
    c. The number of predators in the area varies with population density.
    d. Territorial disputes led to injury and death of some males.

    Textbook Reference: *54.3 What Factors Influence Population Densities? pp. 1173–1174*

10. In the graph shown in Question 9, the population size at which the curves for the birth and death rates intersect is
    a. an estimate of the environmental carrying capacity.
    b. an estimate of the intrinsic rate of increase.
    c. the point at which density-dependent regulation begins.
    d. the point at which density-independent regulation begins.

    Textbook Reference: *54.3 What Factors Influence Population Densities? pp. 1173–1174*

11. Which of the following would probably *not* be true of a population whose dynamics are primarily influenced by density-independent factors?
    a. The population's growth pattern is similar to the logistic growth curve.

    b. The birth rate of the population is dependent on the nutritional status of its adult females.
    c. The most important source of mortality in the population is unfavorable weather conditions.
    d. Both a and b

    Textbook Reference: *54.3 What Factors Influence Population Densities? pp. 1173–1174*

12. A species will tend to have high population fluctuations if its members
    a. have long life spans.
    b. have a high reproductive rate.
    c. eat a varied diet.
    d. All of the above

    Textbook Reference: *54.3 What Factors Influence Population Densities? p. 1175*

13. Which of the following would *not* tend to be associated with the ability of a species to attain high population densities?
    a. Large body size
    b. Utilization of abundant resources
    c. Complex social organization
    d. Recent introduction into a new region

    Textbook Reference: *54.3 What Factors Influence Population Densities? pp. 1175–1176*

14. The best way to decrease the population size of a pest species is to
    a. poison it.
    b. introduce additional predators.
    c. decrease the carrying capacity of the habitat for it.
    d. add competitors.

    Textbook Reference: *54.5 How Can We Manage Populations? pp. 1180–1181*

15. Which of the following factors is significant in setting the present carrying capacity of Earth for the human population?
    a. Earth's ability to absorb by-products of human consumption of fossil fuel energy
    b. Availability of water suitable for drinking, irrigation, and other human uses
    c. Human willingness to cause the extinction of other species to expand the human share of Earth's resources
    d. All of the above

    Textbook Reference: *54.5 How Can We Manage Populations? p. 1181*

Application Questions

1. Draw and explain survivorship curves characteristic of humans in a wealthy country, wild birds, and annual plants.
    Textbook Reference: *54.1 How Do Ecologists Study Populations? p. 1170*

2. The zebra mussel was not found in North America before 1985, yet it is now an abundant pest species in the Great Lakes region and Mississippi River drainage. What principles regarding species abundance apply to the explosive increase of the zebra mussel?

Textbook Reference: 54.3 What Factors Influence Population Densities? pp. 1176–1177

3. How does the division of some populations into discrete subpopulations relate to the process known as the rescue effect? How is the rescue effect influenced by barriers to immigration?
Textbook Reference: 54.4 How do Spatially Variable Environments Influence Population Dynamics? p. 1178

4. How do the demographic traits of fish and whales influence how these marine resources respond to overexploitation?
Textbook Reference: 54.5 How Can We Manage Populations? p. 1180

5. As an alternative to the use of chemical pesticides, ecologists have often recommended biological control: the introduction of predators or parasites of pests (especially exotic species) to bring an undesirable population under control. What are the risks of this strategy?
Textbook Reference: 54.5 How Can We Manage Populations? pp. 1180–1181

Answers

Knowledge and Synthesis Answers

1. **d.** The aspects of population structure studied by ecologists include the spacing and age distribution of the members of the population but not the distribution of genotypes.

2. **c.** If we assume that the population of bluegills randomly mixed in the interval between the capture of the first sample and the capture of the second sample, and if marked and unmarked individuals were equally likely to survive, then the proportion of tagged bluegills in the second sample can be used to estimate the total bluegill population. Because 8 of the 40 bluegills in the second sample were tagged (i.e., one-fifth of the sample), the best estimate of total population of bluegills is 5×40 (i.e., the size of the first sample) = 200.

3. **a.** The decrease in consecutive age classes is least between 0 and 1 year, so the survivorship is greatest during that interval.

4. **a.** When a population's birth rate and death rate are both high, the age distribution is dominated by young individuals. A similar age distribution would occur in a population with a high birth rate and a low death rate, but such a population would not be stable in size.

5. **b.** Both curves show exponential growth, but the rate of exponential growth for curve 2 is less than the rate for curve 1. Because the rate of growth is determined by the expression $(b - d)$, increasing d, the death rate, would cause curve 1 to be more like curve 2.

6. **Curve 2.** Because $r_{max} = (b - d)$, you would expect the growth curve for species A to correspond to the curve with the smaller rate of exponential growth.

7. **c.** The curve showing the relationship between population size and time is steepest at point c, so the growth rate would be greatest at that point.

8. **e.** The growth rate at e would be zero. This is the point called the environmental carrying capacity, the point at which the birth and death rates are equal.

9. **b.** The graph indicates that the birth rate is independent of population size, so event b would not be true. The graph shows that the death rate is density-dependent.

10. **a.** The environmental carrying capacity is the equilibrium population size when the birth rate equals the death rate. The birth rate equals the death rate at the intersection of the two curves.

11. **d.** The logistic growth curve describes density-dependent growth because the rate of increase in the size of the population steadily decreases as the carrying capacity is approached. The nutritional status of the females in the population would depend on the availability of food and hence would be density-dependent. Unfavorable weather, on the other hand, generally occurs without any predictable relationship to the size of a population and thus is a density-independent factor.

12. **b.** Species with long-lived individuals that have low reproductive rates generally have more stable populations than species with short-lived individuals that have high reproductive rates. Species that depend on only a few kinds of foods and that have a fragmented population structure tend to be less stable than those that use a greater variety of resources and that have larger and less-fragmented populations.

13. **a.** Figure 54.11 shows the strong tendency for small species to reach greater population densities than species with larger body sizes.

14. **c.** The best way to control a pest species is to reduce the carrying capacity of the habitat for the species by removing the resources it depends on.

15. **d.** All three of these factors contribute to setting the present carrying capacity of Earth for humanity, though water availability is not a factor everywhere.

Application Answers

1. See Figure 54.3 of the textbook. All of the graphs depict the survivorship of a group of individuals born at the same time (a cohort) from their time of birth until all members of the group have died. In the human population in the United States, death rates are low until old age. In many species of wild birds, the probability of an individual's dying at a particular age is almost the same over most of its lifetime. In annual plants (and in other organisms that produce many offspring), poor survivorship of young individuals is followed by much lower death rates in the middle part of the life span.

2. The high population density of zebra mussels in this continent illustrates that species introduced into a new region often reach levels of abundance above those found in their native ranges because their normal diseases and predators are absent (see Figure 54.12).

3. The rescue effect occurs if a subpopulation that has undergone a large decline is saved from extinction by the immigration of individuals from other subpopulations. As one would predict, barriers to immigration reduce the rescue effect (see Figure 54.15).

4. Although stocks of both many fish and many whale species have been overexploited, the recovery of fish populations is potentially faster for two reasons: first, most species of fish have high reproductive capacities; second, populations of commercially valuable fish are often located in coastal waters that are under the legal jurisdiction of a single country. In contrast, whales, as marine mammals, reproduce at very low rates and are therefore slower to recover from overexploitation than are stocks of most fish. Furthermore, because whales have a wide geographical distribution, they are an international resource requiring the cooperation of many countries to be successfully managed (see Figure 54.17).

5. One risk of the biological control strategy is simply that the introduced parasite or predator will fail to control the pest. A more serious hazard is that the introduced species will attack or outcompete other species that are considered valuable. Two instances of this outcome— the cases of the moth *Cactoblastis cactorum* and the toad *Bufo marinus*—are described in this chapter (see Figure 54.18).

CHAPTER 55 Community Ecology

Important Concepts

An ecological community consists of all the species that live and interact in a particular area.

- Communities are loose assemblages of species. In most cases, species enter and drop out of communities independently over environmental gradients (see Figure 55.1). Where environmental conditions change abruptly, however, the ranges of many species may end at the same place.

- The organisms in a community use diverse sources of energy. Organisms are grouped into trophic levels according to the number of steps through which energy passes to get to them. Energy passes, in sequence, from autotrophic primary producers (photosynthesizers) to heterotrophs, beginning with primary consumers (herbivores) followed by secondary consumers (which feed on herbivores), and so on. Detritivores (decomposers) feed on the dead bodies and waste products of other organisms. Omnivores obtain their food from more than one trophic level (see Table 55.1).

- A food chain is a sequence of feeding linkages beginning with a primary producer. Food chains in a community are usually interconnected in a complex food web (see Figure 55.2).

- Most communities have only three to five trophic levels, partly due to the loss of energy as it moves up the food chain.

- A pyramid of energy shows how energy decreases in moving from lower to higher trophic levels. A pyramid of biomass shows the mass of organisms existing at different trophic levels at a given moment of time. For the same community, these pyramids usually have similar shapes. Major exceptions are most aquatic communities, in which a small mass of rapidly dividing unicellular photosynthesizers can support a much larger biomass of herbivores (see Figure 55.3).

- Detritivores are essential for the continued productivity of ecosystems because they release nutrients from dead bodies that can be taken up again by plants.

The properties of ecological communities are influenced by how organisms affect one another as they seek food.

- Interactions of organisms in a community fall into five general categories (see Table 55.2):
  1. Mutually beneficial (mutualism).
  2. Beneficial to one participant while leaving the other unaffected (commensalism).
  3. Harmful to one participant while leaving the other unaffected (amensalism).
  4. Beneficial to one participant but harmful to the other (predator–prey, parasite–host interactions).
  5. Mutually harmful (competition).

- Predation and parasitism are universal processes. Parasites are typically smaller than their hosts and live in or on their bodies. Microparasites include viruses, bacteria, and protists. Predators are typically larger than and live outside the bodies of their prey.

- Predator–prey interactions often lead to cyclic fluctuations in both predator and prey populations. Oscillations in the population of an herbivorous prey species may be influenced by both food supply and predators (see Figure 55.5).

- Predators may restrict the habitat and geographic range of their prey.

- Mimicry is an adaptation of prey to predation. In Batesian mimicry, a palatable prey species evolves a resemblance to an unpalatable or noxious one. In Müllerian mimicry, two or more unpalatable or noxious species converge to resemble one another (see Figure 55.7).

- Microparasite infections can persist only if, on average, each infected host individual transmits the infection to one other individual. As a result, rates of infection by a microparasite species typically rise, then fall, and do not rise again until a sufficiently dense population of susceptible hosts has reappeared.

- Competition occurs when individuals reduce the ability of other individuals to access shared resources, either by interfering with their activities (interference competition) or by reducing the available resources (exploitation competition). Intraspecific competition occurs among individuals of the same species and is a major density-dependent mechanism of population regulation. Interspecific competition occurs among individuals of different species and can result in competitive exclusion, in which one species prevents all members of another species from using a habitat (see Figure 55.9).

- A competitively superior species may restrict the geographical range of another species (see Figure 55.10).

- Commensal and amensal relationships between species are widespread.

- Mutualistic interactions are both very common and often critically important to the participating species. Examples are the associations of plants and nitrogen-fixing bacteria, of corals and photosynthetic protists, and of plants and their pollinators.

A single species may cause a progression of indirect effects on a community.

- In a trophic cascade, one predator affects many different species at successively lower trophic levels (see Figure 55.13). Because individuals of many species move from one habitat to another, a trophic cascade may affect more than one ecosystem.

- Ecosystem engineers are organisms that build structures that create environments for other species.

- Keystone species exert an influence on ecological communities out of proportion to their abundance. Examples are predatory species that increase species richness by feeding on dominant competitors (see Figure 55.15) and plant species that serve as food for many different animals when food is otherwise scarce.

All ecological communities are subjected to a variety of disturbances.

- A disturbance is an event that changes the survival rate of one or more species in a community. Small disturbances are much more common than large ones, but a few large events may cause most of the changes in a community.

- Ecological succession is a change in community composition following a disturbance. The patterns and causes of succession are varied, but species that first colonize a site often alter the conditions so that they become favorable for other species that come after them.

- Primary succession begins with the establishment of organisms on newly available sites that previously had no organisms (see Figure 55.17).

- Secondary succession begins when organisms reestablish themselves on disturbed sites where some organisms survived the disturbance. This type of succession may also begin with the dead parts of organisms (see Figure 55.18).

- According to the intermediate disturbance hypothesis, communities subjected to very high or very low levels of disturbance have fewer species than those that experience intermediate levels. The hypothesis suggests that in areas with high disturbance levels, only species with great dispersal abilities and high reproductive rates can persist, whereas in areas with low disturbance, species richness declines because competitively dominant species displace others (see Figure 55.19).

- Species that have already become established may facilitate or inhibit colonization by other species.

Several properties of an ecological community influence, or are influenced by, its species richness.

- Species richness—the number of species living in a community—shows a geographical pattern: fewer species are found in high-latitude than in low-latitude regions (see Figure 55.20); fewer are found in relatively flat areas than in mountainous regions; and fewer are found on islands and peninsulas than in an equivalent area on the nearest mainland.

- Species richness increases with primary productivity up to a point, after which it declines (see Figure 55.21). One hypothesis to explain this decline postulates that interspecific competition becomes more intense when productivity is higher, resulting in competitive exclusion of some species.

- Communities with more species tend to be more productive and more stable than communities with fewer species (see Figure 55.22).

The Big Picture

- As you learned in previous chapters, some mutualistic relationships have extraordinary evolutionary or ecological significance. Examples are the role of endosymbiosis in the evolution of eukaryotic cells and the association of nitrogen-fixing bacteria with their host plants.

- In urging a cautious approach to human manipulation of the natural environment, ecologists like to say that "You can never do just one thing." The descriptions in this chapter of the effects of keystone species and of other complex community interactions illustrate the meaning of this saying.

- As discussed in Chapter 57, if we wish to preserve a rare species, in many cases the most important step is to protect its habitat. This often requires an understanding of (and manipulation of) ecological succession to prevent changes in habitat unfavorable to the rare species.

Common Problem Areas

- Students sometimes have difficulty understanding the difference between pyramids of energy and pyramids of biomass. Pyramids of energy can *never* be inverted because there must always be a lower rate of energy flow through any given trophic level than through the trophic level below it. Pyramids of biomass *can* be inverted because they do not represent energy flow rates, but rather the amount of living matter present in the various trophic levels of the community at a given moment in time.

Study Strategies

- This chapter contains descriptions of a number of studies of community structure, including several that relate species richness to other community characteristics. As you study these descriptions, focus on what question the study was designed to answer and how the results support the conclusions drawn from the study.

- Review the following animated tutorial and activities on the Companion Website/CD:
  Tutorial 55.1 Predator-Prey Simulation
  Tutorial 55.2 Primary Succession on a Glacial Moraine
  Activity 55.1 The Major Trophic Levels
  Activity 55.2 Ecosystem Energy Flow

Test Yourself

Knowledge and Synthesis Questions

1. You are studying the ecology of a pond that contains populations of many species of algae, protists, invertebrates, fish, and amphibians. The ecological community of the pond comprises the populations of
   a. all the species living in the pond and the abiotic environment (water, sunlight, etc.) that supports the populations.
   b. all the species living in the pond.
   c. the invertebrates and vertebrate animals living in the pond.
   d. None of the above
   Textbook Reference: 55.1 What Are Ecological Communities? p. 1185

2. Examine the following food web. Organism 9 is a(n)

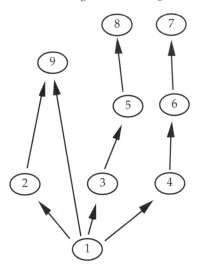

   a. herbivore.
   b. primary carnivore.

   c. secondary carnivore.
   d. omnivore.
   Textbook Reference: 55.1 What Are Ecological Communities? pp. 1186–1187

3. The food web shown in Question 2 has _____ trophic levels.
   a. three
   b. two
   c. four
   d. five
   Textbook Reference: 55.1 What Are Ecological Communities? pp. 1186–1187

4. In which of the following ecosystems would you expect to observe the lowest biomass of primary consumers relative to the biomass of the primary producers?
   a. Tropical evergreen forest
   b. Temperate grassland
   c. Freshwater lake
   d. Open ocean
   Textbook Reference: 55.1 What Are Ecological Communities? pp. 1187–1188

5. Which vertical line in the following graph represents the time at which the predator population is increasing and the prey population is decreasing?

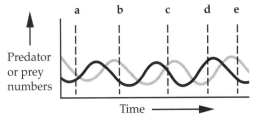

   Textbook Reference: 55.2 What Processes Influence Community Structure? pp. 1189–1190

6. Which of the following statements about mimicry systems is true?
   a. Mimicry systems are either Batesian or Müllerian, but never a combination of the two.
   b. In Müllerian mimicry systems, all species may be regarded as mimics of one another.
   c. In Batesian mimicry systems, directional selection causes the unpalatable species to evolve away from their mimics.
   d. Both b and c
   Textbook Reference: 55.2 What Processes Influence Community Structure? pp. 1190–1191

7. Which of the following statements about microparasites is true?
   a. Microparasite infection rates typically rise only when a sufficiently dense population of susceptible host individuals is present.
   b. Microparasites include viruses, bacteria, protists, and parasitic worms.
   c. For a microparasite infection to persist in a population, more than one new host individual, on average, must become infected with the microparasite before each infected host dies.

d. Both a and c

Textbook Reference: 55.2 What Processes Influence Community Structure? p. 1191

8. A bird eats the fruit of a plant species. The seeds are not digested and germinate in the bird's excrement at some distance from the parent plant. This is an example of
   a. predation.
   b. competition.
   c. commensalism.
   d. mutualism.

Textbook Reference: 55.2 What Processes Influence Community Structure? p. 1194

9. Certain birds follow swarms of foraging army ants and prey upon the insects that the ants flush. The relationship between these birds and the ants is an example of
   a. competition.
   b. commensalism.
   c. mutualism.
   d. Both a and b

Textbook Reference: 55.2 What Processes Influence Community Structure? pp. 1192–1194

10. Which of the following is a possible outcome of competition involving two species?
    a. Some individuals of both species experience reduced growth and reproductive rates.
    b. One species excludes the other from a habitat it would occupy in the absence of competition.
    c. One species restricts the geographic range of the other species.
    d. All of the above

Textbook Reference: 55.2 What Processes Influence Community Structure? pp. 1192–1193

11. If a species of plant that is trampled by an animal eventually evolves sharp spines that prevent trampling, we can say that its association with the animal has changed from _____ to _____.
    a. amensalism; competition.
    b. amensalism; commensalism.
    c. commensalism; competition.
    d. commensalism; mutualism.

Textbook Reference: 55.2 What Processes Influence Community Structure? pp. 1192–1194

12. Which of the following is most characteristic of a keystone species?
    a. Very abundant
    b. Removal having a great effect on community structure
    c. Herbivore
    d. Uniform distribution

Textbook Reference: 55.3 How Do Species Interactions Cause Trophic Cascades? pp. 1196–1197

13. If the following list of stages of ecological succession occurring on glacial moraines in Alaska were put in correct chronological sequence, which would be third?
    a. Arrival of willows and alders
    b. Arrival of lichens
    c. Arrival of conifers
    d. Increase in soil nitrogen content

Textbook Reference: 55.4 How Do Disturbances Affect Ecological Communities? pp. 1197–1198

14. Which of the following would represent primary succession?
    a. The development of an aquatic community in a newly excavated farm pond, eventually followed by the filling in of the pond with sediment and its conversion to a forest
    b. The gradual establishment of a forest following a fire that destroyed the previous community
    c. The invasion of weeds followed by shrubs and trees on an abandoned farm field
    d. Both a and b

Textbook Reference: 55.4 How Do Disturbances Affect Ecological Communities? p. 1197

15. Species richness is
    a. positively correlated with stability of annual productivity.
    b. often greatest in communities with an intermediate level of primary production.
    c. typically greatest in communities experiencing the least amount of disturbance.
    d. Both a and b

Textbook Reference: 55.5 What Determines Species Richness in Ecological Communities? pp. 1200–1201

16. Given that the following terrestrial regions are equal in size, in which would you expect the greatest species richness?
    a. A mountainous region on the mainland of equatorial South America
    b. A mountainous island located near the equator
    c. A flat, lowland region on the mainland of equatorial South America
    d. A mountainous region in the western United States

Textbook Reference: 55.5 What Determines Species Richness in Ecological Communities? p. 1200

Application Questions

1. Some ecologists have regarded a community (especially a plant community) as a tightly integrated "super-organism," whereas other ecologists have argued that each species in a community is individually distributed according to its particular interactions with the environment. How has this question been investigated? Which view of communities appears to be better supported by the evidence?

Textbook Reference: 55.1 What Are Ecological Communities? pp. 1185–1186

2. Where do biomass pyramids like the one shown below occur, and what are the conditions that create them?

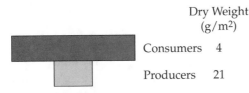

Dry Weight
(g/m²)

Consumers   4

Producers   21

Textbook Reference: 55.1 What Are Ecological Communities? pp. 1187–1188

3. Review the description of Connell's classic experiment on competition between *Balanus* and *Chthamalus* barnacle populations (see Figure 55.9). Do you think that the interaction between these species is an example of interference competition or exploitation competition? Why?
Textbook Reference: 55.2 What Processes Influence Community Structure? pp. 1192–1193

4. Describe how you would assess the degree of interspecific competition that exists between two rodent species living in the same habitat. How would you interpret the results?
Textbook Reference: 55.2 What Processes Influence Community Structure? pp. 1192–1193

5. Wolves were reintroduced to Yellowstone National Park in 1995. Describe the ecological effects within the park of this politically controversial decision.
Textbook Reference: 55.3 How Do Species Interactions Cause Trophic Cascades? pp. 1195–1196

Answers

Knowledge and Synthesis Answers

1. **b.** An ecological community comprises all the populations living in an ecosystem but excludes its nonliving components.

2. **d.** Because organism 9 eats from both the primary producer level (1) and the herbivore level (2), it would be an omnivore.

3. **c.** Four trophic levels are depicted. The levels are primary producer (1), herbivore (2, 3, 4), primary carnivore (5, 6), and secondary carnivore (7, 8). The omnivore (9) occupies either the herbivore or primary carnivore level, depending on whether it is feeding on a primary producer or on an herbivore.

4. **a.** Compared to grasslands or aquatic ecosystems, forests typically have a lower ratio of primary consumer biomass to primary producer biomass because trees (the dominant primary producers) store energy for long periods of time in difficult-to-digest forms.

5. **c.** The key to answering this question is realizing that prey cycles lead predator cycles, so the black line indicates the prey, and the gray line, the predator.

6. **d.** Only answer a is false. Many mimicry systems include both Batesian and Müllerian mimics (see Figure 55.7).

7. **a.** Microparasites include viruses, bacteria, and protists—all of which are far smaller than their host—but not multicellular parasites such as worms. For a microparasite infection to persist, there must be a sufficiently dense host population for at least one (but not necessarily more than one) new host individual to become infected, on average, before each infected host dies.

8. **d.** Both species have benefited; the bird dispersed and provided fertilizer for the plant's seed, and the plant provided food for the bird.

9. **d.** Both ants and birds are competing for the same food supply; the birds also benefit from the ants' activity in flushing the insects, but the ants might not be affected.

10. **d.** It is a generally true of competition, whether intraspecific or interspecific, that it causes reduced growth and reproductive rates of some individuals participating in the competitive interaction. Interspecific competition in particular may lead either to exclusion from a habitat of one of the competing species (competitive exclusion) or to range restriction.

11. **a.** Initially the plant is harmed and the animal is unaffected (amensalism). After the evolution of sharp spines, the two species are competitors for space.

12. **b.** By definition, removal of a keystone species has a larger effect on the community than would be expected based on its abundance.

13. **d.** Of the stages listed, the arrival of lichens would be first, followed by the arrival of willows and alders, an increase in soil nitrogen content (the correct answer), and finally the arrival of conifers.

14. **a.** Succession that begins on a newly available site, like a freshly excavated pond, is primary. Succession that occurs after disturbances such as fire or the conversion of a natural community to farmland is considered secondary.

15. **d.** Of the first three choices, only c is incorrect. The highest levels of species richness are associated with intermediate levels of disturbance.

16. **a.** Equatorial regions have more species than comparable regions at higher latitudes. Mountainous regions have more species than lowland areas at the same latitude. And mainland areas have more species than comparable islands.

Application Answers

1. The question of whether plant communities are "superorganisms" or loose assemblages of species has been investigated by studying how the species composition of communities changes along environmental gradients, such as soil moisture. The results show that each species of plant has its own unique response to environmental factors. Thus the plant community found at a particular location is best regarded not as a "superorganism," but rather as a loose assemblage of species that happen to be adapted to the environmental conditions found there (see Figure 55.1).

2. In most aquatic ecosystems, the dominant primary producers are unicellular algae. Their populations multiply so rapidly and are cropped so efficiently by slower-growing herbivores that they frequently support a larger herbivore mass than their own. Note that the diagram depicts an inverted pyramid of biomass. A pyramid of energy can never be inverted.

3. Ecologists regard the interaction between the barnacle populations as an instance of interference competition. *Balanus* individuals smother, crush, or dislodge *Chthamalus* individuals but do not directly reduce the food resources available to them.

4. A removal experiment could be done in which one or the other species is removed in two similar areas. One then looks for changes in the abundance of the remaining species. The species with the largest increase in population density was most affected by competition with the removed species.

5. In the Lamar Valley of the park, the reintroduction of wolves resulted in a reduction of the population of elk, the wolves' principle prey. In a trophic cascade, the diminution in the number of elk led to increased tree reproduction of aspens and willows, both of which had been heavily browsed by elk. In turn, the increase in the population of streamside willows resulted in an increase in the number of beaver colonies in the valley.

CHAPTER 56 Ecosystems and Global Ecology

Important Concepts

The Earth's system is composed of cycles of materials, inputs of solar energy, and interactions between living organisms and the physical environment.

- Earth is essentially a closed system with respect to atomic matter, but it is an open system with respect to energy.

- Energy from the sun, combined with the energy of radioactive decay in Earth's interior, drives the processes that move materials around the planet.

- Earth is an unusual planet in having life, oceans, a moderate surface temperature, continental drift, and a large moon. The effects of the moon include: stabilizing the tilt of Earth's axis and thereby influencing climate, helping to produce tides, and slowing Earth's rotation.

- Earth's atmosphere is not in chemical equilibrium. Without life on Earth, oxygen and other components of the atmosphere would form nitric acid that would dissolve in the oceans.

- Energy flows through ecosystems unidirectionally and is dissipated as heat. The elements on which life depends cycle among the four compartments of Earth's physical environment: oceans, freshwaters, atmosphere, and land (see Figure 56.1).

- Concentrations of mineral nutrients are very low in most ocean waters because most elements that enter the oceans gradually sink to the seafloor. Rates of photosynthesis in the oceans are highest in zones of upwelling adjacent to continents, where nutrient-rich cold bottom water rises to the surface (see Figure 56.2).

- Most mineral nutrients that enter freshwater are released by the weathering of rocks and are carried to lakes and rivers via groundwater or by surface flow.

- In lakes, surface waters tend to become depleted of nutrients as organisms die and sink to the bottom, whereas deeper waters become depleted of O_2 as organic matter decomposes. Vertical movements called turnover bring nutrients and dissolved CO_2 to the surface and O_2 to deeper water. In deep lakes found in temperate climates, turnover occurs in the spring and fall when the water temperature is uniformly 4°C (see Figure 56.3). In tropical lakes, turnover is completely dependent on strong winds.

- The lowest layer of the atmosphere is the troposphere, where almost all water vapor is located and where most global air circulation takes place. Above the troposphere is the stratosphere, where incoming ultraviolet radiation is absorbed by ozone (O_3) (see Figure 56.4).

- The most abundant gases in the atmosphere are N_2 (about 78 percent) and O_2 (about 21 percent). Although CO_2 constitutes only 0.03 percent of the atmosphere, it is the source of the carbon used by terrestrial photosynthetic organisms and of the dissolved carbonate used by marine producers. The greenhouse gases (such as carbon dioxide, water vapor, and ozone) trap outgoing infrared (heat) radiation emitted by Earth and thus raise its surface temperature.

- About one-fourth of Earth's surface is covered by land. The type of soil that forms in an area and the elements it contains depend on the underlying rock, as well as on climate, topography, the organisms living there, and the length of time that soil-forming processes have been acting.

Energy flows through the global ecosystem unidirectionally.

- With very few exceptions, energy flow in ecosystems originates with photosynthesis, which captures about 5 percent of the solar energy that arrives on Earth.

- Gross primary productivity is defined as the *rate* at which plants assimilate energy through photosynthesis; the accumulated energy is called gross primary production.

- Net primary production is the portion of assimilated energy left over after subtracting the energy plants use to power their own metabolism. All of the other organisms in an ecosystem directly or indirectly derive their energy from net primary production, which takes the form of plant growth and reproduction (see Figure 56.5).

- The major limits on primary production in terrestrial ecosystems are low temperatures and lack of moisture.

In aquatic systems, production is limited by light, nutrients, and temperature (see Figures 56.6 and 56.7).

- Human activities are having an increasing effect on the flow of energy through the global ecosystem. Humans are now appropriating about 20 percent of the annual net primary production on Earth, but the percentage varies widely among regions.

Elements cycle through the global ecosystem.

- The pattern of movement of a chemical element through organisms and compartments of the physical environment is called its biogeochemical cycle.

- The hydrological cycle is driven by evaporation, most of it from ocean surfaces. Underground aquifers (sedimentary rocks) contain large amounts of water, but because this water has a long residence time underground, it plays a small role in the hydrological cycle (see Figure 56.8). Human activities, such as dam construction and pumping of groundwater for irrigation, have altered the flux of water from the land to the oceans (see Figure 56.9). If current water consumption trends continue, about half of the world's population will have an inadequate supply of water by 2025.

- Fires consume the energy of, and release chemical elements from the vegetation they burn. Biomass burning also contributes a significant quantity of carbon dioxide and other greenhouse gases to the atmosphere. Humans deliberately start most fires and can take steps to reduce their undesirable consequences.

- Nearly all the carbon in organisms come from carbon dioxide in the atmosphere or dissolved bicarbonate (HCO_3^-) in water. Carbon is incorporated into organic molecules by photosynthesis; respiration and combustion break down these organic molecules and return carbon to the atmosphere and water (see Figure 56.10).

- Fossil fuels exist because in the remote past, large quantities of organic carbon were removed from the carbon cycle by burial of organisms in sediments lacking oxygen. The ever-increasing rate of burning of fossil fuels over the past 150 years has elevated the CO_2 concentration of the atmosphere (see Figure 56.11).

- Less than half of the CO_2 released into the atmosphere by human activities remains there; most of the rest dissolves in the oceans, which contain 50 times the amount of dissolved inorganic carbon as the atmosphere.

- Several physical and biological processes act to lower atmospheric CO_2 levels. These include: sedimentation of the calcium carbonate shells of marine organisms; the transfer of carbon to deep water via the sinking of dense, saline water in the North Atlantic Ocean; and increased carbon storage in terrestrial vegetation, principally in forests and savannas.

- Historical records and computer models indicate that Earth is warmer when atmospheric carbon dioxide levels are higher and cooler when they are lower (see Figure 56.12). The global warming caused by a doubling of the atmospheric CO_2 concentration would increase mean annual temperatures worldwide and disrupt current precipitation patterns. It might also melt the Greenland and Antarctic ice caps, thereby causing the flooding of coastal regions. Global warming is also expected to increase the incidence of many diseases.

- Nitrogen gas (N_2) makes up 78 percent of Earth's atmosphere, but nitrogen can be converted into biologically useful forms by only a few species of microorganisms that can carry out nitrogen fixation. Thus, nitrogen is often in limited supply in ecosystems. Microorganisms also perform denitrification, which removes nitrogen from the biosphere and returns it to the atmosphere (see Figure 56.14).

- Total nitrogen fixation by humans as a result of the use of fertilizers and burning of fossil fuels equals global natural nitrogen fixation (see Figure 56.15). Human-caused perturbations of the nitrogen cycle have a variety of adverse effects, including groundwater contamination, increased air pollution, reductions in species richness, and eutrophication.

- Sulfur is released as the gases sulfur dioxide and hydrogen sulfide by volcanoes and fumaroles. Sulfur is always abundant enough to meet the needs of organisms.

- The burning of fossil fuels produces emissions that form sulfuric and nitric acid in the atmosphere. The resulting acid precipitation affects all industrialized countries and has been shown to have harmful effects on lake ecosystems and on terrestrial plants (see Figure 56.17).

- The phosphorus cycle differs from the cycles of carbon, nitrogen, and sulfur in that it lacks a gaseous phase. On land, most phosphorus becomes available through the weathering of phosphorus-containing rocks (see Figure 56.19).

- Phosphorus is often a limiting nutrient in soils and lakes. However, adding phosphorus (mostly as phosphates derived from detergents and fertilizers) to lakes is a major cause of eutrophication (nutrient enrichment). Algal blooms and the depletion of oxygen in affected lakes are among the undesirable consequences of this process (see Figure 56.16).

- Recovery and recycling of phosphorus from sewage and animal wastes could reduce the amount of phosphorus entering lakes and rivers while supplying much of the needs of the fertilizer and detergent industries.

- Organisms may be deficient in any of several minerals that they require in very small amounts. Over much of the ocean, a scarcity of dissolved iron limits the rate of photosynthesis. In some terrestrial regions, iodine, cobalt, and selenium are not available in the concentrations necessary to meet the needs of endothermic vertebrates.

- Because biochemical cycles interact in significant ways, alterations in any one cycle affect the others. Scientists are constantly discovering previously unknown interactions (see Figure 56.20).

Ecosystems provide human society with indispensable goods and services.

- Among the goods and services provided by ecosystems are food, clean water, clean air, flood control, soil stabilization, pollination, climate regulation, spiritual fulfillment, and aesthetic enjoyment. It would be either impossible or prohibitively expensive to replace these benefits.

- Modifications of Earth's ecosystems have contributed to human welfare, but the benefits have not been equally distributed. Moreover, short-term increases in some ecosystem goods and services have resulted in long-term degradation of others.

- The Millennium Ecosystem Assessment (MA) was established in 2001 to provide a global assessment of Earth's ecosystems, determine trends in the services they provide, and assess their importance for human well-being (see Figure 56.21).

- The most important cause of alterations in ecosystems has been changes in land use as natural ecosystems have been converted to other, more intensive uses, especially cropland (see Figure 56.22). Though such altered ecosystems provide many benefits (such as food production), often they also have resulted in the degradation of other services (such as the ability to provide clean water and protection from floods).

Many options exist for the sustainable management of ecosystems.

- The economic value of a sustainably managed ecosystem is often higher than that of a converted or intensively exploited ecosystem (see Figure 56.23).

- Because ecosystem services are considered "public goods" that have no market value, it may take government action to create incentives encouraging sustainable ecosystem management. In addition, education programs are needed to increase public awareness of how much human activities affect ecosystem sustainability.

The Big Picture

- Keep in mind the different ways in which energy and matter move through ecosystems. Energy *flows unidirectionally* from producers to higher trophic levels and is ultimately dissipated as heat, whereas, elements and water continually *cycle* through ecosystems.

- The productivity of ecosystems, both terrestrial and aquatic, is relevant to the problem of providing food for the exploding human population (discussed in Chapter 54). The increasing share of global primary production appropriated by the expanding human population is also relevant to the problem of maintaining biodiversity (discussed in Chapter 57).

- Knowledge of how materials move through biogeo-chemical cycles is crucial for understanding and predicting the effects of human alterations of these cycles. Global warming and eutrophication of lakes are two important examples.

- The important roles of nitrogen fixers and other microorganisms in the nitrogen cycle illustrate the sometimes overlooked ways in which rather obscure members of the biotic community may be essential for the healthy functioning of ecosystems.

Commom Problem Areas

- Why eutrophication—the "enrichment" of a body of water with nutrients—often results in a "dead zone" may be puzzling to you. The reason is that the additional growth of photosynthetic algae and bacteria caused by the addition of nutrients such as phosphorous and nitrogen will later result in an increased rate of aerobic respiration by microorganisms as they decompose the dead photosynthesizers. This process may lead to oxygen depletion in part or all of a body of water, such as the deeper waters of a lake.

Study Strategies

- In studying the material on biogeochemical cycles, focus on the following basic questions: Where are the abiotic reserves of an element located? How does the element leave the reserve and enter living organisms? Why do living organisms need the element? How does the element return to its abiotic reserve?

- Review the following animated tutorials and activity on the Companion Website/CD:
  Tutorial 56.1 The Ocean Conveyor Belt
  Tutorial 56.2 The Global Hydrologic Cycle
  Tutorial 56.3 The Global Carbon Cycle
  Tutorial 56.4 The Global Nitrogen Cycle
  Activity 56.1 Ecosystem Energy Flow

Test Yourself

Knowledge and Synthesis Questions

1. Zones of upwelling near the shores of land masses have high primary production because such zones
   a. bring nutrients from the seafloor up to the surface.
   b. are characterized by clear water that permits light to penetrate to an unusually great depth.
   c. bring warm water to the surface.
   d. trap nutrients washed into the ocean from nearby land masses.
   Textbook Reference: *56.1 What Are the Compartments of the Global Ecosystem? p. 1206*

2. Which of the following features would be characteristic of oceans but *not* of freshwater ecosystems?
   a. Receives material from land mostly via groundwater
   b. Seasonal mixing of materials
   c. Elements buried in bottom sediments for long periods of time
   d. Bottom waters frequently lacking oxygen
   Textbook Reference: *56.1 What Are the Compartments of the Global Ecosystem? p. 1206*

3. In the following diagram, the temperature of a freshwater lake is plotted against depth.

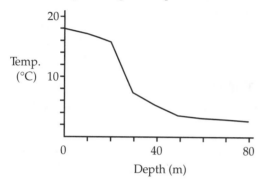

   This temperature profile would be most characteristic of the lake during the _____.
   a. winter
   b. spring
   c. summer
   d. fall
   Textbook Reference: *56.1 What Are the Compartments of the Global Ecosystem? pp. 1207–1208*

4. Based on data in the graph from Question 3, the thermocline for the lake is located between _____ and _____ meters.
   a. 0; 20
   b. 20; 30
   c. 30; 50
   d. 50; 80
   Textbook Reference: *56.1 What Are the Compartments of the Global Ecosystem? pp. 1207–1208*

5. Which of the following features would be characteristic of the stratosphere but *not* of the troposphere?
   a. Most ultraviolet radiation is absorbed here.
   b. Most water vapor resides here.
   c. Most of the mass of the atmosphere lies here.
   d. Circulation of this layer influences ocean currents.
   Textbook Reference: *56.1 What Are the Compartments of the Global Ecosystem? p. 1208*

6. Which of the following compartments of the global ecosystem is characterized by very slow movement of materials within it?
   a. Oceans
   b. Freshwaters
   c. Atmosphere
   d. Land
   Textbook Reference: *56.1 What Are the Compartments of the Global Ecosystem? p. 1208*

7. Which environmental factor frequently limits primary production in both terrestrial and aquatic ecosystems?
   a. Temperature
   b. Moisture
   c. Light
   d. Nutrient supply
   Textbook Reference: *56.2 How Does Energy Flow Through the Global Ecosystem? p. 1210*

8. A plant in the dark uses 0.02 ml of O_2 per minute. The same plant in sunlight releases 0.14 ml of O_2 per minute. A correct estimate of its rate of *gross* primary production is _____ ml of O_2 per minute.
   a. 0.02
   b. 0.12
   c. 0.14
   d. 0.16
   Textbook Reference: *56.2 How Does Energy Flow Through the Global Ecosystem? p. 1209*

9. Which of the following statements about biogeochemical cycles is *not* true?
   a. Most elements remain longest in the living portion of their cycle.
   b. Gaseous elements cycle more quickly than elements without a gaseous phase.
   c. You may have some atoms in your body that were once part of a dinosaur.
   d. Biogeochemical cycles all include both organismal and nonliving components.
   Textbook Reference: *56.3 How Do Materials Cycle Through the Global Ecosystem? pp. 1211–1219*

10. Next to each of the following features, place a *c, n, p,* or *s* if it is characteristic of the biogeochemical cycles of carbon, *n*itrogen, *p*hosphorus, or *s*ulfur. (Note: Some features may apply to more than one cycle.)
    _____ Major reservoir is atmospheric
    _____ Major reservoir is in sedimentary rock
    _____ Often in short supply in ecosystems
    _____ Fossil fuel reserve is part of this cycle
    _____ Involved in cloud formation
    _____ Lacks a gaseous phase
    _____ Includes a form that is a greenhouse gas
    _____ Most fluxes involve organisms
    Textbook Reference: *56.3 How Do Materials Cycle Through the Global Ecosystem? pp. 1211–1219*

11. Which of the following statements is *not* a major concern about our altering of the carbon cycle as it relates to climate change?
    a. The Greenland and Antarctic ice caps may melt if global warming continues.
    b. The burning of fossil fuels adds compounds of sulfur to the atmosphere.
    c. The increase in atmospheric CO_2 exceeds the ability of the oceans to absorb the increase.
    d. CO_2 is a gas that traps infrared radiation.
    Textbook Reference: *56. How Do Materials Cycle Through the Global Ecosystem? pp. 1213–1218*

12. Which of the following statements concerning acid precipitation is *false*?
    a. Acids that enter the atmosphere primarily affect ecosystems located less than 100 kilometers from the source of the pollution.
    b. Though lakes are very sensitive to acidification, their pH can return rapidly to normal values.
    c. Regulation of emission sources has raised the pH of precipitation in many parts of the eastern United States in the last two decades.
    d. Sulfuric acid and nitric acid from the burning of fossil fuels are the major causes of acid precipitation.
    Textbook Reference: 56.3 How Do Materials Cycle Through the Global Ecosystem? p. 1218

13. The process by which a lake ecosystem is altered by eutrophication involves several stages, each causing the next. Which of the following stages would occur *second* in the chain of causation?
    a. Algal blooms occur.
    b. Oxygen levels drop in deeper water.
    c. Phosphorus input from sewage and agricultural runoff increases.
    d. Respiratory demand from decomposers increases.
    Textbook Reference: 56.3 How Do Materials Cycle Through the Global Ecosystem? p. 1219

14. The availability of which of the following micro-nutrients limits the rate of photosynthesis over much of the ocean?
    a. Cobalt
    b. Iron
    c. Selenium
    d. Iodine
    Textbook Reference: 56.3 How Do Materials Cycle Through the Global Ecosystem? p. 1219

15. Of the following ecosystem types, which is being converted to cropland at the greatest rate at the present time?
    a. Deserts
    b. Temperate forests
    c. Tropical and subtropical deciduous forests
    d. Chaparral
    Textbook Reference: 56.4 What Services Do Ecosystems Provide? p. 1222

16. Which of the following is a major obstacle to the sustainable management of ecosystems?
    a. Many ecosystem services are considered "public goods" that have no market value.
    b. The total economic value of a sustainably managed ecosystem is almost always lower than that of an intensively exploited ecosystem.
    c. Most people are not aware of how much human activities affect the functioning of ecosystems.
    d. Both a and c
    Textbook Reference: 56.5 What Options Exist to Manage Ecosystems Sustainably? p. 1223

Application Questions

1. Diagram a generalized biogeochemical cycle, showing the compartments in which elements may reside and the processes by which they move from one compartment to another.
    Textbook Reference: 56.1 What Are the Compartments of the Global Ecosystem? p. 1206

2. If the dry biomass of the primary producers of an ecosystem is 1,000 g per square meter and the efficiency of transfer between trophic levels is 20 percent, what dry biomass weight would you expect at the tertiary consumer level?
    Textbook Reference: 56.2 How Does Energy Flow Through the Global Ecosystem? p. 1209

3. How do farming methods that require the large-scale use of irrigation alter the hydrological cycle?
    Textbook Reference: 56.3 How Do Materials Cycle Through the Global Ecosystem? pp. 1211–1212

4. How may human-set fires be contributing to global warming? What steps can be taken to reduce the frequency and intensity of fires started by humans?
    Textbook Reference: 56.3 How Do Materials Cycle Through the Global Ecosystem? pp. 1212–1213

5. Why may global warming lead to the spread of human disease?
    Textbook Reference: 56.3 How Do Materials Cycle Through the Global Ecosystem? p. 1216

6. How do farming methods that require the large-scale use of manufactured fertilizers contribute to eutrophication and other adverse environmental effects?
    Textbook Reference: 56.3 How Do Materials Cycle Through the Global Ecosystem? pp. 1217–1219

Answers
Knowledge and Synthesis Answers

1. **a.** Low concentrations of mineral nutrients limit the primary production of most ocean waters. In zones of upwelling, nutrient-rich water is brought to the surface from the ocean bottom, thereby enhancing the growth of photosynthesizing plankton.

2. **c.** Most elements remain in bottom sediments of the ocean for millions of years, until they are elevated above sea level by movements of Earth's crust.

3. **c.** The steep thermocline evident in this temperature profile would typically be established only by mid- to late-summer.

4. **b.** The range of depths over which the temperature changes most abruptly is 20 to 30 meters.

5. **a.** The ozone that shields the surface of Earth from most incoming ultraviolet radiation is located in the stratosphere.

6. **d.** Unlike their behavior in air and water, elements on land move slowly and usually only short distances.

7. **a.** Temperature, nutrient supply, and light can limit production in aquatic ecosystems; temperature and moisture most often limit production on land.

8. **d.** The amount of production used in maintenance and biosynthesis is added to net primary production to determine gross primary production. Photosynthetic release of O_2 in the light is an estimate of net production; O_2 use in the dark is an estimate of maintenance and biosynthesis costs. You would add 0.02 ml to 0.14 ml to get an estimate of gross primary production of 0.16 ml per minute.

9. **a.** Most elements cycle through organisms more quickly than they cycle through the nonliving world.

10. **n.** Major reservoir is atmospheric
    c., p. Major reservoir is in sedimentary rock
    n., p. Often in short supply in ecosystems
    c. Fossil fuel reserve is part of this cycle
    s. Involved in cloud formation
    p. Lacks a gaseous phase
    n., c. Includes a form that is a greenhouse gas
    n. Most fluxes involve organisms

11. **b.** The addition of compounds of sulfur to the atmosphere has more to do with acid precipitation than it does with global warming.

12. **a.** Acid precipitation is a regional, not a local, environmental problem. Acids may travel hundreds of kilometers before they settle to Earth in precipitation or as dry particles.

13. **a.** The correct sequence is:

    1. Phosphorus input from sewage and agricultural runoff increases.

    2. Algal blooms occur.

    3. Respiratory demand from decomposers increases.

    4. Oxygen levels drop in deeper water.

14. **b.** Because iron is insoluble in oxygenated water, it rapidly sinks to the ocean floor and therefore may be very scarce in the surface waters in which photosynthesis occurs.

15. **c.** Conversion to cropland is rapidly reducing the extent of a number of tropical and subtropical ecosystem types, including deciduous forests.

16. **d.** Two significant barriers to sustainable ecosystem management are the lack of market value of many ecosystem services and the lack of public understanding of how human activities may harm ecosystems.

Application Answers

1. See Figure 56.1 in the text.

2. The answer is 8 grams. The flow of energy is from primary producers, to primary consumers (herbivores), to secondary consumers, to tertiary consumers. If each of these three energy transfers is only 20 percent efficient, then the biomass of tertiary consumers should be:

   $$1000 \text{ g} \times 0.2 = 200 \text{ g}$$
   $$200 \text{ g} \times 0.2 = 40 \text{ g}$$
   $$40 \text{ g} \times 0.2 = 8 \text{ g}$$

3. The demand for water for irrigation often results in the building of dams, canals, and reservoirs. These structures cause more water to be evaporated from the land and less water to flow to the ocean. Dams also cause seasonal fluctuations of water flows to be lessened, with possible undesirable effects on seawater salinity at the mouth of the dammed river. Excessive pumping of groundwater for irrigation results in the depletion of this resource.

4. Fires that burn biomass, most of which are deliberately started, contribute large fractions of the annual production of several greenhouse gases, particularly carbon dioxide, carbon monoxide, ammonia, and tropospheric ozone. Steps that we can take to reduce the frequency and severity of fires include controlled burning of fire-prone vegetation (to reduce the risk of high-intensity canopy fires), improved productivity of tropical agriculture (to reduce the rate at which forests are burned to create more agricultural land), and greater use of solar cookers (to reduce the demand for firewood).

5. Cold temperatures in winter kill many pathogens. The milder winters associated with global warming will allow increasing numbers of pathogens to survive; hence the diseases they cause will become more common. An example of this process is the spread of dengue fever, a tropical human disease that is expanding its range to higher latitudes. Some diseases of plants and other organisms are also expected to spread as a consequence of global warming.

6. If farmers apply fertilizers to croplands in quantities in excess of the ability of the plants to absorb them, some of the excess nitrates and phosphates leave the soil and enter bodies of water. This process contributes to eutrophication of lakes and the creation of "dead zones" in the sea, such as the one in the Gulf of Mexico around the mouth of the Mississippi River. Excessive use of nitrogen-containing fertilizers has also caused contamination of groundwater and has been linked to outbreaks of toxic dinoflagellates in estuaries on the Atlantic coast.

CHAPTER 57 Conservation Biology

Important Concepts

Conservation biology is the scientific study of how to preserve the diversity of life.

- The science of conservation biology draws on concepts and knowledge from population genetics, evolution, ecology, biogeography, wildlife management, economics, and sociology. The development of this science has been spurred by the accelerating pace of human-caused extinctions of species.

- Conservation biology is a normative science, meaning that it embraces certain values and applies scientific methods to the goal of achieving these values.

- Conservation biology is guided by three basic principles: evolution is the process that unites all of biology, the ecological world is dynamic, and humans are a part of ecosystems.

- Throughout the history of life on Earth, many organisms have changed the environment, creating conditions that favored some existing organisms but negatively affected others. For thousands of years human beings have caused extinctions (see Figure 57.1). Nevertheless, the current situation is unique because a single species is causing all the major environmental changes—a species that has awareness of the changes it is causing.

- There are many reasons why people value biodiversity.

  - Many individual species are important because they supply useful products such as food, fiber, and medicinal drugs.

  - Species are necessary for the functioning of ecosystems and the many benefits and services those ecosystems provide.

  - Extinctions lessen the aesthetic pleasure many people derive from interacting with other organisms.

  - Extinctions deprive us of opportunities to study the structure and functioning of ecological communities and ecosystems.

  - Extinctions of other species caused by human activities raise ethical issues because those species are judged to have intrinsic value.

Biologists use several methods to predict changes in biodiversity.

- Accurate estimates of the number of extinctions that will occur in the next century are impossible for four reasons: we do not know how many species live on Earth today; we do not know the range of most species; it is difficult to determine when a species becomes extinct (see Figure 57.2); and we cannot predict the future with certainty.

- Estimates of current rates of extinction worldwide are based primarily on species–area relationships and rates of tropical deforestation (see Figure 57.3).

- To estimate the risk of extinction of a particular population, biologists develop models that incorporate information about its size, its genetic variation, and the morphology, physiology, and behavior of its members.

- Endangered species are those that are in imminent danger of extinction over all or much of their range. Threatened species are those that are likely to become endangered in the near future. Populations with only a few individuals confined to a small range are especially vulnerable to extinction.

Human activities that threaten the survival of species include habitat destruction, overexploitation, introduction of exotic species, and climate change.

- In the United States, the most important cause of species extinction today is habitat loss, degradation, and fragmentation, especially for species that live in fresh waters (see Figure 57.4). As habitats are fragmented into small patches, populations of species that require large areas cannot be maintained.

- The proportion of a habitat patch subject to detrimental edge effects increases as patch size decreases (see Figure 57.5). For example, species from surrounding habitats may invade the edges of the patch to compete with or prey upon the patch inhabitants. The persistence of species in small patches may be improved if the patches are connected by corridors of suitable habitat through which individuals can disperse (see Figure 57.7).

- Overhunting was, until recently, the most important cause of extinction. Overexploitation for commercial reasons (such as use in traditional medicine, and trade in pets, ornamental plants, and tropical hardwoods) continues to threaten many species.

- Exotic pests, predators, and competitors introduced by humans have driven many species to extinction and threaten many others (see Figure 57.9).

- As rapid global warming occurs, species with poor dispersal abilities may become extinct because of their inability to keep pace with climate change by shifting their ranges. High sea surface temperatures attributed to global warming are already causing increased mortality of reef-forming corals (see Figure 57.10).

Conservation biologists use many strategies to maintain biological diversity.

- Protected areas preserve habitat, prevent over-exploitation, and may serve as nurseries for the replenishment of populations in unprotected areas. In selecting areas most worthy of protection, biologists use such criteria as species richness, endemism, taxonomic uniqueness, unusual ecological or evolutionary phenomena, and global rarity (see Figures 57.11 and 57.12).

- Conservation biologists have identified 595 "centers of imminent extinction." These are concentrated in tropical forests, on islands, and in mountainous regions.

- Practitioners of restoration ecology attempt to return degraded habitats to their natural state. An example is the current project to restore an extensive prairie ecosystem in Montana (see Figure 57.14).

- Humans often reduce the frequency and intensity of disturbances such as fires and floods and thereby endanger species dependent on these disturbances. Reestablishment of historic disturbance patterns (for example, by controlled burning in forests), may help preserve such species.

- In the case of wetlands, it is possible to create new habitats to substitute for the ones being destroyed by development. Restoration is most successful when wetlands are planted with a mixture of species instead of just one or two (see Figure 57.16).

- Certification programs have been devised to ensure that forest and marine species are harvested in ways that both avoid overexploitation and protect biodiversity and ecosystem productivity.

- Truly endangered species cannot withstand any rate of harvest. The Convention on International Trade in Endangered Species (CITES) determines which species are banned in international trade. For example, CITES has instituted a ban on international trade in elephant ivory.

- Controlling invasions of exotic species is often important in preserving biodiversity. Using the traits found in most invasive species, conservation biologists have developed a decision tree to help them determine whether an exotic species should be introduced into North America (see Figure 57.17).

- In many instances, ecologists and economists have been able to demonstrate that biodiversity is profitable, thereby encouraging industries and government agencies to protect endangered species and ecological communities. For example, intact ecosystems may be a reliable source of water (see Figure 57.18), may attract ecotourists, and may improve agricultural productivity by providing pollination services (see Figure 57.19).

- Reconciliation ecology is the practice of using land in which people live and extract resources in ways that sustain biodiversity. In some instances, achieving this goal only requires simple changes in such matters as lawn care or construction methods.

- For a small fraction of endangered species, captive propagation can help to prevent extinction by maintaining the species during critical periods, and by providing individuals for reintroduction into the wild (see Figure 57.20).

- Science can supply information about the accelerating rate of species extinctions, but it is up to society as a whole to determine what rate of species loss due to human activities is acceptable.

The Big Picture

- Conservation biology provides an excellent example of the importance of the synthetic approach to science. In this case, knowledge and concepts derived from different areas of biology, as well as the social sciences, are being applied to the solution of the problem of species and ecosystem preservation.

- Preserving biological diversity is considered by many informed individuals to be one of the most important challenges facing humanity in the twenty-first century.

Common Problem Areas

- It may not be apparent to you why a fragmented habitat is unlikely to support as many species as similar unbroken habitat of equal extent. One reason is that some species may require large expanses of a particular habitat in order to survive. A second reason is that smaller habitat patches are more susceptible to edge effects, as illustrated by Figure 57.5.

Study Strategies

- This chapter refers to and relies on a number of concepts introduced in earlier chapters on ecology, such as: invasive species, habitat fragmentation, and the management of populations (Chapter 54); species richness and disturbance (Chapter 55); and species–area relationships and island biogeography (Chapter 52). You may find it useful to review those concepts as you study the material in this chapter.

- Review the following animated tutorial and activity on the Companion Website/CD:
  Tutorial 57.1 Edge Effects
  Activity 57.1 Concept Matching

Test Yourself

Knowledge and Synthesis Questions

1. The concepts of conservation biology come mainly from all of the following fields *except*
   a. ecology.
   b. evolutionary biology.
   c. population genetics.
   d. immunology.
   Textbook Reference: *57.1 What Is Conservation Biology? p. 1227*

2. The most likely cause of the extinctions of many large mammals in North America and Australia in the last 20,000 years is
   a. rapid climate change associated with glaciation.
   b. overhunting by human beings.
   c. the formation of land bridges to neighboring continents that allowed many new species of competitors and predators to invade these regions.
   d. massive volcanism that caused destruction of the food supply for these mammals.
   Textbook Reference: *57.1 What Is Conservation Biology? p. 1228*

3. Based on species–area relationships, ecologists predict that
   a. about one million tropical evergreen forest species may become extinct in the next hundred years.
   b. a 90 percent loss of habitat will result in loss of about 9 percent of the species living there.
   c. the area required by most species will need to be reduced.
   d. extinction is inevitable.
   Textbook Reference: *57.2 How Do Biologists Predict Changes in Biodiversity? p. 1230*

4. In the United States, the groups of organisms with the highest proportion of endangered or extinct species live in
   a. grasslands.
   b. the deciduous forest biome.
   c. freshwater habitats.
   d. deserts.
   Textbook Reference: *57.3 What Factors Threaten Species Survival? p. 1231*

5. In the following graph, select the curve (*a, b,* or *c*) that correctly shows the expected relationship between habitat patch area and the proportion influenced by edge effects.

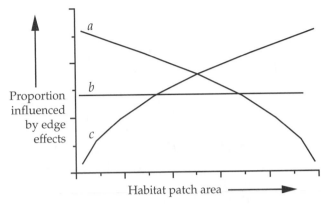

Textbook Reference: *57.3 What Factors Threaten Species Survival? p. 1231*

6. Which of the following is *not* currently a major cause of the global reduction in biodiversity?
   a. Overexploitation
   b. Global warming
   c. Habitat destruction
   d. Introduction of foreign predators and disease
   Textbook Reference: *57.3 What Factors Threaten Species Survival? pp. 1231–1234*

7. Why do conservation biologists believe that global warming may lead to extensive decimation of species?
   a. Because little change in plant community composition has occurred in the past, we cannot expect present communities to adapt to climate change.
   b. The magnitude of climate change will be much greater than past periods of climatic change.
   c. Many sedentary species may not be able to shift their ranges at the same pace as the northern movement of temperature zones.
   d. All of the above
   Textbook Reference: *57.3 What Factors Threaten Species Survival? p. 1234*

8. "Bleaching" of coral
   a. occurs when corals lose their dinoflagellate symbionts.
   b. is most severe in areas with cloudy water.
   c. is associated with high surface water temperatures.
   d. Both a and c
   Textbook Reference: *57.3 What Factors Threaten Species Survival? p. 1234*

9. A defect of the "hotspot" approach to conserving biodiversity is that it does not direct attention to
   a. regions of low species richness that may nevertheless contain unique species or ecological communities.
   b. marine regions.
   c. regions with a high number of endemic species.
   d. Both a and b
   Textbook Reference: *57.4 What Strategies Do Conservation Biologists Use? p. 1235*

10. Which of the following efforts to preserve biodiversity best exemplifies *restoration ecology*?
    a. A campaign to discourage pesticide use on lawns
    b. A project to convert ranch land into a natural prairie
    c. Designation of the habitat of an endangered species as a protected area
    d. Elimination of commercial trade in products derived from endangered and threatened species
    Textbook Reference: *57.4 What Strategies Do Conservation Biologists Use? pp. 1236–1237*

11. Of the answer choices listed for Q. 10, which best exemplifies *reconciliation ecology*?
    Textbook Reference: *57.4 What Strategies Do Conservation Biologists Use? pp. 1241–1242*

12. Experiments on wetland restoration have demonstrated that planting a richer mixture of species is associated with
    a. faster accumulation of belowground nitrogen.
    b. more complex vegetation structure.
    c. more rapid development of vegetation cover.
    d. All of the above
    Textbook Reference: *57.4 What Strategies Do Conservation Biologists Use? pp. 1238–1239*

13. Which of the following is *not* likely to be a characteristic of an invasive plant species?
    a. Specific root symbionts and seed dispersers
    b. Large range in its native continents
    c. Short generation time
    d. Small seeds
    Textbook Reference: *57.4 What Strategies Do Conservation Biologists Use? p. 1240*

14. The fynbos shrub community in South Africa
    a. is threatened by introduced species of taller, faster-growing plants.
    b. is a fire-adapted community.
    c. helps to maintain a regional supply of high-quality water.
    d. All of the above
    Textbook Reference: *57.4 What Strategies Do Conservation Biologists Use? pp. 1240–1241*

15. Researchers have found that coffee production is highest in plantations located closest to forest patches. The explanation for this correlation is that the forests
    a. are the habitat of predators that reduce the populations of herbivores that feed on the coffee plants.
    b. help to maintain soil moisture and to prevent erosion and loss of soil nutrients.
    c. are the habitat of pollinators of the coffee flowers.
    d. cause wind velocities and daily temperature fluctuations to be decreased in adjacent habitats.
    Textbook Reference: *57.4 What Strategies Do Conservation Biologists Use? pp. 1241–1242*

16. Which of the following about the California Condor is *false*?
    a. It is being introduced into regions that were not part of its historic geographical range as part of the effort to save the species.
    b. It became endangered partly because of high mortality that resulted from its eating carcasses containing lead shot.
    c. It has experienced an increase in population size in the last decade as a result of captive propagation.
    d. All of the above
    Textbook Reference: *57.4 What Strategies Do Conservation Biologists Use? p. 1243*

Application Questions

1. Aside from the aesthetic benefits of biodiversity, list three economic reasons people should care about species extinctions.
   Textbook Reference: *57.1 What Is Conservation Biology? pp. 1228–1229*

2. Under the Endangered Species Act of 1973, some species are listed as endangered, whereas others are deemed threatened. What is the distinction between these two categories?
   Textbook Reference: *57.2 How Do Biologists Predict Changes in Biodiversity? p. 1230*

3. What are corridors, and how have experiments demonstrated that they help species persist in patchy environments?
   Textbook Reference: *57.3 What Factors Threaten Species Survival? pp. 1231–1232*

4. Why are exotic species frequently a threat to the biological diversity of the region where they are introduced?
   Textbook Reference: *57.3 What Factors Threaten Species Survival? p. 1233*

5. In recent years, severe forest fires in many parts of the western United States have raised controversy concerning the use of controlled burning as a forest management tool. What are the benefits of this practice?
   Textbook Reference: *57.4 What Strategies Do Conservation Biologists Use? pp. 1237–1238*

Answers
Knowledge and Synthesis Answers

1. **d.** Although immunology is sometimes a useful tool in assessing the amount of genetic variation that exists in a population, its importance is less than the other fields listed.

2. **b.** Because these extinctions followed human colonization of these regions, most biologists favor the hypothesis that overhunting by humans was their most likely cause.

3. **a.** Based on estimates of current and future disappearance of tropical evergreen forests, about one million species that live in these communities could become extinct. This loss is not inevitable if we reduce the rate at which these forests are converted to pasture and cropland.

4. **c.** Habitat destruction and pollution have caused extinction or endangerment of a very high proportion of aquatic freshwater species.

5. **a.** Because the edge of a habitat patch equals the perimeter of the patch, it is proportionally greater for smaller habitat patches. Therefore, the relationship between edge effects and habitat patch areas is an inverse relationship, as shown by curve *a*.

6. **b.** Global warming is predicted to be a major cause of extinction in the future, but is not yet having an impact on populations.

7. **c.** Although the expected magnitude of the climate change due to global warming may be similar to past climate changes, the rate of warming will be greater. This may make it impossible for many species to extend their ranges at the same rate as the northward movement of the temperature zones.

8. **d.** High surface water temperatures cause coral to become "bleached" when the coral polyps lose their endosymbiotic dinoflagellates (which contain photosynthetic pigments).

9. **d.** The "hotspot" concept emphasizes the preservation of regions of high species richness and endemism. It overlooks marine regions and terrestrial regions of relatively low species richness.

10. **b.** Restoration ecology focuses on the reestablishment of entire ecosystems to their natural state.

11. **a.** Reconciliation ecology focuses on ways in which biodiversity can be sustained in landscapes in which people live.

12. **d.** When a species-rich mixture is introduced, wetland restoration is more successful for all of the reasons listed.

13. **a.** A species that requires specific mutualists, such as root symbionts and seed dispersers, is not likely to be successful when introduced to a new geographical region because the species on which it is dependent are unlikely to occur in its new location.

14. **d.** See Figure 57.18 in the textbook.

15. **c.** Although forests might plausibly have all of the effects mentioned, experiments indicate that native pollinators (bees) whose habitat is the forest are responsible for the better quantity and quality of coffee seeds produced on adjacent plantations.

16. **a.** The California condor's historic range included the regions of California and Arizona, where it is currently being reintroduced.

Application Answers

1. A list of economic reasons people should care about species extinctions would include the importance of natural products, ecosystems services, and nature tourism.

2. Endangered species are in imminent danger of extinction over all or a significant portion of their range. Threatened species are those likely to become endangered in the near future.

3. Corridors are relatively thin strips of habitat of a particular type that connect larger patches of the same type of habitat. Their importance in permitting individuals to disperse from one patch to another was demonstrated in experiments described in Figures 54.15 and 57.7 in the textbook.

4. Native species that live in a particular community have evolved to cope successfully with the particular predators, competitors, and diseases that are part of its environment. Introduced species represent a change in the environment for which native species may not have been prepared by natural selection.

5. In ecosystems that naturally experience periodic fires, controlled burning prevents the excessive accumulation of fuel (dead branches, leaf litter, etc.) and thereby lessens the danger that catastrophic, tree-consuming canopy fires will occur. In addition, because many species in such ecosystems require periodic fires for successful establishment and reproduction, controlled burning may be necessary to maintain species richness.

NOTES

NOTES

NOTES

NOTES

NOTES